THE FISH IMMUNE SYSTEM

Organism, Pathogen, and Environment

This is Volume 15 in the

FISH PHYSIOLOGY series
Edited by William S. Hoar, David J. Randall, and Anthony P. Farrell

A complete list of books in this series appears at the end of the volume.

THE FISH IMMUNE SYSTEM

Organism, Pathogen, and Environment

Edited by

GEORGE IWAMA
University of British Columbia
Vancouver, British Columbia
Canada

TERUYUKI NAKANISHI
National Research Institute of Aquaculture
Inland Station
Tamaki-cho, Watarai Mie 519-04
Japan

ACADEMIC PRESS
San Diego London Boston New York Sydney Tokyo Toronto

Front cover photograph: Autograft and allograft in rainbow trout, one month post-grafting (see Chapter 4, Figure 1 for more details).

This book is printed on acid-free paper. ♾

Academic Press, Inc.
525 B Street, Suite 1900, San Diego, California 92101-4495, USA
http://www.apnet.com

Academic Press Limited
24-28 Oval Road, London NW1 7DX, UK
http://www.hbuk.co.uk/ap/

Library of Congress Catalog Card Number: 76-84233

PRINTED IN THE UNITED STATES OF AMERICA
96 97 98 99 00 01 EB 9 8 7 6 5 4 3 2 1

CONTENTS

CONTRIBUTORS

Numbers in parentheses indicate the pages on which the authors' contributions begin.

DOUGLAS P. ANDERSON *(289), Salmon Bay Biologicals, Seattle, Washington 98117*

AKIRA CHIBÁ *(1), Department of Biology, Nippon Dental University, Niigata 951, Japan*

TREVOR P. T. EVELYN *(339), Pacific Biological Station, Department of Fisheries and Oceans, Nanaimo, British Columbia, Canada V9R 5K6*

STEPHEN L. KAATTARI *(207), School of Marine Science, Virginia Institute of Marine Science, College of William and Mary, Gloucester Point, Virginia 23062*

MARGARET J. MANNING *(159), Department of Biological Sciences, University of Plymouth, Plymouth, Devon PL4 8AA, United Kingdom*

TERUYUKI NAKANISHI *(159), National Research Institute of Aquaculture, Inland Station, Watarai Mie 519-04, Tamaki-cho, Japan*

JON D. PIGANELLI *(207), University of Colorado Health Sciences Center, Denver, Colorado 80262*

CARL B. SCHRECK *(311), Oregon Cooperative Fishery Research Unit, U. S. Department of Interior, Oregon State University, Corvallis, Oregon 97331*

C. J. SECOMBES *(63), Department of Zoology, University of Aberdeen, Aberdeen AB9 2TN, United Kingdom*

MARY F. TATNER *(255), Division of Infection and Immunity, University of Glasgow, Glasgow G12 8QQ, United Kingdom*

ALBERTO VARAS *(1), Department of Cell Biology, Faculty of Biology, Complutense University, 28040 Madrid, Spain*

TOMOKI YANO *(105), Faculty of Agriculture, Kyushu University, Fukuoka 812, Japan*

AGUSTÍN G. ZAPATA *(1), Department of Cell Biology, Faculty of Biology, Complutense University, 28040 Madrid, Spain*

PREFACE

Fish health affects many facets of our lives. The production of high-quality fish through aquaculture for both food and wild-stock enhancement, the maintenance of optimum conditions for all aspects of fish research, and the fact that fish health reflects the quality of our aquatic environments are only some of the practical concerns. The importance of fish health is not limited to the professionals who work in fisheries sciences. From a phylogenetic point of view, fish serve as a good model for studying the vertebrate immune system because they have a relatively simple system. The expansion and development of the study of fish health to encompass other disciplines will yield greater appreciation and knowledge of how host and pathogens interact to define a state of health, or disease, and how those systems evolved.

This book was born out of both need and interest. There was a need to update many areas of the field of fish immunology. It has been a decade since the publication of *Fish Immunology,* edited by Manning and Tatner (1985), and more than 25 years since the publication of the review chapter on fish immunology by John Cushing in Volume IV of this series. New techniques in molecular biology have enabled a degree of resolution hardly imaginable at the time of Cushing's review. Furthermore, research in many different species during this time has increased our knowledge of the diversity and phylogenetic trends among fish species. In addition, there is growing interest in the fish physiology community, and even in medical groups, in having more detail about fish health and the fish immune system. We have attempted to give both the fish health professional and the fish physiologist not directly associated with fish immunology a critical and comprehensive review of the fish immune system in the contexts of both phylogeny and overall fish health.

In our view of the defense system of fishes, the primary line of defense is the nonspecific, or natural, immune system. If the antigen successfully penetrates that first line of defense, it is then dealt with by various components of both the natural and the acquired, or specific, immune systems. It is not surprising, perhaps, that invertebrates rely on nonspecific types of defense mechanisms and that higher vertebrates have both natural and acquired immune systems. This phylogenetic pattern is reflected in the ontogenetic pattern of fishes, and we have incorporated these concepts into the design of this book. We have considered the subject of fish immunology as part of a larger picture encompassing interactions among the fish, the pathogen, and the environment in which they both live. We have included reviews that shed light on the interactions among the three components of this relationship. The chapters begin with a description of the tissues and organ systems that are most directly related to the fish immune system. This is followed by several reviews of the basic knowledge of the fish immune system, including an overview of current topics in research and development. Descriptions of how the main components of the immune system are modulated by various endogenous and exogenous factors are presented after such basic information. Despite the great amount of knowledge that has been acquired about the immune system of fishes, as well as about particular fish diseases, there are large gaps in knowledge about the details of topics such as the site for antigen entry and the site for pathogen proliferation. Some of these important topics are addressed in the final chapter of this book. We hope that the new directions in research that each of the authors outlines will stimulate, direct, and challenge scientists working in all facets of fish health.

We are deeply grateful to each of the authors who participated in this project. We appreciate the dedication that has resulted in chapters of the highest quality, despite the heavy demands on each author's time. We are grateful to Professors David Randall and Anthony Farrell for their vision for this volume and for their encouragement. We are also thankful to Dr. Charles Crumly for advice about managing the editorial duties. We thank Dr. Yasuo Inui for reading all the chapters and for his comments regarding the organization of the book. We are indebted to Dr. M. Ototake, Ms. Grace Cho, and A. Matsumoto for their assistance in various aspects of the preparation of the book. We also thank Drs. Edwin L. Cooper, Joanne S. Stolen, and Willem B. Muiswinkel for reviewing the manuscripts.

GEORGE IWAMA
TERUYUKI NAKANISHI

1

CELLS AND TISSUES OF THE IMMUNE SYSTEM OF FISH

AGUSTÍN G. ZAPATA
AKIRA CHIBÁ
ALBERTO VARAS

I. INTRODUCTION: THE PHYLOGENETIC POSITION OF FISH

The first fossils recognizable as vertebrates are the Ostracoderms. They corresponded to a type of Agnathan fish that, apart from the lack of jaws, showed more primitive features than the modern fish, including the lack of paired fins and the presence of only a single, dorsally-located nostril. Nevertheless, modern Cyclostomes (i.e., myxinoids and lampreys) share many of their anatomical characteristics.

Jawed fishes arose nearly 400 million years ago during the Silurian period. Their descendants, the bony fish (Osteichthyes) and cartilaginous fish (Chondrichthyes) are the predominant forms in the current seas and

THE FISH IMMUNE SYSTEM:
ORGANISM, PATHOGEN, AND ENVIRONMENT

freshwaters (Fig. 1). These two groups of fishes diverged from distinct Placoderm lines during the late Silurian or early Devonian period. The cartilaginous fishes, which are morphologically very similar to their Devonian ancestors, represent in a sense the end of an evolutionary line. In contrast, the bony fish constitute a progressive group that includes two large subdivisions, the ray-finned fishes (Actinopterygii) and the fleshy-finned fishes (Sarcopterygii). A small group of the latter, the Crossopterygians, were directly ancestral to amphibians. Another group of Sarcopterygii, the Dipnoi or lungfish, possess various morphological and embryological similarities to amphibians but are not directly in the line of emergence of higher vertebrates.

II. FISH LEUKOCYTES

Because several recent reviews have reported on the structural aspects of fish leukocytes in detail, and other chapters of this volume discuss fish blood cells, mainly monocytes and macrophages, we will describe only some of their more relevant features as an introduction to the principal aim of the chapter, the histophysiology of fish lymphoid organs.

A. Cyclostomes

In hagfish, only a single population of granulocytes has been reported (Tomonaga *et al.,* 1973b; Mattisson and Fänge, 1977). These cells are probably heterophilic granulocytes of gnathostomes and constitute about half of the leukocytes in the blood of *Myxine glutinosa* (Mattisson and Fänge, 1977). They are spherical in shape with motile pseudopodia and contain a lobate nucleus. Their cytoplasmic granules are ovoid or rod shaped (Mattisson and Fänge, 1977). No peroxidase activity is found in the hagfish granulocytes (Johansson, 1973). These cells show amoeboid movement, are found extravascularly in various tissues, and at least in *M. glutinosa,* have phagocytic capacity (Mattisson and Fänge, 1977).

In contrast to the hagfish, different subpopulations of granulocytes are distinguishable in lampreys (Rowley *et al.,* 1988). Among these subpopulations, the heterophilic granulocytes are found consistently and constitute about half of the leukocytes in the blood of adult *Lampetra fluviatilis* with a single subpopulation of the granulocytes. However, there are relatively low ratios of this subpopulation (8%) in the blood of the larval lampreys in which heterophilic granulocytes and acidophils are observed. On the other hand, acidophilic granulocytes are found in the blood of both larval and adult *Lampetra* spp and also both in the intestinal and renal lymphohe-

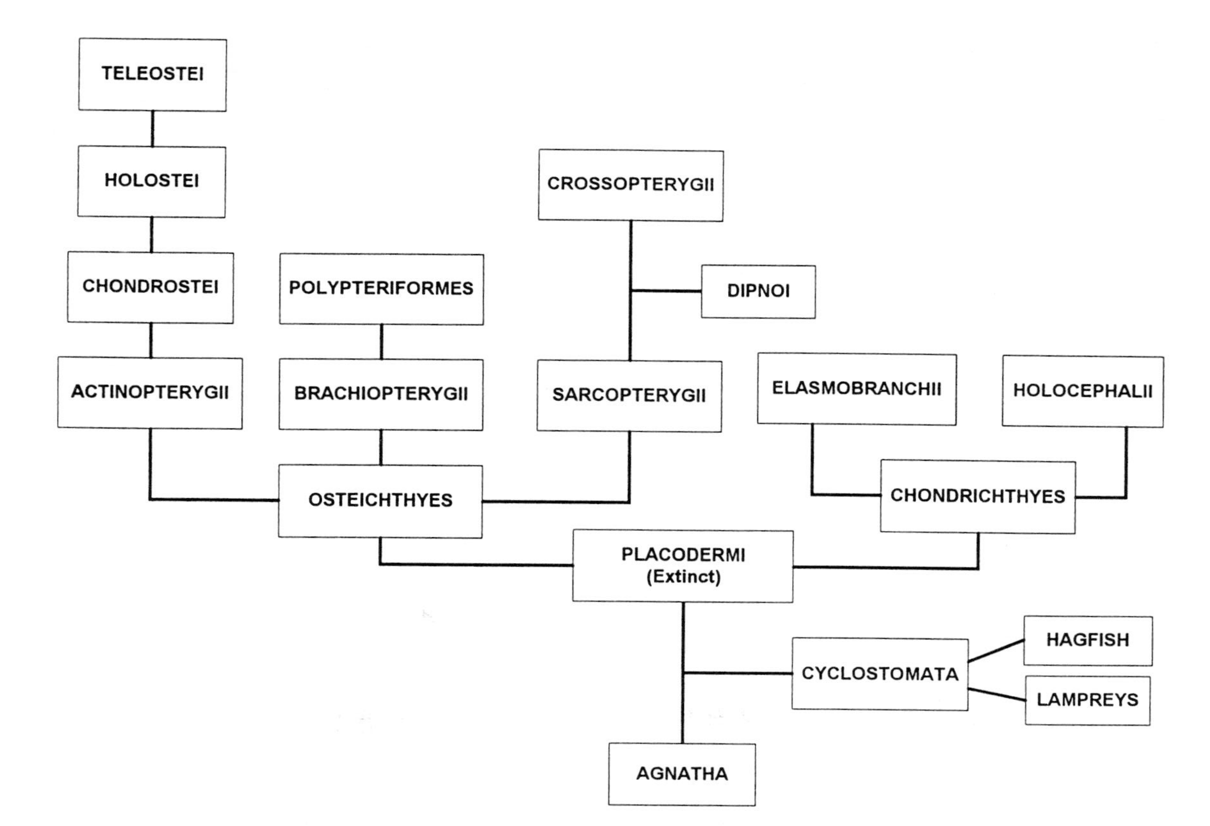

Fig. 1. A phylogenetical tree of fish.

mopoietic tissue of larval lampreys (Rowley *et al.,* 1988). They are more common in the blood of ammocoetes than in adults. Basophilic granulocytes are reported only from *Lampetra planeri* (Fey, 1966) and *Petromyzon marinus* (Fey, 1966). Heterophilic granulocytes have a bilobed or trilobed nucleus and an azurophilic cytoplasm. Cytochemical studies show that they are acid phosphatase, esterase, and periodic acid–Schiff (PAS) positive, but peroxidase negative (Rowley *et al.,* 1988). Electronmicroscopically, cytoplasmic granules are suggested to be in fact part of a single maturation series. *In* the lamprey, *L. fluviatilis,* heterophilic granulocytes phagocytose antibody-coated sheep erythrocytes *in vitro* (Fujii, 1981). The acidophilic granulocytes of *P. marinus* are round cells with minor cytoplasmic processes and contain an eccentric and irregularly shaped nucleus. Their cytoplasmic granules are electron dense, membrane bound, and homogeneous. Histochemical data on the peroxidase in the acidophilic granulocytes are few and inconsistent, for example, producing a negative reaction in *Lampetra* (Rowley *et al.,* 1988) or variable reactions in *L. planeri* (Fey, 1966). Cytological characterization of the basophilic granulocytes is insufficient due to the paucity of data on this type of granulocyte.

B. Chondrichthyes

Little is known about holocephalan granulocytes (Hine and Wain, 1988a; Mattisson *et al.,* 1990). Blood of *Chimaera monstrosa* contains two types of granulocytes referred to as coarse acidophilic granulocytes and fine granulocytes. The former has numerous coarse granules that stain bright red with Giemsa, whereas the latter is full of very fine, red granules. The granulocytes in the cranial lymphohemopoietic tissue also have been studied electronmicroscopically and have been designated as acidophils and heterophils. Cytoplasmic granules of acidophils are electron dense and homogeneous, whereas those of heterophils are heterogeneous in their internal structure (Mattisson *et al.,* 1990).

Granulocytes of elasmobranchs have been studied in several species but their identification and terminology are somewhat confusing due to the enormous heterogeneity (Rowley *et al.,* 1988). Mainwaring and Rowley (1985) reported four types of these cells in the blood of *Scyliorhinus canicula* and referred to them as G1, G2, G3, and G4 granulocytes. Of these cell types, G1, G3, and G4 are classified as acidophils and account for about 27, 3, and 9%, respectively, of the leukocyte population. G2 granulocytes, constituting 1.3% of the total leukocytes, correspond to heterophils. No basophilic granulocytes are found in the *Scyliorhinus* peripheral blood. In contrast, only two types of granulocytes, termed G1 and G2, were described in two species of rays, *Raja clavata* and *R. microcellata.* Both are categorized as acidophils. Among the *Scyliorhinus* granulocyte subpopulations, only G1 shows phagocytic activity and localization at sites of inflammation. Amoeboid activity and chemokinesis/chemotaxis to leukotriene-B_4 are found not only for G1 but also G3 granulocytes. Strong acid phosphatase activity has been reported in G1, G2, and G3 granulocytes, but no peroxidase activity is demonstrated in any of the granulocyte subpopulations (Mainwaring and Rowley, 1985; Rowley *et al.*, 1988). For more detailed information on cytological characteristics such as fine structure and developmental series of the elasmobranch granulocytes (see Ainsworth, 1992 and Hine, 1992).

C. Actinopterygii, Brachiopterygii, and Sarcopterygii

Except for teleosts, the largest group of fishes, little information is available about the granulocytes of bony fish. Hine and Wain (1988b) described three types of granulocytes, heterophils, acidophils, and basophils, in the cranial myeloid tissue of the sturgeon, *Acipenser brevirostrum.* The *Acipenser* heterophils contain two types of cytoplasmic granules: type I granules are positive for alkaline phosphatase but negative for peroxidase, and type II granules are negative for all enzyme tests applied. On the other hand, the acidophils have homogeneous, electron-dense granules which are negative for alkaline phosphatase but positive for peroxidase. Holosteans have two kinds of granulocytes, tentatively designated as acidophils and basophils (Scharrer 1944).

In the myeloid tissues of the kidney and spleen of the bichir, *Polypterus senegalus,* a brachyopterygian, two types of granulocytes occur: one contains electron-dense homogeneous granules in the cytoplasm and the other has less electron-dense, ovoid granules with a heterogeneous or fibrillar content (Chibá, 1994).

In dipnoans (lungfish), granulocytes have been studied in all surviving genera (Hine *et al.*, 1990; Bielek and Strauss, 1993). Owing to the diversity or heterogeneity of the granulocytes among the species studied, there is no universally accepted classification of the lungfish granulocytes. Hine *et al.* (1990) examining *Neoceratodus forsteri* described four types of granulocytes: basophilic, acidophilic, heterophilic, and neutrophilic granulocytes. On the other hand, Bielek and Strauss (1993) investigating *Lepidosiren paradoxa* reported three types of granulocytes, based on their ultrastructure and peroxidase cytochemistry: acidophilic I, acidophilic II, and basophilic granulocytes. The acidophilic I type is the most numerous and contains granules showing variable peroxidase reactions. The characterization of the cell types in lungfish resembles that described in studies on the cell composition of elasmobranchs (Mainwaring and Rowley, 1985; Rowley *et al.,* 1988) rather than that typical of higher vertebrates (Bielek and Strauss, 1993).

In teleosts, three types of granulocytes, heterophils, acidophils, and basophils, have been reported (Ellis, 1977; Rowley *et al.,* 1988; Hine, 1992; Ainsworth, 1992). However, there is enormous variation within the teleosts in both relative abundance and staining reaction of the granulocytes. For example, in the carp, *Cyprinus carpio,* all three types of granulocytes are found in the blood (Rowley *et al.,* 1988). Among them, the heterophils and basophils are the least numerous and constitute 1% of the total leukocyte count. They have kidney-shaped nuclei and two types of cytoplasmic granules: small, peroxidase negative granules, and large, peroxidase positive granules. The acidophilic granulocytes are rather abundant and constitute 8% of the total leukocyte count. The acidophilic granules are peroxidase positive and contain round or irregular granules with heterogeneous contents. In contrast, basophilic granulocytes are peroxidase negative and contain round granules. In salmonids, heterophilic granulocytes predominate with acidophils and basophils either absent or present in very low numbers (Rowley *et al.,* 1988). In the plaice, *Pleuronectes platessa* (Ellis, 1976) and the eel (Kusuda and Ikeda, 1987), only one type of granulocyte, heterophils, has been reported. In the gilthead seabream, *Sparus auratus* (Meseguer *et al.,* 1994a), acid phosphatase activity is evaluated as a cytochemical marker to differentiate the acidophils from the heterophils. Acidophils of the loach, *Misgurunus anguillicaudatus,* have a unique feature, the presence of one large acidophilic granule (Ishizeki *et al.,* 1984).

III. THE LYMPHOID ORGANS OF CYCLOSTOMES

As mentioned above, Agnatha (hagfish and lampreys) were the first living vertebrates to appear in the fossil record. There is considerable discrepancy about the immunological capacity of these primitive fish. Classical reports indicated that cyclostomes were able to respond, albeit poorly, to T-dependent and T-independent antigens, although the molecular mechanisms involved in these responses are unclear (see review by Du Pasquier, 1993). Neither major histocompatibility complex (MHC) molecules (Kronenberg *et al.,* 1994) nor immunoglobulins have been found in hagfish. In contrast, a member of the C3-like complement protein family appears to be involved in nonanticipatory responses, which is related more to invertebrates than the immune response of vertebrates (Ishiguro *et al.,* 1992). Lampreys contain immunoglobulin (Ig) heavy and light chains but they are not covalently joined; the nature of the light chains needs confirmation, and immunoglobulin genes have yet to be characterized (Litman *et al.*, 1992). Although morphologically identifiable lymphocytes appear in the peripheral blood, and lymphohemopoietic tissue occurs in various locations, including both the nephros and the intestinal lamina propria, lampreys do not have true lymphoid organs.

Neither hagfish nor lampreys possess a thymus, and secondary lymphoid organs consist of lymphohemopoietic tissues morphologically and functionally equivalent to the bone marrow. As we will discuss later, fishes contain numerous organs morphologically and functionally equivalent to the bone marrow, which appears for the first time during vertebrate phylogeny in the urodelans of the Plethodontidae family (Zapata and Cooper, 1990; Zapata *et al.*, 1995). In this regard we have emphasized recently that in fish, as in other vertebrates without bone marrow, any organ provided with an adequate stromal cell microenvironment can house and differentiate blood cell progenitors (Zapata *et al.*, 1995). In the Atlantic hagfish, *Myxine glutinosa,* the pronephros was formerly considered to be equivalent to the thymus (Fänge, 1966). A pronephros is formed embryologically in all vertebrate groups producing urine during a brief larval period but, except in a few teleosts, it is missing or largely transformed in adult animals. In adult mixinoids, it consists of ciliated ducts, called nephrostomes, the distal ends of which penetrate the wall of a vein and join a "central mass" suspended in the blood. Our ultrastructural study of the central mass of different-sized Atlantic hagfish demonstrated, confirming previous histological studies by Holmgren (1950), that the central mass is an epithelial filtering organ that contains some areas of active erythropoiesis, macrophages and plasma cells (Fig. 2) (Zapata *et al.*, 1984). The occurrence of plasma cells in mixinoids is, however, a matter of discussion because Tomonaga and Fujii (1994) have failed to find them in the Pacific hagfish, *Eptatetrus stoutii,* and, as mentioned previously, immunoglobulins are absent in these primitive vertebrates. Furthermore, lymphocyte-like cells were described in the muscle–velum complex of Pacific hagfish as well as a thymus equivalent (Riviere *et al.*, 1975). We think, however, that these really represent satellite cells of the skeletal muscle.

In addition, intestinal lymphohemopoietic accumulations that follow the walls of plexiform veins along the entire length of the gut were described as a phylogenetic precursor of the spleen (Good *et al.*, 1966) but in fact these bear mere structural resemblance to the bone marrow of higher vertebrates (Fig. 3). Ultrastructurally, they consist of mature and developing granulocytes scattered in cell clusters between large fat cells in a supporting network of fibroblastic reticular cells (Tanaka *et al.*, 1981). Other authors reported the presence in these areas of erythrocytes in all stages of development (Good *et al.*, 1966), but Tomonaga *et al.* (1973a) did not find synthesis

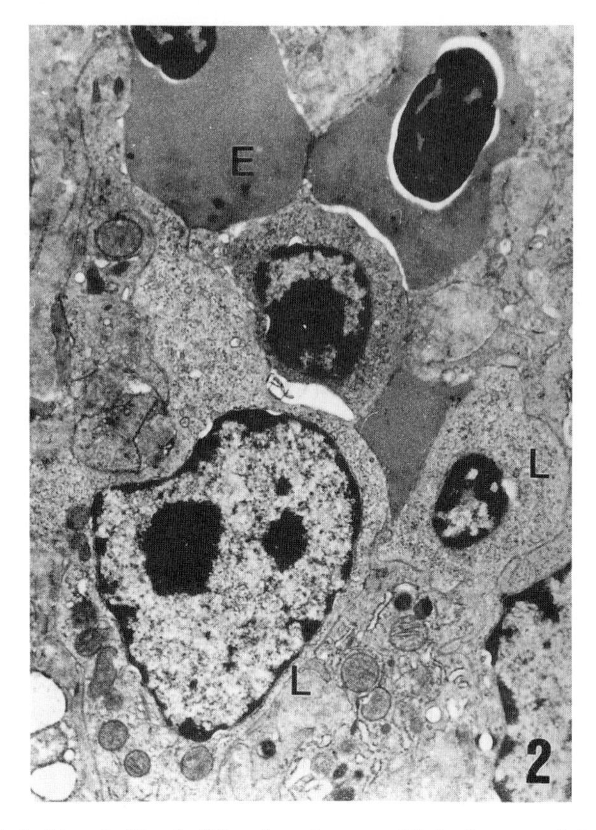

Fig. 2. Foci of lymphoid cells (L) and erythrocytes (E) in the pronephric central mass of the Atlantic hagfish, *Myxine glutinosa.* ×6000.

of hemoglobin in the *Eptatetrus* gut. We suggest that hagfish erythroid cells mature and differentiate outside the intestine in the peripheral blood and/or the pronephros.

On the other hand, migrating leukocytes in the basal mucosa and submucosa, as well as in the epithelium but not lymphoid aggregates, have been detected in hagfish intestine (Fichtelius *et al.,* 1968). Finally, the liver and the perivascular spaces of the gill (Tomonaga 1973) have been described as sites of blood cell production, although they seem to appear as a consequence of blood cell migration from other lymphohemopoietic foci.

In lampreys, lymphoid accumulations that appear in the branchial area have been described as homologous to the thymus of gnathostomes (Good *et al.,* 1972). However, ultrastructural analysis confirmed that these regions represent filtering sites in which phagocytic endothelia of blood vessels and/or circulating macrophages trap both antigenic and nonantigenic mate-

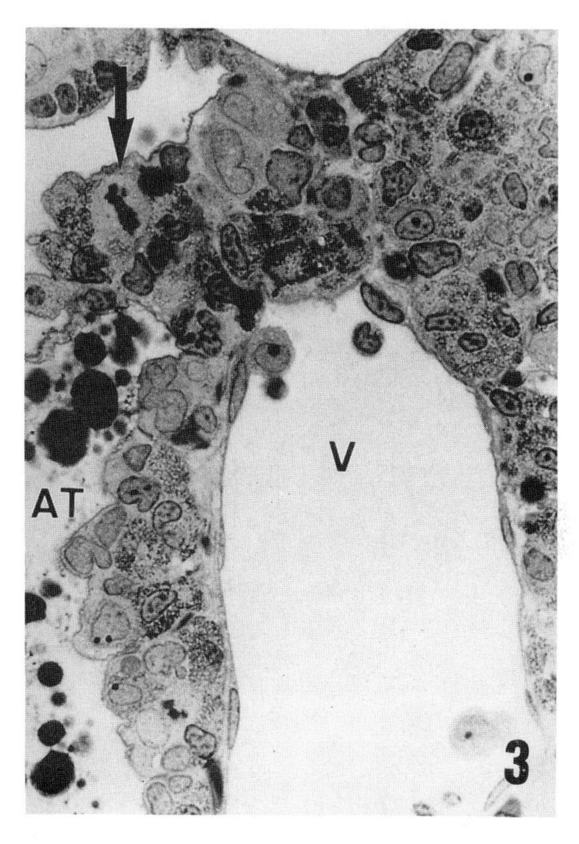

Fig. 3. Lymphohemopoietic tissue of the intestinal lamina propria of the Pacific hagfish, *Eptatretus burgeri.* Granulocytes and blast cells, some of them in division (arrow) appear between large fat cells (AT) and enlarged blood vessels (V). ×1000.

rials from the pharyngeal lumen (Page and Rowley, 1982; Ardavín and Zapata, 1988).

In the absence of a histologically recognizable thymus, the lamprey lymphoid organs consist of lymphohemopoietic tissue located in different organs throughout the complex life cycle of these animals. These changes in the location of lymphohemopoietic tissue indirectly suggest the relevance of the inductive cell microenvironment in the functioning of lymphohemopoietic organs of primitive vertebrates (Tanaka *et al.,* 1981; Ardavín *et al.,* 1984; Ardavín and Zapata, 1987). The main lymphohemopoietic foci of the ammocoetes, the larval form of lampreys, include the thyphlosole, a fold of the midgut formerly described as a primitive spleen (Tanaka *et al.,* 1981), the nephric fold plus the larval opisthonephros, and the neighboring adipose tissue (Ardavín *et al.,* 1984; Ardavín and Zapata, 1987). All of these have the same histological organization which, on the other hand, resembles that of the bone marrow. It consists of cell cords that house all the blood cell lineages, including lymphocytes and plasma cells, in a meshwork of sinusoidal blood vessels and, in the case of nephros, renal tubules (Fig. 4). During metamorphosis, lampreys undergo profound modifications including the formation of the intestinal folds, and the degeneration of the larval opisthonephros in association with the formation of the adult kidney. The loose connective tissue that provides a suitable cell microenvironment for the lymphohemopoietic activity of these organs is then replaced with fibroblasts, macrophages, and dense masses of collagenous fibers. As a consequence, these organs lose their hemopoietic capacity, which is transferred to the supraneural body and, to a less extent, to the adult opisthonephros

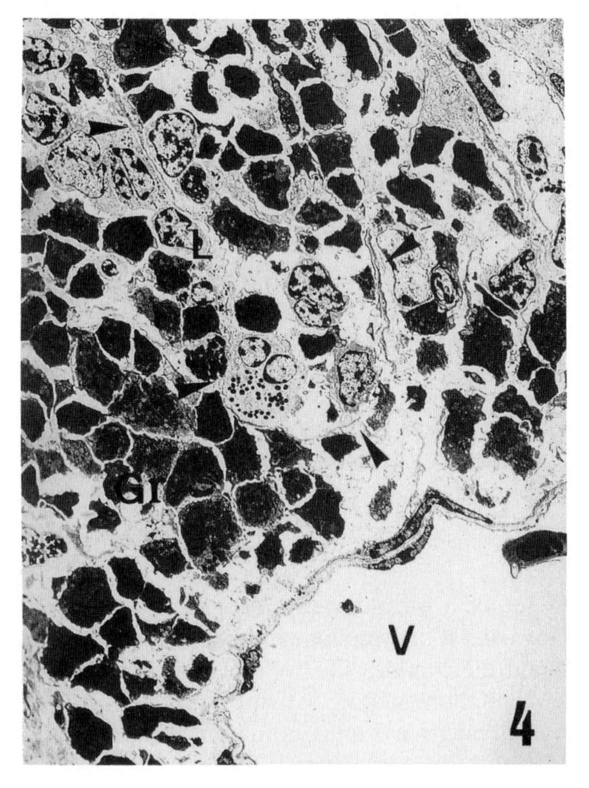

Fig. 4. Developing and mature granulocytes (Gr), mainly heterophils and acidophils, and lymphoid cells (L) occupy the typhlosolar lamina propria of an ammocoete of the sea lamprey, *Petromyzon marinus.* Fibroblastic reticular cell processes (arrowheads), central vein (V). ×1500.

organized just after metamorphosis. Similar changes throughout the life cycle have been reported in the above-described lymphohemopoietic masses of the myxinoid gut (Tomonaga *et al.,* 1973a). Table I summarizes these changes in the sea lamprey, *Petromyzon marinus.*

The supraneural body is a fat column along the lamprey central nervous system which, prior to metamorphosis, does not contain lymphohemopoietic tissue but in postmetamorphic lampreys houses blood cell progenitors, becoming the most important lymphohemopoietic organ of adult lampreys. Its histological organization is remarkably similar to that of mammalian bone marrow, with cell cords arranged among large fat cells. Presumably during larval life, the blood pressure is insufficient to promote the adequate inductive cell microenvironment in the supraneural body, but during metamorphosis the increased pressure due to the development of the heart, and considerable lipid depletion due to the starvation associated with metamorphosis, favor the occurrence of lymphohemopoiesis in the organ.

The liver and the intestinal epithelium of lampreys, as described above for hagfish, contain a few infiltrated lymphoid cells (Ardavín *et al.,* 1984), and in the posterior gut the epithelium takes up macromolecules (Langille and Youson, 1985).

Unfortunately, little information is available on the immunological functions of the lymphoid organs of lampreys. Macrophages from typhlosole, cavernous bodies, nephros, and the supraneural body appear to be involved in trapping carbon and bacteria (Page and Rowley, 1984; Ardavín and Zapata, 1988). Immunization with foreign erythrocytes induces the appearance of antigen-binding cells in both the typhlosole and the blood 4 days later (Fujii *et al.,* 1979), as well as increased numbers of plasma cells in typhlosole, opisthonephros, and the supraneural body (Fujii, 1981; Hagen *et al.,* 1983). In addition, stimulation with bacillus Calmette-Guerin in Freund's complete adjuvant (FCA) promotes cell proliferation in the supra-

Table I
Lymphohemopoietic Foci throughout the Life Span of the Sea Lamprey, *Petromyzon marinus*

Ammocoetes (Stages I–III)	Typhlosole; nephric fold and larval opisthonephros; adipose tissue (only Stages I and II)
Premetamorphosing ammocoetes (Stages IV)	Typhlosole; supraneural body
Macrophthalmia (Stage V)	Supraneural body
Parasitic adult (Stage VI)	Adipose tissue; adult opisthonephros; supraneural body

neural body, as demonstrated by the *in vivo* uptake of tritiated thymidine (Good *et al.*, 1972).

IV. THE LYMPHOID ORGANS OF MODERN FISH

In all modern fish, including both Chondrichthyes and Osteichthyes, thymus, spleen, and somewhat-developed gut lymphoid aggregates appear. In addition, a myriad of different organs, ranging from the primitive meninges to the kidney, contain lymphohemopoietic tissue, which is a typical condition in the vertebrates that lack bone marrow (Table II).

A. The Primary Lymphoid Organs

1. The Bone Marrow Equivalents

In primitive fishes that do not possess a lymphohemopoietic bone marrow, blood cell formation occurs in distinct organs that share structural and functional resemblances to the bone marrow of higher vertebrates.

a. Chondrichthyes: Holocephali (Chimaeriformes). Although Elasmobranchii and Holocephalii are both Chondrichthyes, the most primitive vertebrates containing true lymphoid organs, there are important differences between the two groups which are reflected in their respective lymphohemopoietic organs. In the orbital and the subcranial region of the rabbit-

Table II
Bone Marrow Equivalents in Gnathostomate Fish

Chondrichthyes	
Holocephali	Orbic and cranium
Elasmobranchi	Epigonal organ, Leydig's organ, meninges
Osteichthyes	
Actinopterygii	
Chondrostei	Cranium, heart, kidney, gonads
Holostei	Cranium, heart, kidney
Teleostei	Kidney
Brachiopterygii	
Polypteriformes	Olfactory sac, meninges, kidney
Sarcopterygii	
Dipnoi	Spiral valve, kidney, gonads
Crossopterygii	To be studied

fish (*Chimaera, Hydrolagus*) there are lobed masses first described as lymph nodes without efferent ducts, and later as lymphoid or myeloid tissue. Recently, both histological and electron microscopical studies (Mattisson *et al.,* 1990) have confirmed the blood cell–forming nature of the organs of these area, which in *Chimaera monstrosa* occur in the orbit, the preorbital canal of the cranial cartilage, and a depression of the basis cranii. Their structure resembles that of the epigonal and Leydig's organs of the elasmobranchs, containing, principally, mature and developing granulocytes and, to a lesser extent, lymphocytes and plasma cells, arranged in a network of ramified reticular cells (Fig. 6). Apart from the evident lymphohemopoietic capacity of these areas, we know, however, nothing about their possible immunological significance.

No other important lymphohemopoietic foci appear in *Chimaera,* although it has been suggested that certain lymphopoietic activity exists in the intestinal spiral fold richly infiltrated by lymphocytes, and erythropoiesis has been reported in the cardiac epithelium of larval ratfish.

b. Chondrichthyes: Elasmobranchii. The occurrence of lymphohemopoietic tissue in the walls of the esophagus (Leydig's organ) and the parenchyma of both testes and ovaries (epigonal organs) of most elasmobranchs has been known for a long time. More recently, lymphohemopoietic tissue has been reported in the primitive meninges of at least some elasmobranchs (Chibá *et al.,* 1988; Torroba *et al.,* 1995). It is presumably homologous to that occurring in the meninges of ganoids (see later) and in the brain of the urodeles *Ambystoma* and *Megalobathracus japonicus.* On the other hand, the adult kidney in only a few elasmobranch species contains lymphohemopoietic tissue, although blood cell formation occurs in this organ during the early stages of embryonic life, and the nephrogenic mesenchyme of the postcardinal vein of early dogfish embryos contains presumptive blood cell progenitors (see Zapata *et al.,* 1996).

A gland-like tissue in the ray esophagus, named Leydig's organ, has been known since the seventeen century. It is found in most elasmobranchs although it is poorly developed in some species and others lack it altogether. Also, a "milt-like" structure was reported many years ago in the testes of elasmobranchs and later in the ovaries. The organ is not present in all the species studied, leading some authors to suggest a reciprocal relationship between the Leydig's and epigonal organs (Fänge, 1984). However, many species possess both structures (Zapata, 1981a). The presence or absence of the epigonal organs seems to be related to the developmental pattern of the genital ridge (Fänge, 1987). An epigonal organ will not develop if either the genital ridge develops into gonads extending over the whole length of the peritoneal cavity, or if its anterior part forms the gonad while the posterior part regresses. In contrast, in those species where the anterior part of the genital ridge gives rise to the gonad, the posterior part develops to a variable degree into lymphohemopoietic tissue. This may constitute anything from small masses at the surface or the anterior pole of the gonad, to an enormous organ that surrounds and covers the liver and extends to the rectal gland. Although the epigonal organ is normally bilateral, elasmobranchs containing only one ovary have larger epigonal organs.

The histological characteristics of both the Leydig's and epigonal organs are extensively known but there are few ultrastructural studies (Zapata *et al.,* 1996) and, apart from their assumed lymphohemopoietic capacity, almost no information is available on their function.

The Leydig's organ consists, in all studied species of lymphohemopoietic masses, of both the dorsal and the ventral part of the gut submucosa from the oral cavity to the stomach. Histologically, it exhibits a series of anastomosing lobes penetrated by a few arteries and capillaries, and separated by large, sinusoidal blood vessels. Irregular venous sinuses provided with thin endothelial walls and a poor arterial supply have been reported in the dogfish epigonal organ. Such vascular sinusoids are clearly identified by scanning electron microscopy in the Leydig's organ of *Triakis scyllia* (Fig. 5). The sinusoids occupy preferently both the dorsal and ventral zones of the organ, whereas the inner, central part is comparatively poorly vascularized. In the transitional zone, capillaries are connected to the peripheral sinusoids by large cavities.

Both organs, on the other hand, are histologically similar, exhibiting a stroma of fibrocytes and collagenous fibers arranged between the sinusoidal blood vessels, which contain mature and developing granulocytes, mainly corresponding to heterophils and acidophils, and a variable amount of lymphocytes and plasma cells (Fig. 6) (Zapata *et al.,* 1996). A few high molecular weight and low molecular weight Ig-positive cells appear in the Leydig's organ of *Raja kanojei* (Tomonaga *et al.,* 1984). Matthews (1950) reported active erythropoiesis in the Leydig's organ of the basking shark, *Cetorhinus maxime,* and after splenectomy, the organ contained a compensatory erythropoietic activity (Fänge and Johansson-Sjöbeck, 1975), although in elasmobranchs erythro- and thrombopoiesis are largely confined to the spleen.

The immunological capacity of these organs is, unfortunately, unknown. On the basis of their histological organization (mainly the abundance of sinusoids and the presence of lymphoid cells), Manning (1984) proposed that they may provide an idoneous microenvironment for trapping and processing antigens. Ellis (1977) found that eosinophils from both the spleen and the Leydig's organ of *R. naevus* trapped immune complexes via surface receptors. However, the injection of distinct materials, including dextran,

Fig. 5. Transendothelial cell migration (arrows) from the lymphohemopoietic parenchyma into the sinusoidal lumen in the Leydig's organ of *Triakis scyllia.* ×1700.

carbon, bacteria, latex spheres, and erythrocytes, revealed that little of this appears in the epigonal and Leydig's organs, whereas both spleen and liver contain significant amounts (Rowley *et al.,* 1988). On the other hand, the abundance in hydrolytic enzymes (acid phosphatases, glycosidases, chitinase), due, in turn, to the presence of granulocytes in these organs, suggested to Fänge and colleagues their involvement in immunodefense mechanisms (Fänge, 1982).

There is little information on the ontogenetical development of these organs. In *Torpedo marmorata,* Drzewina (1910) described the primordium of the Leydig's organ in 2-cm-long embryos, and more recently, Navarro (1987) found a few granulopoietic cells in the primitive gonadal parenchyma and among the mesenchymal cells of the esophageal submucosa in 3.5-cm-long dogfish embryos. These data were confirmed by Lloyd-Evans (1993) for the Leydig's organ but, according to this author, the epigonal organ does

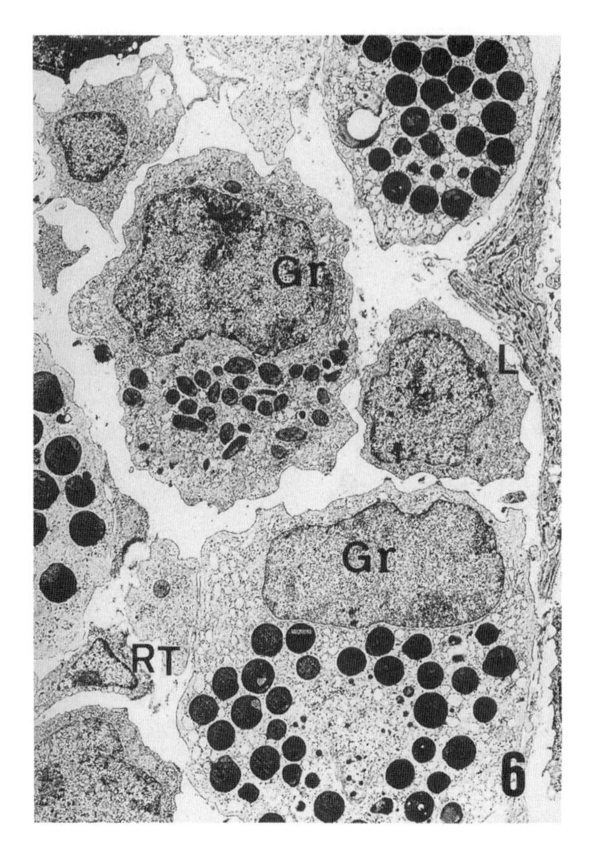

Fig. 6. Mature and developing granulocytes (Gr) and lymphoid cells (L) among the cell processes of reticular cells (RC) in the Leydig's organ of *Raja radiata*. (Courtesy of Dr. Arthur Mattisson, Univ. of Göteborg.) ×3500.

not appear until the egg case splitting stage (6 months). At the end of stage II (external gills) the amount of granulopoietic tissue is increased, both organs are richly vascularized, the first lymphoid cells appear, and mature cells begin to migrate through the sinusoidal walls. From 4 months onward a few Ig-positive cells appear in the Leydig's organ (Lloyd-Evans, 1993). The Leydig's organ reaches the adult condition before the epigonal organ, a fact also pointed out by Lloyd-Evans (1993), in parallel with the development of the vascular pattern.

Many years ago Vialli (1933) reported occasional hemopoiesis in the meninges of *Pristiurus*. More recently we confirmed (Chibá *et al.*, 1988) that some elasmobranchs contain lymphohemopoietic tissue in the "meninx primitiva" throughout the central nervous system. In other primitive fish, such as Chondrostei, Holostei, and *Polypterus,* but also in some amphibians and even in early human embryos, there are lymphohemopoietic foci in both meninges and choroid plexuses (for a review see Chibá *et al.,* 1988). Fänge hypothesized that this presence of lymphohemopoietic tissue in some current elasmobranchs as well as in the cranial cartilage of Holocephali reflects its occurrence in extinct Devonian fishes (Fänge, 1984).

In the stingray, *Dasyatis akajei,* meningeal lymphohemopoietic masses appear predominantly in the telencephalon, diencephalon, and mesencephalon. Our ultrastructural study (Chibá *et al.,* 1988) demonstrated that they occupy three distinct microenvironments: one containing developing and mature granulocytes in a stroma of fibroblasts and collagenous fibers; a second consisting of lymphoid cells, reticular cells, and fibers; and an outer region formed by fibrocytes and collagenous fibers. Although, apart from its hemopoietic capacity, the function of this tissue is unclear, two features point out its relevance in the defense mechanisms of elasmobranchs. Macrophage–lymphocyte cell clusters occur in these meningeal masses and resemble those found during the mammalian immune response (Fig. 7). In addition, recent studies indicate that the meningeal lymphohemopoietic tissue is a source of lymphoid cells, macrophages, antigen-presenting cells, and plasma cells that can gain access to brain ventricles where they form cell clusters, presumably in response to pathogens present in the cerebrospinal fluid (Torroba *et al.,* 1995). Serial sectioning of the hypothalamus of some specimens of *T. scyllia* and *S. torazame* confirmed the occurrence of a lymphoid tissue that extended from the meninge to the ventricular lumen. The lymphoid tissue in these areas does not invade the brain parenchyma, but some macrophages migrate through the ependymal cell layer into and out of the ventricular lumen.

Finally, although most investigators have not found lymphohemopoietic tissue in the kidney of adult elasmobranchs (see Zapata and Cooper, 1990; Zapata *et al.,* 1995), it has been described in a few species (Fänge, 1982) and occurs during embryonic life (Navarro, 1987; Hart *et al.,* 1986; Lloyd-Evans, 1993). Primitive hemopoietic precursors occur in the nephrogenic mesenchyme of the postcardinal vein of 2- to 2.5-cm-long dogfish embryos, and the embryonic kidney is the first peripheral tissue to become lymphohemopoietic. Mature and developing granulocytes are identified in the renal parenchyma of 2.5- to 3.5-cm-long embryos; lymphoid cells, most of which express surface Ig (Lloyd-Evans, 1993), emerge soon after. Later, the renal lymphohemopoietic activity declines, disappearing completely in the last stages of embryonic life and the postnatal period. As mentioned previously for lamprey lymphohemopoietic organs, this evolution of renal hemopoiesis in embryonic dogfish indirectly reflects the relevance of cell microenviron-

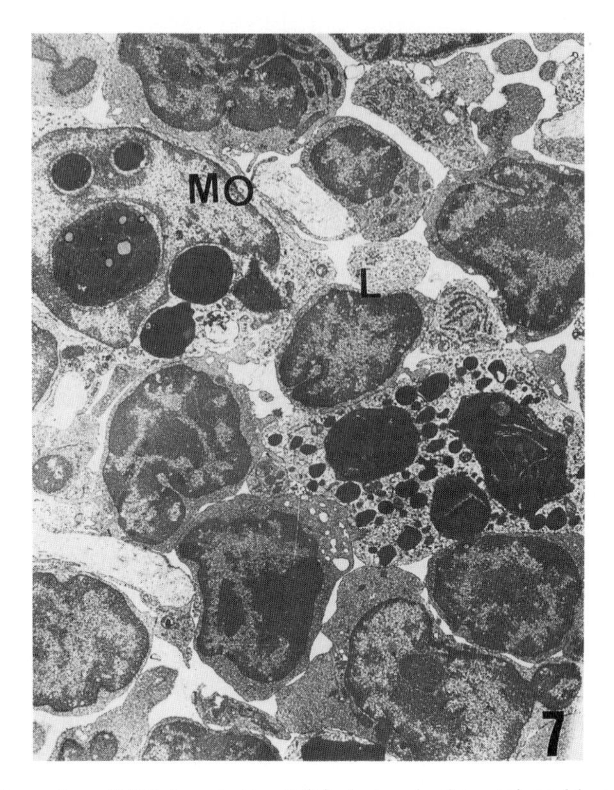

Fig. 7. Macrophage (MO)–lymphoid cell (L) clusters in the meningeal lymphoid tissue of a specimen of *Triakis scyllia.* ×4500.

ments in determining the hemopoietic capacity of vertebrates (Zapata *et al.,* 1995).

c. Actinopterygii: Chondrostei and Holostei. As mentioned above for some elasmobranchs, the main site of granulopoiesis in both chondrosteans and holosteans is a mass, situated in the cranial cavity and associated with the meninges, that in young sturgeons covers the medulla oblongata and the anterior part of the spinal cord (Fänge, 1984). Histologically, these lymphohemopoietic areas consist of closely packed granulocytes and lymphocytes arranged in a meshwork of venous and lymphatic spaces. In addition, the heart of both sturgeons and paddlefish is covered by lymphoid tissue (Clawson *et al.,* 1966). The tissue overlying the base of the heart and extending down over both the atrium and ventricle is similar in organization to mammalian bone marrow, consisting of lymphocytes, reticular cells, granulocytes, and a few, scattered macrophages arranged in cell cords among sinusoidal blood vessels (Fänge, 1984). In both chondrosteans and holosteans, the kidney is an important hemopoietic organ which contains a large number of blast cells as well as developing erythrocytes, granulocytes, lymphocytes, and macrophages, and in *Amia calva* there is granulopoietic tissue in the ovary similar to that found in the elasmobranch epigonal organ (Fänge, 1984).

Primary and secondary immune responses have been reported in *Polyodon spathula* immunized with *Brucella abortus* and *Salmonella parathyphi,* or against protein antigens. They are able to subacutely reject primary and secondary skin allografts with a component of specific memory (for a review see Zapata, 1983), but the role played by the lymphoid organs in immunoreactivity is unknown. Extensive immunization of paddlefish has been reported to result in increased numbers of plasma cells in the spleen and in the pericardial hemopoietic tissue but there is no information on other organs (Good *et al.,* 1966). There is more data on the immunological capacities of holosteans but, still, the functional role of distinct lymphoid organs has not been studied. They do phagocytose bacteria, yeast, and sheep erythrocytes, produce antibodies against different antigens (Bradshaw *et al.,* 1969), and acutely reject allografts (McKinney *et al.,* 1981). Although McKinney *et al.* (1981) failed to demonstrate mixed lymphocyte reaction (MLR) in gars, recently, *in vitro* mitogen-mediated and MLR-induced responses have been reported (Luft *et al.,* 1994).

d. Actinopterygii: Teleostei. The kidney is an important lymphoid organ in teleosts, and consists of two distinct, although structurally similar, segments: the anterior, cephalic or head kidney; and the middle and posterior, trunk kidney. Both regions exhibit hemopoietic capacity but it is greater in the head kidney in which renal function has disappeared (Ellis and de Sousa, 1974; Zapata, 1979a, 1981b). Although some lymphocyte cell clusters seem to occupy defined areas in the kidney of some teleosts (Zapata, 1979a), in general, the lymphohemopoietic cells are scattered at random throughout a stroma of histoenzymatically heterogeneous fibroblastic reticular cells (Quesada *et al.,* 1990; Press *et al.,* 1994) and sinusoidal blood vessels, both with phagocytic capacity (Zapata, 1979a). Every hemopoietic cell lineage seems to be differentiated from cell progenitors (Al-Adhami and Kunz, 1976; Zapata, 1981b) by an important lympho- and plasmacytopoietic capacity (Smith *et al.,* 1967; Zapata, 1979a, 1981b). Despite the fact that some authors have claimed some resemblance of the teleost kidney to lymph nodes (Smith *et al.,* 1967) its capacity for housing and differentiating blood cell precursors supports its phylogenetical relationship to the bone marrow of higher vertebrates (Zapata, 1979a), and in general, it is considered to be a postembryonic source of hemopoietic stem cells (Al-Adhami and Kunz, 1976).

Evidence supports the presence of antigen-presenting cells, and T-like and B-like lymphocytes in the teleost kidney, and accordingly, a role for the renal lymphoid tissue in the defense mechanisms of teleosts. Antigen-binding cells and antibody-producing cells have been found in the lymphohemopoietic tissue of the teleost kidney (see Zapata and Cooper, 1990). Recently, Press *et al.* (1994) demonstrated a higher number of Ig-positive cells in the head kidney than in the spleen. They occur as scattered cells or forming small pyroninophilic cell clusters, especially after antigenic stimulation (van Muiswinkel *et al.,* 1991; Press *et al.,* 1994). Changes affecting the renal lymphoid tissue after immunization have been analyzed with light and electron microscopic immunohistochemistry in the carp head kidney after injection of alum-precipitated bovine serum albumin (BSA) (Imagawa *et al.,* 1991). These authors describe three different Ig-positive cells, identified as blast cells, plasma cells of large lymphoid type, and cells of plasmacytoid type. The former predominated in control, nonimmunized carp and in treated animals 7 days after immunization, whereas the plasmacytoid type cells appear at day 14 postimmunization. They concluded that the cells of the pyroninophilic clusters corresponded to the plasma cell lineage with Ig-producing capacity, and to the large lymphoid type plasma cells which gradually differentiate in these cell clusters. Supernatants, either from cultures of pronephric leukocytes activated with PHA or mixed leukocyte cultures, induce specific proliferation of mitogen-activated leukocytes (Caspi and Avtalion, 1984).

e. Brachiopterygii: Polypteriformes. The polypterids represented by *Polypterus* and *Calamoichthys* are primitive bony fishes belonging to Brachiopterygii or Polypteriformes. Despite the growing interest in the immune responses of primitive vertebrates, no data are available in the literature on the immune system of these fish, except for one study by Waldschmidt (1887) on the existence of a meningeal lymphohemopoietic tissue in *Polypterus,* and the histological description carried out by Yoffey (1929) of the spleen of *Calamoichthys.* Recently, we examined the lymphoid organs of 6 adult bichirs, *Polypterus senegalus,* of both sexes. Our preliminary results (Chibá, 1994) confirm the occurrence (in addition to well-developed thymus, spleen, and gut-associated lymphoid tissue, the structure of which is described below) of lymphohemopoietic foci in the meninges as well as in the olfactory sac. Furthermore, the kidney, including the pronephros and the trunk kidney, contains important masses of lymphoid tissue. The meningeal tissue, mainly granulopoietic in nature, resembles that previously reported in other primitive fish, including some elasmobranchs. The renal lymphohemopoietic tissue is probably organized in cell cords between enlarged blood sinusoids and, in the mesonephros, the renal tubules, as described for teleost kidney. The lymphoid tissue of the olfactory sac occupies the mucosa and consists of different-sized lymphocytes, blast cells, plasma cells, and large melanomacrophages arranged between the components of the connective tissue (Fig. 8).

f. Sarcopterygii: Dipnoi. The lymphoid organs of lungfish have been studied mainly by Jordan and Speidel (1931) and later by Minura and Minura (1977), but information on their immune capacities is lacking. And, in general, available data are sparse. Apart from the thymus and the spleen, the lymphoid organs of Dipnoi include a granulopoietic tissue that develops within the spiral intestine and a similar tissue covering the kidney and the

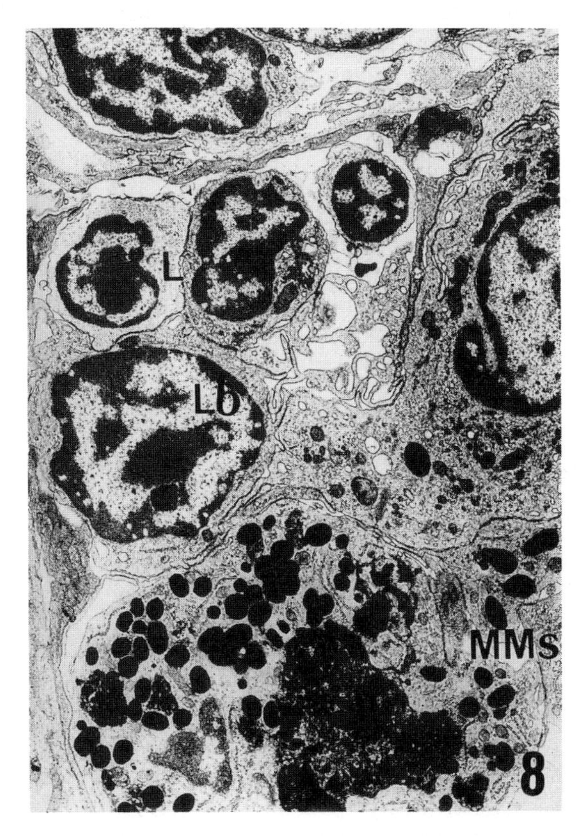

Fig. 8. Lymphoid tissue of the olfactory sac of the bichir, *Polypterus senegalus.* Different sized lymphocytes (L), lymphoblasts (Lb), and large melanomacrophages (MMs) occupy the mucosa arranged between the components of the connective tissue (arrows). ×5000.

gonads (Jordan and Speidel, 1931; Good *et al.,* 1966). The tissue contains a large amount of pigment, presumably melanomacrophages, and abundant eosinophils and lymphocytes.

g. Crossopterygii. Despite their phylogenetic importance we know little about the immune system of Crosopterygii. Extensive lymphomyeloid masses have been reported in the viscera of the coelacanth, *Latimeria chalumnae* (Millot *et al.,* 1978), but any histological description is lacking.

2. The Thymus

Chondrychthyes are the first vertebrates with a histologically identifiable thymus. In general, there are no important structural differences in this regard among the distinct groups of fishes, except in the case of teleosts. The thymus in most teleosts is remarkable for its location near the gill cavity and its permanent continuity, in adult fish, with the pharyngeal epithelium (Fig. 9). In the angler fish, *Lophius piscatorius,* the thymus is, however, located far from the branchial cavity (Fänge and Pulsford, 1985), and *Sicyases sangineus* has a pair of thymus glands in each gill chamber: one gland occupies a superficial position and the second is located close to the gill epithelium (Gorgollón, 1983). In addition, the teleost thymus does not exhibit the typical corticomedullary demarcation found in the thymus gland of all vertebrates. In contrast, two, three, four, and even six regions have been described in distinct teleost species. Their relationship to the thymic cortex and medulla of other vertebrates is controversial (see Castillo *et al.,* 1990; Zapata and Cooper, 1990; Chilmonczyk, 1992). Presumably, the continuity of the thymus with the pharyngeal epithelium, and the lack of a clear corticomedullary regionalization reflect an incomplete migration of the thymic primordium from the gill buds (where it originated) to the underlying mesenchyme during ontogeny. Chilmonczyk (1992) has proposed that fish displaying a more internalized thymus might represent the first step of the internalization process that occurs during vertebrate evolution. These variations from the "common pattern" do not, however, represent a lack of functional capabilities of teleost thymus.

In other nonteleostean fishes in which the histological structure of the thymus has been studied, two clearly defined areas, the cortex and the medulla, have been found (Zapata and Cooper, 1990). Nevertheless, available information on the structure and function is very limited. Accordingly, most data in this section refer to elasmobranch and/or teleost thymus.

The organ is surrounded by a connective tissue capsule consisting of fibroblasts, macrophages, and collagenous fibers that project into the thymic parenchyma by means of connective tissue trabeculae that carry blood vessels and nerves (discussed later). When the gland is large, it divides it

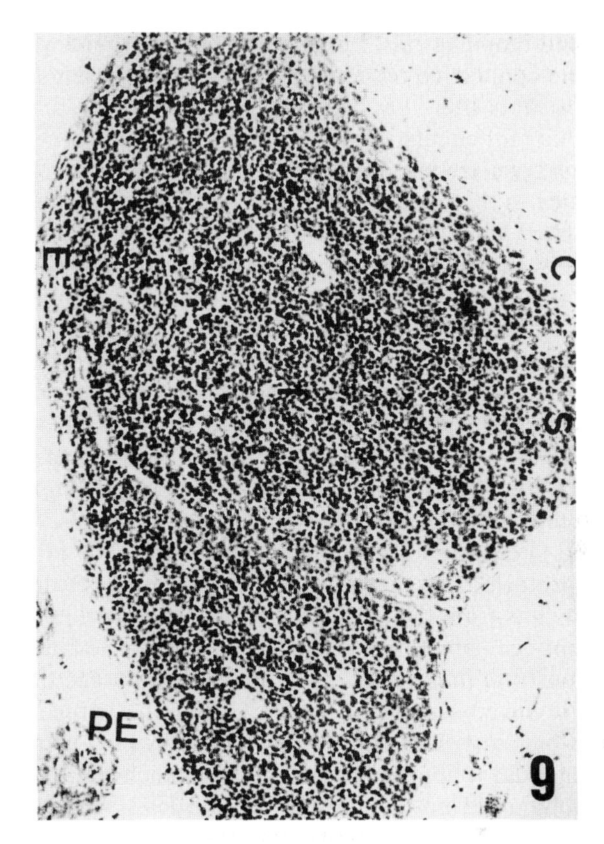

Fig. 9. Thymus gland of a 60-day-old juvenile rainbow trout, *Oncorhynchus mykiss*. Connective tissue capsule (C), subcapsulary region (S), inner zone (I) and outer zone (E). Note the continuity with the pharyngeal epithelium (PE). (Courtesy of Dr. Ana Castillo, Univ. of Orense.) ×200.

into several lobules. The thymic parenchyma consists of a meshwork of epithelial cells that houses free cells, mainly different-sized lymphocytes, but also macrophages, presumptive interdigitating/dendritic cells, myoid cells and, in fewer numbers, other cell types. Both ultrastructural and immunohistochemical studies have demonstrated the heterogeneity of the thymic epithelial component in both elasmobranchs and teleosts, confirming that the most primitive thymus tissues contain cell types very similar to those described in mammalian thymus tissues.

In elasmobranchs, a vast majority of thymic epithelial cells constitute a supporting tridimensional network in both cortex and medulla (Zapata *et al.,* 1996). It extends under the connective tissue elements of capsule, trabeculae, and blood vessels, forming a sheath that separates the thymocytes from the connective tissue with a continuous basement membrane. This "histological barrier" has been related in other vertebrates to a selective permeability for circulating molecules, although the recent demonstration of a transcapsulary route in the mammalian thymus. (Nieuwenhuis *et al.*, 1988) indicates that the problem of access into the thymic parenchyma is more complex.

In the teleost thymus, furthermore, the ultrastructural evidence of fenestrated vascular endothelia, gaps in the epithelial cells surrounding the capillaries, and mainly, the occurrence of pores on the thymic surface reported in young trout, led to Chylmonczyk (1992) to suggest that the thymic barrier, at least in some teleosts, may be of variable permeability. Unpublished observations by this author indicate that only small molecules, such as peroxidase (MW = 40 kD), reach the thymic parenchyma of rainbow trout, *Oncorhynchus mykiss,* after intravenous injection of materials of varied size or molecular weight (cited by Chylmonczyk, 1992). Previously, Tatner and Manning (1982) had reported that the thymus of rainbow trout is relatively unprotected and that antigens can gain entrance directly into the embryonic thymus from the gill cavity. However, experiments using trout fry of different ages immersed in a 0.5% solution of ferritin for different periods demonstrate that the gills, but not the embryonic thymus, are the main organs involved in ferritin trapping during the ontogeny of rainbow trout. Only when very immature (4-day-old) fry were exposed to ferritin for a long time, did it appear in the thymic parenchyma. In this case, the thymic epithelium did not contain ferritin particles, suggesting that they could passively gain access through either the basement membrane under the thymic connective tissue capsule in which the tracers accumulate or the extracellular spaces adjacent to the border between the thymic primordium and the pharyngeal epithelium (Castillo and Zapata, unpublished observations). Presumably, in physiological conditions, antigenic and nonantigenic materials present in the gill cavity of developing trout are, trapped by the gill cells before they can gain access into the thymic parenchyma.

These supporting epithelial cells, on the other hand, show moderate to high electron density and contain cytoplasmic tonofilaments and desmosomes (Fig. 10). In the cortex, they are elongated elements, whereas in both deep cortex and medulla they are wider and exhibit fewer desmosomes and filaments. Although we claimed certain phagocytic capacity for these cells in the thymus of *Torpedo marmorata* and *Raja clavata,* most thymic phagocytosis is in fact associated with macrophages (Zapata, 1980).

Large, pale, secretory-like cells also appear in the elasmobranch thymus (Fig. 10). They contain rough endoplasmic reticulum, a well-developed Golgi complex, and moderately electron-dense secretory vesicles. Surface

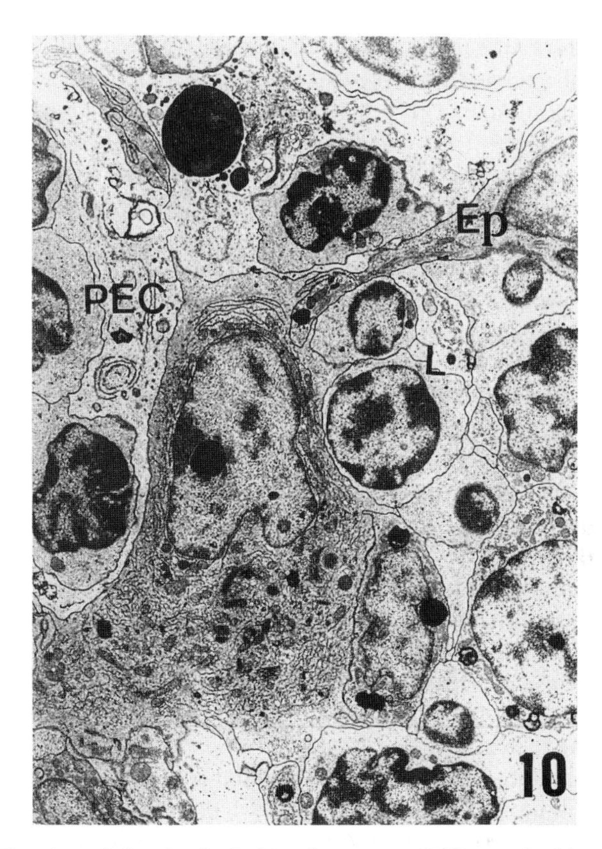

Fig. 10. Thymic medulla of a shark, *Mustelus manazo*. Different sized lymphocytes (L) appear arranged between the supporting epithelial cells (Ep) and the cell processes of pale epithelial cells (PEC). ×4100.

membranous interdigitations, desmosomes, and bundles of tonofilaments confirm the epithelial nature of these cells. Medullary epithelial cysts are frequently observed in the thymus of elasmobranchs. They occur as intracytoplasmic small cavities or as large multicellular complexes that border a big cavity empty or filled with cell debris and/or amorphous materials. Short microvilli or cilia project into the cavity lumen whereas electron-dense granules and numerous filaments accumulate close to the cyst. Hassall's bodies are lacking in the fish thymus, although occasionally they have been reported in light microscopy studies (Good *et al.*, 1966). In the elasmobranch thymus, however, epithelial cells containing masses of degenerated fibrous material that resemble the keratin bundles found in the unicellular Hassall's bodies of mammalian thymus have been observed (Zapata *et al.*, 1996).

The structure of teleost thymus has been well studied, although few ultrastructural analyses have been carried out and there is considerable controversy about the degree of heterogeneity of the thymic epithelium (reviewed in Zapata, 1983; Chylmonczyk, 1992; Zapata *et al.,* 1996). One single type of epithelial cell that constitutes the supporting framework of the whole organ was described for *Rutilus rutilus* and *Gobio gobio* (Zapata, 1981c) and *Scyaces sanguineus* (Gorgollón, 1983). More recently, two different epithelial cell types were identified in the thymus of *Lophius piscatorius* (Fänge and Pulsford, 1985) and *Mugil auratus* (Marinova, 1986). One corresponded to the stellate cells described previously in the cortex of elasmobranch thymus, and the other to a hypertrophic secretory cell similar to the pale cells that appear in the thymic medulla of elasmobranchs.

In addition, epithelial cells but not Hassall's bodies, as already mentioned for elasmobranchs, appear in the teleost thymus, although O'Neill (1989) denies the presence of cell cysts, Hassall's corpuscles, and myoid cells in the thymus of *Harpagifer antarticus.* Both thymulin (Frohely and Deschaux, 1986) and thymosin fraction V–producing epithelial cells have been immunodetected in the teleost thymus, although their relationships to the ultrastructurally identified epithelial cell subtypes as well as their functional relevance, if any, remain unresolved.

Lymphocyte–epithelial cell associations *in situ,* which resemble the so-called nurse cell complexes found in the mammalian thymus, have been observed in the teleost thymus (Pulsford *et al.,* 1991; Alvarez, 1993), whereas thymic macrophage–lymphocyte multicellular complexes occur in both elasmobranchs (Zapata, 1980; Navarro, 1987) and teleosts (Zapata, 1981c; Fänge and Pulsford, 1985). Recently, the presence of thymic nurse-like cells has been confirmed by *in vitro* isolation and characterization of enriched cell suspensions from adult trout thymus (Flaño *et al.,* 1995).

In order to further characterize the thymic epithelial component of rainbow trout we divided the stromal cells into two groups according to their capacity to be stained with antisera raised specifically to vertebrate cytokeratins (Castillo *et al.,* 1990). Among the cytokeratin-negative cells were fibroblasts from thymic capsula and trabeculae, and macrophages from trabeculae and thymic parenchyma. The cytokeratin-containing cells were then classified according to their respective enzyme–histochemical pattern and the location they occupied in the thymus gland (Table III). The main stromal epithelial cell type corresponds to the previously mentioned stellate cells that forms the stroma meshwork in the area referred to by Castillo and colleagues as the inner zone of the trout thymus. These cells are presumably similar to the cortical epithelial cells of mammalian thymus (Nabarra and Adrianarison, 1987), although the latter are negative for alkaline phosphatase. Limiting and peritrabecular epithelial cells are mor-

Table III
Cytokeratin-Containing Cells of the Thymus of Rainbow Trout, *Oncorhynchus mykiss* (modified from Castillo *et al.* (1990) *Thymus* **15,** 153)

	Alkaline phosphatase	Acid phosphatase	Nonspecific esterase	5′-Nucleotidase
Limiting and peritrabecular epithelial cells	+	−	++	−
Large ECs of subcapsulary zone	+	++	+/−	+
Inner zone				
Stellate ECs	+++	−	−	−
Ovoid ECs	−	++	+	+
Outer zone				
Acidophilic ECs	−	−	+/−	−
Cystic cells	−	+++	+++	++

phologically similar to the stellate cells but exhibit a slightly different enzymatic pattern, especially as regards the strong expression of nonspecific esterase. Ovoid cells of the inner zone resemble morphologically the hypertrophied epithelial cells of *Mugil auratus* (Marinova, 1986) and *Lophius piscatorius* (Fänge and Pulsford, 1985), and seem to be involved in the formation of epithelial cysts as proposed by Fänge and Pulsford (1985) and confirm our own results (Castillo *et al.,* 1990). Other keratin-positive cells correspond to large epithelial cells of the subcapsular zone, a region that houses medium and large lymphoid cells, and acidophilic epithelial cells of the outer zone. The enzymatic pattern of these latter ones differs from the other epithelial cells of the thymic stroma and are presumably more related to the epithelial components of the pharynx.

The ultrastructure of thymic epithelial cell microenvironment was also analyzed during the ontogeny of rainbow trout (Castillo *et al.,* 1991), and the results obtained were recently confirmed in juvenile sea bass, *Dicentrarchus labrax* (Abelli *et al.,* 1994). The histogenesis of the gland includes three important steps: the lymphocyte colonization of the paryngeal epithelium, the development of connective tissue trabeculae, and the organization of distinct thymic regions. On hatching, only a few thymocytes appear interspersed among three different epithelial cell types: those cells adjacent to the connective tissue capsule; ramified, electron-dense cells, which seem to be equivalent to the stellate epithelial cells of the inner zone of adult thymus; and electron-lucent, secretory-like cells similar to the pale, ovoid cells of the inner zone of adult thymus. Four days after hatching, the thymus enlarges and numerous gaps appear between the cell processes of the epithelial elements. In 21-day-old trout, thymic trabeculae carrying blood vessels develop and a subcapsular zone appears containing lymphoblasts and large subcapsular epithelial cells. In 30-day-old trout, the outer zone which consists of spindle-shaped epithelial cells and a few, small epithelial cysts appears.

With respect to the embryological origin of these epithelial cell subpopulations, Castillo *et al.* (1991) suggest that epithelial cells of both the outer zone and subcapsulary region may arise from the epithelium lining the digestive part of the pharynx and/or from that covering the inner sides of the operculum. Both are derived from the ectoderm and, remarkably, resemble histologically the teleost epidermis. On the other hand, the epithelial elements of the inner zone might derive from the branchial pharynx, which is considered to be of endodermal origin. This dual origin of the thymic epithelial cell components is assumed for most higher vertebrates, including humans (Crouse *et al.,* 1985).

Macrophages are common in both the cortex and the medulla of fish thymus. Ultrastructurally, they are characterized by the engulfed material that occupies their cytoplasm. Monocytes (Castillo *et al.,* 1990), multinucleated giant cells (Pulsford *et al.,* 1991), and melanomacrophages (Gorgollón, 1983; Pulsford *et al.,* 1991) have been reported in teleost thymus. In the thymus of *O. mykiss,* macrophages are strongly positive for acid phosphatase, 5′-nucleotidase, and nonspecific esterase, and appear scattered throughout the limits between the inner and outer zones, the outer zone, and into the pharyngeal epithelium (Castillo *et al.,* 1990). Those occurring in the inner and subcapsular zone may be involved in the clearance of nonviable thymocytes (Castillo *et al.,* 1990).

No clear evidence is available on the occurrence of interdigitating/dendritic cells in fish thymus. We identified presumptive interdigitating cells which appeared occasionally in the medulla and corticomedullary border of the dogfish, *Mustelus manazo* (Zapata *et al.,* 1996). Other authors have suggested, but not demonstrated, their presence in the thymus of young dogfish (Navarro, 1987) and wild brown trout (Alvarez, 1993). Ultrastructurally they resemble the medullary electron-lucent epithelial cells of elasmobranch thymus already described but they do not contain either tonofilaments or desmosomes. They show few membranous organelles, including short profiles of rough endoplasmic reticulum, tubules of smooth endoplasmic reticulum, and elongated mitochondria. Prominent surface foldings and interdigitations also appear on their surface.

Myoid cells are present in the fish thymus (Zapata and Cooper, 1990), although some authors have been unable to observe them (Pulsford *et al.,* 1984 in *S. canicula;* O'Neill, 1989 in *H. antarticus*). There are important variations in the number of myoid cells reported in different classes of vertebrates with only very few recorded for some fish species (Gorgolloń, 1983). On the other hand, their functional significance still remains obscure, even in higher vertebrates. They appear as round or oval, large cells with a lightly electron-dense nucleus and a cytoplasm containing myofilaments organized in sarcomere-like structures around the nucleus. Degenerated myoid cells have also been found as effected cells within macrophages (Fänge and Pulsford, 1985).

Occasionally, we have found small numbers of cells containing electron-dense cytoplasmic granules that resemble those of neuroendocrine cells in the thymus of various elasmobranchs (Zapata *et al.,* 1996). These have in fact been described in the thymus of all vertebrates (Zapata and Cooper, 1990), although their function within the gland is unknown.

Presence of mature plasma cells and/or Ig-positive cells in the thymus gland has also been demonstrated in all vertebrates including fish, although their origin and function is a matter of discussion. Classical studies using anti-Ig antisera showed numerous Ig-expressing cells in the thymus of both elasmobranchs (Ellis and Parkhouse, 1975) and teleosts (van Loon *et al.,* 1981). Later, the use of specific anti-Ig monoclonal antibodies confirmed the existence of small numbers of these cells (de Luca *et al.,* 1983; Secombes *et al.,* 1983). Furthermore, *in vitro* studies confirmed the appearance of plaque-forming cells in teleost thymus; and intrathymic plasma cells were ultrastructually identified (Zapata, 1981b; Pulsford *et al.,* 1991). They could develop *in situ* in the thymic cell microenvironment if the gland contains B-cell precursors, or they could migrate there occasionally (an unresolved issue), or they might be related to the pool of circulating mature B lymphocytes which in a small proportion migrate into and out of the thymus, at least in mammals.

It is well known that both the structure and function of vertebrate lymphoid organs undergo seasonal variations, a subject described in other chapters of this volume, and repeatedly reported by our group (Zapata and Cooper, 1990; Zapata *et al.,* 1992; Zapata, 1996). In teleosts, as in other vertebrates and especially wild birds, a transitional intrathymic erythropoiesis has been found (Alvarez *et al.,* 1994). This process, which in birds coincides with conditions of high blood demand and can be induced by hemorrhage, has been related to seasonal variations in circulating hormone levels, mainly the sex steroids and thyroid hormones (Fonfría *et al.,* 1983), although we could not establish such a correlation in trout thymus (Alvarez *et al.,* 1994).

Little information is available on thymic vascularization in fish. Chilmonczyk (1992) reported that thymic vascularization in teleosts seems to be derived from the gill vascular system and, therefore, should be solely of arterial origin. Accordingly, blood supply to the trout thymus is achieved by the thymic artery, a branch of the segmental artery coming from the second and third afferent branchial arteries. After entering the cephalic part of the organ, the artery splits repeatedly to form a dense capillary network throughout the entire organ. These capillaries are gathered into the thymic vein and further collected into the anterior vena cava.

Few data are currently available on the innervation of fish thymus. Lagabrielle (1938) demonstrated that innervation of the teleost thymus is by the sympathetic system. Apparently, depending on the location of the thymus in the epithelium of the gill chamber, the fourth (*Salmo salar*) or the fifth (*Corvina umbra*) sympathetic ganglion innervates the thymus.

In general, the fish thymus, like that of higher vertebrates, involutes with age, although the histological details of the process remain largely obscure, and few systematic studies are available. Several years ago we used light microscopy to analyze the histological changes that occur in the lymphoid organs, including the thymus, of an annual small killifish, *Notobranchius guentheri,* throughout its life span (Cooper *et al.,* 1983). The earliest changes, which began to appear in the thymus at four months, consisted principally of an increase of the connective tissue. By 6 months, the thymus showed increased numbers of epithelial cysts and a concomitant decrease in lymphocytes. By 12 months, the thymic tissue of several fish was completely degenerated. There were fewer lymphocytes which appeared masked by massive amounts of connective tissue. At the same time, there were marked tumoral growths apparent in the oral cavity, connective tissue, and mainly kidney and liver. More recently, Ellsaesser *et al.* (1988) studied the thymic involution in the channel catfish. The organ remained constant from 3 to 10 months, sharply increasing in size between 11 and 12 months and then began to decrease after 13 months. At 16 months after hatching, the thymus consisted of a thin epithelioid layer without lymphocytes, and at 18 months was no longer macroscopically visible.

Data on the functionality of fish thymus are restricted to teleosts. Indirect evidence comes from experiments on either the migration of labeled thymocytes to peripheral lymphoid organs or the effects of early thymectomy on the maturation of the immune system (reviewed in Zapata and Cooper, 1990; Chilmonczyk, 1992; Manning, 1994). It suggests that teleost thymus, despite its striking morphology, has the same function as in higher vertebrates, that is, it is the main source of immunocompetent T cells.

Rainbow trout injected intrathymically with tritiated thymidine show radiolabeled lymphocytes in both the spleen and kidney (Tatner, 1985). In addition, trout-labeled blood lymphocytes migrate through the thymus before reaching the spleen and kidney (Tatner and Findlay, 1991). In contrast, plaice lymphocytes collected from the neural duct and reinjected into the same fish after *in vitro* labeling with ^{3}H-uridine did not significantly colonize the thymus (Ellis and de Sousa, 1974).

The effects of thymectomy on the maturation of the fish immune system are discussed in another chapter of this volume. In general, the available data support a correlation between the histological maturation of the teleost thymus, appearance of the lymphocytes in peripheral lymphoid organs, and development of the cell-mediated immune responses.

B. The Secondary Lymphoid Organs

In all vertebrates, secondary peripheral lymphoid organs provide the structural organization for trapping antigens and processing them in suitable ways for the cells that are capable of reacting to them. Although the lymphoid organs in fish, that are the equivalent of bone marrow (apart from their blood-forming function) presumably participate in the immune responses and could therefore be categorized as secondary lymphoid organs, both the spleen and the gut-associated lymphoid tissue (GALT) constitute the main peripheral lymphoid organs. Lymph nodes appear for the first time in birds, but lymphoid aggregates associated with the lymphatic system have been described in both amphibians and reptiles (Zapata and Cooper, 1990). In fish, the secondary circulatory system has remarkable similarities with the lymphatic system of other vertebrates, although little is known about its function. Indirect evidence suggests that the secondary circulatory system of plaice contains all types of leukocytes, but not erythrocytes, found in the primary circulation (Ellis and de Sousa, 1974; Ishimatsu *et al.*, 1992), and recently, it has been speculated that it may be involved in fish immune reactivity (Ishimatsu *et al.*, 1995).

1. The Spleen

The spleen is the major peripheral lymphoid organ of all gnathostomus vertebrates. It is a large, blood-filtering organ that undergoes increasing structural complexity in order to augment its efficiency in trapping and processing antigens. Nevertheless, the amount of lymphoid tissue in the spleen of distinct vertebrates varies greatly, reflecting mainly the pattern of blood circulation and/or the occurrence of other peripheral lymphoid organs. In fish, teleosts that contain abundant lymphohemopoietic tissue in the kidney have a poorly developed splenic lymphoid tissue, whereas in elasmobranchs, which, as already mentioned, lose their renal lymphohemopoietic tissue in adult life, the spleen is an important lymphoid organ. Information on the structure, cell content, and immune functions of the splenic lymphoid tissue in other groups of fish is, however, very limited.

The strong reduction of mesenteries in both *Chimaera* and *Hydrolagus* keeps the spleen of holocephalans free from the stomach and intestine, although it is strongly associated with the pancreas (Fänge, 1984). In general terms, its histological organization resembles that of elasmobranch spleen (Fänge, 1984; Fänge and Nilsson, 1985; Mattisson, pers. comm.). Lymphoid foci constitute the white pulp, and mature and developing erythroid cells and thrombocytes occupy the red pulp. In addition, large ellipsoids appear as pale structures throughout the splenic parenchyma.

In selachians, the spleen is an elongated, lobed organ close to the duodenum. In rays and skates, in contrast, it appears as a round, compact, only slightly lobed mass. The vascularization of elasmobranch spleen was briefly reviewed by Fänge and Nilsson (1985) and Zapata and Cooper (1990), and more recently by Tanaka (1993). According to Fänge and Nilsson (1985), arterial supply to the spleen of a dogfish, *Squalus acanthias,* is achieved by the lienogastric artery originating from the dorsal aorta. In the splenic parenchyma, this artery produces smaller branches without anastomosing. The terminal branches, usually referred to as capillaries, bear thickened walls or sheaths, often named ellipsoids. In semithin sections of dogfish (*Triakis scyllia*) spleen fixed by perfusion, arteries, ellipsoids, and veins are clearly distinguished (Fig. 11). Unmyelinated nerve fibers and their endings are demonstrable by electron microscopy of the smooth muscle layer of the arterial wall. These fibers are considered to be branches of the middle splanchnic nerve (Fänge and Nilsson, 1985). The examination of perfused spleen and splenic vascular corrosion cast by scanning electron microscopy demonstrates that the ellipsoids open into the splenic cords, which are a three-dimensional meshwork of reticular cells where several kinds of blood cells are housed. This meshwork is gradually gathered into the splenic veins and finally joins the hepatic portal system. Ellipsoids appear in many fishes as well as in other vertebrates, but they may be indistinct or lacking in certain species (Yoffey, 1929). In the *Triakis* spleen, ellipsoids are morphologically distinct. In the terminal region of the central capillaries, the endothelial cells occasionally show a discontinuous basement membrane and often appear perforated or loosely connected. These minute spaces lead into the surrounding macrophage sheath, which is generally considered to function as part of the filter or trap for disfunctioning free cells and antigens or immune complexes (see later). In the ellipsoid, adjoining macrophages sometimes seem to be firmly associated with each other by interdigitations seen typically in the epithelial cells (Fig. 12). This condition suggests that macrophages in ellipsoid walls could act as sessile but also fixed cells. In some elasmobranchs, ellipsoidal macrophages contain abundant lipid inclusions and cellular debris (Yoffey, 1929; Zapata, 1980), con-

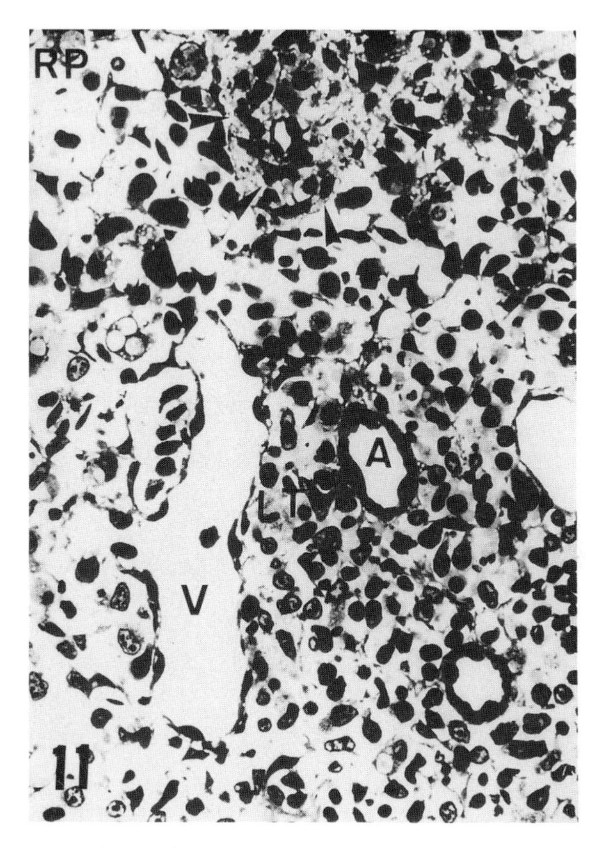

Fig. 11. Arteries (A), veins (V), and ellipsoidal blood vessels (arrowheads) in the splenic parenchyma of a dogfish, *Triakis scyllia.* Lymphoid tissue (LT), red pulp (RP). ×500.

firming Dustin's opinion (1975) that in some species of higher vertebrates, the ellipsoids could function as lipid storage depots.

In the elasmobranch spleen, large masses of lymphoid tissue, which surround central arteries and peripheral ellipsoids, constitute the white pulp (Fänge and Nilsson, 1985; Zapata and Cooper, 1990). It contains different sized lymphocytes, numerous developing and mature plasma cells, and macrophages in a supporting network of fibroblastic reticular cells. Macrophages, lymphocytes, and plasma cells form cell clusters that are possibly involved in immunological exchange. In addition, immunofluorescence double staining has revealed two types of antibody-producing cells in the spleen of some elasmobranchs (Tomonaga *et al.,* 1984). In some species, the limits between the lymphoid white pulp and red pulp, which houses developing and mature erythroid cells and thrombocytes, are poorly

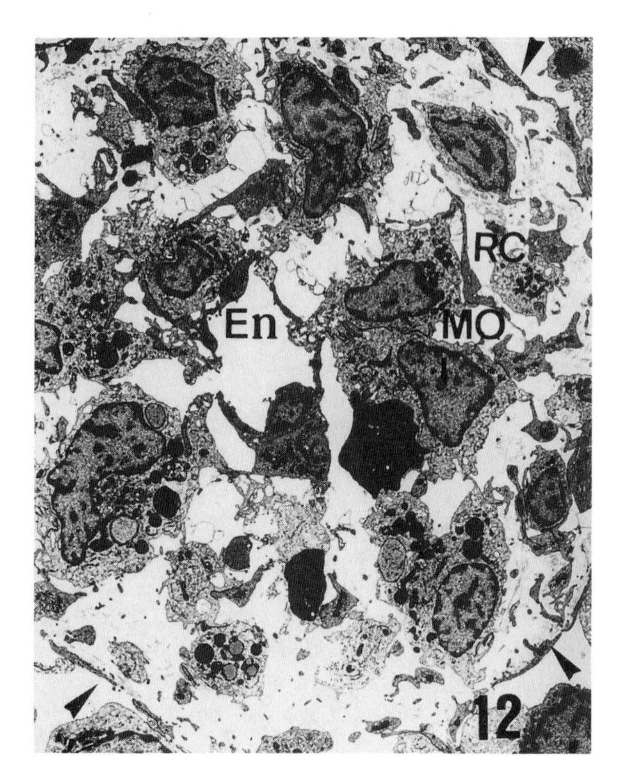

Fig. 12. Ellipsoidal blood vessel of the spleen of *T. scyllia.* Note the endothelial cells (En) surrounded by macrophages (MO) in a meshwork of reticular fibers and cell processes (RC) which define a thin reticular capsule in the outermost of the ellipsoid (arrowheads). ×2000.

defined. However, especially after antigenic stimulation, clearly distinguishable lymphoid aggregates appear (Morrow, 1978). Together, these data indirectly support a primary role for the spleen in antibody synthesis in elasmobranchs and, therefore, in immune responsiveness (Morrow, 1978; Zapata, 1980; Tomonaga *et al.*, 1984, 1985, 1992; Pulsford and Zapata, 1989), despite classical reports of the immune response failing to decrease in splenectomized sharks (Ferren, 1967).

In ganoids (chondrosteans and holosteans), the spleen comprises a red and a white pulp (Good *et al.*, 1966). The white pulp of holosteans consists of small lymphoid aggregates scattered throughout the splenic parenchyma and contain numerous lymphocytes and plasma cells, the number of which increase markedly after extensive immunization (Good *et al.*, 1966). In sturgeons, the white pulp is organized as follicle-like periarterial masses containing lymphocytes, granulocytes (mainly acidophils), and scattered macrophages (Clawson *et al.,* 1966; Fänge, 1984). Ellipsoidal blood vessels are well developed and are surrounded by lymphoid tissue rather then red pulp.

Numerous studies have focused on the histology of the teleost spleen (see reviews of Zapata, 1983; Zapata and Cooper, 1990) although there is little ultrastructural data (Zapata, 1982; Quesada *et al.,* 1990; Zapata and Cooper, 1990; Press *et al.,* 1994). The lymphoid tissue is poorly developed in the teleost spleen. It surrounds small arteries, appears diffuse in the splenic parenchyma, and is related to the so-called melanomacrophage centers (see later). After antigenic stimulation, increased amounts of lymphoid tissue does appear, in the spleen of teleosts. The lymphoid tissue of teleost spleen consists of lymphoid cells, mainly small, medium, and large lymphocytes arranged in a supporting reticular cell meshwork. Some recent data suggest that these stromal reticular cells of the teleost spleen represent a truly heterogeneous cell population. Quesada *et al.* (1990) reported ultrastructural differences between the reticular cells in the red pulp and those of the white pulp of the sea bass spleen, and Press *et al.* (1994) observed variations in the intensity of enzyme reactivity for alkaline phosphatase and 5′-nucleotidase within the spleen, which could reflect differences between distinct reticular cell populations. On the other hand, macrophages appear in both the red and the white pulp, and in the latter, macrophage–lymphocyte–plasma cell clusters have been demonstrated ultrastructurally (Zapata 1982).

In the teleost spleen, ellipsoidal blood vessels are less developed than in the spleen of elasmobranchs, but are organized according to the same pattern with terminal capillaries that show a thin endothelial cell layer surrounded by a sheath of reticular fibers and macrophages (Yoffey, 1929). Remarkably, the splenic arterioles react for nonspecific esterase and alkaline phosphatase, but this reactivity disappears as the vessels enter the ellipsoids, reflecting the special nature of ellipsoidal endothelia (Press *et al.,* 1994). These authors also analyzed the histoenzymatical pattern of ellipsoidal macrophages from teleost spleen, remarking on its similarities to the marginal zone macrophages of the mammalian spleen. In fact, a marginal zone limiting the white and the red pulp is lacking in the teleost spleen, and according to scanning electron microscopic studies on vascular corrosion casts, the splenic circulation is open to the arterial capillaries that end in the reticular meshwork of the red pulp (Kita and Itazawa, 1990).

Antigen-binding and/or antibody-producing cells have been detected, and trout splenocytes can be stimulated by LPS, PPD, and ConA (Zapata, 1983; Zapata and Cooper, 1990), indirectly suggesting the presence of T-like and B-like cells in the teleost spleen. Functions of the splenic lymphoid tissue of teleosts remain, however, controversial, although its role in antigen processing (discussed later) seems to be certain. Splenectomy has no effect on the humoral responses to BSA in some teleosts (Ferren, 1967) although in other species the spleen apparently represents a major lymphoid organ (Yu *et al.,* 1970). Accordingly, Tatner (1985) found a preferential migration of trout thymocytes into the spleen, a fact previously reported by Ellis and de Sousa (1974) in the plaice, and there was a greater involvement of carp spleen in secondary immune responses. Primary intraperitoneal administration of human γ-globulin (HGG) in saline or FCA induced few changes in the spleen. In contrast, a secondary immunization with HGG in FCA, but not in saline, generated numerous pyroninophilic cells closely associated with splenic ellipsoids. More recently, large clusters of Ig-positive cells have been observed in *Salmo salar* 3 months after vaccination (Press *et al.,* 1994), and increased numbers of pyroninophilic cells appear around ellipsoids and melanomacrophage centers after primary and secondary immunization (van Muiswinkel *et al.,* 1991).

The structural organization of the spleen of *Calamoichthys,* as described by Yoffey (1929), is remarkably primitive. Splenic arteries and veins run parallel to each other within lymphoid tissue surrounded by red pulp. In *Polypterus senegalus,* an elongated, voluminous spleen appears closely associated with the intestine. It is histologically poorly developed with unclearly delimited red and white pulp. Together with lymphocytes, clusters of both mature and immature plasma cells and striking granulopoietic foci occupy the splenic parenchyma (Fig. 13).

The spleen of dipnoans is divided into two nonconnected areas within the alimentary canal (Fig. 14). Saito (1984) described anatomically the correlation between the development of the spleen and the enteric blood vessels in the Australian lungfish, *Neoceratodus forsteri,* showing the relationships between the vascular dynamics of the foregut and the yolk sac, and the formation of the spleen. Thus, the splenic primordium first appears as a mesenchymal condensation supplied by the third and fourth vitelline arteries. Gradually, the development of splenic sinuses within the primordium and the formation of the gastric and enteric splenic portal systems occur, with the organ finally growing along the anterior extremity of the spinal valve. In the South American lungfish, *Lepidosiren paradoxa,* it forms a compact organ within the wall of the stomach and the anterior part of intestine, in which red and white regions appear clearly (Good *et al.,* 1966; Fänge, 1982). Jordan and Speidel (1931) described histologically three regions in the spleen of *Protopterus ethiopicus;* the central one consisted of lymphoid cells surrounded by a region, active in erythropoiesis, of cell cords and blood sinuses, and a thin, peripheral capsular zone.

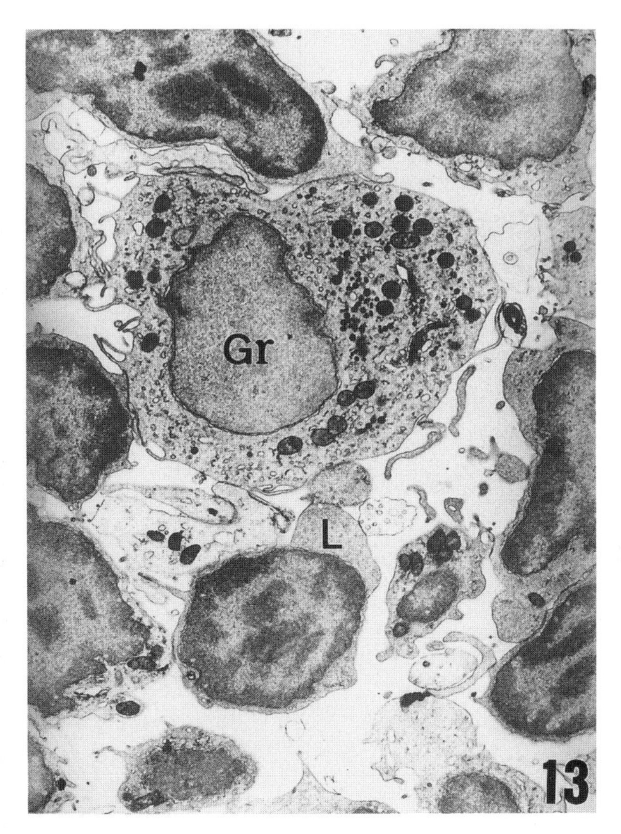

Fig. 13. Granulocytes (Gr), small and medium lymphocytes (L) in the splenic parenchyma of *Polypterus senegalus.* ×3000.

2. Gut-Associated Lymphoid Tissue (GALT) and the Mucosal Immune System

All vertebrates, including Agnatha, contain lymphoid cells isolated in the lamina propria and the intestinal epithelium, but well-organized lymphoid aggregates appear for the first time in Chondrichthyes (Fichtelius *et al.,* 1968; Tomonaga *et al.,* 1986; Hart *et al.,* 1988; Zapata and Cooper, 1990; Zapata *et al.,* 1996). Apart from considerable species-specific variations in size, the histological organization of fish gut-associated lymphoid aggregates is similar in all species studied. It consists of nonencapsulated lymphoid accumulations that contain mainly lymphocytes, macrophages, and plasma cells, as well as different types of granulocytes.

Small lymphoid aggregates had been observed in the spiral valve and/or duodenum of several elasmobranch species (see Zapata *et al.,* 1996), but

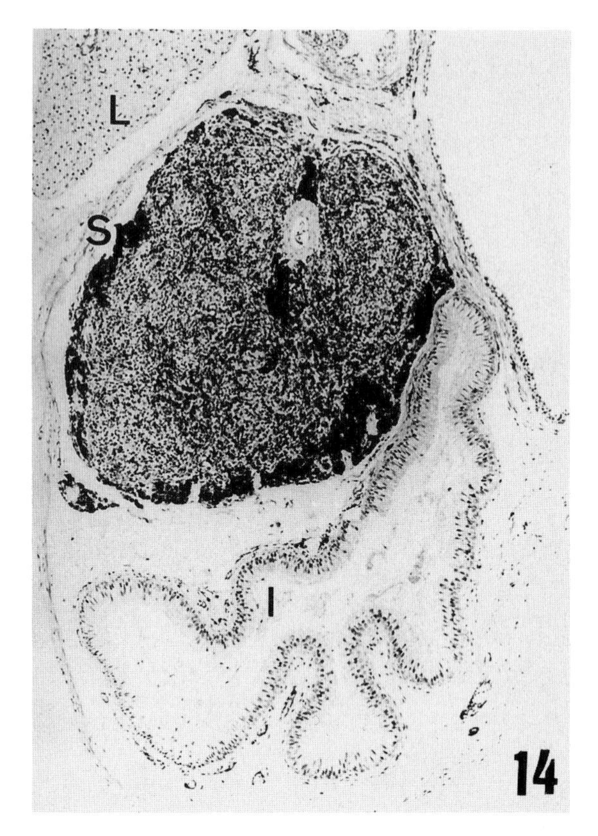

Fig. 14. Spleen (Sp) of a lungfish, *Lepidosiren paradoxa,* closely associated with the intestinal walls (I) at the caudal part of the liver (L). ×30.

more recently, massive infiltrations of lymphoid tissue have been reported in the spiral valve of several sharks, rays, and dogfish (Hart *et al.,* 1986, 1988; Tomonaga *et al.,* 1986). Although Tomonaga *et al.* (1986) remarked that some lymphoid cells present in the intestinal epithelium of elasmobranchs contained cytoplasmic Ig, Hart *et al.* (1988) have emphasized that plasma cells are restricted to lamina propria, whereas granular cells, macrophages, and lymphocytes also appear within the epithelium. In this location, lymphocytes are electron-lucent cells devoid of most cytoplasmic organelles (Fig. 15). As in other lymphoid organs, both high molecular weight (HMW) and low molecular weight (LMW) Ig-forming cells appear in the gut lamina propria of the skate *Raja kenojei* (Tomonaga *et al.,* 1984).

In chondrosteans, mainly in the paddlefish, *Polyodon spathula,* which are frequently parasitized, but not in *Scapirhynchus plathorhynchus,* which

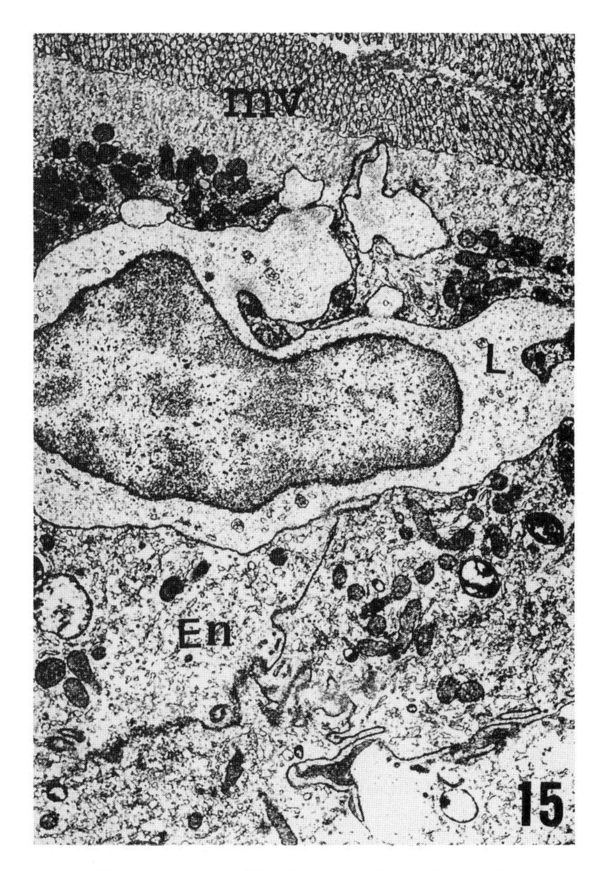

Fig. 15. Intraepithelial lymphocyte (L) of the intestine of *Mustelus manazo*. Enterocytes (En), microvilli (mv). ×6250.

remains uninfected, there is a rich development of the GALT with dense lymphoid accumulations in the region of the ileocecal valve (Good *et al.*, 1966). Lymphoid tissue has also been observed in the gut of sturgeons (Fänge, 1984). In garfish, *Lepisosteus platostomus,* there are no intestinal lymphoid accumulations, but both garfish and bowfin (*Amia calva*) contain considerable numbers of intraepithelial lymphocytes (Fichtelius *et al.*, 1968). Also, the American lungfish, *Lepidosiren paradoxa,* possesses GALT (Good *et al.*, 1966).

In all teleosts, diffuse accumulations of lymphoid tissue appear along the gut (Doggett and Harris, 1991), although in those species that exhibit two distinct intestinal segments, the second one seems to be specialized for antigen trapping and processing (Hart *et al.*, 1988). Apart from different-sized lymphocytes, plasma cells (although not abundant) have been identified in the GALT of numerous teleosts (Rombout *et al.,* 1986; Temkin and McMillan, 1986), increasing in numbers after intraoral antigenic challenge (Pontius and Ambrosius, 1972). Macrophages appear in the intestinal epithelium as well as in the lamina propria and could be involved in scavenging and/or antigen presentation. Intestinal macrophages of the tilapia, *Oreochromis mossambicus,* contain particulate debris even in nonstimulated fish, and 1 h after ferritin administration their number increases in the basal region of epithelium, frequently closely associated with intraepithelial lymphocytes (Doggett and Harris, 1991). Twenty-four hours after intubation, mainly ferritin-containing macrophages occupy the lamina propria. Furthermore, Rombout and colleagues have demonstrated that intestinal macrophages in carp express antigen determinants on their outer surface (Rombout *et al.,* 1986, 1989; Rombout and van den Berg, 1989). This same group reported that isolated, mainly intraepithelial macrophages from carp gut exhibit striking differences to macrophages isolated from other lymphoid organs (Rombout *et al.,* 1993). Like intestinal macrophages of mammals, they showed poor or little adherence to glass or plastic and possibly possess Ig binding capacity. In addition, they form frequent cell clusters with lymphocytes, as previously described *in situ* for goldfish (Temkin and McMillan, 1986).

Rombout *et al.* (1993) also analyzed the phenotypical characteristics of intestinal lymphoid cells using monoclonal antibodies specific either to carp Ig or all leukocytes. Previously, Bielek (1988) had indicated that most, if not all, carp intraepithelial lymphocytes were Ig negative, probably representing T cells or NK cells. Rombout *et al.* (1993) confirmed that apparently Ig-positive cells are restricted primarily to the lamina propria and represent B lymphocytes and/or plasma cells. Various types of granulocytes have also been described in the GALT of teleosts (Zapata, 1979b; Davina *et al.,* 1982; Temkin and Mc Millan, 1986). Especially remarkable is the occurrence of PAS-positive (Vallejo and Ellis, 1989) and eosinophilic granular cells (Bergeron and Woodward, 1982), which might mediate hypersensitive reactions, although IgE is lacking in fish and their mast cells apparently do not contain histamine (Lamas *et al.,* 1991).

As regards the histological evidence of lymphoid tissue associated with the gut and other mucosae including the reproductive tract, skin, and gills, numerous reports in recent years have supported the existence of a common mucosal immune system in fish, mainly in teleosts but also in elasmobranchs (Hart *et al.,* 1988; Zapata and Cooper, 1990). We describe the evidence supporting this occurrence as well as some histophysiological characteristics of the system.

Antigens administered into the gut lead to an increase in the number of intraepithelial leukocytes (Davina *et al.,* 1982), induce the production of specific antibodies in the mucosae and bile but not in the serum (Hart *et al.,* 1988; Davidson *et al.,* 1993), and can elicit a protective immune response in skin mucus (Hart *et al.,* 1988). Although the opposite situation has also been described, generally, oral immunization that induces specific antibodies in the serum but not in the mucus subsequently leads to immunosuppression (Rombout *et al.,* 1989). Davidson *et al.* (1994) also observed a partial or total suppression of circulating anti-HGG antibody titers after oral or anal immunization, respectively. However, when the same oral route was used to administer *A. salmonicida,* instead of observing suppression of the systemic immune system, an enhanced response appeared (Davidson *et al.,* 1994).

IgM has been identified in the gut mucus and bile of both elasmobranchs and teleosts (Hart *et al.,* 1988). However, except for the biliary Ig of sheepshead, *Archosargus probatocephalus,* which appears as a dimer and differences in the molecular weight of the heavy chain of biliary and serum Ig (Lobb and Clem, 1981a), only slight differences have been found between mucosal and serum Igs (Hart *et al.,* 1988; Rombout *et al.,* 1993; Fuda *et al.,* 1992). In the bester, a sturgeon, skin mucus Ig has a lower molecular weight but is antigenically similar to serum Ig (Kintsuji *et al.,* 1994). Furthermore, biliary Ig levels are similar to those of serum in dogfish (Hart *et al.,* 1988), whereas in sheepshead and carp, they are lower in the bile than in the serum (Lobb and Clem, 1981a,b). In this respect, mucosal Igs of *A. probatocephalus* are mainly the result of local synthesis (Lobb and Chem, 1981c). On the other hand, a secretory component has not been identified in the intestine and bile of dogfish (Hart *et al.,* 1988) and some teleosts (Lobb and Clem, 1981c; Rombout *et al.,* 1993). However, it does occur in the skin mucus of sheepshead, and the serum of the nurse shark, *Ginglyomostroma cirratum,* exhibits a high affinity for mammalian secretory component (Underdown and Sockin, 1978).

Bath immunization of catfish, *Ictalurus punctatus,* with dinitrophenyl–horse serum albumin (DNP-HoSA) enhances secretory immunity but is not effective in stimulating the systemic immune system, since the titer of mucosal antibodies increases independently of the systemic humoral response (Lobb, 1987). Davidson *et al.* (1993) indicated recently that the mucosal and systemic immune compartments of rainbow trout are both active although they exhibit different kinetics. After intraperitoneal injection of *A. salmonicida,* the first antibody-secreting cells appeared in the pronephros 2 weeks later, but it was not until week 7 that a significant response was recorded in the intestine. Conversely, antibody-secreting cells appeared at the same time (3 weeks postimmunization) in both organs when fish were immunized orally (Davidson *et al.*, 1993).

Although the mechanisms involved remain to be elucidated, larval dogfish enterocytes internalize and process intact protein molecules (Hart *et al.*, 1988). In teleosts, numerous data support the uptake in the gut of bacterial antigens, whole bacteria, and viruses (Hart *et al.*, 1988). Administered proteins have been found in intercellular spaces within intraepithelial lymphocytes and macrophages, and free in the systemic circulation. Furthermore, enterocytes in the second gut segment of carp have been claimed to be analogous to the M cells of mammalian gut and to transfer antigens to mucosal lymphoid accumulations (Rombout *et al.*, 1993). Other studies demonstrate, however, that the proteins are totally digested within enterocytes (see review by Hart *et al.*, 1988). On the other hand, the absorption capacity apparently varies throughout the development. In general terms, enterocytes of larval or young fish absorb more readily than those of adult animals (Hart *et al.*, 1988). Likewise, the route of administration is important for determining the local and/or systemic immune response. Although various bacteria or viruses are trapped after both oral and anal administration, anal intubation in trout produces a better protective response than either oral or immersion delivery (Hart *et al.*, 1988).

C. Ontogeny of the Fish Lymphoid Organs

The ontogeny of fish lymphoid organs has been reported for several teleosts and in the dogfish, *Scyliorhinus canicula.* Unfortunately, the development of lymphoid tissues has only occasionally been correlated with the onset of distinct immune capacities.

1. Ontogeny of Dogfish Lymphoid Organs

The ontogenetical development of dogfish lymphoid organs has been studied by light and electron microscopy (Hart *et al.*, 1986; Navarro, 1987; Lloyd-Evans, 1993) but no correlation has as yet been shown with the functional maturation of the immune system. This section is devoted to the histological development of the thymus, spleen, and GALT. The ontogeny of other elasmobranch lymphoid organs, such as the Leydig's organ, epigonal organ, and kidney has already been reviewed.

The thymic primordium appears in 1.5-cm-long dogfish embryos as dorsal, paired excrecences of pharyngeal epithelium on both sides of the gill arches, formed by homogeneous, electron-lucent, polygonal epithelial cells. First, lymphoid cells house the thymic anlagen when it is still closely associated to the pharyngeal epithelium. In 3.5-cm-long embryos the thymus gland is already an isolated, lobed organ, the epithelial cells of which show certain morphological heterogeneity. Limits of the cortex–medulla are established throughout stage III, and most histological components found in the thymus of young dogfish (stage IV) appear before hatching. Apart from the supporting epithelial cells there are presumptive secretory cells, some of them containing abundant vesicles and intracytoplasmic cysts, macrophages, and obviously, lymphocytes. Interdigitating cells have been described in dogfish embryos after hatching (Navarro, 1987).

The splenic primordium appears as an isolated organ along the ventral surface of the gut in 3-cm-long dogfish embryos (Navarro, 1987). In this developmental stage, the organ consists of closely joined mesodermal cells with some blood sinuses. The mesodermal cells evolve to a loose reticular network which, at the end of the external gill stage, is colonized by lymphoid cells. In the following stages, the lymphoid tissue increases around the splenic arteries to form an incipient white pulp at the prehatching stage (Navarro, 1987; Lloyd-Evans, 1993). Occasional Ig-positive cells, blasts, and plasma cells appear in the embryonic spleen (Lloyd-Evans, 1993). The first ellipsoids appear in stage III embryos as sheaths of both reticular cells and macrophages that embrace the final region of capillaries.

The ontogeny of the lymphoid tissue of the spiral valve of dogfish was studied by light microscopy by Hart *et al.* (1986). Tomonaga *et al.* (1986) reported that gut lymphoid accumulations appear prior to parturation in viviparous sharks, whereas in ovoviviparous dogfish they develop before feeding (Hart *et al.*, 1986). At the end of stage II, when the thymus is already organized, the spiral valve contains some lymphoid cells and macrophages in the lamina propria. In the following stage, the lymphoid tissue of the spiral valve reaches the adult form, consisting of small lymphoid aggregates that gradually increase in size and cell content. Intraepithelial lymphocytes appear during stages IV and V, whereas both granular cells and plasma cells are evident only 6 months after hatching.

With respect to the parameters regulating GALT development in elasmobranchs, in particular the possible influence of early antigenic stimulation, Tomonaga *et al.* (1986) showed that, as in mammals, although antigens can accelerate the development of intestinal lymphoid tissue, it is already developed in stages considered to be antigen free.

2. Ontogeny of Teleost Immune System

It is difficult to establish a concrete, common pattern of immune system development in different teleosts, principally because of the different methods used for determining the developmental stages in each species and the dependence of these stages on environmental parameters such as temperature, photoperiod, and so forth. For example, lymphoid colonization of the thymus has been dated to occur 22 days prehatch in *Salmo salar* (Ellis,

1977), 5 days prehatch in *Oncorhynchus mykiss* (Manning, 1994), 4 weeks posthatch in the Antarctic fish *Harpagifer antarcticus* (O'Neill, 1989), and in 4 months posthatch in the channel catfish (Grizzle and Rogers, 1976). Nevertheless, some general patterns can be observed.

Most authors agree that the thymus is the first organ to become lymphoid in teleosts (Zapata and Cooper, 1990; Chilmonczyk, 1992; Pulsford *et al.*, 1994), although Chantanachookhin *et al.* (1991) described that the kidney developed first, followed by the spleen and the thymus. In the marine rockfish, *Sebasticus marmoratus,* Nakanishi (1986, 1991) has reported that the thymus develops after the kidney and the spleen. Nevertheless, presumptive lymphohemopoietic stem cells appear in the kidney before the onset of lymphoid differentiation in the thymus (Zapata and Cooper, 1990; Pulsford *et al.,* 1994), although mature lymphocytes colonize the kidney later. It can, therefore, be concluded that the kidney contains hemopoietic stem cells before mature cells appear in the thymus, but mature lymphocytes are present in the kidney only after they are found in the thymus. Although Ellis (1977) reported no evidence of lymphopoietic foci in the yolk sac walls of *S. salar,* erythrocytes and macrophages are present before development of the kidney. The yolk sac is therefore probably the earliest organ exhibiting certain hemopoietic capacities, in teleost embryos, as indicated for dogfish. This must still be confirmed experimentally, and the possible relationships between the different hemopoietic loci found throughout embryonic life remain unknown.

By using monoclonal antibodies either against serum Ig or thymocytes in carp, T-marker-expressing cells appear first in thymus and later in the kidney, whereas Ig-positive cells appear quite late in the kidney (Secombes *et al.,* 1983). Also, Ellis (1977) reported the late appearance of IgM-positive cells in *S. salar.* According to his results, IgM-positive cells and capacity for MLR appear simultaneously (by day 48 posthatch) and coincide with the onset of larval feeding. In contrast, van Loon *et al.* (1981) demonstrated an earlier appearance (days 14–21 postfertilization) of cytoplasmic and surface Ig-positive cells in both thymus and kidney. The early presence of Ig-positive cells in the carp pronephros was recently confirmed by flow cytometry using a monoclonal antibody reactive with the heavy chain of carp Ig (Koumans–van Diepen *et al.,* 1994). Plasma cells did not appear, however, before one month of age, coinciding with an important increase of serum Ig levels. In rainbow trout, Ig-positive lymphocytes were identified immunohistochemically on sections of kidney 4 days after hatching (Razquin *et al.,* 1990) but on renal cell suspensions 12 days before hatching (Castillo *et al.,* 1993). In accordance, serum Ig values peaked earlier in trout (Castillo *et al.,* 1993) than in carp (van Loon *et al.,* 1981). In both, traces of Ig appear in eggs; we discuss this finding in the following section.

In carp, the Ig concentration remains low until day 21, increasing then gradually to reach adult values at 5–8 months of age (van Loon *et al.*, 1981). In rainbow trout, the levels of IgM remain similar to those of unfertilized eggs until Vernier's stage 28, in which surface-IgM-positive cells appear. From this stage they increase gradually, reaching a peak at hatch, coinciding with the occurrence of IgM-positive cells in pronephros, and then diminish (Castillo *et al.,* 1993).

The spleen matures long after the thymus and kidney develop (Ellis, 1977). Van Loon *et al.* (1981) describe a small spleen in carp 14 days postfertilization that acquires adult histological features by the time 6- to 8-week-old carp are examined, and Pulsford *et al.* (1994) demonstrate the absence of both ellipsoids and melanomacrophage centers in the spleen of young *Platichthys flesus,* which are very prominent in adult flounders. As already mentioned, apart from its minor immunological relevance in adult teleosts, the spleen seems to mature slowly.

D. *In Vivo* Antigen Trapping in Teleosts. The Structure and Functional Significance of Melanomacrophage Centers (MMCs)

The trapping and processing of antigens are preliminary, essential steps for the induction of most immune responses. Largely analyzed in teleosts, these steps involve various organs such as gills, skin, gut, spleen, and kidney as well as the so-called melanomacrophage centers (see Zapata, 1983; Lamers, 1985; Zapata and Cooper, 1990). Knowledge of the route of entry of pathogens and/or immunogenic materials in the natural environment in which fish live is of special importance. In these conditions, gills, skin or lateral line, and gut have been claimed to be the main, obviously nonexclusive, routes for *in vivo* antigen uptake in teleosts (Zapata and Cooper, 1990).

The capacity of different components of the GALT, mainly enterocytes and macrophages, for trapping and processing antigens has already been discussed, thus we will now analyze the role played by gills. Presence of carbon in gills has been noted in young trout and dogfish (Manning, 1994) but not in carp and plaice (Ellis *et al.,* 1976). Chilmonczyk and Monge (1980) pointed out the capacity of gill pillar cells to trap foreign materials, including viral hemorrhagic septicemia virus. That data and our ultrastructural studies (Zapata *et al.,* 1987) showed that *Yersinia ruckeri* O antigen was located in the thin mucous layer and adhered to the gill pavement cells. Later, the antigen appeared in vacuoles of the underlying phagocytes, presumably due to the internalization of antigens. The specificity of uptake was later demonstrated immunohistochemically in gill sections by using a specific anti–*Y. ruckeri* O-antigen antiserum (Torroba *et al.,* 1993). A similar trapping pattern has also been described by other authors (Goldes *et al.*, 1986).

On the other hand, nonantigenic material is trapped mainly in three locations: splenic ellipsoids (Zapata and Cooper, 1990), sinusoidal blood vessels, macrophages and reticular cells of the kidney (Herráez, 1988; Dannevig *et al.*, 1990, 1994; Zapata and Cooper, 1990; Press *et al.*, 1994), and ventricular endothelial cells and atrialendocardial macrophages of the heart (Nakamura and Shimozawa, 1994). Later, cells filled with engulfed materials migrate to the MMCs of the kidney and spleen (Herráez and Zapata, 1986, 1991; Herráez, 1988).

Evidence available on antigen trapping in the peripheral lymphoid organs of teleosts, including the cells involved and the underlying mechanisms, is controversial (for a review see Zapata, 1983; Zapata and Cooper, 1990). Soluble antigens (i.e., HGG) were extracellularly detected as immune complexes associated with the reticular fibers of splenic ellipsoidal walls. In the kidney, apparently, pyroninophilic cells and/or macrophages trapped the antigens intracellularly. In addition, a second challenge with HGG resulted in clusters of pyroninophilic cells, mainly in the kidney, which, according to the authors, could become MMCs. Several years ago we remarked that extracellular deposits of immune complexes induced in teleost spleen by the administration of soluble antigens, could be due to the open blood circulation of the organ. Antigens could, therefore, diffuse through the end of ellipsoids to the splenic parenchyma rather than being retained on ellipsoidal reticular fibers (Zapata, 1983). Other authors have demonstrated the relative importance of teleost kidney and spleen in the clearance of particulate and soluble blood-borne substances. MacArthur *et al.* (1983) comparatively analyzed the levels of captation of different organs of plaice injected with ^{51}Cr-labeled turbot erythrocytes and demonstrated that the kidney contained a higher count due to its larger size, although, per gram, the spleen was more active. Nevertheless, there were notable species-specific differences. The involvement of the sinusoidal endothelium of the pronephros in the clearance of circulating materials has been emphasized by several authors (Dannevig and Berg, 1985). Others have demonstrated that it is in fact macrophages associated with venous sinusoids that act as phagocytes and capture bacteria introduced into the blood circulation (Ferguson, 1984). Recently, Dannevig *et al.* (1994) have shown that endothelial cells from rainbow trout pronephros possess endocytic receptors involved in the clearance of circulating soluble ligands. In order to clarify the mechanisms involved in antigen trapping in teleost lymphoid organs and the role, if any, played by the MMCs (the structure of which is described later), we analyzed ultrastructurally the trapping of either sheep erythrocytes (SRBC) (Herráez and Zapata, 1986; Herráez, 1988) or formalin-killed *Y. ruckeri* (Herráez and Zapata, 1987; Herráez, 1988) by goldfish, *Carassius auratus*.

MMCs appear as isolated melanomacrophages and/or large cell clusters in all the fish studied, including Agnatha, but also in amphibians and reptiles (see reviews by Agius, 1980; Herráez, 1988; Zapata and Cooper, 1990; Wolke, 1992). They are aggregates of closely packed, histoenzymatically heterogeneous macrophages that contain diverse inclusions, the most frequent being lipofucsin, melanin, and hemosiderin (Fig. 16). Lipofucsins may be derived from the oxidation of polyunsaturated fatty acids (Agius, 1980, 1985), a molecular species usually present in fish. Hemosiderin is one of the breakdown products of hemoglobin. Phylogenetically, MMCs seem to evolve from the isolated melanomacrophages observed in Agnatha and elasmobranchs into the organized centers found in all bony fish, except

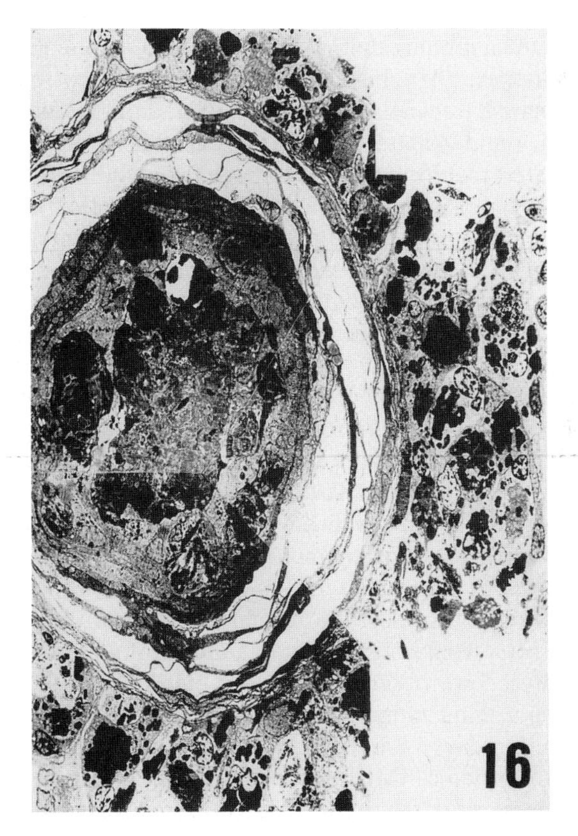

Fig. 16. Large MMC in the pronephros of a goldfish, *Carassius auratus,* 21 days after intraperitoneal secondary challenge with sheep erythrocytes. ×1000.

salmonids (Agius, 1980; Herráez and Zapata, 1986, 1991; Ardavín and Zapata, 1987; Zapata and Cooper, 1990). Furthermore, while they are found in the liver of Agnatha, elasmobranchs, and primitive Osteichthyes, they occur primarily in the main lymphoid organs, kidney and spleen, of teleosts. From an ontogenetical point of view, their appearance, according to Agius (1981), coincides with the first feeding, a fact that, as we have indicated repeatedly, seems to be important for many of the physiological processes of fish.

Teleost MMCs often appear in the axillary of the ellipsoidal branches of the spleen, totally (Agius, 1980, 1985) or partially encapsulated (Herráez and Zapata, 1986, 1991; Meseguer *et al.,* 1991, 1994b) and infiltrated with a few granulocytes and pyroninophilic cells, some of which express Ig (Press *et al.,* 1994). Frequently, a lymphocyte cuff that surrounds the entire arterial system is intimately associated with MMCs (Herráez and Zapata, 1986), although this association has not been observed in some species (Meseguer *et al.,* 1994b). In salmonids the MMCs are poorly defined and developed (Zapata and Cooper, 1990). In fact, there is an opposite correlation between numbers of isolated, free melanomacrophages and the development of large MMCs (Herráez and Zapata, 1986; Meseguer *et al.,* 1994b), although it is unclear if isolated macrophages move to preexisting MMCs, as proposed by Mori (1980), or form new aggregates (Ellis *et al.,* 1976). This probably depends on the amount of engulfed materials and the number of MMCs present (Herráez and Zapata, 1986). In the teleost kidney, the MMCs are distributed randomly throughout the lymphohemopoietic tissue, with those containing melanin being more abundant in the pronephros than in the mesonephros.

After administration of sheep red blood cells (SRBC) erythrocytes first appear in the cytoplasm of ellipsoidal macrophages in the spleen (Fig. 17) and in the endothelial cells of the kidney. Later, macrophages in the spleen and phagocytic reticular cells and/or macrophages in the kidney transport the engulfed erythrocytes to the MMCs (Fig. 16) (Herráez and Zapata, 1986). A similar pattern of trapping was observed by Lamers (1985) after injecting *Aeromonas hydrophila* in *Cyprinus carpio,* but he was unable to demonstrate antigen transport into the MMCs. When the changes undergone by the MMCs from both spleen and kidney of goldfish were morphometrically analyzed we observed increased numbers of splenic centers 5 and 7 days after immunization. Significant differences in the size of MMCs became evident on days 14 and 21. After a second challenge, the numbers but not the size of splenic MMCs underwent another increase. In the kidney, especially after secondary immunization, the changes were faster and stronger than in the spleen and occurred principally in the pronephros (Herráez and Zapata, 1986).

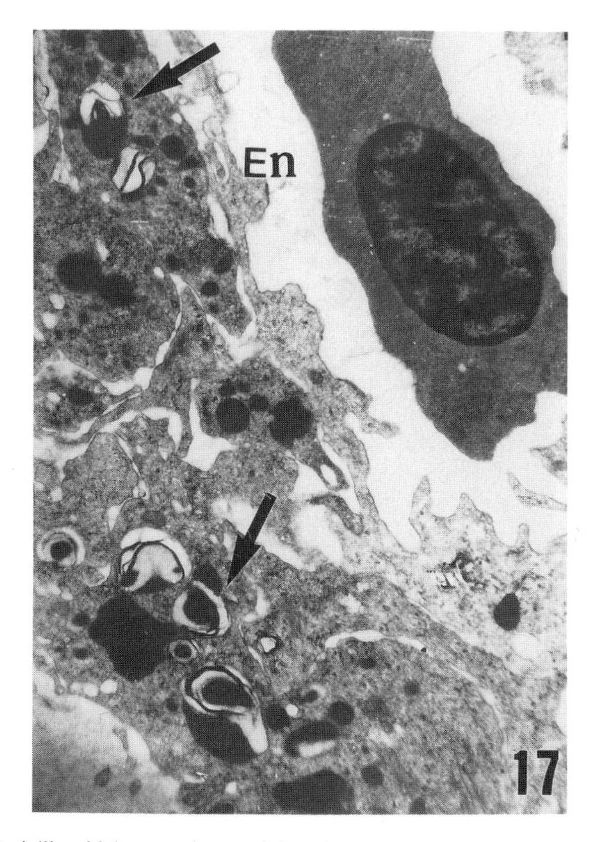

Fig. 17. Periellipsoidal macrophages of the spleen of a goldfish, *C. auratus,* intraperitoneally immunized with formalin-killed *Yersinia ruckeri.* Engulfed bacteria (arrows) appear in the phagocyte cytoplasm. Endothelial cells (En). ×11500.

When goldfish were primarily or secondarily immunized intraperitoneally with a single injection of 0.1 ml formalin-killed *Y. ruckeri,* the pattern of antigen trapping and processing in both spleen and kidney was similar to that found in SRBC-immunized fish (Fig. 17), but remarkably, the morphometrical analysis indicated no variation in number, size, or area occupied by MMCs in any of the lymphoid organs analyzed (Herráez and Zapata, 1987). According to our ultrastructural evidence, the bacteria but not the erythrocytes were completely engulfed and digested in the cytoplasm of phagocytic MMCs and, thus, did not change in either number or size (Fig. 18).

These results suggest that MMCs are merely scavengers in the fish lymphoid organs and not the phylogenetical precursors of the germinal

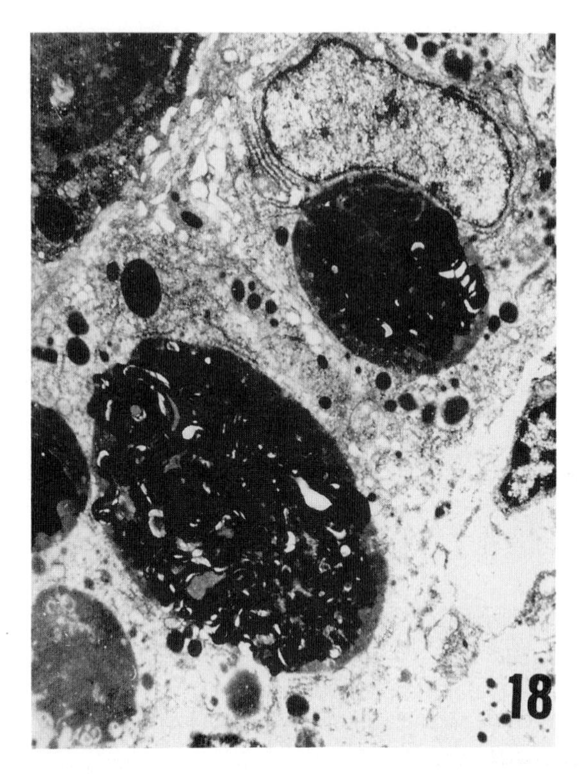

Fig. 18. Large groups of bacteria appear engulfed in the cytoplasm of macrophages of a MMC of the pronephros of *C. auratus* 21 days after intraperitoneal injection of formalin-killed *Y. ruckeri.* ×11500.

centers of higher vertebrates, as proposed by some authors (Lamers, 1985; Zapata and Cooper, 1990; Press *et al.,* 1994), in which macrophages filled with degraded materials, including antigens, accumulate. We have recently discussed that the absence of germinal centers in lower vertebrates is related to the lack of an efficient mechanism for selecting the B-cell mutants generated throughout the immune response, but not to the lack of somatic hypermutation in these vertebrates (Zapata *et al.,* 1995). Other authors have emphasized, in agreement with our arguments, that the general function of the MMCs is the "centralization" of endogenous and exogenous materials for destruction, detoxification, or recycling (Vogelbein *et al.,* 1987). These same authors compared MMCs with chronic inflammatory lesions derived from the accumulation of monocyte-like cells (Vogelbein *et al.,* 1988), and Tsujii and Seno (1990) pointed out that the formation of large MMCs in the fish lymphoid organs could be related to a poor lysosomal enzyme repertoire of fish macrophages, a fact also reported by other authors

(Herráez and Zapata, 1986, 1991). In this respect, we concluded several years ago that phagocytic cells of the largest MMCs are residual elements that have almost exhausted lysosomal activity (Herráez and Zapata, 1991).

Apart from defence mechanisms, MMCs have been proposed to be involved in numerous processes including aging, tissue breakdown, and pathological conditions (Agius, 1980; Herráez 1988; Wolke, 1992). A relationship between iron metabolism and MMCs has been emphasized by various authors and some have claimed that their most important function is erythrocyte destruction (see Herráez, 1988). Furthermore, under starvation and in diseased fish, increased iron deposits occur in the splenic centers (Wolke, 1992; Manning, 1994). Supporting this breakdown function of erythrocytes, we demonstrated that phenylhydrazine-induced anemia results in a rapid increase of both the number and size of MMCs. Furthermore, 5 days after treatment, MMCs degenerated by the time of onset of erythroblasts, suggesting a relationship between the iron necessary for erythropoiesis and the rapid disappearance of MMCs (Fig. 19). (Herráez and Zapata, 1986; Herráez, 1988).

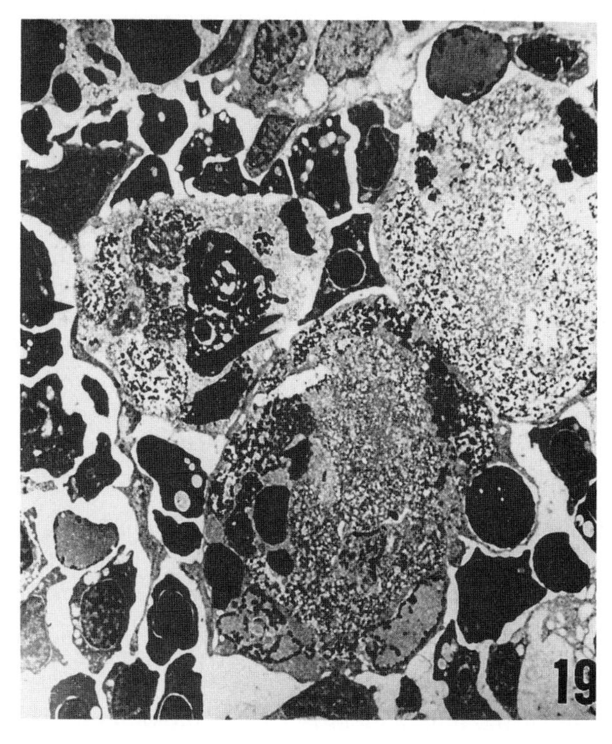

Fig. 19. Degenerated MMCs of the spleen of an anemic goldfish 10 days after phenylhydrazine administration. ×5600.

Another controversial aspect of the biology of fish MMCs is the origin and functional significance, if any, of the melanin some of them contain. Melanin is a complex polymer that can absorb and neutralize free radicals, cations, and other potentially toxic agents. Edelstein (1971) has suggested that it could contribute to the killing of bacteria in macrophages, since it produces hydrogen peroxide from the oxidation of NADH, which in turn can be used in the bactericidal iodination system. In this respect, most authors agree that fish macrophages from MMCs engulf the melanin from melanocytes and melanophores (Agius, 1985). Some authors, however, have indicated that fish macrophages can synthesize melanin because they exhibit tyrosinase activity (Zuasti *et al.*, 1989). Ultrastructural results demonstrate, however, that macrophages from MMCs do not possess the necessary cell machinery (Herráez and Zapata, 1991). Zuasti and colleagues analyzed the tyrosinase activity in cell fractions derived from total kidney but not from isolated macrophages or MMCs. In addition, recent data have confirmed that isolated pronephros monocyte/macrophages engulf melanin associated with cell debris from the same origin (Meseguer *et al.*, 1994b). This indirectly suggests that under physiological conditions, these cell types phagocytose both premelanosomes and melanosomes, in which tyrosinase activity occurs, and thus, this activity remains in both isolated melanomacrophages and MMCs.

V. FURTHER DIRECTIONS

We have reviewed the current knowledge about the cellular components of the fish immune system which consists mainly of circulating leukocytes and lymphoid organs. Although little information on this matter is available we have tried to comment on the histophysiological aspects that provide a better knowledge of the immune capabilities of these vertebrates. Because many aspects remain unresolved and must be elucidated in the next few years, we would like to summarize some, which in our opinion, deserve special attention:

1. Production of specific reagents that will permit better phenotypical characterization of the fish lymphocytes and their cell subsets
2. Analysis of the role played by distinct types of granulocytes in the defence mechanisms of fish
3. Clarification of the true immunological capacities of both hagfish and lampreys
4. Analysis of the immunological role of the different bone marrow-equivalent lymphoid organs that appear in primitive fish

5. A better characterization of the cell components, functional capabilities, and ontogenetical appearance of the mucosae-associated lymphoid tissue
6. Study of the ontogenetical correlation between the appearance of fish lymphoid organs and maturation of immune capacities
7. *In vivo* analysis of the mechanisms of uptake and processing of antigens; cells and mechanisms of transport involved from the site of trapping to the lymphoid organs
8. Development of new approaches to study the engima of the lack of germinal centers in fish and, in general, in ectothermic vertebrates

REFERENCES

Abelli, L., Romano, N., Scapigliati, G., and Mastrolia, L. (1994). Cytological observations on thymic stroma of sea bass, *Dicentrarchus labrax* L. *Dev. Comp. Immunol.* **18,** Suppl. 1, S62.

Agius, C. (1980). Phylogenetic development of melanomacrophage centers in fish. *J. Zool.* (London) **191,** 11–31.

Agius, C. (1981). Preliminary studies on the ontogeny of the melanomacrophages of teleost haemopoietic tissues and age-related changes. *Dev. Comp. Immunol.* **5,** 597–606.

Agius, C. (1985). The melanomacrophages centers of fish: A review. *In* "Fish Immunology" (M. J. Manning and M. F. Tatner, eds.), pp. 85–105. Academic Press, London.

Ainsworth, A. J. (1992). Fish granulocytes: Morphology, distribution, and function. *Annu. Rev. Fish Dis.* **2,** 123–148.

Al-Adhami, M. A., and Kunz, Y. W. (1976). Hemopoietic centers in the developing angelfish *Pterophyllum scalare* (Curier and Valenciennes). *Wilheim Roux's Arch.* **179,** 393–401.

Alvarez, F. (1993). Immunobiología de salmónidos. Análisis morfológico de los mecanismos de defensa de *Salmo trutta* contra la infección por *Saprolegnia* sp. Estudio del tegumento y los órganos linfoides. Doctoral Thesis, University of León.

Alvarez, F., Flaño, E., Villena, A. J., Zapata, A. G., and Razquin, B. E. (1994). Seasonal intrathymic erythropoietic activity in trout. *Dev. Comp. Immunol.* **18,** 409–420.

Ardavín, C. F., and Zapata, A. G. (1987). Ultrastructure and changes during metamorphosis of the lymphohemopoietic tissue of the larval anadromous sea lamprey *Petromyzon marinus. Dev. Comp. Immunol.* **11,** 79–93.

Ardavín, C. F., and Zapata, A. G. (1988). The pharyngeal lymphoid aggregates of lampreys: A morphofunctional equivalent of the vertebrate thymus? *Thymus* **11,** 59–65.

Ardavín, C. F., Gomariz, R. P., Barrutia, M. G., Fonfría, J., and Zapata, A. G. (1984). The lymphohemopoietic organs of the anadromous sea lamprey *Petromyzon marinus.* A comparative study throughout its life span. *Acta Zool.* (Stockholm) **65,** 1–15.

Bergeron, T., and Woodward, B. (1982). Ultrastructure of the granule cells in the small intestine of the rainbow trout (*Salmo gairdneri*) before and after stratum granulosum formation. *Can. J. Zool.* **61,** 133–138.

Bielek, E. (1988). Ultrastructural analysis of leucocyte interaction with tumour targets in a teleost, *Cyprinus carpio* L. *Dev. Comp. Immunol.* **12,** 809–821.

Bielek, E., and Strauss, B. (1993). Ultrastructure of the granulocytes of the South American lungfish, *Lepidosiren paradoxa:* Morphogenesis and comparison to other leucocytes. *J. Morph.* **218,** 29–41.

Bradshaw, C. M., Clem, L. W., and Sigel, M. M. (1969). Immunologic and immunochemical studies on the gar, *Lepisosteus platyrhincus*. *J. Immunol.* **103,** 496–504.

Caspi, R. R., and Avtalion, R. R. (1984). Evidence for the existence of an IL-2–like lymphocyte growth promoting factor in a bony fish *Cyprinus carpio*. *Dev. Comp. Immunol.* **8,** 51–60.

Castillo, A., Razquin, B. E., Löpez-Fierro, P, Alvarez, F., Zapata, A. G., and Villena, A. J. (1990). Enzyme and immunohistochemical study of the thymic stroma in the rainbow trout, *Salmo gairdneri,* Richardson. *Thymus* **36,** 159–173.

Castillo, A., López-Fierro, P., Zapata, A. G., Villena, A. J., and Razquin, B. E. (1991). Posthatching development of the thymic epithelial cells in the rainbow trout, *Salmo gairdneri:* An ultrastructural study. *Am. J. Anat.* **190,** 299–307.

Castillo, A., Sánchez, C., Dominguez, J., Kaatari, S., and Villena, A. J. (1993). Ontogeny of IgM and IgM-bearing cells in rainbow trout. *Dev. Comp. Immunol.* **17,** 419–424.

Chantanachookhin, C., Seikai, T., and Tanaka, M. (1991). Comparative study of the ontogeny of the lymphoid organs in three species of marine fish. *Aquaculture* **99,** 143–155.

Chibá, A. (1994). Light and electron microscopic observation of the hemopoietic organs of the bichir, *Polypterus senegalus* (Brachiopterygii). *Acta Anat. Nippon.* **69,** 99, (Abstract).

Chibá, A., Torroba, M., Honma, M., and Zapata, A. G. (1988). Occurrence of lymphohaemopoietic tissue in the meninges of the stingray *Dasyatis akajei* (Elasmobranchii, Chondrichthyes). *Am. J. Anat.* **183,** 268–276.

Chilmonczyk, S. (1992). The thymus in fish: Development and possible function in the immune response. *Annu. Rev. Fish Dis.* **2,** 181–200.

Chilmonczyk, S., and Monge, D. (1980). Rainbow trout gill pillar cells: Demonstration of inert particle phagocytosis and involvement in viral infection. *J. Reticuloendothel. Soc.* **28,** 327–332.

Clawson, C. C., Finstad, J., and Good, R. A. (1966). Evolution of immune response. V. Electron microscopy of plasma cells and lymphoid tissue of the paddlefish. *Lab. Invest.* **15,** 1830–1847.

Cooper, E. L., Zapata, A. G., Barrutia, M. G., and Ramirez, J. A. (1983). Aging changes in lymphopoietic and myelopoietic organs of the annual cyprinodont fish, *Notobranchius guentheri. Exp. Gerontol.* **18,** 29–38.

Crouse, D. A., Turpen, J. B., and Sharp, J. G. (1985). Thymic nonlymphoid cells. *Surv. Immonol. Res.* **4,** 120–134.

Dannevig, B. H., and Berg, T. (1985). *In vitro* degradation of endocytosed protein in pronephros cells of the char (*Salmo alpinus* L.). The effects of temperature and inhibitors. *Dev. Comp. Immunol.* **9,** 231–240.

Dannevig, B. H., Struksnaes, G., Skogh, T., Mork Kindberg, G., and Berg, T. (1990). Endocytosis via the scavenger- and the mannose-receptor in rainbow trout (*Salmo gairdneri*) pronephros is carried out by nonphagocytic cells. *Fish Physiol. Biochem.* **8,** 229–238.

Dannevig, B. H., Lauve, A., Press, C. M. L., and Landsverk, T. (1994). Receptor-mediated endocytosis and phagocytosis by rainbow trout head kidney sinusoidal cells. *Fish Shellfish Immunol.* **4,** 3–18.

Davidson, G. A., Ellis, A. E., and Secombes, C. J. (1993). Route of immunization influences the generation of antibody secreting cells in the gut of rainbow trout, *Oncorhynchus mykiss* (Walbaum 1792). *Dev. Comp. Immunol.* **17,** 373–376.

Davidson, G. A., Ellis, A. E., and Secombes, C. J. (1994). A preliminary investigation into the phenomenon of oral tolerance in rainbow trout (*Oncorhynchus mykiss,* Walbaum 1792). *Fish Shellfish Immunol.* **4,** 141–151.

Davina, J. H. M., Parmentier, H. K., and Timmermans, L. P. M. (1982). Effect of oral administration on *Vibrio* bacteria on the intestine of cyprinoid fish. *Dev. Comp. Immunol. Suppl.* **2,** 157–166.

de Luca, D., Wilson, M., and Warr, G. W. (1983). Lymphocyte heterogeneity in the trout *Salmo gairdneri,* defined with monoclonal antibodies to IgM. *Eur. J. Immunol.* **13,** 546–551.

Doggett, T. A., and Harris, J. E. (1991). Morphology of the gut-associated lymphoid tissue in *Oreochromis mossambicus* and its role in antigen absorption. *Fish Shellfish Immunol.* **1,** 213–228.

Drzewina, A. (1910). Sûr l'organe lymphoïde et la muqueuse de l'oesophage de la torpille. *Arch. d'Anat. Micr.* **12,** 1–18.

Du Pasquier, L. (1993). Evolution of the immune system. *In* "Fundamental Immunology" (W. E. Paul, ed.), pp. 199–233. Raven Press, New York.

Dustin, P. (1975). Ultrastructure and function of the elipsoids of the spleen. Their relationship with fat metabolism and red blood cells. *Haematologica* **60,** 136–154.

Edelstein, L. M. (1971). Melanin: A unique biopolymer. *In* "Pathobiology Annual" (H. L. Ioachim, ed.) Vol. 1, pp. 309–324. Butterworth, London.

Ellis, A. E. (1976). Leucocytes and related cells in the plaice *Pleuronectes platessa. J. Fish Biol.* **8,** 143–156.

Ellis, A. E. (1977). The leucocytes of fish: A review. *J. Fish Biol.* **11,** 453–491.

Ellis, A. E., and de Sousa, M. (1974). Phylogeny of the lymphoid system. I. A study of the fate of circulating lymphocytes in plaice. *Eur. J. Immunol.* **4,** 338–343.

Ellis, A. E., and Parkhouse, R. M. E. (1975). Surface immunoglobulins on the lymphocytes of the skate *Raja naevus. Eur. J. Immunol.* **5,** 726–728.

Ellis, A. E., Munro, A. L. S., and Roberts, R. J. (1976). Defense mechanisms in fish. I. A study of the phagocytic system and the fate of intraperitoneally injected particulate material in the plaice *Pleuronectes platessa* L. *J. Fish Biol.* **8,** 67–78.

Ellsaesser, C. F., Bly, J. E., and Clem, L. W. (1988). Phylogeny of lymphocyte heterogeneity. The thymus of the channel catfish. *Dev. Comp. Immunol.* **12,** 787–799.

Fänge, R. (1966). Comparative aspects of excretory and lymphoid tissue. *In* "Phylogeny of Immunity" (R. T. Smith, P. A. Miescher, and R. A. Good, eds.), pp. 140–145. University of Florida Press, Gainesville.

Fänge, R. (1982). A comparative study of lymphomyeloid tissue in fish. *Dev. Comp. Immunol. Suppl.* **2,** 22–33.

Fänge, R. (1984). Lymphomyeloid tissues in fishes. *Vidensk. Meddr. Dansk. Naturh. Foren.* **145,** 143–162.

Fänge, R. (1987). Lymphomyeloid system and blood cell morphology in elasmobranchs. *Arch. Biol. (Bruxelles)* **98,** 187–208.

Fänge, R., and Johansson-Sjöbeck, M.-L. (1975). The effect of splenectomy on the hematology and on the activity of 6-amino-levulinic acid dehydratase (ALA-D) in hemopoietic tissue of the dogfish *Scyliorhinus canicula* (Elasmobranchii). *Comp. Biochem. Physiol.* **52A,** 577–580.

Fänge, R., and Nilsson, S. (1985). The fish spleen: Structure and function. *Experientia* **41,** 152–152.

Fänge, R., and Pulsford, A. (1985). The thymus of the angler fish *Lophius piscatorius* (Pisces: Teleostei): A light and electron microscopic study. *In* "Fish Immunology" (M. J. Manning and M. F. Tatner, eds.), pp. 293–311. Academic Press, London.

Ferguson, H. W. (1984). Renal portal phagocytosis of bacteria in rainbow trout (*Salmo gairdneri* Richardson): Ultrastructural observations. *Can. J. Zool.* **62,** 2505–2511.

Ferren, F. A. (1967). Role of the spleen in the immune response of teleosts and elasmobranchs. *J. Florida Med. Assoc.* **54,** 434–437.

Fey, F. (1966). Vergleichende hämozytologie niederer Vertebraten. III. Granulozyten. *Folia Haematol.* **86,** 1–20.

Fichtelius, K. E., Finstad, J., and Good, R. A. (1968). Bursa equivalents of bursaless vertebrates. *Lab. Invest.* **19,** 339–351.

Flaño, E., Alvarez, F., López-Fierro, P., Razquin, B. E., Villena, A. J., and Zapata, A. G. (1996). *In vitro* and *in vivo* characterization of fish thymic nurse cells. *Develop. Immunol.,* **5,** 17–24.

Fonfría, J., Barrutia, M. G., Garrido, E., Ardavín, C. F., and Zapata, A. G. (1983). Erythropoiesis in the thymus of the spotless starling *Sturnus unicolor. Cell Tiss. Res.* **232,** 445–455.

Frohely, M. F., and Deschaux, P. A. (1986). Presence of tonofilaments and thymic serum factor (FTS) in thymic epithelial cells of a freshwater fish (Carp, *Cyprinus carpio*) and a seawater fish (Bass, *Dicentrarchus labrax*). *Thymus* **8,** 235–244.

Fuda, H., Hara, A., Yamazaki, F., and Kobayashi, K. (1992). A peculiar immunoglobulin M (IgM) identified in eggs of chum salmon (*Oncorhynchus keta*). *Dev. Comp. Immunol.* **16,** 415–423.

Fujii, T. (1981). Antibody-enhanced phagocytosis of lamprey polymorphonuclear leucocytes against sheep erythrocytes. *Cell Tiss. Res.* **219,** 41–51.

Fujii, T., Nakagawa, H., and Murakawa, S. (1979). Immunity in lamprey. II. Antigen-binding responses to sheep erythrocytes and hapten in the ammocoete. *Dev. Comp. Immunol.* **3,** 609–620.

Goldes, S. A., Ferguson, H. W., Daoust, P. Y., and Moccia, R. D. (1986). Phagocytosis of the inert suspended clay kaolin by the gills of rainbow trout, *Salmo gairdneri* Richardson. *J. Fish Dis.* **9,** 147–151.

Good, R. A., Finstad, J., Pollara, B., and Gabrielsen, A. E. (1966). Morphological studies on the evolution of the lymphoid tissues among the lower vertebrates. *In* "Phylogeny of Immunity" (R. T. Smith, P. A. Miescher, and R. A. Good, eds.), pp. 149–167. University of Florida Press, Gainesville.

Good, R. A., Finstad, J., and Litman, J. (1972). Immunology. *In* "The Biology of Lampreys" (M. V. Hardisty and I. C. Potter, eds.) Vol. I, pp. 127–206. Academic Press, London.

Gorgollón, P. (1983). Fine structure of the thymus in the adult cling fish *Sicyases sanguineus* (Pisces Gobiesocidae). *J. Morph.* **177,** 25–40.

Grizzle, J. M., and Rogers, R. C. (1976). Anatomy and histology of the channel catfish. Auburn University Agricultural Experiment Station, Auburn, AL, 94 pp.

Hagen, M., Filosa, M. F., and Youson, J. H. (1983). Immunocytochemical localization of antibody-producing cells in adult lamprey. *Immunol. Lett.* **6,** 87–92.

Hart, S., Wrathmell, A. B., and Harris, J. E. (1986). Ontogeny of gut-associated lymphoid tissue (GALT) in the dogfish *Scyliorhinus canicula* L. *Vet. Immunol. Immunopathol.* **12,** 107–116.

Hart, S., Wrathmell, A. B., Harris, J. E., and Grayson, T. H. (1988). Gut immunology in fish: A review. *Dev. Comp. Immunol.* **17,** 241–248.

Herráez, M. P. (1988). Estructura y función de los centros melanomacrofágicos de *Carassius auratus.* Doctoral Thesis. University of León.

Herráez, M. P., and Zapata, A. G. (1986). Structure and function of the melanomacrophage centers of the goldfish *Carassius auratus. Vet. Immunol. Immunopathol.* **12,** 117–126.

Herráez, M. P., and Zapata, A. G. (1987). Trapping of intraperitoneally-injected *Yersinia ruckeri* in the lymphoid organs of *Carassius auratus:* The role of melanomacrophage centers. *J. Fish Biol.* **31,** Suppl. A, 235–237.

Herráez, M. P., and Zapata, A. G. (1991). Structural characterization of the melanomacrophage centers (MMC) of goldfish *Carassius auratus. Eur. J. Morph.* **29,** 89–102.

Hine, P. M. (1992). The granulocytes of fish. *Fish Shellfish Immunol.* **2,** 79–98.

Hine, P. M., and Wain, J. M. (1988a). Observations on the composition and ultrastructure of holocephalan granulocytes. *N. Z. J. Mar. Freshwater Res.* **22,** 63–73.

Hine, P. M., and Wain, J. M. (1988b). Ultrastructural and cytochemical observations on the granulocytes of the sturgeon, *Acipenser brevirostrum* (Chondrostei). *J. Fish Biol.* **33,** 235–245.

Hine, P. M., Lester, R. J. G., and Wain, J. M. (1990). Observations on the blood of Australian lungfish. *Neoceratodus forsteri* Klefft. I. Ultrastructure of granulocytes, monocytes, and thrombocytes. *Aust. J. Zool.* **38,** 131–144.

Holmgrem, N. (1950). On the pronephros and the blood in *Myxine glutinosa. Acta Zool.* (*Stockholm*) **31,** 233–348.

Imagawa, T., Hashimoto, Y., Kon, Y., and Sugimura, M. (1991). Immunoglobulin-containing cells in the head kidney of carp (*Cyprinus carpio* L.) after bovine serum albumin injection. *Fish Shellfish Immunol.* **1,** 173–185.

Ishiguro, A., Kobayashi, K., Suzuki, M., Titani, K., Tomonaga, S., and Kurosawa, Y. (1992). Isolation of a hagfish gene that encodes a complement component. *EMBO J.* **11,** 829–837.

Ishimatsu, A., Iwama, G. K., Bentley, T. B., and Heisler, N. (1992). Contribution of the secondary circulatory system to acid–base regulation during hypercapnia in rainbow trout (*Oncorhynchus mykiss*). *J. Exp. Biol.* **170,** 43–56.

Ishimatsu, A., Iwama, G. K., and Heisler, N. (1995). Physiological roles of the secondary circulatory system in fish. *In* "Advances in Comparative and Environmental Physiology" (N. Heisler, ed.), Vol. 21, in press.

Ihsizeki, K., Nawa, T., Tachibana, Y., Sakahu, Y., and Iida, S. (1984). Hemopoietic sites and development of eosinophil granulocytes in the loach, *Misgurunus anguillicaudatus. Cell Tiss. Res.* **235,** 419–426.

Johansson, M.-L. (1973). Peroxidase in blood cells of fishes and cyclostomes. *Acta Reg. Soc. Sci. Litt. Gotheburg Zool.* **8,** 53–56.

Jordan, H. E., and Speidel, C. C. (1931). Blood formation on the African lungfish, under normal conditions and under conditions of prolonged estivation and recovery. *J. Morph. Physiol.* **51,** 319–371.

Kintsuji, H., Kawahara, E., Nomura, S., Kusuda, R. (1994). Characterization of serum and skin mucus immunoglobulins of the bester. *Dev. Comp. Immunol.* **18,** IV, (Abstract).

Kita, J., and Itazawa, Y. (1990). Microcirculatory pathways in the spleen of the rainbow trout *Oncorhynchus mykiss. Jpn. J. Ichthyol.* **37,** 265–272.

Koumans–van Diepen, J. C. E., van de Lisdonk, M. H. M., Taverne-Thiele, A. J., Verburg–van Kemenade, B. M. L., and Rombout, J. H. W. M. (1994). Characterization of immunoglobulin-binding leucocytes in carp (*Cyprinus carpio* L.). *Dev. Comp. Immunol.* **18,** 45–56.

Kronenberg, M., Brines, R., and Kaufman, J. (1994). MHC evolution: A long-term investment in defense. *Immunol. Today* **15,** 4–6.

Kusuda, R., and Ikeda, Y. (1987). Studies on classification of eel leucocytes. *Nippon Suisan Gakkaishi* **53,** 205–209.

Lagabrielle, J. (1938). Contribution à l'étude anatomique, histologique et embryologique du thymus chez les téléostéens. Doctoral Thesis, University of Bordeaux.

Lamas, J., Bruno, D. W., Santos, I., Anadón, R., and Ellis, A. E. (1991). Eosinophilic granular cell response to intraperitoneal injection with *Vibrio anguillarum* and its extracellular products in rainbow trout, *Onkorhynchus mykiss. Fish Shellfish Immunol.* **1,** 187–194.

Lamers, C. H. J. (1985). The reaction of the immune system of fish to vaccination. Doctoral Thesis, Agricultural University of Wageningen.

Langille, R. M., and Youson, J. H. (1985). Protein and lipid absorption in the intestinal mucosa of adult lampreys (*Petromyzon marinus*) following induced feeding. *Can. J. Zool.* **63,** 691–702.

Litman, G. W., Rast, J. P., Hulst, M. A., Litman, R. T., Shamblott, M. J., Haire, R. N., Hinds-Frey, K. R., Buell, R. D., Margittai, M., Ohta, Y., Zilch, A. C., Good, R. A., and Amemiya,

C. T. (1992). Evolutionary origins of immunoglobulin gene diversity. *In* "Progress in Immunology" (J. Gergely, M. Benczur, A. Erdei, A. Falus, G. Y. Füst, G. Medgyesi, G. Y. Petranyi, and E. Rajnavölgyi, eds.), Vol. VIII, pp. 107–114. Springer-Verlag, Budapest.

Lloyd-Evans, P. (1993). Development of the lymphomyeloid system in the dogfish, *Scyliorhinus canicula. Dev. Comp. Immunol.* **17,** 501–514.

Lobb, C. J. (1987). Secretory immunity induced in catfish. *Ictalurus punctatus,* following bath immunization. *Dev. Comp. Immunol.* **11,** 727–738.

Lobb, C. J., and Clem, L. W. (1981a). Phylogeny of immunoglobulin structure and function. XI. Secretory immunoglobulins in the cutaneous mucus of the sheepshead *Archosargus probatocephalus. Dev. Comp. Immunol.* **5,** 587–596.

Lobb, C. J., and Clem, L. W. (1981b). Phylogeny of immunoglobulin structure and function. XII. Secretory immunoglobulins in the bile of the marine teleost *Archosargus probatocephalus. Mol. Immunol.* **18,** 615–619.

Lobb, C. J., and Clem, L. W. (1981c). Phylogeny of immunoglobulin structure and function. X. Humoral immunoglobulins of sheepshead. *Archosargus probatocephalus. Dev. Comp. Immunol.* **5,** 271–282.

Luft, J. C., Clem, L. W., and Bly, J. E. (1994). *In vitro* mitogen-induced and MLR-induced responses of leucocytes from the spotted gar, a holostean fish. *Fish Shellfish Immunol.* **4,** 153–156.

MacArthur, J. I., Fletcher, T. C., and Thomson, A. W. (1983). Distribution of radiolabeled erythrocytes and the effect of temperature on clearance in the plaice (*Pleuronectes platessa* L.). *J. Reticuloendothel. Soc.* **34,** 13–21.

Mainwaring, G., and Rowley, A. F. (1985). Studies on granuloctye heterogeneity in elasmobranches. *In* "Fish Immunology" (M. J. Manning and M. F. Tatner, eds.), pp. 57–69. Academic Press, Orlando.

Manning, M. J. (1984). Phylogenetical origins and ontogenetic development of immunocompetent cells in fish. *Dev. Comp. Immunol. Suppl.* **3,** 61–68.

Manning, M. J. (1994). Fishes. *In* "Immunology. A comparative approach" (R. J. Turner, ed.), pp. 69–100. John Wiley & Sons, Chichester.

Marinova, T. E. (1986). Comparative observations of thymic epithelial cell ultrastructure in some teleost fishes and mammals. *Verh. Anat. Gessel.* **805,** 633–668.

Matthews, L. H. (1950). Reproduction in the basking shark *Cetorhinus maximus. Phil. Trans. R. Soc. London (B)* **234,** 247–315.

Mattisson, A. G., and Fänge, R. (1977). Light and electron microscopic observations on the blood cells of the Atlantic hagfish. *Myxine glutinosa* (L.). *Acta Zool.* (*Stockholm*) **58,** 205–221.

Mattisson, A., Fänge, R., and Zapata, A. G. (1990). Histology and ultrastructure of the cranial lymphohaemopoietic tissue in *Chimaera monstrosa* (Pisces, Holocephali). *Acta Zool.* (*Stockholm*) **71,** 97–106.

McKinney, E. C., McLeod, T. F., and Sigel, M. M. (1981). Allograft rejection in a holostean fish, *Lepisosteus platyrhincus. Dev. Comp. Immunol.* **5,** 65–75.

Meseguer, J., Esteban, M. A., and Agulleiro, B. (1991). Stromal cells, macrophages, and lymphoid cells in the head kidney of sea bass (*Dicentrarchus labrax* L.). An ultrastructural study. *Arch. Histol. Cytol.* **54,** 299–309.

Meseguer, J., López-Ruiz, A., and Esteban, M. A. (1994a). Cytochemical characterization of leucocytes from the seawater teleost gilthead seabream (*Sparus aurata* L.). *Histochemistry* **102,** 37–44.

Meseguer, J., López-Ruiz, A., and Esteban, M. A. (1994b). Melanomacrophages of the seawater teleosts, sea bass (*Dicentrarchus labrax*) and gilthead seabream (*Sparus auratus*): Morphology, formation, and possible function. *Cell Tiss. Res.* **277,** 1–10.

Millot, J., Anthony, J., and Robineau, D. (1978). "Anatomie de Latimeria chalumnae. III". Éditions de Centre National de la Recherche Scientifique: Paris. pp. 133–134.
Minura, O. M., and Minura, I. (1977). Timo de *Lepidosiren paradoxa* (Fitz, 1836). Peize Dipnóico. *Bol. Fisiol. Anim. Univ. S. Paulo* **1,** 29–38.
Mori, M. (1980). Studies on the phagocytic system in goldfish. I. Phagocytosis of intraperitoneally injected carbon particles. *Fish Pathol.* **15,** 25–30.
Morrow, W. J. W. (1978). The immune response of the dogfish *Scyliorhinus canicula L.* Doctoral Thesis, Plymouth Polytechnic University.
Nabarra, B., and Adrianarison, I. (1987). Ultrastructural studies of thymic reticulum. I. Epithelial component. *Thymus* **9,** 95–121.
Nakamura, H., and Shimozawa, A. (1994). The heart as a host defense organ in fish. *Dev. Comp. Immunol.* **18,** Suppl., 1, S61 (Abstract).
Nakanishi, T. (1986). Seasonal changes in the hormonal immune response and the lymphoid tissues of the marine teleost *Sebasticus marmoratus. Vet. Immunol. Immunopathol.* **12,** 213–223.
Nakanishi, T. (1991). Ontogeny of the immune system in *Sebasticus marmoratus:* Histogenesis of the lymphoid organs and effects of thymectomy. *Env. Biol. Fish.* **30,** 135–145.
Navarro, R. (1987). Ontogenia de los órganos linfoides de *Scyliorhinus canicula.* Estudio ultraestructural. Master's Thesis. University Complutense, Madrid.
Nieuwenhuis, P., Stot, R. J. M., Wagenaar, J. P. A., Wubbena, A. S., Kampinga, J. and Karrenbeld, A. (1988). The transcapsular route: a new way for (self) antigens to bypass the blood–thymus barrier? *Immunol. Today* **9,** 372–375.
O'Neill, J. (1989). Ontogeny of the lymphoid organs in an antarctic teleost, *Harpagifer antarcticus* (Notothenioidei: Perciformes). *Dev. Comp. Immunol.* **13,** 25–33.
Page, M. and Rowley, A. F. (1982). A morphological study of pharyngeal lymphoid accumulations in larval lampreys. *Dev. Comp. Immunol. Suppl.* **2,** 35–40.
Page, M. and Rowley, A. F. (1984). The reticulo-endothelial system of the adult river lamprey, *Lampetra fluviatilis* L.: The fate of intravascularly injected colloidal carbon. *J. Fish Dis.* **7,** 339–353.
Pontius, H., and Ambrosius, H. (1972). Contribution to the immune biology of poikilothermic vertebrates. IX. Studies on the cellular mechanism of humoral immune reactions in perch *Perca fluviatilis L. Acta Biol. Med. Ger.* **29,** 319–339.
Press, C. M. L., Dannevig, B. H., and Landsverk, T. (1994). Immune and enzyme histochemical phenotypes of lymphod and nonlymphoid cells within the spleen and head kidney of Atlantic salmon (*Salmo salar* L.). *Fish Shellfish Immunol.* **4,** 79–93.
Pulsford, A., Morrow, W. J. W., and Fänge, R. (1984). Structural studies of the thymus of the dogfish *Scyliorhinus canicula L. J. Fish Biol.* **25,** 353–360.
Pulsford, A., and Zapata, A. G. (1989). Macrophages and reticulum cells in the spleen of the dogfish, *Scyliorhinus canicula. Acta Zool.* (*Stockholm*) **70,** 221–227.
Pulsford, A., Fänge, R., and Zapata, A. G. (1991). The thymic microenvironment of the common sole, *Solea solea. Acta Zool.* (*Stockholm*) **72,** 209–216.
Pulsford, A., Tomlinson, M. G., Lemaire-Goni, Glynn, P. J. (1994). Development and immunocompetence of juvenile flounder *Platichthys flesus* L. *Fish Shellfish Immunol.* **4,** 63–68.
Quesada, J., Villena, M. L., and Agulleiro, B. (1990). Structure of the spleen of the sea bass (*Dicentrarchus labrax*): A light and electron microscopic study. *J. Morph.* **206,** 273–281.
Razquin, B. E., Castillo, A., López-Fierro, P., Alvarez, F., Zapata, A. G., and Villena, A. J. (1990). Ontogeny of IgM-producing cells in the lymphoid organs of rainbow trout, *Salmo gairdneri* Richardson: An immuno- and enzyme-histochemical study. *J. Fish Biol.* **36,** 159–173.

Riviere, H. B., Cooper, E. L., Reddy, A. L., and Hildemann, W. H. (1975). In search of the hagfish thymus. *Am. Zool.* **15,** 39–49.

Rombout, J. W., and van den Berg, A. A. (1989). Immunological importance of the second gut segment of carp. I. Uptake and processing of antigens by epithelial cells and macrophages. *J. Fish Biol.* **35,** 13–22.

Rombout, J. W., Block, L. J., Lamers, C. H., and Egberts, E. (1986). Immunization of carp (*Cyprinus carpio*) with *Vibrio anguillarum* bacteria: Indications for a common mucosal immune system. *Dev. Comp. Immunol.* **10,** 341–351.

Rombout, J. W., van den Berg, A. A., Berg, C. T. G. A., Witte, P., and Egberts, E. (1989). Immunological importance of the second gut segment of the carp. III. Systemic and/or mucosal immune responses after immunization with soluble and particulate antigens. *J. Fish Biol.* **35,** 179–186.

Rombout, J. H. W. M., Taverna-Thiele, A. J., and Villena, M. (1993). The gut-associated lymphoid tissue (GALT) of carp (*Cyprinus carpio* L.): An immunocytochemical analysis. *Dev. Comp. Immunol.* **17,** 55–66.

Rowley, A. F., Hunt, T. C., Page, M., and Mainwaring, G. (1988). Fish. *In* "Vertebrate Blood Cells" (A. F. Rowley and N. A. Ratcliffe, eds.), pp. 19–127. Cambridge University Press, Cambridge.

Saito, H. (1984). The development of the spleen in the Australian lungfish *Neoceratodus forsteri* Kreft, with special reference to its relationship to the gastro-enteric vasculature. *Am. J. Anat.* **169,** 337–360.

Scharrer, E. (1944). The histology of the meningeal lymphoid tissue in the ganoids *Amia* and *Lepisosteus. Anat. Rec.* **88,** 291–310.

Secombes, C. J., van Groningen, J. J. M., van Muiswinkel, W. B., and Egberts, E. (1983). Ontogeny of the immune system in the carp (*Cyprinus carpio*). The appearance of antigenic determinants on lymphoid cells detected by mouse anti-carp thymocyte monoclonal antibodies. *Dev. Comp. Immunol.* **7,** 455–464.

Smith, A. M., Potter M., and Merchant, E. B. (1967). Antibody-forming cells in the pronephros of the teleost *Lepomis macrochirus. J. Immunol.* **99,** 876–882.

Tanaka, Y. (1993). "The Spleen: Origin and Evolution". Kyuh. Print., Tokyo.

Tanaka, Y., Saito, Y., and Gotoh, H. (1981). Vascular architecture and intestinal hematopoietic nests of two cyclostomes, *Eptatretus burgeri* and ammocoetes of *Entosphenus reissneri:* A comparative morphological study. *J. Morph.* **170,** 71–93.

Tatner, M. F. (1985). The migration of labeled thymocytes to the peripheral lymphoid organs in the rainbow trout, *Salmo gairdneri* Richardson. *Dev. Comp. Immunol.* **9,** 85–91.

Tatner, M. F., and Findlay, C. (1991). Lymphocyte migration and localization patterns in rainbow trout *Oncorynchus mykiss,* studies using the tracer sample method. *Fish Shellfish Immunol.* **1,** 107–117.

Tatner, M. F., and Manning, M. J. (1982). The morphology of the trout, *Salmo gairdneri,* thymus. Some practical and theoretical considerations. *J. Fish Biol.* **21,** 27–32.

Temkin, R. J., and McMillan, D. B. (1986). Gut-associated lymphoid tissue (GALT) of the goldfish, *Carassius auratus. J. Morph.* **190,** 9–26.

Tomonaga, S. (1973). Study on the blood cells and hemocytopoietic tissues of the hagfish, *Eptatretus burgeri.* III. Lymphoid cells within the epithelial layer. *Yamaguchi Igaku* **22,** 37–43.

Tomonaga, S., and Fujii, R. (1994). Phylogenetic emergence of immunoglobulin-producing cells. *Acta Anat. Nippon.* **69,** 100, (Abstract).

Tomonaga, S., Hirokane, H., and Awaya, K. (1973a). The primitive spleen of the hagfish. *Zool. Mag.* **82,** 215–217.

Tomonaga, S., Shinohara, H., and Awaya, K. (1973b). Fine structure of the peripheral blood cells of the hagfish. *Zool. Mag.* **82,** 215–217.

Tomonaga, S., Kobayashi, K., Kajii, T., and Awaya, K. (1984). Two populations of immunoglobulin-forming cells in the skate, *Raja kenojei:* Their distribution and characterization. *Dev. Comp. Immunol.* **8,** 803–812.

Tomonaga, S., Kobayashi, K., Hgiwara, K., Sasaki, K., and Sezaki, K. (1985). Studies on immunoglobulin and immunoglobulin-forming cells in *Heterodontus japonicus,* a cartilaginous fish. *Dev. Comp. Immunol.* **9,** 617–626.

Tomonaga, S., Kobayashi, K., Hagiwara, K., Yamaguchi, K., and Awaya, K. (1986). Gut-associated lymphoid tissue in the elasmobranchs. *Zool. Sci.* **3,** 453–458.

Tomonaga, S., Zhang, H., Kobayashi, K., Fujii, R., and Teshima, K. (1992). Plasma cells in the spleen of the Aleutian skate, *Bathyraja aleutica. Arch. Histol. Cytol.* **55,** 287–294.

Torroba, M., Anderson, D. P., Dixon, O. W., Casares, F., Varas, A., Alonso, L., Gómez del Moral, M., and Zapata, A. G. (1993). *In vitro* antigen trapping by gill cells. A fluorescence and immunohistochemical study. *Histol. Histopathol.* **8,** 363–369.

Torroba, M., Chibá, A., Vicente, A., Varas, A., Sacedón, R., Jimenez, E., Honma, Y., and Zapata, A. G. (1995). Macrophage–lymphocyte cell clusters in the hypothalamic ventricle of some elasmobranch fish. Ultrastructural analysis and possible functional significance. *Anat. Rec.* **242,** 400–410.

Tsujii, T., and Seno, S. (1990). Melanomacrophage centers in the aglomerular kidney of the sea horse (teleosts): Morphologic studies on their formation and possible function. *Anat. Rec.* **226,** 460–470.

Underdown, B. J., and Socken, D. J. (1978). A comparison of secretory component–immunoglobulin interactions amongst different species. *Adv. Exp. Med. Biol.* **107,** 503–511.

Vallejo, A. N., and Ellis, A. E. (1989). Ultrastructural study of the response of eosinophil granule cells to *Aeromonas salmonicida* extracellular products and histamine liberators in rainbow trout, *Salmo gairdneri* Richardson. *Dev. Comp. Immuno.* **13,** 133–148.

van Loon, J. J. A., van Oosterom, R., and van Muiswinkel, W. B. (1981). Development of the immune system in carp (*Cyprinus carpio*). *In* "Aspects of Developmental and Comparative Immunology" (J. B. Solomon, ed.), Vol. I, pp. 469–470. Pergamon Press, New York.

van Muiswinkel, W. B., Lamers, C. H. J., and Rombout, J. H. W. M. (1991). Structural and functional aspects of the spleen in bony fish. *Res. Immunol.* **142,** 362–366.

Vialli, M. (1933). Formazioni linfoidi meningee e perimeningee nei selaci *Bol. Zool.* (*Italy*) **4,** 123–134.

Vogelbein, W. K., Fournie, J. W., and Overstreet, R. M. (1987). Sequential development and morphology of experimentally induced hepatic melanomacrophage centres in *Rivulus marmoratus. J. Fish Biol.* **31,** 145–153.

Vogelbein, W. K., Fournier, J. W., and Overstreet, R. M. (1988). Ultrastructural features of hepatic macrophage center induction by the apicomplexam, *Calyptospora funduli. Proc. Int. Fish Health Conf.* p. 95, Vancouver. (Abstract).

Waldschmidt, J. (1887). Beitrag zur Anatomie des Zentralnervesystem und des Geruchsorgan von *Polypterus bichir. Anat. Anz.* **2,** 308–322.

Wolke, R. E. (1992). Piscine macrophage aggregates: A review. *Annu. Rev. Fish Dis.* **2,** 91–108.

Yoffey, M. (1929). A contribution to the study of the comparative histology and physiology of the spleen, with reference chiefly to its cellular constituents. *J. Anat.* **63,** 314–344.

Yu, M.-L., Sarot, D. A., Filazzola, R. J., and Perlmutter, A. (1970). Effects of splenectomy on the immune response of the blue gourami, *Trichogaster trichopterus,* to infectious pancreatic necrosis (IPN) virus. *Life Sci.* G. Pt. II, 749–755.

Zapata, A. G. (1979a). Ultrastructural study of the teleost fish kidney. *Dev. Comp. Immunol.* **3,** 55–65.

Zapata, A. G. (1979b). Ultraestructura del tejido linfoide asociado al tubo digestivo (GALT) de *Rutilus rutilus. Morf. Norm. Patol. Sec. A,* **3,** 23–29.

Zapata, A. G. (1980). Ultrastructure of elasmobranch lymphoid tissue. I. Thymus and spleen. *Dev. Comp. Immunol.* **4,** 459–471.

Zapata, A. G. (1981a). Ultrastructure of elasmobranch lymphoid tissue. II. Leydig's and epigonal organs. *Dev. Comp. Immunol.* **5,** 43–52.

Zapata, A. G. (1981b). Lymphoid organs of teleost fish. II. Ultrastructure of renal lymphoid tissue of *Rutilus rutilus* and *Gobio gobio. Dev. Comp. Immunol.* **5,** 685–690.

Zapata, A. G. (1981c). Lymphoid organs of teleost fish. I. Ultrastructure of the thymus of *Rutilus rutilus. Dev. Comp. Immunol.* **5,** 427–436.

Zapata, A. G. (1982). Lymphoid organs of teleost fish. III. Splenic lymphoid tissue of *Rutilus rutilus* and *Gobio gobio. Dev. Comp. Immunol.* **6,** 87–94.

Zapata, A. G. (1983). Phylogeny of the fish immune system. *Bull. Inst. Pasteur* **81,** 165–186.

Zapata, A. G. (1996). Periodic cycles and immunity. *In* "The Physiology of Immunity" (J. A. Marsh and M. D. Kendall, eds.). pp. 377–393. CRC Press, Boca Raton.

Zapata, A. G., and Cooper, E. L. (1990). "The Immune System: Comparative Histophysiology." John Wiley and Sons, Chichester.

Zapata, A. G., Fänge, R., Mattisson, A. G., and Villena, A. (1984). Plasma cells in adult Atlantic hagfish, *Myxine glutinosa. Cell Tiss. Res.* **235,** 691–693.

Zapata, A. G., Torroba, M., Alvarez, F., Anderson, D. P., Dixon, O. W., Wisniewski, M. (1987). Electron microscopic examination of antigen uptake by salmonid gill cells after bath immunization with a bacterin. *J. Fish. Biol.* **31,** Suppl. A, 209–217.

Zapata, A. G., Torroba, M., and Varas, A. (1992). Seasonal variations in the immune system of lower vertebrates. *Immunol. Today* **13,** 142–147.

Zapata, A. G., Torroba, M., Vicente, A., Varas, A., Sacedón, R., Jiménez, E. (1995). The relevance of cell microenvironments for the appearance of lymphohaemopoietic tissues in primitive vertebrates. *Histol. Histopathol.* **10,** 761–778.

Zapata, A. G., Torroba, M., Sacedón, R., Varas, A., and Vicente, A. (1996). Structure of the lymphoid organs of elasmobranchs. *J. Exp. Zool.* **275,** 125–143.

Zuasti, A., Jara, J. R., Ferrer, C., Solano, F. (1989). Occurrence of melanin granules and melano synthesis in the kidney of *Sparus auratus.* Pigment *Cell Res.* **2,** 93–99.

2

THE NONSPECIFIC IMMUNE SYSTEM: CELLULAR DEFENSES

C. J. SECOMBES

I. INTRODUCTION

A variety of leukocyte types are involved in nonspecific cellular defenses of fish, and include monocytes/macrophages, granulocytes, and nonspecific cytotoxic cells (NCCs). Macrophages and granulocytes are mobile phagocytic cells found in the blood and secondary lymphoid tissues (see Chapter 1) and are particularly important in inflammation, which is the cellular response to microbial invasion and/or tissue injury leading to the local

THE FISH IMMUNE SYSTEM:
ORGANISM, PATHOGEN, AND ENVIRONMENT

accumulation of leukocytes and fluid. Indeed, in 1891 Metchnikoff wrote, "We might turn our attention to the class of fishes where we find inflammatory processes similar to those that are known from the study of higher animals." Less mobile tissue granulocytes, termed eosinophilic granular cells (EGCs), are also involved in the host response to bacterial and helminth pathogens at mucosal sites such as the gills and gut. These cells can degranulate, releasing immunopharmacological agents in a manner analogous to mammalian mast cells. On the other hand, virus-infected host cells and protozoan pathogens may be the target for NCCs also present in the blood, lymphoid tissues, and mucosal sites, which can spontaneously kill cells via an apoptic and necrotic mechanism.

An important feature of all of these events is their lack of specificity, allowing large numbers of cells to be mobilized quickly. However, unlike specific defences (see Chapters 4 and 5), there is no memory component, and thus, subsequent exposure to the same pathogen does not lead to higher and faster "secondary" responses. Nevertheless, the cells involved in nonspecific immunity are able to interact with the cells of the specific immune system and can be stimulated by them and their products. This is particularly true during delayed or chronic inflammatory reactions, which occur if inflammatory stimuli persist.

II. MORPHOLOGY AND ISOLATION

A number of morphological, physical, and functional characteristics can be used to distinguish cells involved in nonspecific defenses. Differences in surface molecules are also the basis for generating specific antibodies able to recognize individual cell types. Some of these characteristics and, where available, specific antibodies can be used to isolate particular cell types. In addition, some progress toward establishing fish leukocyte cell lines has been made.

A. Macrophages

Macrophages can be easily isolated from a variety of sources, including blood (monocytes), lymphoid organs (especially the kidney), and the peritoneal cavity. Leukocyte preparations can be enriched for macrophages in a variety of ways. First, it is possible to separate macrophages from lymphocytes and blast cells by density gradient centrifugation, since macrophages are larger and more granular (Secombes, 1990). However, it is more difficult to separate macrophages from granulocytes with this method. It is also possible to enrich for macrophages by eliciting them into the peritoneal cavity with phlogistic agents able to induce inflammation. Since granulocytes are highly mobile, they tend to predominate during the first few days after injection, with macrophages increasing later on. Thus, by choosing an appropriate time to harvest the cells, a good purity can be obtained. For more pure preparations, it is common to use two other features of macrophages: their ability to adhere firmly to substrates and to live for weeks in culture. Adherance to glass or plastic culture dishes is a rapid way to substantially increase the purity of macrophage-enriched suspensions, and can be achieved in 2–3 h. Immediately postadherance some contamination with granulocytes can still occur, but due to the relatively short life span of the latter, within a day or two of culture, purities of >95% macrophages can be obtained.

Isolated macrophages can be identified in a variety of ways. Typically, they are mononuclear, nonspecific esterase positive, and peroxidase negative (see Chapter 1). Functionally, they can act as accessory cells for lymphocyte responses, are avidly phagocytic, can secrete oxygen and nitrogen free radicals, and can kill a variety of pathogens (bacteria, helminth larvae), although differences do exist between macrophages from different sources (Secombes, 1990). While little is known about their surface markers, it is clear that they possess both antibody (Fc) and complement receptors (Secombes and Fletcher, 1992), and express class II MHC molecules (Secombes, 1994a). However, to date, no fish macrophage-specific antibodies have been generated, although antibodies to a related cell type present in the brain (glial cells) do exist (Dowding *et al.*, 1991). Despite the absence of specific antibodies for fish macrophages, flow cytometry can be used to help identify macrophages after phagocytosis of FITC-labeled particles (Thuvander *et al.*, 1992).

B. Granulocytes

Granulocytes can be subdivided into neutrophils, eosinophils, and basophils (see chapter 1). Neutrophils and eosinophils are the most common types, with basophils being absent in most species. As with macrophages, granulocytes can be isolated from blood, lymphoid tissues, and the peritoneal cavity. Leukocyte suspensions can be enriched for granulocytes by density gradient centrifugation and by collection of peritoneal cells within a few days of injecting an eliciting agent. Granulocyte numbers in the blood can be greatly increased within 24 h of stressing fish, although how similar they are functionally to cells from unstressed fish is not clear. Granulocytes will adhere to culture vessels, especially if precoated with celloidine (Lamas and Ellis, 1994a). However, after 48 h in culture their viability decreases dramatically.

The most conspicuous feature of isolated granulocytes are the granules present in their cytoplasm. These granules will stain positively for many dyes (e.g., Sudan black) and enzymes (e.g., peroxidase), and this can be used to identify the cells. In some species, as in salmonids, neutrophils are polymorphonuclear which allows simple detection of these cells. More recently, neutrophil-specific monoclonal antibodies (MoAb) have been raised for several fish species, including channel catfish (*Ictalurus punctatus*) (Bly *et al.*, 1990; Ainsworth *et al.*, 1990) and Atlantic salmon (*Salmo salar*) (Pettersen *et al.*, 1995). These MoAb can be used in a variety of methods, such as magnetic cell sorting (MACS) or flourescent-activated cell sorting (FACS), to obtain very pure neutrophil populations for functional assays. FACS analysis also allows their size/granularity profiles to be determined, where they can be seen as larger, more granular cells than other blood leukocytes (Fig. 1). Isolated granulocytes (especially neutrophils) are highly mobile, phagocytic, and produce reactive oxygen species but their bactericidal activity is often relatively poor compared with macrophages (see following). As with macrophages, they appear to possess both Fc and complement receptors, as evidenced in opsonization studies. Eosinophilic granular cells (EGSs) found in the stratum granulosum of the gut, gills, skin, meninges, and surrounding major blood vessels, are not considered to be eosinophils but rather mast cells (Vallejo and Ellis, 1989). The granules

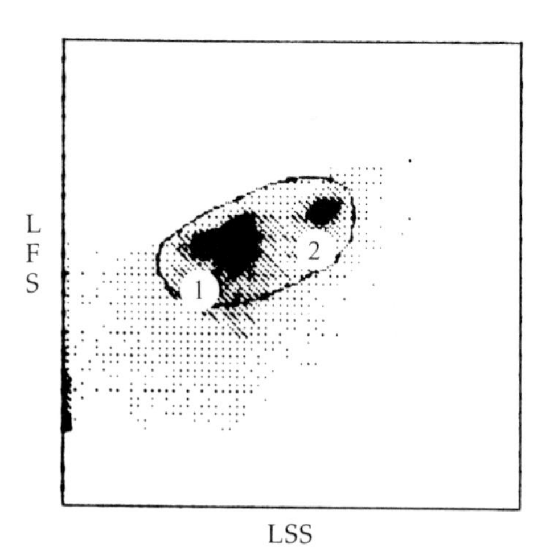

Fig. 1. Flow cytometric analysis of rainbow trout blood leucocytes. Two populations can be visualized that differ in their size (*y* axis) by granularity (*x* axis) profiles. Lymphocytes (1), which form the majority of the blood leukocyte population, are smaller and less granular compared with the larger and more granular neutrophils (2). LFS, log forward light scatter; LSS, log side light scatter.

stain with basic dyes such as toluidine blue and astra blue, contain phospholipids and acid mucopolysaccharides, and are positive for alkaline phosphatase, arylsulphatase, and 5′-nucleotidase activity. Acid phosphatase activity has also been associated with EGCs but is confined to lysosomal structures. EGCs have been isolated from the gut by digestion of the tissue with collagenase followed by density gradient centrifugation and adherance (Dorin *et al.*, 1993). Adherance ranged from 21% of cells after a 30-min incubation period to 67% after 180 min. Functionally, such cells have been shown to endocytose foreign proteins, and to possess the capacity to degrade them via cathepsin D activity.

C. Nonspecific Cytotoxic Cells

Nonspecific cytotoxic cells (NCC) in fish are considered to be equivalent functionally to mammalian natural killer (NK) cells. They can lyse a wide spectrum of mammalian tumor cells lines, as well as transformed fish protozoan parasites (Evans and Jaso-Friedmann, 1992). In sharks, the cell type responsible for this spontaneous cytotoxicity is considered to be the macrophage (McKinney *et al.*, 1986) but in bony fish a second, much smaller lymphocyte-like cell type is involved. Unlike mammalian NK cells, these cells do not contain cytoplasmic granules and their nucleus is very pleomorphic. They can be obtained from blood, lymphoid tissues, and gut, with tissue leukocytes having the highest activity. Indeed, in the kidney some 15–20% of leukocytes are considered to be NCCs (Evans *et al.*, 1984).

A variety of approaches have been used to enrich for NCCs, including dissociation from tumor target cells, density gradient centrifugation, and flow cytometry. Flow cytometry has been most effective, giving a four- to six-fold increase in NCC activity. However, size microheterogeneity exists in this population, giving activity in more than one sorted fraction (Evans *et al.*, 1987).

Recently, a "function-associated molecule" (FAM) has been described on NCCs, which is involved in recognition of target cells. This FAM is a vimentin-like molecule, and antibodies raised to it inhibit NCC activity (Harris *et al.*, 1992). Thus, this putative antigen receptor is also a good marker for the identification and isolation of fish NCCs. In addition, anti-rat transferrin receptor MoAb (OX-26) cross-react with catfish NCCs (Evans and Jaso-Friedmann, 1992).

D. Cell Lines

To date very few leukocyte cell lines exist from fish. However, some progress in generating cell lines has been made, particularly in channel catfish and Japanese flounder (*Paralichthys olivaceus*). Remarkably, spontaneous development of long-term leukocyte cell lines has been achieved in some species, as for example with a monocyte-like cell line in catfish (Vallejo *et al.,* 1991) and goldfish (*Carassius auratus*) (Wang *et al.,* 1995), and a myelopoietic culture in rainbow trout (*Oncorhynchus mykiss*) (Diago 1994). In catfish, the monocyte-like cells are nonspecific esterase and peroxidase positive (but negative for Sudan black), are phagocytic, and can function as efficient accessory cells in autologous cultures. Typically, the cells grow in clumps and contain high numbers of blasts. In goldfish, the cells are also nonspecific esterase positive, phagocytic, and capable of releasing both oxygen and nitrogen free radicals after appropriate stimulation.

It has also been possible to establish catfish leukocyte cell lines with transient stimulation with phorbol ester and calcium ionophore (Miller *et al.,* 1994). Such cells continue to proliferate in the absence of further stimulation, feeder cells, or exogenous factors, in contrast to mammalian leukocyte cell lines. Long-term, cloned B-cell lines, T-cell lines, and monocyte/macrophage cell lines have been established in this manner, with a success rate of >95% of tested fish.

Finally, fish leukocyte cell lines can be established by immortalization with oncogene transfection (Tamai *et al.,* 1993). A variety of plasmid constructs containing the oncogenes have been tested using electroporation for transfection. Cotransfection resulted in the highest efficiences, especially using a combination of *c-fos* and c-Ha-*ras* with a CMV promoter. The cell lines typically contain 40–45% lymphocytes, 5–25% neutrophils and 30–55% platelets. While these cell lines have been used primarily to study cytokine secretion, the phagocytic activity of immortalized neutrophils has also been established.

III. INFLAMMATION

Three major events happen during inflammatory responses (Suzuki and Iida, 1992). First there is an increased blood supply to the infected area. This is followed by an increased capillary permeability, and lastly there is a migration of leukocytes out of the capillaries and into the surrounding tissue. Once in the tissue they migrate toward the site of infection, attracted by a variety of host- and pathogen-derived molecules. Thus, when a pathogen gains entry to the tissues of a host it quickly encounters a network of phagocytic cells with potent microbicidal activity, which limit its spread or removes it altogether.

A. Cellular Events

1. Acute Information

Acute information in fish has been studied following induction by a large variety of natural and experimental stimuli. These include injection with phlogistic agents such as bacteria, exposure to metazoan parasites, subcutaneous inoculation with fungi, intrapulmonary stimulation (lungfish) with carbon or latex, and wounding (Roberts, 1989; Suzuki and Iida, 1992; Woo, 1992). In all cases a common acute inflammatory response is elicited, characterized by neutrophilia and monocytosis in the blood, and an accumulation of neutrophils and macrophages at the site of injury or infection (Roberts, 1989; Suzuki and Iida, 1992). The cellular response is typically biphasic, especially in response to potentially pathogenic organisms, with the increase in blood neutrophils and their extravasation preceeding the appearance of monocytes and macrophages (Ellis, 1986). Neutrophilia occurs within an hour of giving an inflammatory stimulus, and commonly reaches a peak after 48 h. The lack of mitotic figures in the infiltrating cells strongly suggests that they arrive by migration and are not the result of perivascular proliferation at the site of inflammation. In support of this, serum from fish injected with stimulants such as lipopolysaccharide (LPS) or muramyl dipeptide (MDP) is capable of stimulating colony formation of head kidney leucocytes (especially macrophages), and this activity peaks at 1–2 days postinjection (Kodama *et al.,* 1994a). Nevertheless, changes do occur in the migrating cell populations. For example, in neutrophils increases in a range of enzyme activities can occur (Park and Wakabayashi, 1991), including peroxidase, Mg-ATPase, malate dehydrogenase, and glucose-6-phosphate–dependent glycogen synthetase, resulting in increases in glycogen in the latter case (readily identifiable with periodic acid Schiff, PAS). Additionally, the activity of phosphorylases that metabolize glycogen also increase, presumably to sustain the cellular activities occurring in these cells. In some situations an increase in immature neutrophils is seen in the blood (i.e., with a range of weaker enzyme activities), possibly as a result of excessive migration of mature cells to the inflammed site.

Neutrophils are phagocytic and, at the peak of the response, most of these cells possess phagosomes containing ingested material. Nevertheless, it is the macrophages that usually have the largest phagocytic capacity and are able to ingest many more particles per cell (Suzuki, 1984). Indeed, as the response progresses, their cytoplasm appears "foamy" due to extensive phagocytosis and cytoplasmic vacuolation. However, since neutrophil numbers can far exceed macrophage numbers, their relative contribution to bacterial clearance may be substantial. In addition to phagocytosis of the inflammatory agent, if degeneration of muscle fibers occurs during the response (as following wounding) then phagocytosis of the necrotic muscle also occurs (Roberts, 1989). The majority of phagocytosis is seen during the first 3–4 days, after which the phagocytes return to their resting state and their numbers decline. This decrease in cell number may be due in part to cell emmigration, but lysis of cells may also occur. The release of intracellular enzymes as a consequence of phagocyte lysis may contribute to the hostile environment for the pathogen, and it has even been suggested that neutrophils actively disgorge their enzymes as an extracellular killing mechanism (Ellis, 1986). Similarly, the inability of filamentous fungi to grow after attachment of brown trout (*Salmo trutta*) inflammatory cells may be through such a mechanism (Wood *et al.*, 1986). Extracellular enzymes will also damage host tissue, possibly contributing to the hemorrhagic liquifaction of host tissue commonly seen during bacterial infections. However, purulent lesions associated with enzyme release from degenerating neutrophils are not seen in fish.

After cellular infiltration and phagocytosis comes tissue repair, particularly in the case of skin and muscle lesions (Roberts, 1989). Where there is a breach of the epidermal integrity, such as with a wound, in addition to the risk of secondary infection there is also an immediate osmotic imbalance in fish. Epidermal healing is very rapid, and within hours a 2- to 3-cell-thick epidermis can cover the wound. This is as a result of migration of preexisting Malpighian cells from adjacent normal skin, which consequently shows a marked reduction in thickness. Following the clearance of necrotic muscle from lesions, replacement fibrosis and muscle fiber regeneration begin. Elongate fibroblasts appear within the first week after the inflammatory insult and undergo fibroplasia during the second week. Muscle and scale regeneration are more apparent during the third week, and by the end of the fourth week increased cellularity of the dermis together with an increased number of melanophores in the lesion (causing a darkening of the wound) are all that allow it to be distinguished from surrounding normal tissue.

The inflammatory response to metazoan parasites is basically similar to the above, but there are some significant differences relating to the pathogen being too large to be phagocytosed, and commonly being able to survive unharmed as long as the host lives. Parasite infections are generally chronic and this can lead to a chronic inflammatory response (see below) or encapsulation following on from the initial acute inflammatory response (Pulsford and Matthews, 1984; Hoole, 1994). However, inflammatory responses do not always occur, particularly if the parasite only penetrates the epidermis with anchor processes or enters through the skin. In these circumstances, only a tissue response may be elicited, consisting of hyperplasia of skin epidermal cells or gill epithelial cells coupled with an increase in mucus production. Such reactions are typical of infections with protozoan ectoparasites, with glochidia larvae of molluscs, with monogeneans, and at the site of epidermal penetration by cercariae of various digeneans (e.g., *Cryptocotyle lingua*). In the latter case, encystment in the skin or muscle may follow penetration but still without eliciting a true inflammatory reaction. The cysts are composed of a delicate membrane of parasite origin, surrounded by a host capsule of fibroblasts and the occasional melanophore giving a characteristic "black-spot" (McQueen *et al.,* 1973). Where an inflammatory response is elicited it is commonly associated with a tissue response. Neutrophils, eosinophils, and macrophages may accumulate at a site of epidermal hyperplasia, or form a second component of the host reaction during encystment. Thus, in the latter case, the host response in established infections often consists of an inner leukocyte layer and an outer connective tissue layer.

The role of eosinophils in fish inflammatory responses is not clear. While they are known to be involved in antiparasite responses in a few species (Cone and Wiles, 1985; Reimschuessel *et al.,* 1987), in most they are absent from the response. However, eosinophilic granular cells (EGCs) may have an important role. Degranulation of EGCs in the intestine of rainbow trout occurs 1 h after an intraperitoneal injection of bacterial exotoxin (ECP) from *Aeromonas salmonicida* or *Vibrio anguillarum* (Powell *et al.,* 1993) coincident with a fall in tissue histamine levels, a transient appearance of histamine in the blood, and widespread vasodilation. In addition, EGCs appear in the blood, kidney, and spleen of injected fish, suggesting they are a mobile cell population in certain situations. In *Catostomus commersoni,* extragastric histamine levels are higher in tissues containing an equivalent cell type, the PAS-positive granular leukocyte (PAS-GL), compared with tissues where they are absent (Barber and Westermann, 1978). Injection of *C. commersoni* and rainbow trout with the secretagogue compound 48/80, a histamine liberator, causes cytoplasmic vacuolation, loss of granule definition, and degranulation of PAS-GL/EGC within 24 h. Ultrastructurally, the response resembles the anaphylactic granule extrusion of mammalian mast cells (Vallejo and Ellis, 1989). This degranulation is inhibitable by antihistamines (promethazine and cimetidine), in contrast to mammalian mast cells where such agents can only block histamine receptors on target cells. Degranulation is not a cytotoxic event, and regeneration of cytoplasmic granules occurs, associated with an increased cytoplasmic activity. Thus, EGC and PAS-GL appear to be functionally akin to mast cells in addition to their morphological and locational similarities.

2. Chronic Inflammation

In situations where inflammatory stimuli are not eliminated during an acute inflammatory response, a chronic inflammatory response may follow (Roberts, 1989). Granulomas are typical of such responses and consist of organized collections of mature mononuclear phagocytes within a fibrous tissue stroma. They can be induced *in vivo* by a wide range of bacterial, fungal, or parasitic diseases, by diet-related diseases, by injection with adjuvants, or by experimental autoimmunity, and probably represent an attempt to isolate and destroy pathogens evading the acute inflammatory response.

Lymphocytes often appear early on in the chronic inflammatory infiltrate, followed by a large influx of macrophages and a concurrent monocytosis. As the response progresses, the macrophages aggregate together and transform into a granuloma consisting of epithelioid cells and multinucleated giant cells (MGCs), although in some circumstances the lesion may consist of macrophages surrounded by a zone of epithelioid cells. Extensive melanization and fibrosis are characteristic of established granulomas. The delayed nature of the granulomatous response coupled with the early appearance of lymphocytes suggests that fish can mount delayed-type hypersensitivity reactions, as do mammals (Thomas and Woo, 1990; Ramakrishna *et al.*, 1993).

Such *in vivo* responses demonstrate the remarkable maturation sequence of mononuclear phagocytes. As in acute inflammatory responses, blood monocytes migrate to the inflammatory site where they develop into tissue macrophages. The macrophages in turn transform into epithelioid cells or MGCs. Epithelioid cells are large, polygonal cells, typically closely associated and with poorly defined margins. They have oval nuclei and granular cytoplasm due to the presence of numerous free ribosomes and dense lysosomal vesicles, possibly reflecting a more mature cell type. Multinucleated giant cells, as their name suggests, are polykaryons and are believed to be derived from the fusion of macrophages or epithelioid cells. The nuclei are of approximately equal size and may vary in number from two to many hundreds within one polykaryon. They are arranged in the form of a ring or arc, the Langhans giant cell (Fig. 2), or are more randomly dispersed throughout the cytoplasm, the foreign body giant cell. Numerous vesicles, mitochondria, and even melanin granules also occur in the cytoplasm of MGCs, and are present within the ring of nuclei in the Langhans type indicating a greater degree of cellular organization than in the foreign body type. Multinucleated giant cells are phagocytic but less so than macrophages. That MGCs can be derived from monocytes and macrophages has been confirmed *in vitro* using purified cell populations (Secombes, 1985).

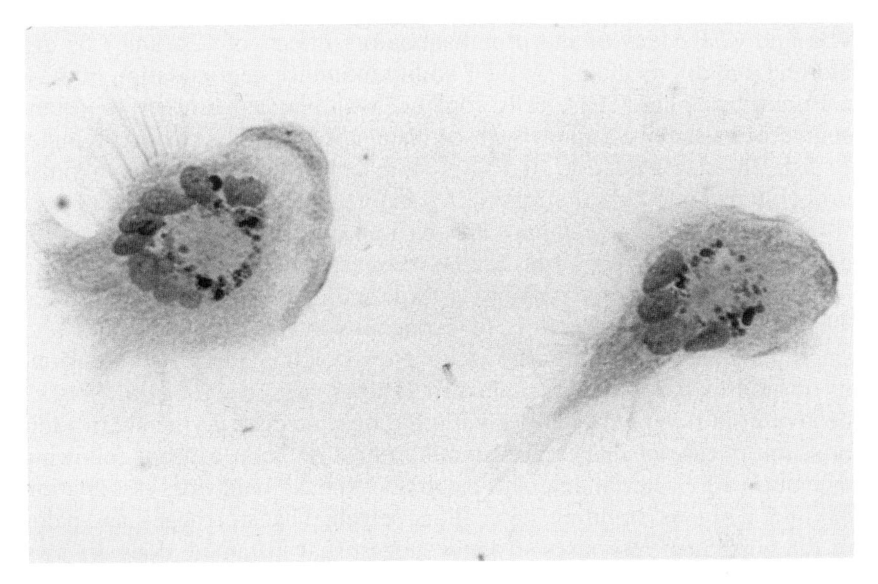

Fig. 2. A light micrograph of rainbow trout Langhans-type giant cells in culture (×700). Note the multiple nuclei arranged in an arc. From Secombes, 1985. With permission.

B. The Control of Inflammation

The development of inflammatory reactions is controlled by a number of mediators, including cytokines, eicosanoids, complement factors, and other vasoactive compounds released from phagocytes, EGCs/PAS-GL, and thrombocytes. Fast-acting, preformed mediators such as vasoactive amines initiate the response. Later, newly synthesized molecules such as eicosanoids serve to attract and activate leukocytes. On arrival at the site of infection, leukocytes themselves release mediators that regulate the response. However, it is the persistence of an antigen/pathogen that exerts the ultimate control.

1. Vasoactive Amines or Proteins

A number of major blood enzyme systems have a role in inflammation by producing vasoactive molecules. These include the clotting system, the fibrinolytic system, the kinin system, and the complement system. While little is known of the relative contribution of these systems to inflammation in fish, it is clear that they share many similarities with their mammalian counterparts (Alexander and Ingram, 1992; Olson, 1992; Sakai, 1992). For example, it is known that molecules equivalent to complement factors C3 and C5 exist in fish, which breakdown to give the anaphylactic compounds

C3a and C5a. Many of the proinflammatory effects of C3a and C5a are indirect and are mediated by their ability to induce degranulation of mast cells and basophils. These cells, together with platelets, are an important source of vasoactive amines such as histamine and 5-hydroxytryptamine. In fish, the degranulation of EGCs in response to bacterial exotoxin is probably mediated via the release of complement factors in this manner.

Although vasoactive amines are released during inflammatory-type reactions in fish, the role of histamine has been questioned in view of the low tissue levels in comparison with birds and mammals (Suzuki and Iida, 1992), and the variable effect of exogenous histamine on vascular and smooth muscle of different fish species. However, immediate hypersensitivity responses can be induced in fish (Jurd, 1987), resulting in "shock" behavior (fin clamping and disorientation) or immediate erythema. In addition, the release of intestinal histamine into the blood of trout following injection with bacterial exotoxin results in "shock" behavior, vasodilation of visceral organs, vomiting, defecation, petechial haemorrhages, and palor of the gills. Such responses strongly suggest that histamine does act as a mediator of inflammatory responses in fish, although it is probably not the only mediator released. Interestingly, 5-hydroxytryptamine (serotonin) has been found in "polymorphous" granular cells in the gills of rainbow trout (Nilsson and Holmgren, 1992) and may have an involvement.

2. Eicosanoids

Eicosanoids are a group of lipid mediators derived from eicosapolyenoic acids (polyunsaturated fatty acids with 20 carbon atoms), especially arachidonic acid, that have potent proinflammatory effects. Eicosanoids are not stored in cells but released soon after they are produced following cell stimulation and mobilization of phospholipases. They include prostaglandins (PGs) and thromboxanes (TXs) derived from cyclooxygenase activity, and leukotrienes (LTs) and lipoxins (LXs) derived from lipoxygenase activity. All of these molecules are known to be released by fish leukocytes, including purified macrophages (Pettitt *et al.,* 1991), neutrophils (Tocher and Sargent, 1987) and thrombocytes (Lloyd-Evans *et al.,* 1994).

Generation of lipoxygenase products by leukocytes has been particularly well studied in fish because they are derived from a relatively limited number of cell types, mainly phagocytes, in contrast to PGs which can be released from most cells. For example, rainbow trout macrophages are able to secrete LTs and LXs following stimulation with calcium ionophore or opsonized zymosan, although in contrast to mammalian macrophages they secrete more LXs than LTs. Indeed, indications that lymphocytes do not release LTs and LXs come from *in vitro* studies showing that the regression line obtained by plotting the amount of LT and LX produced as a function of the concentration of adherent leukocytes (mainly macrophages) present in a mixed leukocyte culture, crosses the *x*-axis rather than the *y*-axis (Fig. 3; Rowley *et al.,* 1995). The main isomers produced are LTB_4 and LXA_4, although this can vary depending on dietary intake. Thus, feeding fish diets rich in *n*-3 fatty acids increases the relative amount of eicosapentaenoic acid present, resulting in a shift to higher levels of LTB_5 and LXA_5 (Ashton *et al.,* 1994). LXs are produced more slowly than LTs and require the collaboration of 5- and 12-lipoxygenase enzymes (Rowley *et al.,* 1994). The possible activity of 15-lipoxygenase activity has also been demonstrated recently (Knight *et al.,* in press).

LTs and LXs have a number of activities on nonspecific and specific immune functions. With respect to the former, they are able to augment phagocytosis (Rainger *et al.,* 1992) and to act as potent chemoattractants for neutrophils (covered in a subsequent section). In addition, injection of nordihydroguaiaretic acid (NDGA), a lipoxygenase inhibitor, significantly reduces the relative numbers of macrophages and neutrophils elicited into the peritoneal cavity of rainbow trout by an intraperitoneal injection of formalin-killed *A. salmonicida* relative to numbers in control fish given

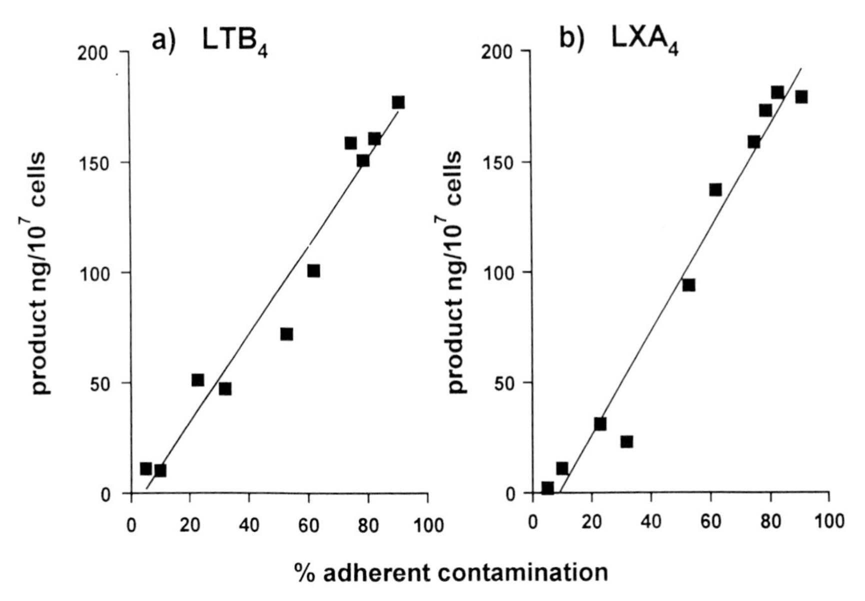

Fig. 3. Regression analysis of released (a) LTB_4 and (b) LXA_4 from isolated rainbow trout leukocytes stimulated with calcium ionophore, as a function of the concentration of adherent cells present in the culture. That the regression line crosses the *x*-axis indicates that only adherent (predominantly macrophages) leukocytes are able to release these products. From Rowley *et al.,* 1995. With permission.

A. salmonicida alone (Rainger *et al.*, 1992). Similarly, in plaice (*Pleuronectes platessa*), erythema induced by intradermal injection of certain fungal extracts can be abrogated by inhibitors of LT release, whereas inhibitors of cyclooxygenase activity (indomethacin) have no effect (Baldo and Fletcher, 1975). Interestingly, in addition to the role of LTB_4 in inflammation, LTC_4, LTD_4, and LTE_4 are functionally critical components of the "slow reacting substance of anaphylaxis" (SRS-A). Slow reacting substance of anaphylaxis induces contraction of smooth muscle with a uniquely slow onset and relaxation that distinguishes it from histamine and other vasoactive factors, and is associated with early inflammatory events.

Recently, the interest in PG release has been rekindled with the discovery that there are two forms of cyclooxygenase enzymes in mammals, with one form being inducible (COX2) in macrophages and other immunologically active cells by the action of cytokines, lipopolysaccharide (LPS), and phorbol esters (O'Sullivan *et al.*, 1992). While it remains to be determined whether two forms exist in fish, it has been demonstrated that stimulation of trout head kidney macrophages with LPS induces higher PG production than is obtained from control, unstimulated cells, suggesting that COX activity has been increased in some manner (Rowley *et al.*, 1995). In mammals, PGs can interact with histamine and bradykinin to induce vasodilation and increase vascular permeability, but down regulate a number of phagocyte functions. In fish, PGs have been shown to inhibit macrophage respiratory burst activity (Novoa *et al.*, in press) as well as suppress lymphocyte proliferation (Secombes *et al.*, 1994) and antibody production (Knight and Rowley, 1995).

3. Cytokines

A number of cytokines have an involvement in inflammatory reactions. These include tumor necrosis factor α (TNFα), interleukin 1 (IL-1), IL-6, and a variety of chemokines. Indeed, TNFα is the principal mediator of the host response to Gram negative bacteria and is able to induce the release of other cytokines such as IL-1, IL-6, and chemokines. In fish there is good evidence for the biological activity of IL-1-like molecules (Ellsaesser and Clem, 1994; Verburg van Kemenade *et al.*, 1995) and chemokines (Howell, 1987). Antigenic cross-reactivity of antisera to mammalian TNF-α, IL-1, and IL-6 (Ahne, 1993; Robertsen *et al.*, 1994; Verburg van Kemenade *et al.*, 1995) has also been reported, as has biological cross-reactivity of mammalian IL-1 (Clem *et al.*, 1991) and TNFα (Hardie *et al.*, 1994a). The role of such cytokines in inflammatory reactions of fish has still to be determined, although it is clear that human rTNFα is a good chemoattractant for trout neutrophils (Jang *et al.*, 1995a) and will be discussed further under Modulation of Killing Mechanisms.

IV. PHAGOCYTE MIGRATION

Locomotor behavior of leukocytes is a key factor allowing their accumulation at sites of infection during inflammation. Indeed, it is well known in fish that injection of phlogistic agents can elicit leukocytes to a site, especially macrophages and neutrophils. More recently, evidence for the migration of EGCs *in vivo* has been obtained (Lamas *et al.,* 1991). However, the use of *in vitro* assays has allowed more precise analysis of leukocyte migration. From such studies, two main types of locomotory behavior have been described in fish: an increase in the speed of migration (chemokinesis) and an increase in directional migration (chemotaxis) (Sharp *et al.,* 1991a).

A variety of assays have been used to study leukocyte locomotion in fish. These have included migration under agarose assays (Griffin, 1984), leukocyte polarization assays (Wood and Matthews, 1987; Taylor and Hoole, 1993), and various modified Boyden chamber techniques where the cells and chemoattractants are separated by a filter (Weeks *et al.,* 1986; Sharp *et al.,* 1991a). Both macrophages and granulocytes are clearly migratory in such assays, with neutrophils and a second type of granulocyte (L1) migrating in some species (Taylor and Hoole, 1993). While most studies have failed to distinguish between chemokinesis and chemotaxis, a few have. For example, by tracing the paths of individual cells in the migration under agarose assay it has been possible to show that rainbow trout leukocytes respond to fetal calf serum in a random way but respond to trout serum in a unidirectional way (Fig. 4). Similarly, using the Boyden chamber technique, it has been possible to perform checkerboard assays where the concentration of the chemoattractant is varied above and below the filter, allowing increases in concentration in the presence and absence of a gradient. Such assays have shown that the responses to bacterial products (Nash *et al.,* 1986), parasite extracts (Sharp *et al.,* 1991a), and lipoxins (Sharp *et al.,* 1992) are random, whereas migration to leukotrienes, mammalian C5a, and LPS-activated plasma is directional but often with a random component (Sharp *et al.,* 1992; Newton *et al.,* 1994).

Many host- and pathogen-derived factors have been shown to be chemoattractants for fish leukocytes. Of the host-derived factors, lipoxygenase products are particularly potent chemoattractants, with lipoxins inducing responses some three- to four-fold higher than leukotrienes (Sharp *et al.,* 1992). Little evidence of stereospecificity is apparent using a number of 4- and 5-series LXs. However, eicosanoid-rich supernatants from ionophore-challenged macrophages do show differences in their ability to induce *in vitro* locomotion of trout neutrophils, depending upon the diet given to the fish from which the macrophages were derived (Ashton *et al.,* 1994).

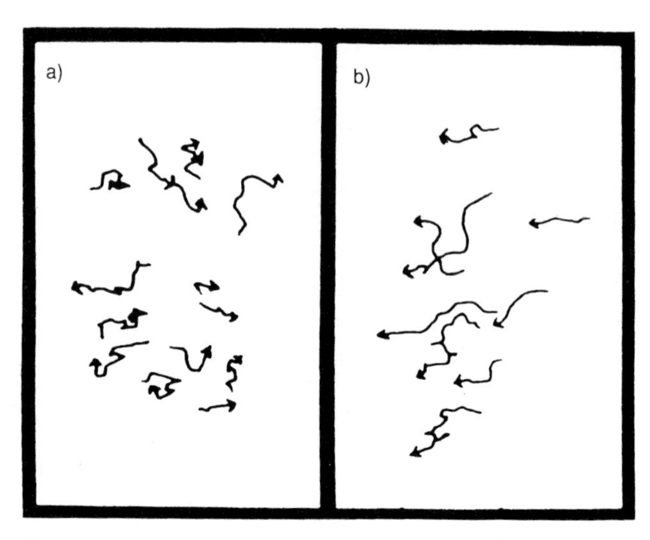

Fig. 4. Migration paths of rainbow trout blood leukocytes in response to (a) fetal calf serum or (b) normal trout serum (×500). Note the random vs. unidirectional migration paths in response to these chemoattractants. Reprinted from *Dev. Comp. Immunol.* B. R. Griffin, Random and directed migration of trout (*Salmo gairdneri*) leukocytes: Activation by antibody, complement, and normal serum components, 589–597, copyright 1984, with kind permission from Elsevier Science Ltd, The Boulevard, Langford Lane, Kidlington OX5 1GB, UK.

Fish fed diets rich in *n*-3 polyunsaturated fatty acids (PUFA) give more chemo-attractive supernatants than those from fish fed diets rich in *n*-6 PUFA. Serum-derived factors are also potent chemoattractants in fish, especially after appropriate activation (e.g., with LPS, zymosan, etc.), as are factors in inflammatory exudate fluid (MacArthur *et al.,* 1985). Confirmation that complement factors in serum are important chemoattractants has been obtained using mammalian C5a (Obenauf and Hyder Smith, 1992; Newton *et al.,* 1994). Finally, some cytokines are also chemoattractants for leukocytes, especially the chemokines (Van Damme, 1994). Generation of possible chemokines from fish leukocytes has been demonstrated in a few instances (Howell, 1987; Bridges and Manning, 1991), and human rTNFα has been shown to attract rainbow trout neutrophils in a dose-dependent manner, inhibitable with anti-TNF receptor MoAb (Jang *et al.,* 1995a). Not all host molecules induce locomotion of leukocytes, and there are many examples of migration-inhibitory substances released from fish leukocytes (Smith *et al.,* 1980; Song *et al.,* 1989).

Pathogen-derived chemoattractants include extracts of digenean (Wood and Matthews, 1987) and cestode parasites (Sharp *et al.,* 1991a; Taylor and Hoole, 1993), acanthocephalan parasites (Hamers *et al.,* 1992), and bacterial products (Nash *et al.,* 1986; Weeks *et al.,* 1988; Lamas and Ellis, 1994a). It has also been shown that the combination of host factors (e.g., normal serum) and bacterial-derived factors can significantly enhance migration (Lamas and Ellis, 1994a; Newton *et al.,* 1994).

In mammals, leukocyte migration *in vivo* is in large part determined by the expression of adhesion molecules such as selectins and integrins. For example, the appearance of neutrophils at sites of acute inflammation require the expression of the selectin ELAM on the endothelium in these areas (Roitt *et al.,* 1993). To date, virtually nothing is known about the role of adhesion molecules in fish. The main exception is a recent report in catfish, where PBL cytotoxicity toward allogeneic cells can be inhibited with an MoAb that appears to recognize an integrin-like molecule similar to LFA-1, present on all catfish leukocytes (Yoshida *et al.,* 1995). Similarly, there are few experimental studies on adhesion of leukocytes in fish. One exception is leukocyte adherance to cestode parasites (Sharp *et al.,* 1991b; Hoole, 1994), where the involvement of antibody, complement, and acute-phase proteins (e.g., C-reactive protein, CRP) has been the focus of attention (Fig. 5).

V. PHAGOCYTOSIS

Phagocytosis is the process whereby cells internalize, kill, and digest invading microorganisms. It can be divided into three main phases: attachment of the particle to the cell surface, ingestion involving the formation of a phagosome, and lastly, breakdown of the particle within the phagosome. *In vivo* and *in vitro* studies have shown that monocytes/macrophages and granulocytes (neutrophils and in some cases eosinophils) are phagocytic and will ingest a wide range of inert and antigenic particles (Ainsworth, 1992; Secombes and Fletcher, 1992; Steinhagen and Jendrysek, 1994), and soluble ligands (Dorin *et al.,* 1993; Dannevig *et al.,* 1994). Thrombocytes have also been described as phagocytic, although their phagocytic capacity is very low and it is not clear if these cells are capable of intracellular digestion. Within tissues, endothelial cells in the kidney and parenchymal cells in the liver can also endocytose molecules in a receptor-dependent manner (Dannevig *et al.,* 1994).

Attachment of a particle to the phagocyte membrane is a prerequisite for uptake, and is a relatively passive process. Nevertheless, fish phagocytes have the capacity to discriminate between targets, suggesting the involvement of surface receptors. Since phagocytosis can proceed *in vitro* in the absence of serum, it is likely that a number of lectin-like receptors are present on macrophages. This has been shown to be the case in tilapia (*Oreochromis spilurus*), where preincubation of macrophages with L-fucose, D-galactose, D-glucose, D-mannose, α-methyl mannoside, and *N*-acetyl-D-

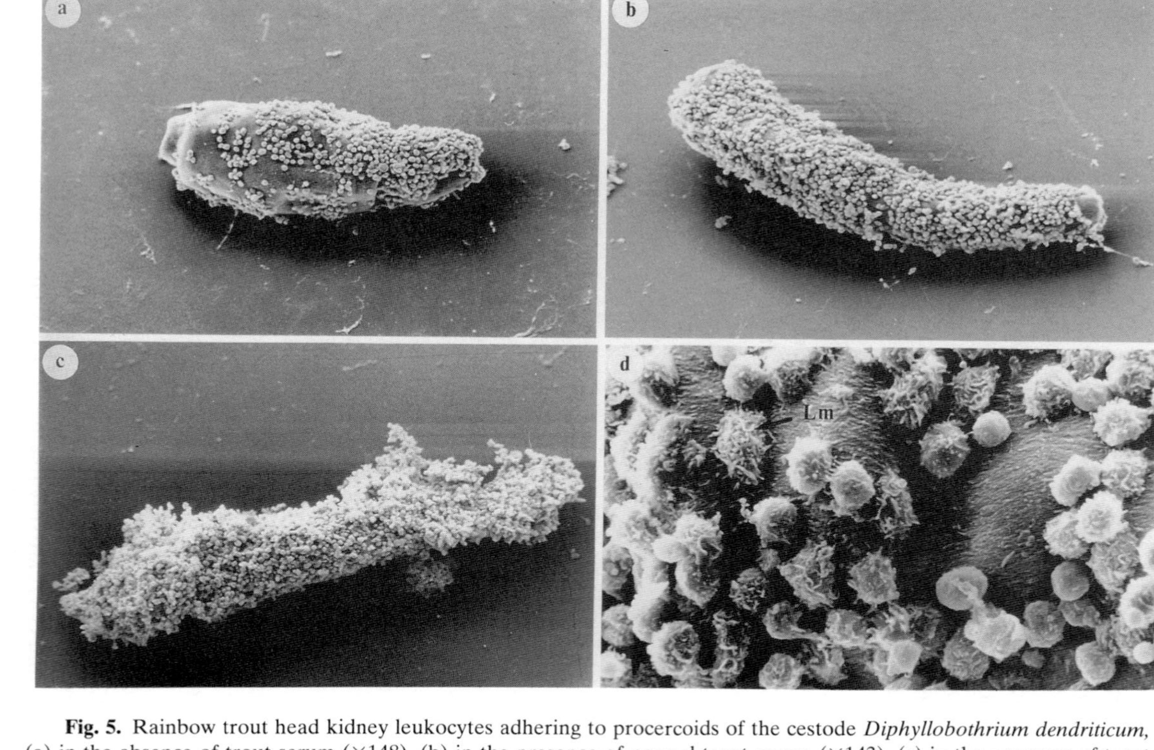

Fig. 5. Rainbow trout head kidney leukocytes adhering to procercoids of the cestode *Diphyllobothrium dendriticum*, (a) in the absence of trout serum (×148), (b) in the presence of normal trout serum (×143), (c) in the presence of trout immune serum (×130), or (d) higher magnification of (c) showing lamellipodial extensions (Lm) from the cells to the parasite surface (×1271). From Sharp *et al.*, 1991b. With permission.

glucosamine all significantly inhibit phagocytosis (Saggers and Gould, 1989). Atlantic salmon macrophages have also been shown to possess receptors for β-glucans (Engstand and Robertsen, 1994), present in the cell wall of most yeast and mycelial fungi.

It is well known that opsonization of particles with hemolytically active normal serum greatly increases adherence to macrophages and neutrophils, and subsequent ingestion (Matsuyama *et al.*, 1992; Rose and Levine, 1992). Heat-inactivated normal serum is far less active, suggesting that complement is the vital factor. This is supported by the finding that incubation of opsonized particles with antisera to fish C3 abolishes the opsonizing effect. In addition, purified mammalian complement components have been used successfully to opsonize particles (Johnson and Smith, 1984). Activation of complement via both the classical and alternative pathways is effective for opsonization. Interestingly, it has been suggested that differences in the uptake of bacterial species/strains may relate to their relative ability to activate complement (Lamas and Ellis, 1994a). Pretreatment of phagocytes with trypsin greatly decreases uptake of opsonized particles, suggesting it is a receptor-mediated event (Matsuyama *et al.*, 1992). Indeed, Fc receptors for antibody have been demonstrated on fish phagocytes (Haynes *et al.*, 1988). However, in general, opsonization with heat-inactivated antisera gives poor uptake. This is possibly due to a low receptor number on resting cells, since activated cells show enhanced uptake of antibody-opsonized particles (Secombes and Fletcher, 1992). Finally, another serum component, CRP, has also been demonstrated to act as an opsonin in fish (Nakanishi *et al.*, 1991).

Ingestion of a particle is an active process and can occur by engulfment or enfoldment. During engulfment, pseudopodia are extended, encircle the particle, and fuse. The resulting phagosome is drawn into the cell. In enfoldment, as seen in the gar (*Lepisosteus platyrhincus*), individual pseudopodia encircle the particle and wrap around it several times before fusion and phagosome formation. Such processes require the active participation of cytoskeletal proteins, especially actin. Thus, incubation with substances that inactivate intracellular actin, such as botulinum C_2 toxin, prevents ingestion by fish phagocytes (Kodama *et al.*, 1994b). The divalent cation Ca^{2+} is also essential, and in its absence phagocytosis is blocked. A variety of factors may influence the kinetics of phagocytosis, such as temperature, incubation time, and target-to-effector (E:T) ratio. This has been elegantly demonstrated recently, using a second-order factorial experimental design with turbot (*Scophthalmus maximus*) phagocytes, allowing detection of variable interactions and response-variable maxima to be modeled (Leiro *et al.*, 1995). Response surface plots show that rates of phagocytosis *in vitro* are affected by all three parameters, and that moving away from the within-domain maximum, the response drops in all directions (i.e., there is an optimum time, temperature, and ratio, and the response drops sharply away from this) (Fig. 6). Interestingly, the E:T ratio has the strongest influence on phagocytosis.

Following ingestion, electron microscope studies have shown that cytoplasmic granules and vesicles converge upon and fuse with the phagosomes, discharging their contents into the lumen around the particle (Lamas and Ellis, 1994b). Degranulation is particularly well seen in neutrophils, where staining of the granules for peroxidase activity allows visualization of the reaction product in the phagosomes following granule–phagosome fusion. A similar process is thought to occur in monocytes/macrophages following fusion with primary lysosomes. It has also been suggested that some of the phagocyte's granule contents may be released into the external milieu and contribute to extracellular killing (Lamas and Ellis, 1994b).

While phagocytosis is the most distinctive function of phagocytes, external substances can also be internalized by pinocytosis or fluid-phase endocytosis (Weeks *et al.,* 1987; Lauve and Dannevig, 1993). In mammals this occurs in a receptor-independent manner, by simple membrane invagination. The molecules are then transported to lysosomes for degradation, or to the plasma membrane after partial degradation (see Phagocytes as Accessory Cells). In rainbow trout, uptake of extracellular fluid by macrophages has been estimated to be 7.2 nl/10^6 cells h^{-1}, or 1.4% of the cell volume per hour (Lauve and Dannevig, 1993). If the diameter of newly formed endocytic vesicles is assumed to be similar to those in mammals, some 70% of the macrophage surface is internalized per hour. Uptake is temperature dependent but does not follow a linear time course, suggesting recycling of endocytosed fluid.

VI. PHAGOCYTE KILLING MECHANISMS

Phagocytes are able to kill pathogens using a variety of killing mechanisms that can be broadly categorized as oxygen dependent or oxygen independent. In many instances, where the pathogen is a microorganism, killing by fish phagocytes is an intracellular event and can occur relatively quickly (i.e., within 1 h [Daly *et al.,* 1994]. However, it is quite clear that extracellular killing can also occur, as with killing of helminth larvae by rainbow trout macrophages (Whyte *et al.,* 1989). While little is known about larvacidal mechanisms in phagocytes, progress has been made with respect to bactericidal activity and will be reviewed in a subsequent section.

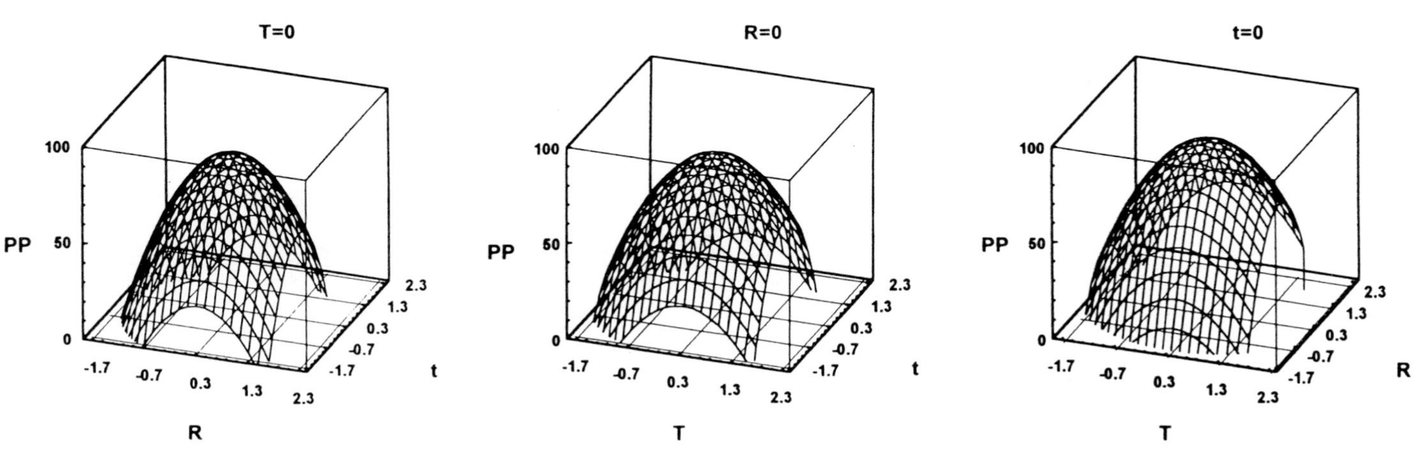

Fig. 6. Response surface plots of percentage phagocytosis (PP) of *Glugea caulleryi* spores by turbot adherent splenic cells, as a second-order function of temperature (T), incubation time (t) and spore-to-phagocyte ratio (R). In each plot, one of the three variables is fixed at its domain-central value (i.e., coded as zero). Note that in each case the response drops sharply away from the maximum in all directions. Reprinted from *Comp. Biochem. Physiol.* 112C, Leiro et al., A factorial experimental design for investigation of the effects of temperature, incubation time and pathogen-to-phagocyte ratio on *in vitro* phagocytosis by turbot adherent cells. 215–220, copyright 1995, with kind permission from Elsevier Science Ltd, Pergamon Imprint, The Boulevard, Langford Lane, Kidlington OX5 1GB, UK.

A. Oxygen-Dependent Mechanisms

When phagocytes ingest particles, there is an increased oxygen uptake that is independent of mitochondrial respiration (i.e., is not prevented in the presence of metabolic inhibitors). For example, rainbow trout head kidney phagocytes consume some 12 nmol O_2/min/10^7 cells after stimulation with zymosan compared with an uptake of 1.3 nmol O_2/min/10^7 cells in resting cells (Nagelkerke *et al.,* 1990). Soluble membrane stimulants such as phorbol esters can also elicit this response. This so-called respiratory burst is used in the generation of a number of oxygen and nitrogen free radicals known to be toxic for bacteria and protozoan parasites.

1. Reactive Oxygen Species (ROS)

It is well known that oxygen free radicals are produced by fish phagocytes during the respiratory burst (Secombes and Fletcher, 1992). Some have been detected directly, as with superoxide anion (O_2^-) and hydrogen peroxide, whereas others are inferred from indirect evidence, as with generation of singlet oxygen which is associated with chemiluminescence. The primary reaction of the respiratory burst is the one electron reduction of molecular oxygen to O_2^- catalyzed by NADPH oxidase. NADPH oxidase is a complex, multicomponent enzyme found in the plasma membrane of phagocytes (Segal and Abo, 1993). It consists principally of a low potential cytochrome *b* and a flavoprotein that act as an electron transport chain using the reducing equivalents provided by NADPH. The NADPH is produced via the hexose monophosphate shunt, making this process also glucose dependent. Evidence exists for the presence of NADPH oxidase in trout phagocytes and the glucose-dependent production of O_2^-, suggesting that a similar situation occurs in fish (Secombes and Fletcher, 1992). Scavenger/inhibitor studies have shown that O_2^- produced from fish macrophages is not particularly toxic for fish bacterial pathogens, but hydrogen peroxide and/or its derivatives are (Sharp and Secombes, 1993; Hardie *et al.,* 1996). Similarly, in cell-free systems, hydrogen peroxide is a potent bactericidal agent (Karczewski *et al.,* 1991; Hardie *et al.,*1996).

Both macrophages and neutrophils can generate ROS, although some differences do exist. For example, in Atlantic salmon the production of O_2^- following stimulation with phorbol esters is greater from neutrophils than macrophages (Lamas and Ellis, 1994a), whereas in channel catfish, neutrophils appear to be relatively poor producers of ROS (Dexiang and Ainsworth, 1991a), although a direct comparison with the response in macrophages was not made in the latter study.

As with phagocytosis, opsonization of particles can lead to augmented generation of ROS. Thus, bacteria opsonized with normal serum or heat-inactivated antiserum enhances ROS production, and bacteria opsonized with both antibody and complement induces the highest responses (Waterstrat *et al.*, 1991; Lamas and Ellis, 1994a). Similarly, the E:T ratio affects ROS production, with an optimal ratio of 1:50–1:100 using Atlantic salmon neutrophils incubated with *A. salmonicida* (Lamas and Ellis, 1994a). Live bacteria are often more stimulatory than killed bacteria (Stave *et al.*, 1984; Lamas and Ellis, 1994a), but high E:T ratios (i.e., >1:100) suppress ROS production. The relative virulence of the strain of bacteria used can also affect the response, although this is complicated by potential differences in the presence of antioxidative defenses (i.e., superoxide dismutase, catalase) between strains. *In vitro* temperature also affects ROS production, which is typically lower at lower temperatures (Hardie *et al.*, 1994b). However, it appears that acclimation *in vivo* to low temperatures can overcome this effect to a large extent (Dexiang and Ainsworth, 1991b; Hardie *et al.*, 1994b).

2. Reactive Nitrogen Species (RNS)

There is increasing evidence that fish are able to generate reactive nitrogen species such as nitric oxide (NO). This has come primarily from studies of the nervous system, where NO acts as an inter- or intracellular messenger. NO is synthesized from arginine via the action of NO synthase (NOS), which hydroxylates the terminal (guanidino) carbon to give citrulline and NO. NOS has been demonstrated in the CNS and brain of Atlantic salmon and rainbow trout in enzyme histochemical (diaphorase) studies (Schober *et al.*, 1993; Ostholm *et al.*, 1994). Interestingly, in mammals, in addition to constitutively expressed NOS, there is an inducible form (iNOS) expressed in phagocytes, particularly macrophages, after stimulation with cytokines. Its expression is crucial for the destruction of certain pathogens, although NO by itself is not sufficiently reactive to initiate deleterious reactions such as lipid peroxidation. However, it can form more potent peroxidizing, nitrating, and nitrosating species, such as hydroxyl radicals ($OH\cdot$), nitrogen dioxide (NO_2), nitrogen trioxide (N_2O_3), and nitronium ions (NO_2^+). In addition, it can react with iron to form nitrosyl heme complexes, and with ROS to form peroxynitrites ($ONOO^-$) (Hibbs, 1992). Recent evidence suggests that fish also possess an iNOS.

NOS activity has been detected in channel catfish head kidney leukocytes following intraperitoneal injection with live *Edwardsiella ictaluri* (Schoor and Plumb, 1994), although the cells responsible for NOS activity were not identified. In goldfish, Neumann *et al.* (1995) have shown that a long-term macrophage cell line and primary cultures of kidney macrophages secrete NO (as detected by nitrite accumulation) after incubation with LPS or supernatants from leukocytes stimulated with Con A and phorbol ester (deemed to contain a macrophage activating factor, MAF). Furthermore, the MAF-containing supernatants synergized with LPS for induction of NO production (Fig. 7). Addition of arginine analogues (N^G-monomethyl-L-arginine or amino-guanidine) to these cultures inhibited induction, confirming the dependence on arginine metabolism for NO production in fish. Lastly, in rainbow trout and goldfish, a partial sequence for i NOS (with 70–75% amino acid homology to mammalian i NOS) has been obtained using cDNA stimulated from i macrophages (Hardie *et al.*, 1994c; Kerry *et al.*, 1996). However, addition of an arginine analogue to MAF-activated trout macrophages had no effect on their *in vitro* killing activity for *Renibacterium salmoninarum* (Hardie *et al.*, 1996), or on macrophage larvacidal activity for postpenetration larvae of *Diplostomum spathaceum* (Chappell *et al.*, 1994), although there was no confirmation that NO production had been induced in these studies. Thus, the relevance of NO production to fish defenses has yet to be established.

B. Oxygen-Independent Mechanisms

Virtually nothing is known about phagocyte oxygen-independent killing mechanisms. It is known that fish phagocytes possess a large array of enzymes (see Chapter 1) that would be potentially bactericidal following lysosomal fusion with phagosomes. They also possess lysozyme, which is known to be bactericidal for a number of fish pathogens (Grinde, 1989)

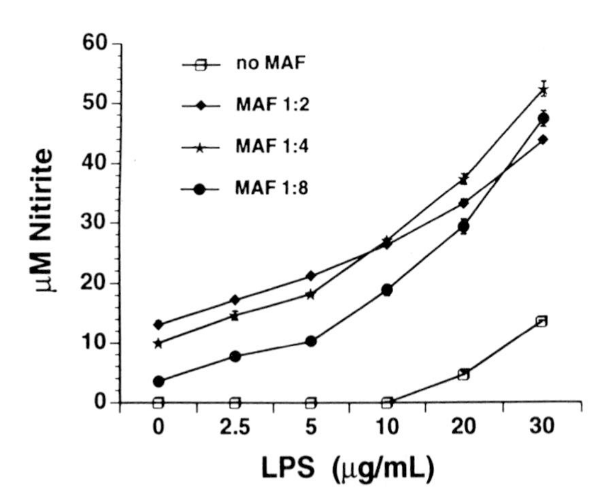

Fig. 7. Induction of nitric oxide, as detected by accumulation of nitrite in tissue culture supernatants, from a goldfish macrophage cell line stimulated with varing concentrations of MAF and LPS. Data are from a representative experiment. Reprinted from *Dev. Comp. Immunol.* 19, Neumann *et al.,* Macrophage activating factor(s) secreted by mitogen stimulated goldfish kidney leukocytes synergize with bacterial lipopolysaccharide to induce nitric oxide production in teleost macrophages, 473–482, copyright 1995, with kind permission from Elsevier Science Ltd, The Boulevard, Langford Lane, Kidlington OX5 1GB, UK.

and can be secreted to allow extracellular killing. Cationic proteins are also known to be bactericidal for a number of fish pathogens (Kelly *et al.*, 1990), but to date, none have been isolated from fish. Cytotoxic cytokines may also be involved in anaerobic killing, especially tumor necrosis factors. While it is known that human rTNF cross-reacts with fish leukocytes (Hardie *et al.*, 1994a) no direct evidence of TNF in fish exists. However, the cross-reactivity of human rTNFα to trout macrophages can be inhibited by preincubation of the cells with MoAb to the 55 kDa TNF receptor (Jang *et al.*, 1995a). Such MoAb also significantly inhibit macrophage responsiveness to supernatants from stimulated leukocytes (macrophages or lymphocytes) which can increase ROS production (Jang *et al.*, 1995b), suggesting this activity may be due to the presence of a fish TNF-like molecule (Fig. 8). Clearly a lot more research is required in this area of fish phagocyte biology.

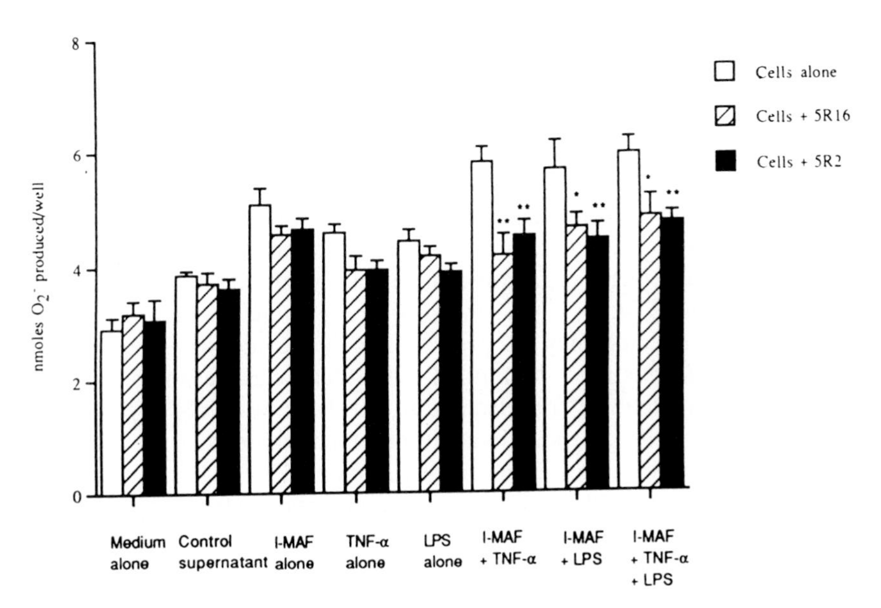

Fig. 8. Respiratory burst activity of rainbow trout head kidney (target) macrophages incubated with medium alone or 1:4 diluted macrophage supernatants, for 24 h prior to the assay. The macrophage supernatants were prepared by incubating head kidney macrophages for 12 h with medium (control supernatant) or various combinations of lymphocyte-derived MAF (1-MAF, diluted 1:4), TNFα 25 iu/ml), and LPS (50 μg/ml). Following this incubation the macrophages were washed and the supernatants harvested 24 h later. In some cases the target macrophages were preincubated with anti-TNF receptor MoAb for 1 h at 6 μg/ml (5R2) or 0.45 μg/ml (5R16) before addition of the macrophage supernatants. Macrophage stimulatory activity is clearly seen in such macrophage supernatants, particularly when using combinations of treatments to obtain the supernatants. In addition, clear inhibition of this activity is apparent ($^{*}p < 0.05$; $^{**}p < 0.01$) in the presence of the MoAb compared with the response of macrophages without MoAb. From Jang *et al.*, 1995b.

C. Modulation of Killing Mechanisms

It is clear that fish phagocyte activity can be modulated by a wide range of endogenous and exogenous factors (Secombes, 1994b). Many of these factors (environmental influences, xenobiotics, immunostimulants, dietary components, stress, pathogens) are covered in other chapters in this volume (see Chapters 7–9) and will not be discussed in detail here. However, as part of the normal functioning of the immune system, phagocyte activity is tightly regulated by cytokines and eicosanoids, and these factors may synergize with secondary signals which are often pathogen-derived. Since the activities of eicosanoids on phagocytes and inflammation have already been described in this chapter, this section will focus on cytokine effects.

That phagocytes, especially macrophages, can be "activated" to a state of heightened killing activity has been known for a decade (Olivier *et al.*, 1985). More recently it has been confirmed that a factor (or factors) generated from fish leukocytes (surface Ig^- lymphocytes) is able to induce a state of macrophage activation *in vitro* (reviewed in Secombes, 1994b). This MAF is predominantly present in a 19-kDa fraction isolated by size exclusion chromatography, cofractionates with interferon activity, and is acid and temperature sensitive. Such characteristics are suggestive of the presence of a type II (γ) interferon in fish. Macrophages are also able to release one or more factors that can increase ROS production in an autocrine fashion. The activity of the factor or factors is significantly inhibited in the presence of MoAb to the 55-kDa TNFα receptor (Jang *et al.*, 1995b), indicating a TNFα-like molecule may be present. Further evidence for a role of TNF-α comes from studies looking at the effects of β-glucans on macrophage ROS production (Robertsen *et al.*, 1994). Low levels of β-glucans (1 μg/ml) enhance ROS production, but this effect is ablated in the presence of a polyclonal rabbit antiserum against human rTNFα. Interestingly, *in vitro* incubation with β-glucans alone does not increase macrophage bactericidal activity, whereas *in vivo* treatment with glucans does, suggesting the requirement for a second factor in macrophage activation, presumably released following *in vivo* treatment. Furthermore, human rTNFα is able to synergize with MAF-containing supernatants to enhance macrophage ROS production (Hardie *et al.*, 1994a).

A hormone with structural similarities to hematopoietic cytokines is growth hormone (GH). It has a four antiparallel helical bundle structure in common with cytokines such as IL-2, IL-4, IL-5, GM-CSF, GCSF, MCSF, and the interferons (Sprang and Bazan, 1993). Injection of rainbow trout with chum salmon (*O. keta*) GH has been shown to enhance kidney leukocyte respiratory burst (chemiluminescence) activity (Sakai *et al.*, 1995). More recently, direct addition of chum salmon GH at 10–100 ng/ml to rainbow trout phagocytes has also been shown to enhance respiratory burst activity (Sakai, pers. comm.).

Recent studies have shown that phagocyte activity can also be down regulated by cytokines (Jang *et al.*, 1994). For example, while MAF-containing supernatants are able to enhance ROS production in rainbow trout macrophages, they decrease 5′ nucleotidase activity in the same cells (Hepkema and Secombes, 1994). Inhibition of ROS production by trout macrophages has been demonstrated using natural bovine transforming growth factor β_1 (TGFβ_1) added to activated cells at 1 ng/ml (Jang *et al.*, 1994). TGFβ is also able to prevent activation of macrophages by coincubation with an activating signal. Evidence that fish cells can produce this macrophage deactivating factor comes from two observations. First, generation of factors capable of enhancing ROS production from macrophages was increased in the presence of anti-TGFβ serum (Jang *et al.*, 1995b), suggesting a suppressive influence had been overcome. In mammals, the mature TGFβ_1 peptide is 99–100% identical (Derynck, 1994) but has not been isolated from other vertebrate groups. Similarly, mature TGFβ_2 peptide is ≥95% identical across amphibians, birds, and mammals. Even TGFβ_5, a TGFβ unique to amphibians, is 75% identical to TGFβ_1. So TGFβs are very conserved and, consequently, polyclonal anti-TGFβ_1 sera show considerable cross-species reactivity. Secondly, a TGFβ has recently been sequenced from macrophage-enriched kidney cDNA, isolated from vaccinated rainbow trout, with 68% predicted amino acid homology to human TGFβ_1 (Secombes *et al.*, 1996; Accession no. X99303). Paradoxically, incubation of resting macrophages with low doses (0.1 ng/ml) of bovine TGFβ_1 has the opposite effect and actually enhances ROS production.

Sympathetic neurotransmitters can also enhance or inhibit rainbow trout kidney respiratory burst activity. Alpha adrenergic receptor agonists (phenylephrine) and cholinergic agonists (carbachol) enhance ROS production, whereas beta adrenergic agonists (isoproterenol) and epinephrine suppress ROS production (Flory and Bayne, 1991).

VII. PHAGOCYTES AS ACCESSORY CELLS

It is well known that accessory cells are required for lymphocyte responses in fish (Clem *et al.* 1991; Vallejo *et al.* 1992a). There are two main ways that phagocytes can function as accessory cells. First, they can take up and process antigens, presenting them on their surface in association with class II MHC molecules, and second, they can secrete soluble mediators involved in lymphocyte activation, such as IL-1. It is clear that both of these activities are performed by fish phagocytes, primarily from studies in channel catfish (Vallejo *et al.,* 1992a; Ellsaesser and Clem, 1994).

A. Antigen Processing and Presentation

It is well known from histological studies on fish lymphoid tissues (spleen, kidney, gut) that macrophages are involved in antigen uptake *in vivo.* However, recent studies *in vitro* have made a significant advance in determining the functional significance of this phenomenon. Studies on antigen presentation in catfish have shown that autologous blood leukocytes can act as efficient stimulators of proliferation and antibody secretion, following a pulse with antigen and subsequent fixation with paraformaldehyde. Peripheral blood leukocytes fixed prior to antigen pulsing cannot act as accessory cells. Genetic restriction appears to be operating since allogeneic blood leukocytes are not efficient antigen presenters, as predicted by the known involvement of MHC molecules in mammalian systems. Lysates from cells incubated with antigen show that significant proteolysis occurs within 4 h after pulsing. In addition, the antigen presenting function can be abolished by treatment of the pulsed cells with substances known to interfere with processing and presentation, such as chloroquine, NH_4Cl, PMSF, leupeptin, or monensin prior to or during antigen pulsing. However, such treatment does not affect antigen uptake. Similarly in Atlantic salmon macrophages, it has been shown that bacterial antigens are highly susceptible to proteolysis, which is inhibitable with NH_4Cl, while LPS is more resistant (Espelid and Jorgensen, 1992). In catfish, B cells as well as monocytes are able to present antigens to lymphocytes, again as predicted by the distribution of MHC class II molecules on mammalian leukocytes, which are known to be transcribed in fish leukocytes (Hordvik *et al.,* 1993; Klein *et al.,* 1993; Glamann, 1995; see Chapter 4).

It has been possible to confirm that antigen is reexpressed on the surface of antigen-presenting cells in fish by fractionation of cell homogenates (Vallejo *et al.,* 1992b). Using radiolabeled antigen, radioactivity is demonstrable in the plasma membrane of cells incubated with antigen at 4°C. Subsequent incubation at 27°C shows a progressive decrease in membrane-bound antigen and a corresponding increase in endosome- or lysosome-associated antigen during a 3-h period. After 5 h, membrane-bound radioactivity increases again with a concomitant decrease in the endosome/lysosome fractions. Membrane preparations from such cells are efficient stimulators of proliferation by autologous leukocytes from antigen-primed fish, supporting the notion that processed antigen is recycled to the cell surface for presentation to lymphocytes.

Temperature studies have shown that antigen processing and presentation can occur in blood leukocytes at low temperatures but require a longer exposure to antigen prior to fixation (i.e., 8 h at 17°C vs. 5 h at 27°C) for optimal stimulation (Vallejo *et al.,* 1992c). This appears to be primarily due to a slower rate of catabolism of the antigen rather than a slow uptake at low temperatures. Indeed, sufficient antigen is taken up within 1–2 h at low temperatures for stimulation of lymphocytes, providing the cells are subsequently cultured at 27°C for 10 h.

B. Interleukin-1–Like Activity

Supernatants from fish phagocytes are able to stimulate proliferation of fish T cells and murine IL-1-dependent cell lines, and the induction of *in vitro* antibody production to thymus-dependent and -independent antigens by fish B cells (Ellsaesser and Clem, 1994; Secombes, 1994c; Verburg-van Kemenade *et al.,* 1995). In carp, epithelial cells, macrophages, and neutrophilic granulocytes secrete these factors, whereas in channel catfish, monocytes are the main source. Stimulation of the cells with LPS or phorbol ester is required to give maximal IL-1-like activity. In catfish, this activity appears to be present in at least two forms: a high molecular weight form (70 kDa) active on fish cells but not mouse cells, and a low molecular weight form (15 kDa) active for mouse cells but not fish cells. In carp, only the 15-kDa species is present. Western blot analysis of such supernatants reveals antigenic cross-reactivity with antisera to mammalian IL-1α and IL-1β, and such antisera also neutralize the biological activity of the supernatants. Furthermore, molecular analysis of mRNA from catfish monocytes with a murine IL-1α cDNA probe reveals the presence of a hybridizing species with a similar mobility to that seen using mouse monocyte mRNA. Such findings strongly suggest that fish cells secrete an IL-1-like molecule that is important in activating lymphocytes.

Supernatants containing natural mammalian IL-1 (murine or human) can also stimulate catfish T-cell proliferation, although such cells do not respond to rIL-1 (Hamby *et al.,* 1986; Ellsaesser and Clem, 1994). Thus, fish lymphocytes also appear to possess a surface receptor that can recognize and respond to IL-1. However, as with the IL-1-like factor itself, further characterization is required to confirm the nature of these molecules.

VIII. NONSPECIFIC CYTOTOXICITY

The leukocytes of several teleost species have been shown to be capable of spontaneous cytotoxic reactions against a wide variety of established fish and mammalian cell lines, virus-infected cells, and against protozoan parasites (see Evans and Jaso-Friedmann, 1992). This constitutes another component of the nonspecific cellular defenses. The killing is spontaneous in that leukocytes from unprimed fish can effect it, and it does not require any apparent induction period. Thus, cytotoxicity starts immediately upon addition of target cells and increases up to a maximum value within some 2–8 h, depending on the species. Direct physical contact is required for killing, with E : T ratios as low as 1.5–2 : 1 able to induce lysis using enriched NCC. Nonspecific cytotoxic cells are present in a number of lymphoid organs (thymus, kidney, spleen), the peritoneal cavity, and blood, although activity is usually lowest in the latter. In addition, NCC-like cells can be found in the liver (nonparenchymal cells) but have little cytolytic activity.

Although NCC cytotoxicity is termed nonspecific it does appear to be selective. Thus, the ability to lyse one type of target does not always correlate with the ability to lyse a second. Cytotoxicity is apparent against established fibroblastic, epithelial, or malignantly transformed cell lines but not against normal resting xenogeneic cells, suggesting a role in protection against neoplasia. Susceptibility to lysis does not appear to be related to the activational state of the targets since Con A–induced blast cells are not lysed. Nonspecific cytotoxic cells also appear to have an important role in resistance to viral and parasitic infection. For example, fish cell lines infected with virus (e.g., IPNV) are more susceptible to killing than uninfected cells (Moody *et al.,* 1985). In addition, NCC can kill protozoan parasites (*Ichthyophthirius multifiliis* and *Tetrahymena pyriformis*), although optimal killing requires target cell immobilization and a cytotoxic period of 10 h or longer (Graves *et al.,* 1985). That the same type of effector cell is responsible for killing cell lines and protozoa can be shown by depletion studies, where following preincubation with one target type a marked decrease in activity is seen in the other.

Morphological studies of isolated teleost NCC (see Morphology and Isolation at the beginning of this chapter) have shown that they are the smallest leukocytes, possessing a very pleomorphic, clefted nucleus and relatively little cytoplasm with no cytoplasmic granules. Villi are present on their surface, and they attach to targets using long, membranous filaments, allowing many binding sites.

Many factors have been shown to influence NCC activity, such as diet (Kiron *et al.,* 1993), temperature (LeMorvan-Rocher *et al.,* 1995), stress (Evans and Jaso-Friedmann, 1994), administration of growth hormone (Kajita *et al.,* 1992), fish strain (Ristow *et al.,* 1994), and age (Faisal *et al.,* 1989). Activity is higher in young animals and at low temperatures, suggesting NCC are particularly important when specific (lymphocyte-mediated) responses are relatively poor. However, poststress NCC activity is suppressed as with lymphocyte responses.

In elasmobranchs a rather different type of spontaneous cytotoxicity occurs during periods of low environmental temperatures (McKinney *et al.*, 1986). Under such conditions, nurse shark (*Ginglymostoma cirratum*) blood leukocytes are able to lyse xenogeneic erythrocytes. The effector cells are glass adherent and phagocytic, suggesting that they are monocytes which are the dominant adherent cell type. Leukocytes taken from fish in relatively warm water (>26°C) do not exhibit spontaneous cytotoxicity, due to the presence of a nonadherent, nonphagocytic regulatory cell type that inhibits this activity (Haynes and McKinney, 1991). The relationship between the cells that effect spontaneous cytotoxicity in elasmobranchs and NCCs in teleosts is not clear.

A. Recognition of Target Cells

Recognition and binding to target cells by teleost NCCs appears to be receptor mediated. Indeed, MoAb generated against purified NCCs are able to inhibit (by 60–65%) lysis of target cells by unfractionated NCCs preincubated with the MoAb. Target cells preincubated with such MoAb have no effect upon lysis. Nonspecific cytotoxic cells preincubated with MoAb also show inhibition of conjugate formation between effector and target cells. The determinants recognized by these MoAb are present on the cell membrane and belong to a single protein with a molecular weight between 40 and 42 kDa, determined by western blot analysis (Jaso-Friedmann *et al.*, 1993). Anti-vimentin MoAb cross-react to this NCC receptor protein, showing that it contains vimentin-like determinants. Interestingly, the anti–NCC receptor MoAb also inhibit mammalian NK cell activity, suggesting that this vimentin-like molecule is evolutionarily conserved (Harris *et al.*, 1992).

Binding of the anti–NCC receptor MoAb to NCCs increases expression of $p56^{lck}$ (a cytoplasmic kinase), inositol lipid turnover (i.e., induces release of inositol phosphates), mobilization of intracellular calcium reserves (Harris *et al.*, 1992), and rapid protein phosphorylation (Jaso-Friedmann *et al.*, 1994). This release of second messengers confirms the molecule is involved in signal transduction. Calcium-activated kinases appear to provide the strongest signal to initiate phosphorylation, although cAMP-cGMP cyclic nucleotide-dependent kinase activity also participates in signaling and regulation of cytotoxicity. Indeed, tyrosine phosphorylation appears to be a prerequisite step in cytotoxicity, as with NK cells. Interestingly, dephosphorylation of membrane proteins by phosphatases also regulates NCC activity, as evidenced by increased activity *in vitro* and *in vivo* in the presence of protein phosphatase inhibitors such as fluoride and/or vanadate (Evans and Jaso-Friedmann, 1994). Maximum augmentation occurs when inherent levels of NCC activity are low. Clearly, multiple second messenger pathways are involved in NCC lysis of target cells, with an equilibrium existing between kinase and phosphatase activities.

Recently, an NCC target cell antigen has been discovered using anti-idiotypic MoAb generated against idiotopes on an MoAb that recognizes the putative NCC receptor (Lester *et al.,* 1994). Thus, preincubation of NCCs with this MoAb inhibits subsequent lysis of target cells by 76% and conjugate formation by 50%. The anti-idiotypic MoAb acts effectively as an NCC receptor and recognizes its ligand on head kidney, spleen, thymus, blood leukocytes, liver, and brain cells. Western blot analysis reveals reactivity with proteins of 54 and 65 kDa on target cell lysates. A MoAb prepared against human Epstein Barr virus–transformed lymphoblastic cells also inhibits lysis of target cells by NCC and, interestingly, recognizes a determinant on the fish protozoan parasite *T. pyriformis* (Leary *et al.,* 1994). The antigen recognized by this MoAb has been subjected to *N*-terminal microsequencing, as have trypsin digests of the antigen. None of the sequences obtained have significant homology to known proteins, suggesting a novel molecule is recognized on *T. pyriformis* by this MoAb.

B. Killing Mechanisms

Little is known about the mechanism by which NCCs lyse target cells. As with NK cells, following conjugate formation, reorientation of organelles occurs in NCCs, resulting in polarization of the bulk of the cytoplasm and the Golgi apparatus toward the target cell contact area. NCCs do not appear to possess azurophilic cytoplasmic granules, so granule exocytosis is unlikely to be a major factor in killing, unlike the situation with NK cells (Evans and Jaso-Friedmann, 1992). This would also suggest that serine esterases (granzymes) are not involved. However, lysis of targets by NCCs is an energy-dependent process requiring an intact cytoskeletal architecture, secretory vesicles, and calcium. Interestingly, NCCs inflict both apoptic (DNA fragmentation) and necrotic lesions in target cells (Greenlee *et al.,* 1991), as seen with mammalian killer cells that utilize two complementary killing mechanisms mediated via binding of Fas with the Fas ligand and perforin (Kagi *et al.,* 1994; Lowin *et al.,* 1994).

ACKNOWLEDGMENTS

Thanks go to Dr. T. C. Fletcher (Dept. Zoology, University of Aberdeen) for critically reading the manuscript. Thanks also go to Mr. S. Hamdani (Dept. Zoology, University of Aberdeen) for Fig. 1.

REFERENCES

Ahne, W. (1993). Presence of interleukins (IL-1, IL-3, IL-6) and the tumor necrosis factor (TNF alpha) in fish sera. *Bull. Eur. Assoc. Fish Pathol.* **13,** 106–107.

Ainsworth, A. J. (1992). Fish granulocytes: Morphology, distribution, and function. *Annu. Rev. Fish Dis.* **2,** 123–148.

Ainsworth, A. J., Dexiang, C., and Greenway, T. (1990). Characterization of monoclonal antibodies to channel catfish, *Ictalurus punctatus,* leucocytes. *Vet. Immunol. Immunopathol.* **26,** 81–92.

Alexander, J. B., and Ingram, G. A. (1992). Noncellular nonspecific defence mechanisms of fish. *Annu. Rev. Fish Dis.* **2,** 249–279.

Ashton, I., Clements, K., Barrow, S. E., Secombes, C. J., and Rowley, A. F. (1994). Effects of dietary fatty acids on eicosanoid-generating capacity, fatty acid composition, and chemotactic activity of rainbow trout (*Oncorhynchus mykiss*) leucocytes. *Biochim. Biophys. Acta* **1214,** 253–262.

Baldo, B. A., and Fletcher, T. C. (1975). Inhibition of immediate hypersensitivity responses in flatfish. *Experientia* **31,** 495–496.

Barber, D. L., and Westermann, J. E. M. (1978). Observations on development and morphological effects of histamine liberator 48/80 on PAS-positive granular leukocytes and heterophils of *Catostomus commersoni. J. Fish Biol.* **13,** 563–573.

Bly, J. E., Miller, N. W., and Clem, L. W. (1990). A monoclonal antibody specific for neutrophils in normal and stressed channel catfish. *Dev. Comp. Immunol.* **14,** 211–221.

Bridges, A. F., and Manning, M. J. (1991). The effect of priming immersions in various human gamma globulin (HGG) vaccines on humoral and cell-mediate immune responses after intraperitoneal HGG challenge in the carp, *Cyprinus carpio* L. *Fish Shellfish Immunol.* **1,** 119–129.

Chappell, L. H., Hardie, L. J., and Secombes, C. J. (1994). Diplostomiasis: The disease and host–parasite interactions. *In* "Parasitic Diseases of Fish" (A. W. Pike and J. W. Lewis, eds.), pp. 59–86. Samara Publishing, Tresaith, Dyfed, UK.

Clem, L. W., Miller, N. W., and Bly, J. E. (1991). Evolution of lymphocyte subpopulations, their interactions and temperature sensitivities. *In* "Phylogenesis of Immune Functions" (G. W. Warr and N. Cohen, eds.), pp. 191–213. CRC Press, Boca Raton, Florida.

Cone, D. K., and Wiles, M. (1985). Trophozoite morphology and development site of two species of *Myxobolus* (Myxozoa) parasitizing *Catastomus commersoni* and *Notemigonus crysoleucas* in Atlantic Canada. *Can. J. Zool.* **63,** 2919–2923.

Daly, J. G., Moore, A. R., and Olivier, G. (1994). Bactericidal activity of brook trout (*Salvelinus fontinalis*) peritoneal macrophages against avirulent strains of *Aeromonas salmonicida. Fish Shellfish Immunol.* **4,** 273–283.

Dannevig, B. H., Lauve, A., McL. Press, C., and Landsverk, T. (1994). Receptor-mediated endocytosis and phagocytosis by rainbow trout head kidney sinusoidal cells. *Fish Shellfish Immunol.* **4,** 3–18.

Derynck, R. (1994). Transforming growth factor-beta. *In* "The Cytokine Handbook," 2nd edition (A. Thomson, ed.), pp. 319–342. Academic Press, London.

Dexiang, C., and Ainsworth, A. J. (1991a). Assessment of metabolic activation of channel catfish peripheral blood neutrophils. *Dev. Comp. Immunol.* **15,** 201–208.

Dexiang, C., and Ainsworth, A. J. (1991b). Effect of temperature on the immune system of channel catfish (*Ictalurus punctatus*). II. Adaptation of anterior kidney phagocytes to 10°C. *Comp. Biochem. Physiol.* **100A,** 913–918.

Diago, M. L., Lopez-Fierro, M. P., Razquin, B., and Villena, A. (1994). Long-term myelopoietic cultures from the renal hematopoietic tissue of the rainbow trout, *Oncorhynchus mykiss* W.: Phenotypic characterization of the stromal cells. *Exp. Hematol.* **21,** 1277–1287.

Dorin, D., Sire, M. -F., and Vernier, J. -M. (1993). Endocytosis and intracellular degradation of heterologous protein by eosinophilic granulocytes isolated from rainbow trout (*Oncorhynchus mykiss*) posterior intestine. *Biol. Cell* **79,** 219–224.

Dowding, A. J., Maggs, A., and Scholes, J. (1991). Diversity amongst the microglia in growing and regenerating fish CNS: Immunohistochemical characterization using FL.1, an anti-macrophage monoclonal antibody. *Glia* **4,** 345–364.

Ellis, A. E. (1986). The function of teleost fish lymphocytes in relation to inflammation. *Int. J. Tiss. React.* **8,** 263–270.

Ellsaesser, C. F., and Clem, L. W. (1994). Functionally distinct high and low molecular weight species of channel catfish and mouse IL-1. *Cytokine* **6,** 10–20.

Engstad, R. E., and Robertsen, B. (1994). Specificity of a β-glucan receptor on macrophages from Atlantic salmon (*Salmo salar* L.). *Dev. Comp. Immunol.* **18,** 397–408.

Espelid, S., and Jorgensen, T. O. (1992). Antigen processing of *Vibrio salmonicida* by fish (*Salmo salar* L.) macrophages *in vitro*. *Fish Shellfish Immunol.* **2,** 131–141.

Evans, D. L., and Jaso-Friedmann, L. (1992). Nonspecific cytotoxic cells as effectors of immunity in fish. *Annu. Rev. Fish Dis.* **2,** 109–121.

Evans, D. L., and Jaso-Friedmann, L. (1994). Role of protein phosphatases in the regulation of nonspecific cytotoxic cell activity. *Fish Shellfish Immunol.* **18,** 137–146.

Evans, D. L., Carlson, R. L., Graves, S. S., and Hogan, K. T. (1984). Nonspecific cytotoxic cells in fish (*Ictalurus punctatus*). IV. Target cell binding and recycling capacity. *Dev. Comp. Immunol.* **8,** 823–833.

Evans, D. L., Smith, E. E., and Brown, F. E. (1987). Nonspecific cytotoxic cells in fish (*Ictalurus punctatus*). VI. Flow cytometric analysis. *Dev. Comp. Immunol.* **11,** 95–104.

Faisal, M., Ahmed, I. I., Peters, G., and Cooper, E. L. (1989). Natural cytotoxicity of tilapia leucocytes. *Dis. Aquat. Org.* **7,** 17–22.

Flory, C. M., and Bayne, C. J. (1991). The influence of adrenergic and cholinergic agents on the chemiluminescent and mitogenic responses of leukocytes from the rainbow trout, *Oncorhynchus mykiss. Dev. Comp. Immunol.* **15,** 135–142.

Glamann, J. (1995). Complete coding sequence of rainbow trout MHC II β chain. *Scand. J. Immunol.* **41,** 365–372.

Graves, S. S., Evans, D. L., and Dawe, D. L. (1985). Antiprotozoan activity of nonspecific cytotoxic cells (NCC) from the channel catfish (*Ictalurus punctatus*). *J. Immunol.* **134,** 78–85.

Greenlee, A. R., Brown, R. A., and Ristow, S. S. (1991). Nonspecific cytotoxic cells of rainbow trout (*Oncorhynchus mykiss*) kill YAC-1 targets by both necrotic and apoptic mechanisms. *Dev. Comp. Immunol.* **15,** 153–164.

Griffin, B. R. (1984). Random and directed migration of trout (*Salmo gairdneri*) leukocytes: Activation by antibody, complement, and normal serum components. *Dev. Comp. Immunol.* **8,** 589–597.

Grinde, B. (1989). Lysozyme from rainbow trout, *Salmo gairdneri* Richardson, as an antibacterial agent against fish pathogens. *J. Fish Dis.* **12,** 95–104.

Hamby, B. A., Huggins, E. M., Jr., Lachman, L. B., Dinarello, C. A., and Sigel, M. M. (1986). Fish lymphocytes respond to human IL-1. *Lymphokine Res.* **5,** 157–162.

Hamers, R., Lehmann, J., Sturenberg, F. -J., and Taraschewski, H. (1992). *In vitro* study of the migratory and adherent responses of fish leucocytes to the eel-pathogenic acanthocephalan *Paratenuisentis ambiguus* (van Cleave, 1921) Bullock et Samuel, 1975 (Eoacanthocephala: Tenuisentidae). *Fish Shellfish Immunol.* **2,** 43–51.

Hardie, L. J., Chappell, L. H., and Secombes, C. J. (1994a). Human tumor necrosis factor α influences rainbow trout *Oncorhynchus mykiss* leucocyte responses. *Vet. Immunol. Immunopathol.* **40,** 73–84.

Hardie, L. J., Fletcher, T. C., and Secombes, C. J. (1994b). Effect of temperature on macrophage activation and the production of macrophage activating factor by rainbow trout (*Oncorhynchus mykiss*) leucocytes. *Dev. Comp. Immunol.* **18,** 57–66.

Hardie, L. J., Grabowski, P., Ralston, S., McGuigan, F., and Secombes, C. J. (1994c). Isolation and partial coding sequence for nitric oxide synthase from rainbow trout macrophages. *Dev. Comp. Immunol.* **18(Suppl. 1),** S88.

Hardie, L. J., Ellis, A. E., and Secombes, C. J. (1996). *In vitro* activation of rainbow trout macrophages stimulates killing of *Renibacterium salmoninarum* concomitant with augmented generation of respiratory burst products. *Dis. Aquat. Org.* **25,** 175–183

Harris, D. T., Kapur, R., Frye, C., Acevedo, A., Camenisch, T., Jaso-Friedmann, L., and Evans, D. L. (1992). A species-conserved NK cell antigen receptor is a novel vimentin-like molecule. *Dev. Comp. Immunol.* **16,** 395–403.

Haynes, L., and McKinney, E. C. (1991). Shark spontaneous cytotoxicity: Characterization of the regulatory cell. *Dev. Comp. Immunol.* **15,** 123–134.

Haynes, L., Fuller, L., and McKinney, E. C. (1988). Fc receptor for shark IgM. *Dev. Comp. Immunol.* **12,** 561–571.

Hepkema, F. W., and Secombes, C. J. (1994). 5′Nucleotidase activity of rainbow trout *Oncorhynchus mykiss* macrophages: Correlation with respiratory burst activity. *Fish Shellfish Immunol.* **4,** 301–309.

Hibbs, J. B., Jr. (1992). Overview of cytotoxic mechanisms and defence of the intracellular environment against microbes. *In* "The Biology of Nitric Oxide. 2. Enzymology, Biochemistry, and Immunology." (S. Moncada, M. A. Marletta, J. B. Hibbs, Jr., and E. A. Higgs, eds.), pp. 201–206. Portland Press, London.

Hordvik, I., Grimholt, U., Fosse, V. M., Lie, O., and Endresen, C. (1993). Cloning and sequence analysis of cDNAs encoding the MHC class II β chain in Atlantic salmon (*Salmo salar*). *Immunogenetics* **37,** 437–441.

Hoole, D. (1994). Tapeworm infections in fish: Past and future problems. *In* "Parasitic Diseases of Fish" (A. W. Pike and J. W. Lewis, ed.), pp. 119–140. Samara Publishing Ltd., Tresaith, Dyfed, UK.

Howell, C. J. stG. (1987). A chemokinetic factor in the carp, *Cyprinus carpio. Dev. Comp. Immunol.* **11,** 139–146.

Jang, S. I., Hardie, L. J., and Secombes, C. J. (1994). Effects of transforming growth factor β_1 on rainbow trout *Oncorhynchus mykiss* macrophage respiratory burst activity. *Dev. Comp. Immunol.* **18,** 315–323.

Jang, S. I., Mulero, V., Hardie, L. J., and Secombes, C. J. (1995a). Inhibition of rainbow trout phagocyte responsiveness to human tumor necrosis factor α (hTNFα) with monoclonal antibodies to the hTNFα 55 kDa receptor. *Fish Shellfish Immunol.* **5,** 61–69.

Jang, S. I., Hardie, L. J., and Secombes, C. J. (1995b). Elevation of rainbow trout *Oncorhynchus mykiss* macrophage respiratory burst activity with macrophage-derived supernatants. *J. Leukocyte Biol.* **57,** 943–947.

Jaso-Friedmann, L., Leary III, J. H., and Evans, D. L. (1993). Nonspecific cytotoxic cells in fish: Antigenic cross-reactivity of a function-associated molecule with the intermediate filament vimentin. *Cell. Immunol.* **148,** 208–217.

Jaso-Friedmann, L., Leary III, J. H., and Evans, D. L. (1994). Pathways of signaling in nonspecific cytotoxic cells: Effects of protein kinase and phosphatase inhibitors and evidence for membrane tyrosine phosphorylation. *Cell. Immunol.* **153,** 142–153.

Johnson, E., and Smith, P. (1984). Attachment and phagocytosis by salmon macrophages of agarose beads coated with human C3b and C3bi. *Dev. Comp. Immunol.* **8,** 623–630.

Jurd, R. D. (1987). Hypersensitivity in fishes: A review. *J. Fish Biol.* **31A,** 1–7.

Kagi, D., Ledermann, B., Burki, K., Seiler, P., Odermatt, B., Olsen, K. J., Podack, E. R., Zinkernagel, R. M., and Hengartner, H. (1994). Cytotoxicity mediated by T cells and natural killer cells is greatly impaired in perforin-deficient mice. *Nature* **369,** 31–37.

Kajita, Y., Sakai, M., Kobayashi, M., and Kawauchi, H. (1992). Enhancement of nonspecific cytotoxic activity of leucocytes in rainbow trout *Oncorhynchus mykiss* injected with growth hormone. *Fish Shellfish Immunol* **2,** 155–157.

Karczewski, J. M., Sharp, G. J. E., and Secombes, C. J. (1991). Susceptibility of strains of *Aeromonas salmonicida* to killing by cell-free generated superoxide anion. *J. Fish Dis.* **14,** 367–373.

Kelly, D., Wolters, W. R., and Jaynes, J. M. (1990). Effect of lytic peptides on selected fish bacterial pathogens. *J. Fish Dis.* **13,** 317–321.

Kiron, V., Gunji, A., Okamoto, N., Satoh, S., Ikeda, Y., and Watanabe, T. (1993). Dietary nutrient–dependent variations on natural killer activity of the leucocytes of rainbow trout. *Gyobyo Kenkyu* **28,** 71–76.

Klein, D., Ono, H., O'Huigin, C., Vincek, V., Goldschmidt, T., and Klein, J. (1993). Extensive MHC variability in cichlid fishes of lake Malawi. *Nature* **364,** 330–334.

Knight, J., and Rowley, A. F. (1995). Immunoregulatory activities of eicosanoids in the rainbow trout (*Oncorhynchus mykiss*). *Immunology* **85,** 389–393.

Knight, J., Holland, J. W., Bowden, L. A., Halliday, K., and Rowley, A. F. (1995). Eicosanoid generating capacities of different tissues from the rainbow trout, *Oncorhynchus mykiss*. *Lipids* **30,** 451–458.

Kodama, H., Mukamoto, M., Baba, T., and Mule, D. M. (1994a). Macrophage colony–stimulating activity in rainbow trout (*Oncorhynchus mykiss*) serum. *Modulat. Fish Immune Responses* **1,** 59–66.

Kodama, H., Baba, T., and Ohishi, I. (1994b). Inhibition of phagocytosis by rainbow trout (*Oncorhynchus mykiss*) macrophages by botulinum C_2 toxin and its trypsinized component II. *Dev. Comp. Immunol.* **18,** 389–395.

Laing, K. J., Grabowski, P. S., Belosevic, M., and Secombes, C. J. (1996). A partial sequence for nitric oxide synthase from a goldfish (*Carassius auratus*) macrophage cell line. *Immunol. Cell Biol.* **74,** 374–379.

Lamas, J., and Ellis, A. E. (1994a). Atlantic salmon (*Salmo salar*) neutrophil responses to *Aeromonas salmonicida. Fish Shellfish Immunol.* **4,** 201–219.

Lamas, J., and Ellis, A. E. (1994b). Electron microscopic observations of the phagocytosis and subsequent fate of *Aeromonas salmonicida* by Atlantic salmon neutrophils *in vitro. Fish Shellfish Immunol.* **4,** 539–546.

Lamas, J., Bruno, D. W., Santos, Y., Anadon, R., and Ellis, A. E. (1991). Eosinophilic granular cell response to intraperitoneal injection with *Vibrio anguillarum* and its extracellular products in rainbow trout, *Oncorhynchus mykiss. Fish Shellfish Immunol.* **1**, 187–194.

Lauve, A., and Dannevig, B. H. (1993). Fluid-phase endocytosis by rainbow trout headkidney macrophages. Fish Shellfish Immunol. **3,** 79–87.

Leary III, J. H., Evans, D. L., and Jaso-Friedmann, L. (1994). Partial amino acid sequence of a novel protozoan parasite antigen that inhibits nonspecific cytotoxic cell activity. *Scand. J. Immunol.* **40,** 158–164.

Leiro, J., Siso, M. I. G., Ortega, M., Santamarina, M. T., Sanmartin, M. L. (1995). A factorial experimental design for investigation of the effects of temperature, incubation time and pathogen-to-phagocyte ratio on *in vitro* phagocytosis by turbot adherent cells. *Comp. Biochem. Physiol.* **112C,** 215–220.

LeMorvan-Rocher, C., Troutaud, D., and Deschaux, P. (1995). Effects of temperature on carp leukocyte mitogen-induced proliferation and nonspecific cytotoxic activity. *Dev. Comp. Immunol.* **19,** 87–95.

Lester III, J. P., Evans, D. L., Leary III, J. H., Fowler, S. C., and Jaso-Friedmann, L. (1994). Identification of a target cell antigen recognized by nonspecific cytotoxic cells using an anti-idiotype monoclonal antibody. *Dev. Comp. Immunol.* **18,** 219–229.

Lloyd-Evans, P., Barrow, S., Hill, D. J., Bowden, L. A., Rainger, G. E., Knight, J., and Rowley, A. F. (1994). Eicosanoid generation and effects on the aggregation of thrombocytes from the rainbow trout, *Oncorhynchus mykiss. Biochim. Biophys. Acta* **1215,** 291–299.

Lowin, B., Hahne, M., Mattmann, C., and Tschopp, J. (1994). Cytolytic T-cell cytotoxicity is mediated through perforin and Fas lytic pathways. *Nature* **370,** 650–652.

MacArthur, J. I., Thomson, A. W., and Fletcher, T. C. (1985). Aspects of leucocyte migration in the plaice, *Pleuronectes platessa* L. *J. Fish Biol.* **27,** 667–676.

Matsuyama, H., Yano, T., Yamakawa, T., and Nakao, M. (1992). Opsonic effect of the third complement component (C3) of carp (*Cyprinus carpio*) on phagocytosis by neutrophils. *Fish Shellfish Immunol.* **2,** 69–78.

McKinney, E. C., Haynes, L., and Droese, A. L. (1986). Macrophage-like effector of spontaneous cytotoxicity from the shark. *Dev. Comp. Immunol.* **10,** 497–508.

McQueen, A., MacKenzie, K., Roberts, R. J., and Young, H. (1973). Studies on the skin of plaice (*Pleuronectes platessa* L.). III. The effect of temperature on the inflammatory response to the metacercariae of *Cryptocotyle lingua* (Creplin, 1825) (Digenea: Heterophyidae). *J. Fish Biol.* **5,** 241–247.

Miller, N. W., Chinchar, V. G., and Clem, L. W. (1994). Development of leukocyte cell lines from the channel catfish (*Ictalurus punctatus*). *J. Tissue Culture Methods* **16,** 1–7.

Moody, C. E., Serreze, D. V., and Reno, P. W. (1985). Nonspecific cytotoxic activity of teleost leukocytes. *Dev. Comp. Immunol.* **9,** 51–64.

Nagelkerke, L. A. J., Pannevis, M. C., Houlihan, D. F., and Secombes, C. J. (1990). Oxygen uptake of rainbow trout, *Oncorhynchus mykiss,* phagocytes following stimulation of the respiratory burst. *J. Exp. Biol.* **154,** 339–353.

Nakanishi, Y., Kodama, H., Murai, T., Mikami, T., and Izawa, H. (1991). Activation of rainbow trout complement by C-reactive protein. *Am. J. Vet. Res.* **52,** 397–401.

Nash, K. A., Fletcher, T. C., and Thomson, A. W. (1986). Migration of fish leucocytes *in vitro:* The effect of factors which may be involved in mediating inflammation. *Vet. Immunol. Immunopathol.* **12,** 83–92.

Neumann, N. F., Fagan, D., and Belosevic, M. (1995). Macrophage activating factor(s) secreted by mitogen stimulated goldfish kidney leukocytes synergize with bacterial lipopolysaccharide to induce nitric oxide production in teleost macrophages. *Dev. Comp. Immunol.* **19,** 473–482.

Newton, R. A., Raftos, D. A., Raison, R. L., and Geczy, C. L. (1994). Chemotactic responses of hagfish (Vertebrata, Agnatha) leucocytes. *Dev. Comp. Immunol.* **18,** 295–303.

Nilsson, S., and Holmgren, S. (1992). Cardiovascular control by purines, 5-hydroxytryptamine, and neuropeptides. *In* "Fish Physiology" (W. S. Hoar, D. J. Randall, and A. P. Farrell, eds.), Vol. XIIB, pp. 301–341. Academic Press, London.

Novoa, B., Figueras, A., Ashton, I., and Secombes, C. J. (in press). *In vitro* studies on the regulation of rainbow trout (*Oncorhynchus mykiss*) macrophage respiratory burst activity. *Dev. Comp. Immunol.*

Obenauf, S. D., and Hyder Smith, S. (1992). Migratory response of nurse shark leucocytes to activated mammalian sera and porcine C5a. *Fish Shellfish Immunol.* **2,** 173–181.

Olivier, G., Evelyn, T. P. T., and Lallier, R. (1985). Immunity to *Aeromonas salmonicida* in coho salmon (*Oncorhynchus kisutch*) induced by modified Freund's complete adjuvant: Its nonspecific nature and the probable role of macrophages in the phenomenon. *Dev. Comp. Immunol.* **9,** 419–432.

Olson, K. R. (1992). Blood and extracellular fluid volume regulation: Role of the renin–angiotensin system, kallikrein–kinin system, and atrial natriuretic peptides. *In* "Fish Physiology" (W. S. Hoar, D. J. Randall and A. P. Farrell, eds.), Vol. XIIB, pp. 136–254. Academic Press, London.

Ostholm, T., Holmqvist, B. I., Alm, P., and Ekstrom, P. (1994). Nitric oxide synthase in the CNS of the atlantic salmon. *Nueroscie. Lett.* **168,** 233–237.

O'Sullivan, M. G., Huggins, E. M. Jr., Meade, E. A., DeWitt, D. L., and McCall, C. E. (1992). Lipopolysaccharide induces prostaglandin H synthase-2 in alveolar macrophages. *Biochem. Biophys. Res. Comm.* **187,** 1123–1127.

Park, S. W., and Wakabayashi, H. (1991). Activities of glycogen phosphorylase and glycogen synthetase in eel neutrophils. *Gyobyo Kenkyu* **26,** 35–43.

Pettersen, E. F., Fyllingen, I., Kavlie, A., Maaseide, N. P., Glette, J., Endresen, C., and Wergeland, H. I. (1995). Monoclonal antibodies reactive with serum IgM and leukocytes from Atlantic salmon (*Salmo salar* L.). *Fish Shellfish Immunol.* **5,** 275–287.

Pettitt, T. R., Rowley, A. F., Barrow, S. E., Mallet, A. I., and Secombes, C. J. (1991). Synthesis of lipoxins and other lipoxygenase products by macrophages from the rainbow trout, *Oncorhynchus mykiss. J. Biol. Chem.* **266,** 8720–8726.

Powell, M. D., Briand, H. A., Wright, G. M., and Burka, J. F. (1993). Rainbow trout (*Oncorhynchus mykiss* Walbaum) intestinal eosinophilic granule cell (EGC) response to *Aeromonas salmonicida* and *Vibio anguillarum* extracellular products. *Fish Shellfish Immunol.* **3,** 279–289.

Pulsford, A., and Matthews, R. A. (1984). An ultrastructural study of the cellular response of the plaice, *Pleuronectes platessa* L., to *Rhipidocotyle johnstonei* nom. nov. (pro-*Gasterostomum* sp. Johnstone, 1905) Matthews, 1968 (Digenea: Bucephalidae). *J. Fish Dis.* **7,** 3–14.

Rainger, G. E., Rowley, A. F., and Pettitt, T. R. (1992). Effect of inhibitors of eicosanoid biosynthesis on the immune reactivity of the rainbow trout, *Oncorhynchus mykiss. Fish Shellfish Immunol.* **2,** 143–154.

Ramakrishna, N. R., Burt, M. D. B., and MacKinnon, B. M. (1993). Cell-mediated immune response of rainbow trout (*Oncorhynchus mykiss*) to larval *Pseudoterranova decipiens* (Nematoda; Ascaridoidea) following sensitization to live sealworm, sealworm extract, and nonhomologous extracts. *Can. J. Fish. Aquat. Sci.* **50,** 60–65.

Reimschuessel, R., Bennett, R. O., May, E. B., and Lipsky, M. M. (1987). Eosinophil granular cell response to a microsporidian infection in a sergeant major fish, *Abudefduf saxatilis* (L.). *J. Fish Dis.* **10,** 319–322.

Ristow, S., Grabowski, L., Young, W., and Thorgaard, G. (1994). Arlee strain of rainbow trout exhibits a low level of nonspecific cytotoxic cell (NCC) activity. *Faseb. J.* **8,** A986.

Roberts, R. J. (1989). "Fish Pathology," 2nd edition. Bailliere Tindall, London.

Robertsen, B., Engstad, R. E., and Jorgensen, J. B. (1994). β-Glucans as immunostimulants in fish. *Modula. Fish Immune Responses* **1,** 83–99.

Roitt, I., Brostoff, J., and Male, D. (1993). "Immunology," 3rd edition. Mosby, London.

Rose, A. S., and Levine, R. P. (1992). Complement-mediated opsonization and phagocytosis of *Renibacterium salmoninarum. Fish Shellfish Immunol.* **2,** 223–240.

Rowley, A. F., Lloyd-Evans, P., Barrow, S. E., and Serhan, C. N. (1994). Lipoxin biosynthesis by trout macrophages involves the formation of epoxide intermediates. *Biochemistry* **33,** 856–863.

Rowley, A. F., Knight, J., Lloyd-Evans, P., Holland, J. W., and Vickers, P. J. (1995). Eicosanoids and their role in immune modulation in fish—a brief overview. *Fish Shellfish Immunol.* **5,** 549–567.

Saggers, B. A., and Gould, M. L. (1989). The attachment of microorganisms to macrophages isolated from tilapia *Oreochromis spilurus* Gunther. *J. Fish Biol.* **35,** 287–294.

Sakai, D. K. (1992). Repertoire of complement in immunological defense mechanisms of fish. *Annu. Rev. Fish Dis.* **2,** 223–247.

Sakai, M., Kobayashi, M., and Kawauchi, K. (1995). Enhancement of chemiluminescent responses of phagocytic cells from rainbow trout, *Oncorhynchus mykiss,* by injection of growth hormone. *Fish Shellfish Immunol.* **5,** 375–379.

Schober, A., Malz, C. R., and Meyer, D. L. (1993). Enzymehistochemical demonstration of nitric oxide synthase in the diencephalon of the rainbow trout (*Oncorhynchus mykiss*). *Neurosci. Lett.* **151,** 67–70.

Schoor, W. P., and Plumb, J. A. (1994). Induction of nitric oxide synthase in channel catfish *Ictalurus punctatus* by *Edwardsiella ictaluri. Dis. Aquat. Org.* **19,** 153–155.

Secombes, C. J. (1985). The *in vitro* formation of teleost multinucleate giant cells. *J. Fish Dis.* **8,** 461–464.

Secombes, C. J. (1990). Isolation of salmonid macrophages and analysis of their killing activity. *Tech. Fish Immunol.* **1,** 137–154.

Secombes, C. J. (1994a). Macrophage activation in fish. *Modulat. Fish Immune Responses* **1,** 49–57.

Secombes, C. J. (1994b). Enhancement of fish phagocyte activity. *Fish Shellfish Immunol.* **4,** 421–436.

Secombes, C. J. (1994c). The phylogeny of cytokines. *In* "The Cytokine Handbook," 2nd edition (A. W. Thomson, ed.), pp. 567–594. Academic Press, London.

Secombes, C. J., and Fletcher, T. C. (1992). The role of phagocytes in the protective mechanisms of fish. *Annu. Rev. Fish Dis.* **2,** 53–71.

Secombes, C. J., Clements, K., Ashton, I., and Rowley, A. F. (1994). The effect of eicosanoids on rainbow trout, *Oncorhynchus mykiss,* leucocyte proliferation. *Vet. Immunol. Immunopathol.* **42,** 367–378.

Secombes, C. J., Hardie, L. J., and Daniels, G. (1996). Cytokines in fish: An update. *Fish Shellfish Immunol.* **6,** 291–304.

Segal, A. W., and Abo, A. (1993). The biochemical basis of the NADPH oxidase of phagocytes. *TIBS* **18,** 43–47.

Sharp, G. J. E., and Secombes, C. J. (1993). The role of reactive oxygen species in the killing of the bacterial fish pathogen *Aeromonas salmonicida* by rainbow trout macrophages. *Fish Shellfish Immunol.* **3,** 119–129.

Sharp, G. J. E., Pike, A. W., and Secombes, C. J. (1991a). Leucocyte migration in rainbow trout (*Oncorhynchus mykiss* Walbaum): Optimization of migration conditions and responses to host and pathogen (*Diphyllobothrium dendriticum* Nitzsch) derived chemoattractants. *Dev. Comp. Immunol.* **15,** 295–305.

Sharp, G. J. E., Pike, A. W., and Secombes, C. J. (1991b). Rainbow trout (*Oncorhynchus mykiss* Walbaum, 1792) leucocyte interactions with metacestode stages of *Diphyllobothrium dendriticum* (Nitzsch, 1824), (Cestoda, Pseudophyllidea). *Fish Shellfish Immunol.* **1,** 195–211.

Sharp, G. J. E., Pettitt, T. R., Rowley, A. F., and Secombes, C. J. (1992). Lipoxin-induced migration of fish leukocytes. *J. Leuk. Biol.* **51,** 140–145.

Smith, P. D., McCarthy, D. H., and Paterson, W. D. (1980). Further studies on furunculosis in fish. *In* "Fish Diseases," 3rd COPRAQ Session (W. Ahne, ed.), pp. 113–119. Springer-Verlag, Berlin.

Song, Y. L., Lin, T., and Kou, G. H. (1989). Cell-mediated immunity of the eel, *Anguilla japonica* (Temminck and Schlegel), as measured by the migration inhibition test. *J. Fish Dis.* **12,** 117–123.

Sprang, S. R., and Bazan, J. F. (1993). Cytokine structural taxonomy and mechanisms of receptor engagement. *Curr. Opinion Struct. Biol.* **3,** 815–827.

Stave, J. W., Roberson, B. S., and Hetrick, F. M. (1984). Factors affecting the chemiluminescent response of fish phagocytes. *J. Fish Biol.* **25,** 197–206.

Steinhagen, D., and Jendrysek, S. (1994). Phagocytosis by carp granulocytes; *in vivo* and *in vitro* observations. *Fish Shellfish Immunol.* **4.** 521–524.

Suzuki, K. (1984). A light and electron microscope study on the phagocytosis of leucocytes in rockfish and rainbow trout. *Bull. Jpn. Soc. Sci. Fish.* **50,** 1305–1315.

Suzuki, Y., and Iida, T. (1992). Fish granulocytes in the process of inflammation. *Annu. Rev. Fish Dis.* **2,** 149–160.

Tamai, T., Shirahata, S., Sato, N., Kimura, S., Nonaka, M., and Hiroki, M. (1993). Purification and characterization of interferon-like antiviral protein derived from flatfish (*Paralichthys olivaceus*) lymphocytes immortalized by oncogenes. *Cytotechnology* **11,** 121–131.

Taylor, M. J., and Hoole, D. (1993). *Ligula intestinalis* (L.) (Cestoda: Pseudophyllidea): Polarization of cyprinid leucocytes as an indicator of host- and parasite-derived chemoattractants. *Parasitology* **107,** 433–440.

Thomas, P. T., and Woo, P. T. K. (1990). *In vivo* and *in vitro* cell-mediated immune responses of rainbow trout, *Oncorhynchus mykiss* (Walbaum), against *Cryptobia salmositica* Katz, 1951 (Sarcomastigophora: Kinetoplastida). *J. Fish Dis.* **13,** 423–434.

Thuvander, A., Johannisson, A., and Grawe, J. (1992). Flow cytometry in fish immunology. *Tech. Fish Immunol.* **2,** 19–26.

Tocher, D. R., and Sargent J. R. (1987). The effect of calcium ionophore A23187 on the metabolism of arachidonic and eicosapentaenoic acids in neutrophils from a marine teleost fish rich in (*n*-3) polyunsaturated fatty acids. *Comp. Biochem. Pysiol.* **87B,** 733–739.

Vallejo, A. N., and Ellis, A. E. (1989). Ultrastructural study of the response of eosinophil granule cells to *Aeromonas salmonicida* extracellular products and histamine liberators in rainbow trout *Salmo gairdneri* Richardson. *Dev. Comp. Immunol.* **13,** 133–148.

Vallejo, A. N., Ellsaesser, C. F., Miller, N. W., and Clem, L. W. (1991). Spontaneous development of functionally active long-term monocytelike cell lines from channel catfish. *In Vitro Cell. Dev. Biol.* **27A,** 279–286.

Vallejo, A. N., Miller, N. W., and Clem, L. W. (1992a). Antigen processing and presentation in teleost immune responses. *Annu. Rev. Fish Dis.* **2,** 73–89.

Vallejo, A. N., Miller, N. W., Harvey, N. C., Cuchens, M. A., Warr, G. W., and Clem, L. W. (1992b). Cellular pathway(s) of antigen processing and presentation in fish APC: Endosomal involvement and cell-free antigen presentation. *Dev. Immunol.* **3,** 51–65.

Vallejo, A. N., Miller, N. W., and Clem, L. W. (1992c). Cellular pathway(s) of antigen processing in fish APC: Effect of varying *in vitro* temperatures on antigen catabolism. *Dev. Comp. Immunol.* **16,** 367–381.

Van Damme, J. (1994). Interleukin-8 and related chemotactic cytokines. *In* "The Cytokine Handbook," 2nd edition (A. W. Thomson, ed.), pp. 186–208. Academic Press, London.

Verburg van Kemenade, B. M., Weyts, F. A. A., Debets R., and Flik, G. (1995). Carp macrophages and neutrophilic granulocytes secrete an interleukin 1–like factor. *Dev. Comp. Immunol.* **19,** 59–70.

Wang, R., Neumann, N. F., Shen, Q., and Belosevic, M. (1995). Establishment and characterization of a macrophage cell line from the goldfish. *Fish Shellfish Immunol.* **5,** 329–346.

Waterstrat, P. R., Ainsworth, A. J., and Capley, G. (1991). *In vitro* responses of channel catfish, *Ictalurus punctatus,* neutrophils to *Edwardsiella ictaluri. Dev. Comp. Immunol.* **15,** 53–63.

Weeks, B. A., Warinner, J. E., Mason, P. L., and McGinnis, D. S. (1986). Influence of toxic chemicals on the chemotactic response of fish macrophages. *J. Fish Biol.* **28,** 653–658.

Weeks, B. A., Keisler, A. S., Warinner, J. E., and Matthews, E. S. (1987). Preliminary evaluation of macrophage pinocytosis as a technique to monitor fish health. *Mar. Environ. Res.* **22,** 205–213.

Weeks, B. A., Sommer, S. R., and Dalton, H. P. (1988). Chemotactic response of fish macrophages to *Legionella pneumophila:* Correlation with pathogenicity. *Dis. Aquat.* Organisms **5,** 35–38.

Whyte, S. K., Chappell, L. H., and Secombes, C. J. (1989). Cytotoxic reactions of rainbow trout macrophages for larvae of the eye fluke *Diplostomum spathaceum* (Digenea). *J. Fish Biol.* **35,** 333–345.

Woo, P. T. K. (1992). Immunological responses of fish to parasitic organisms. *Annu. Rev. Fish Dis.* **2,** 339–366.

Wood, B. P., and Matthews, R. A. (1987). The immune response of the thick-lipped grey mullet, *Chelon labrosus* (Risso, 1826), to metacercarial infections of *Cryptocotyle lingua* (Creplin, 1825). *J. Fish Biol.* **31A,** 175–183.

Wood, S. E., Willoughby, L. G., and Beakes, G. W. (1986). Preliminary evidence for inhibition of *Saprolegnia* fungus in the mucus of brown trout, *Salmo trutta* L., following experimental challenge. *J. Fish Dis.* **9,** 557–560.

Yoshida, S. H., Stuge, T. B., Miller, N. W., and Clem, L. W. (1995). Phylogeny of lymphocyte heterogeneity: Cytotoxic activity of channel catfish peripheral blood leukocytes directed against allogeneic targets. *Dev. Comp. Immunol.* **19,** 71–77.

3

THE NONSPECIFIC IMMUNE SYSTEM: HUMORAL DEFENSE

TOMOKI YANO

THE FISH IMMUNE SYSTEM:
ORGANISM, PATHOGEN, AND ENVIRONMENT

I. INTRODUCTION

The serum, mucus, and eggs of fish contain a variety of substances that nonspecifically inhibit the growth of infectious microorganisms. These substances are predominantly proteins or glycoproteins and many of them are believed to have their counterparts or precursors in the blood and hemolymph of invertebrates. They are specific in that they react with just one chemical group or configuration, but they have been called "nonspecific" because of the substances with which they react are very common, and they do not influence the growth of only one microorganism. In spite of their potential importance in host defense, their biological and physicochemical properties have not yet been fully elucidated. This chapter will introduce such humoral defense factors, concentrating upon lysozyme, complement, interferon, C-reactive protein, transferrin, and lectin (hemagglutinin).

II. LYSOZYME

Lysozyme is found in a wide range of vertebrates (Osserman *et al.*, 1974), and is one of the defensive factors against invasion by microorganisms. It splits the β (1→4) linkages between *N*-acetylmuramic acid and *N*-acetylglucosamine in the cell walls (peptidoglycan layers) of Gram-positive bacteria, thus preventing them from invading (Salton and Ghuysen, 1959). In the case of Gram-negative bacteria, which are not directly damaged by lysozyme, the enzyme becomes effective after complement and other enzymes have disrupted the outer cell wall, thereby unmasking the inner peptidoglycan layer of the bacteria (Glynn, 1969; Neeman *et al.*, 1974; Hjelmeland *et al.*, 1983). In addition to a direct antibacterial effect, lysozyme promotes phagocytosis as an opsonin, or by directly activating polymorphonuclear leukocytes and macrophages (Klockars and Roberts, 1976; Jollès and Jollès, 1984).

According to Salton (1957) and Jollès (1969), 'true' lysozymes have to satisfy the following criteria: (1) The enzyme lyses *Micrococcus lysodeikticus* cells; (2) is readily adsorbed by chitin-coated cellulose; (3) is a low molecular weight protein; and (4) is stable at acidic pH at higher temperatures, but is inactivated under alkaline conditions. Most of the early lysozyme data were obtained from birds, and classical representatives of this enzyme family are therefore called c type (chicken type). Most c-type lysozymes consist of a single amino acid chain of about 129 residues, have a molecular weight of about 14.5 kDa, and have a restricted degrading effect on chitin which is a major component of the cell walls of fungi and the exoskeletons of certain invertebrates. All mammalian lysozymes thus far examined have proved to be of the c type. It is suggested that c-type lysozymes share a common ancestor with g-type (goose-type) lysozymes which have 185 amino acid residues and a molecular weight of about 20.5 kDa (Jollès and Jollès, 1984; Mckenzie and White, 1991).

A. Biological Functions of Fish Lysozyme

In fish, lysozyme is distributed mainly in tissues rich in leucocytes, such as the head kidney, at sites where the risk of bacterial invasion is high, such as the skin, the gills and the alimentary tract, and in the eggs (Fletcher and Grant, 1968; Fletcher and White, 1973; Ourth, 1980; Mochizuki and Matsumiya, 1981; Studnicka *et al.,* 1986; Grinde *et al.,* 1988a; Kawahara and Kusuda, 1988a; Grinde, 1989; Oohara *et al.,* 1991; Yousif *et al.,* 1991; Holloway *et al.,* 1993). This implies that fish lysozymes play an important role in the host defense mechanisms against infectious diseases (Fänge *et al.,* 1976; Murray and Fletcher, 1976; Lundblad *et al.,* 1979; Lindsay, 1986; Lie *et al.,* 1989).

In plaice, lysozyme activity has been identified histochemically in monocytes and neutrophils (Murray and Fletcher, 1976). These cells probably contribute to the serum lysozyme activity since their number increases concomitantly with serum lysozyme levels (Fletcher and White, 1973). Recently, the occurrence of lysozyme in the eggs of coho salmon was reported (Yousif *et al.,* 1991). The eggs were shown to contain high levels of the enzyme, the concentration in the yolk being 1900 μg/ml. Similar levels of lysozyme were found in the eggs of two other species of salmonids, although the source of the lysozyme present in eggs is unknown. It seems likely, however, that it is released from the kidney and other lysozyme-rich tissues of the mother and transported to the developing eggs via the serum (Yousif *et al.,* 1991).

Grinde (1989) investigated the antibacterial effect of two lysozyme variants (types I and II), purified from the kidney of rainbow trout, on seven bacterial strains of Gram-negative species. They found that type I was surprisingly potent, having substantial antibacterial activity on all the strains tested. On the other hand, hen egg-white lysozyme, which was used as a reference, was bactericidal only against the one species that was considered nonpathogenic. These results suggest that fish lysozyme, in contrast to mammalian lysozyme, has substantial antibacterial activity not only against Gram-positive bacteria, but also against Gram-negative bacteria in the absence of complement.

The lysozyme obtained from coho salmon eggs has been shown to be bactericidal to *Aeromonas hydrophila, Aeromonas salmonicida,* and *Carnobacterium piscicola* at a concentration of 700 μg/ml, a concentration approximately one-third of that found in the yolk of most salmonid eggs. However, the kidney disease bacterium *Renibacterium salmoninarum* was not killed when incubated with as much as 1900 μg/ml of the enzyme for 90 minutes (Yousif *et al.,* 1994a). These findings indicate that lysozyme plays a role in preventing the mother-to-progeny (vertical) transmission of some bacterial fish pathogens, and its failure to kill *R. salmoninarum* helps to explain why this organism is readily transmitted vertically. Rainbow trout kidney lysozyme (type I) showed substantial antimicrobial activity (both bactericidal and bacteriolytic) against Gram-negative bacteria (Grinde, 1989). In contrast, coho salmon egg lysozyme was bactericidal but not bacteriolytic (Yousif *et al.,* 1994a), indicating that the mechanism underlying the bactericidal and lytic properties of the enzyme are not always the same, as had been suggested by Iacono *et al.* (1980).

B. Physicochemical Properties of Fish Lysozyme

Grinde *et al.* (1988b) purified two different lysozymes, designated types I and II, from the kidney of rainbow trout (Table I). Type II lysozyme was approximately three times as potent as type I when its activity was tested using *Micrococcus luteus* as the substrate. Both lysozymes were c-type lysozymes and had a molecular weight of 14.4 kDa and isoelectric points of 9.5 and 9.7, respectively. The optimum pH of both enzymes was approximately the same as those reported for tilapia and plaice lysozymes (Sankaran and Gurnani, 1972; Fletcher and White, 1973). Dautigny *et al.* (1991) established the complete 129 amino acid sequences of both type I and type II lysozymes using protein chemistry microtechniques. The two amino acid sequences differed only at position 86, type I having aspartic acid and type II having alanine.

From the skin mucus of ayu fish, two different types of lysozyme were purified (Itami *et al.,* 1992). Both enzymes shared a molecular mass of 18 kDa but had isoelectric points of 9.4 and 9.8, respectively. Only the lysozyme activity of type I was inhibited by heparin (100 units/ml) and both enzymes were inactivated by histamine (10 mM). Bacteriolytic activity of the two enzymes was high for formalin-killed cells of *A. hydrophila* and *Pasteurella piscicida,* but low against those of *Vibrio anguillarum,* a pathogenic agent of ayu fish. Lysozymes were also isolated from the eggs of coho salmon (Yousif *et al.,* 1991) and the skin mucus of carp (Takahashi *et al.,* 1986).

C. Factors Influencing Lysozyme Activity

1. Season, Sex, and Stage of Sexual Maturity

Serum concentration of lumpsucker lysozyme showed seasonal variations and was higher in the male than in the female (Fletcher *et al.,* 1977).

Table I
Physicochemical Properties of Fish Lysozymes[a]

Species	Site	MW (kDa)	PI	Opt. pH	Opt. temp. (°C)	Source[b]
Anguilla japonica						
(Japanese eel)	Skin mucus			6.0	30	(a)
	Serum kidney			8.0	50	
Cyprinus carpio						
(Carp)	Skin mucus	24, 33		7.2, 9.0	40	(b)
Plecoglossus altivelis						
(Ayu fish)	Skin mucus	I: 18	I: 9.4	I: 6.3–6.9	I: 35	(c)
		II: 18	II: 9.8	II: 6.3–6.9	II: 35	
Oncorhynchus mykiss						
(Rainbow trout)	Skin mucus			6.0	30–40	(d)
	Kidney	I: 14.4	I: 9.5	I: 5.5	I: 45	(e)
		II: 14.4	II: 9.7	II: 5.5	II: 45	
	Kidney	14.5, 23				(f)
Oncorhynchus kisutch						
(Coho salmon)	Egg	14.5				(g)
Salmo salar						
(Atlantic salmon)	Kidney	12–13				(f)
Seriola quinqueradiata						
(Yellowtail)	Skin, mucus			7.0	40	(h)
	Serum, kidney			8.0	40	
Scatophagus argus						
(Scat)	Liver, gill			6.2, 9.2		(i)
Tilapia mossambica						
(Tilapia)	Liver			5.4		(i)
Pleuronectes platessa						
(Plaice)	Serum	14–15		5.4		(j)

[a] Abbreviations: MW, molecular weight; PI, isoelectric point; Opt., Optimum; temp., temperature.

[b] (a), Kawahara and Kusuda (1988b); (b), Takahashi *et al.* (1986); (c), Itami *et al.* (1992); (d), Hjelmeland *et al.* (1983); (e), Grinde *et al.* (1988b); (f) Lie *et al.* (1989); (g), Yousif *et al.* (1991); (h), Kusuda *et al.* (1987); (i) Sankaran and Gurnani (1972); (j), Fletcher and White (1973, 1976).

In carp, the highest level of the enzyme occurred in spawners (Studnicka *et al.*, 1986), and in Atlantic salmon and brown trout, the lysozyme activity markedly decreased during smoltification (parr-smolt transformation) (Muona and Soivio, 1992).

2. Water Temperature

Plaice maintained at a low temperature (5°C) for 3 months exhibited a 70% decrease in serum lysozyme level (Fletcher and White, 1976). A similar decrease in serum lysozyme level was observed in carp (Studnicka *et al.*, 1986). In Japanese eel, however, the opposite result was obtained (Kusuda and Kitadai, 1992): eels maintained at 15°C showed greater serum lysozyme activity compared with those maintained at 20 to 30°C.

3. Stress and Infections

Möck and Peters (1990) reported that rainbow trout stressed by transport or acute water pollution had significantly reduced serum lysozyme levels.

Enhanced serum lysozyme activity was observed in carp infected with *Aeromonas punctata* (Vladimirov, 1968; Siwicki and Studnicka, 1987) or the protozoan *Eimeria subepithelialis* (Studnicka *et al.*, 1986), and in Atlantic salmon experimentally challenged with *A. salmonicida* (Møyner *et al.*, 1993). In carp, serum lysozyme activity increased concomitantly with the elevation of antibody titer (Vladimirov, 1968).

III. COMPLEMENT

The complement system is an integral part of the vertebrate immune system. In mammals, it is composed of two distinct pathways, the classical complement pathway (CCP) and the alternative complement pathway (ACP) (Fig. 1). In the CCP, C1 is first activated by an antigen–antibody complex, and this is followed by a cascade of interactions of C4, C2, C3, C5, C6, C7, C8, and C9. The late-acting complement components (C5 to C9) together form a membrane attack complex (MAC), which induces the death (cytolysis) of target cells. In the ACP, C3 is directly activated in the presence of factors B and D by substances such as the lipopolysaccharide (LPS) of Gram-negative bacteria, inulin, zymosan, and rabbit erythrocytes, and this also leads to the formation of a MAC. Activation of either the CCP or the ACP results in the generation of many biologically important peptides that are involved in inflammatory responses. For example, C3a, an anaphylatoxin, causes smooth muscle contraction and increased capillary permeability mostly through the release of histamine from mast cells; C3b, an opsonin, enhances the phagocytosis of particles to which it becomes attached and probably represents the most important complement-derived biologically active fragment in terms of defense against infection; C5a is a potent chemotactic factor for neutrophils and macrophages, and also has

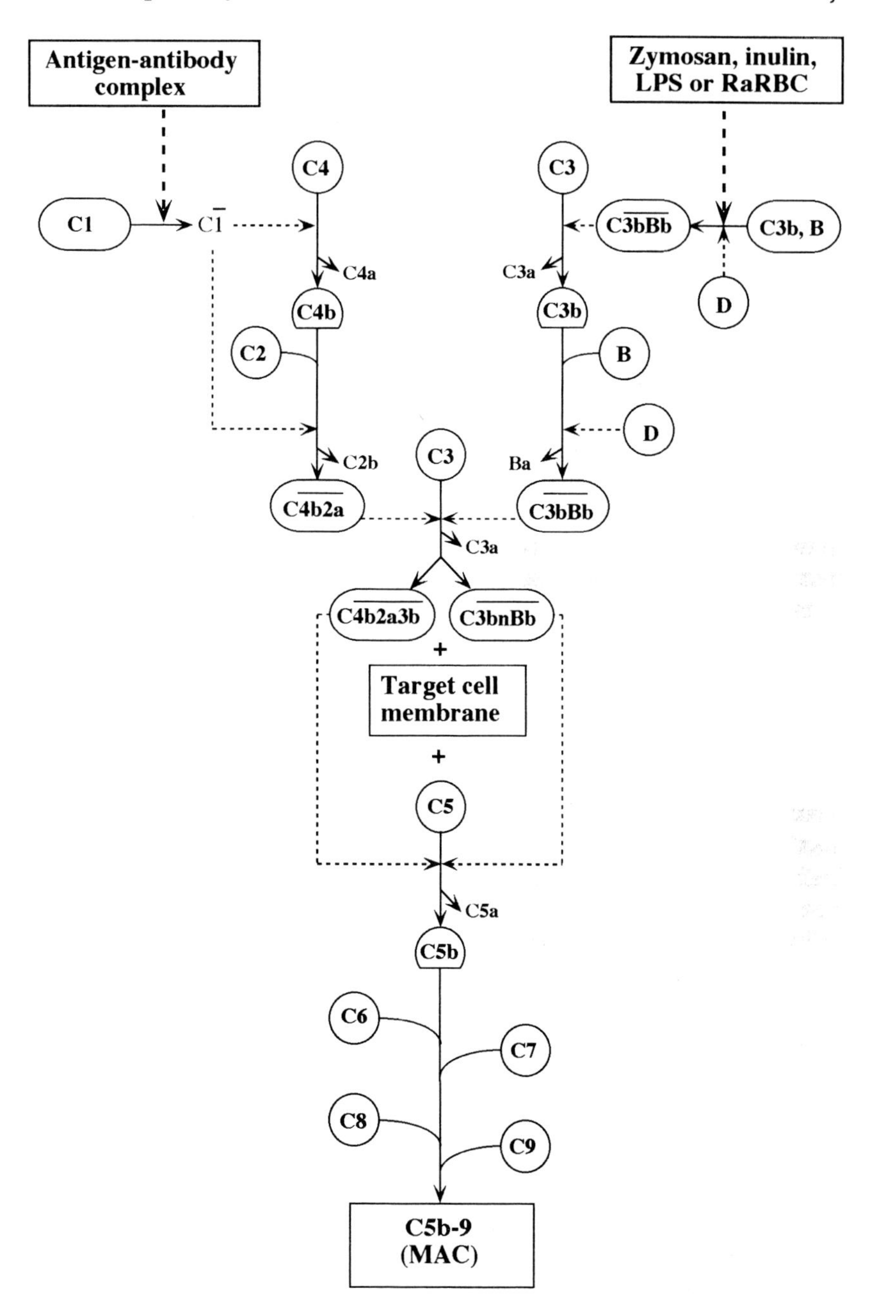

Fig. 1. A scheme showing the classical and alternative complement pathways, generation of biologically active fragments (C4a, C3a, and C5a), and formation of the membrane attack complex in mammals. LPS, lipopolysaccharide; RaRBC, rabbit erythrocytes; B, factor B; D, factor D; MAC, membrane attack complex.

anaphylatoxin activity; and C5b becomes the nucleus for the formation of a MAC.

In mammals, formation and degradation of biologically active fragments are regulated by serum proteins such as C4bp (C4 binding protein), factor I, factor H, and anaphylatoxin inactivator, and by cell surface proteins such as CR1 (complement receptor–1), MCP (membrane cofactor protein), and DAF (decay accelerating factor). Neutrophils and macrophages have receptors for C3b, iC3b, and C5a (CR1, CR3, and C5a receptor), and mast cells have a receptor for both C3a and C4a (C3a/C4a receptor) on their surface membranes (Lambris, 1988; Becherer *et al.,* 1989).

A. The Complement System of Cyclostomes

Cyclostome (lamprey and hagfish) serum lacks cytolytic activity of the complement and has only the alternative pathway of C3 activation (Fujii *et al.,* 1992; Nonaka, 1994).

Proteins that are homologous to mammalian C3 have been isolated from lamprey, *Lampetra japonica* (Nonaka *et al.,* 1984a), and from hagfish, *Eptatretus burgeri* (Fujii *et al.,* 1992; Ishiguro *et al.,* 1992) (Table II). Both

Table II

Complement Components Identified in Fishes, Amphibians, Reptiles, and Birds[a,b]

Animal	Protein	cDNA
Cyclostomes		
Hagfish	C3	C3
Lamprey	C3	C3, B
Cartilaginous fishes		
Nurse shark	C1n, C2n, C3n, C4n, C8n, C9n, MAC	
Bony fishes		
Rainbow trout	C3, C5, MAC	C3, C9
Carp	C1, C4, C2, C3, C5 (C5b), C6, C7, C8, C9, B, D, MAC	
Sand bass	I, H (or C4bp)	
Amphibians		
Bullfrog	C1 (C1q)	
Clawed frog	C4, C3, B, MAC	C3, I
Axolotl	C3	
Reptiles		
Cobra	C3, I, MAC	C3
Birds		
Quail	C3	
Chicken	C1q, C3, B, I	C3

[a] From Yano (1995).

[b] Abbreviations: MAC, membrane attack complex; B, factor B; D, factor D; I, factor I; H, factor H; C4bp, C4-binding protein.

proteins bind to zymosan particles only when incubated in the presence of homologous serum, and when bound, they potentiate phagocytosis by macrophages, indicating that other serum factors are involved in C3 activation and that cyclostome macrophages have C3 receptors on their surface membranes. Lamprey C3 (190 kDa) is composed of three polypeptide chains, the α-chain (84 kDa), β-chain (74 kDa), and γ-chain (32 kDa), linked by disulfide bonds, and has a thiolester bond on the α-chain (Nonaka *et al.,* 1984a). Hagfish C3 also has a three-subunit chain structure (88 kDa, 76 kDa, 25 kDa) and possesses a thiolester bond on the α-chain (Ishiguro *et al.,* 1992). Both C3s show an amino acid sequence closer in identity to mammalian C3 than to mammalian C4, even though they have a three-subunit chain structure similar to mammalian C4 (Ishiguro *et al.,* 1992; Nonaka and Takahashi, 1992). This is not unexpected in view of the widely accepted concept that mammalian C3 and C4 have both derived from a common ancestor, probably during the process of gene duplication (Gorski *et al.,* 1981). From the hagfish *E. burgeri,* C3 with a two-subunit chain structure (115 kDa and 77 kDa) has also been isolated (Fujii *et al.,* 1992), implying that hagfish C3 is present in two different stable forms.

From the Pacific hagfish *Eptatretus stoutii,* C3 (or a C4-like protein) with a three-subunit chain structure (77 kDa, 70 kDa, 30 kDa) has been isolated (Hanley *et al.,* 1992; Raftos *et al.,* 1992). The protein (210 kDa) promoted phagocytosis of yeast by hagfish leukocytes, and EDTA inhibited the binding of C3 to the yeast surface while decreasing the phagocytosis of yeast. Recently, Raison *et al.* (1994) identified a C3 receptor (105 kDa) present on the Pacific hagfish leukocyte surface. Moreover, it was demonstrated that Pacific hagfish granulocytes showed chemotactic migration in response to human C5a and LPS-activated hagfish plasma (Newton *et al.,* 1994). This indicates that specific chemoattractant receptors are present on the surface of hagfish leukocytes, and LPS activation of hagfish plasma generates a potent chemotactic product.

The factor responsible for the lytic activity of lamprey serum has been identified, but has been shown to be unrelated to complement (Gewurz *et al.,* 1965, 1966; Nonaka *et al.,* 1984a). The complete amino acid sequences of lamprey and hagfish C3 (Nonaka and Takahashi, 1992; Ishiguro *et al.,* 1992) and lamprey factor B (Nonaka *et al.,* 1994) have been deduced by sequence analysis of their cDNA clones.

B. The Complement System of Cartilaginous Fishes

Hemolytic complement activity of cartilaginous fishes (lemon shark, nurse shark, and sting ray) was inactivated or depleted by treatment with EDTA, LPS, carrageenan, hydrazine, zymosan, inulin, or heat (48–50°C)

(Legler and Evans, 1967; Day *et al.,* 1970; Culbreath *et al.,* 1991). EDTA-inactivated serum of cartilaginous fish could not have its activity restored by an addition of excess Ca^{2+} and Mg^{2+} (Legler and Evans, 1967; Gigli and Austen, 1971; Koppenheffer, 1987).

The complement system of the nurse shark *Ginglymostoma cirratum* consists of a classical and an alternative pathway (Jensen *et al.,* 1981; Culbreath *et al.,* 1991). The classical pathway is composed of six functionally distinct components named C1n, C2n, C3n, C4n, C8n, and C9n (Ross and Jensen, 1973a,b; Jensen *et al.,* 1981; Hyder Smith and Jensen, 1986) (Table II). C1n corresponds to mammalian C1, and C2n (184 kDa) and C3n correspond to mammalian C4 and C2, respectively. C8n (185 kDa) and C9n (ca. 190 kDa) appear to be analogues of mammalian C8 and C9, respectively (Jensen *et al.,* 1973, 1981; Hyder Smith and Jensen, 1986). However, the correspondence of C4n to any of the mammalian complement components remains unclear.

C1n is activated by EAn (sheep erythrocytes sensitized with nurse shark antibody) and forms EAnC1n which can be lysed by C1-depleted guinea pig serum (C4–C9) (Ross and Jensen, 1973a). Moreover, C8n lyses EAC1gp4-7hu in the presence of C9n, and C9n lyses EAC1gp4-7hu-8gp (Jensen *et al.,* 1973, 1981). These results indicate that C1n, C8n, and C9n are compatible with guinea pig C4, human C7, and guinea pig C8, respectively.

Formation of hemolytic intermediate complexes, EAnC1nC2n and EAnC1nC2nC3n, was also confirmed in the nurse shark. Like their mammalian counterparts, the former was stable while the latter was extremely unstable (Jensen *et al.,* 1981). Activation of nurse shark complement leads to the formation of an MAC on target cells (Jensen *et al.,* 1981). This was confirmed by electron microscopic observation. The average inner diameter of membrane lesions made by the nurse shark MAC was 8.0 nm. Nurse shark leukocytes show chemotactic migration in response to porcine C5a and LPS-activated rat serum, implying that shark leukocytes possess cell surface receptors that can recognize mammalian C5a (Obenauf and Hyder Smith, 1985, 1992).

C. The Complement System of Bony Fishes

Bony fishes, both marine and freshwater, have a CCP and ACP that are directly comparable to those of mammals. To date, the existance of both pathways has been shown in Japanese eel (Iida and Wakabayashi, 1983; Kusuda and Fukunaga, 1987), carp (Yano *et al.,* 1985; Matsuyama *et al.,* 1988a), channel catfish (Ourth and Wilson, 1982a; Lobb and Hayman, 1989), ayu fish (Matsuyama *et al.,* 1988b; Yano *et al.,* 1988b), salmonid fishes (Nonaka *et al.,* 1981a, Ingram, 1987, Røed *et al.,* 1992), porgies (Matsuyama *et al.*, 1988b, Yano *et al.*, 1988b; Sunyer and Tort, 1994), tilapia *Tilapia nilotica* (Matsuyama *et al.*, 1988b; Yano *et al.*, 1988b), and albacore tuna (Giclas *et al.*, 1981). Complement has also been detected in the skin mucus of rainbow trout (Harrell *et al.*, 1976).

In general, bony fish complement shows the highest activity at 15–25°C and retains its activity even at 0–4°C (Day *et al.*, 1970; Gigli and Austen, 1971; Rijkers, 1982; Koppenheffer, 1987; Lobb and Hayman, 1989). This may be a common feature seen in poikiothermic vertebrates, since similar phenomena are observed in amphibians and reptiles (Gigli and Austen, 1971; Koppenheffer, 1987). The complement of most warm water fishes is inactivated when held at 45–54°C for 20 min, whereas that of cold water fishes is inactivated when held at 40–45°C for 20 min (Legler *et al.*, 1967; Dorson *et al.*, 1979; Giclas *et al.*, 1981; Sakai, 1981; Ingram, 1987; Lobb and Hayman, 1989; Røed *et al.*, 1990, 1992; Sunyer and Tort, 1994).

The ACP activity (ACH50) of samples from seven bony fishes (rainbow trout, ayu fish, carp, tilapia, yellowtail, porgy, and flounder) was extremely high compared with that of samples from mammals when measured under optimum conditions (Yano *et al.*, 1988b, Yano, 1992). Similar results have been obtained by Saha *et al.* (1993) and Sunyer and Tort (1994) who measured the ACH50 titers of catfishes and a porgy (sea bream), respectively, together with those of mammals. These results indicate that the role of ACP, which works effectively during the early stages of infection, is much more important in fish than it is in mammals.

1. Biological Functions of Bony Fish Complement

a. Virucidal Activity. Rainbow trout and masu salmon, *O. masou,* fry (4–5 months after hatching) whose sera showed no hemolytic activity were far more susceptible to IHN (infectious hematopoietic necrosis) and IPN (infectious pancreatic necrosis) viruses than were chum salmon, *O. keta,* fry (4–5 months after hatching) whose serum possessed complement activity as high as that of juveniles (Sakai *et al.*, 1994), suggesting that fish complement displays virucidal activity against invading viruses.

b. Bactericidal Activity. Bactericidal activity of complement has been reported for Japanese eel, cyprinid fishes, channel catfish, salmonid fishes, porgy (*Sparus aurata*), and tilapia (*Tilapia nilotica*) (Ourth and Wilson, 1981, 1982a,b; Trust *et al.*, 1981; Munn *et al.*, 1982; Iida and Wakabayashi, 1983; Sakai, 1983; Ourth and Bachinski, 1987a,b; Sugita *et al.*, 1989; Jenkins and Ourth, 1990; Sunyer and Tort, 1995). The bactericidal activity of fish serum is attributable mainly to the activation of the ACP, rather than the CCP (Koppenheffer, 1987).

In general, fish complement displays bactericidal activity against nonvirulent strains of Gram-negative bacteria, but not against Gram-positive bacteria or virulent strains of Gram-negative bacteria (Ourth and Wilson, 1982b; Iida and Wakabayashi, 1983, 1993; Sugita *et al.,* 1989). Ourth and Bachinski (1987b) demonstrated that the catfish ACP is an efficient defense mechanism against nonpathogenic Gram-negative bacteria that contain no sialic acid, but that the catfish ACP is inhibited by the large amount of sialic acid contained in pathogenic Gram-negative bacteria such as *A. salmonicida* and *F. columnaris,* indicating that sialic acid is an important virulence factor for establishing an initial infection. The ability of *A. salmonicida* to resist the bactericidal activity of rainbow trout and channel catfish ACPs could perhaps also be explained by the presence of an A layer and/or LPS, which prevents access of complement proteins to the membranes of the bacteria (Munn *et al.,* 1982; Jenkins *et al.,* 1991). Activation of the channel catfish ACP damaged susceptible *Escherichia coli,* forming an MAC on the cell membrane. The average inner diameter of the lesions was about 9 nm (Jenkins *et al.,* 1991).

c. Parasiticidal Activity. Sera from nonsusceptible rainbow trout lysed *Cryptobia salmositica* (a pathogenic hemoflagellate) via the activation of the ACP (Wehnert and Woo, 1980). The parasites were also lysed when incubated with immune (anti–*C. salmositica*) plasma from salmonids and a smaller amount of homologous complement, indicating that the rainbow trout CCP also works effectively for killing the parasites (Jones and Woo, 1987; Bower and Evelyn, 1988; Woo and Thomas, 1991; Woo, 1992).

d. Opsonic Activity. Opsonic activity of fish complement has been observed in Japanese eel, carp, channel catfish, salmonid fishes, yellowtail (*Seriola quinqueradiata*), and tilapia (Nonaka *et al.,* 1984b; Sakai, 1984a; Honda *et al.,* 1985, 1986; Kusuda and Tanaka, 1988; Moritomo *et al.,* 1988; Saggers and Gould, 1989; Matsuyama *et al.,* 1992; Jenkins and Ourth, 1993). Generally, fish complement exhibits opsonic activity against nonpathogenic bacteria, but not against virulent strains of bacteria (Kusuda and Tanaka, 1988; Moritomo *et al.,* 1988; Iida and Wakabayashi, 1993; Jenkins and Ourth, 1993). It was demonstrated that C3 is the major phagocytosis-promoting factor (opsonin), and C3 receptors are present on the macrophages of salmonid fishes (Johnson and Smith, 1984; Sakai, 1984a), tilapia (*Oreochromis spilurus*) (Saggers and Gould, 1989), and yellowtail (Kusuda and Tanaka, 1988), and on the neutrophils of carp (Matsuyama *et al.,* 1992) and channel catfish (Jenkins and Ourth, 1993).

e. Chemoattracting Activity. It is reported that incubation of rainbow trout and Japanese eel serum with antigen–antibody complex and zymosan, respectively, generated C5a-like chemotactic factors (Griffin, 1984; Suzuki, 1986; Iida and Wakabayashi, 1988).

f. Inactivation of Bacterial Exotoxin(s). Rainbow trout complement is reported to inactivate the toxicity of extracellular products of *A. salmonicida* (Sakai, 1984b).

2. Biochemistry of Bony Fish Complement Components

Nonaka *et al.* (1981a) demonstrated that the CCP of rainbow trout is activated by antigen–antibody complex (sheep erythrocytes sensitized with homologous antibody) in the presence of Ca^{2+} and Mg^{2+}, and that the ACP can be activated by zymosan, inulin, or rabbit erythrocytes in the presence of Mg^{2+} alone. They isolated C3 and C5 proteins from rainbow trout plasma (Nonaka *et al.,* 1981b, 1984b) (Table II). Trout C3 (190 kDa) is composed of two polypeptide chains, the α-chain (128 kDa) and the β-chain (74 kDa), linked by disulfide bonds and has a thiolester bond on the α-chain. Trout C5 (194 kDa) also has a two-subunit structure (133 kDa and 86 kDa) linked by a disulfide bond. Nonaka *et al.* (1981a) isolated an MAC from rabbit erythrocytes lysed by rainbow trout serum, however, they did not identify the constituents of the MAC.

Our laboratory has isolated complement components from C1 to C9 and factors B and D from carp serum (Table III) and by using these purified proteins, we have obtained the following information: (1) C1 and C4 bind to EA (sheep erythrocytes sensitized with carp antibody) in the presence of Ca^{2+}, and C2 binds to EAC14 in the presence of Mg^{2+} (Fig. 2); (2) EA is stable, while EAC142 is quite labile and rapidly inactivated at room temperature; (3) C3 binds to EAC142 even at 0°C, in marked contrast to the case in mammals where the binding reaction is temperature dependent and never occurs at low temperatures; (4) Mg^{2+} is indispensable for the activation of the ACP; (5) C3 (184 kDa) is cleaved into C3a (14 kDa) and C3b (168 kDa) by C3 convertase (C3bBb), and factor B (93 kDa) is split into Ba (34 kDa) and Bb (66 kDa) by factor D; and (6) C5 (C5b), C6, C7, C8, and C9 in molar ratios of 1:1:1:1:4 together form MACs on target cell (rabbit erythrocyte) membranes. (Yano *et al.,* 1985, 1986, 1988a,c; Nakao *et al.,* 1988, 1989; Uemura *et al.,* 1992; Uemura, 1993; Yano and Nakao, 1994; Yano, 1995). From these results it is apparent that the complement system is highly developed at the phylogenetic level of the bony fish.

The degradation pattern of rainbow trout C3 was investigated by incubating trout serum with inulin. The data showed a high degree of homology with the mammalian C3 degradation pattern (Jensen and Koch, 1992). The complete amino acid sequences of rainbow trout C3 (Lambris *et al.,* 1993)

Table III
Molecular Weights of Complement Components Isolated from Carp Serum

	Molecular weight (kDa)[a,b]		
Complement components	Carp	Rainbow trout	Human
C1	1020		794 (22)
C4	170		205 (3)
C2	108 (1)		102 (1)
C3	184 (2)	190 (2)	185 (2)
C5	175 (2)[c]	194 (2)	190 (2)
C6	115 (1)		120 (1)
C7	106 (1)		110 (1)
C8	146 (3)		150 (3)
C9	91 (1)		71 (1)
D	29 (1)		24 (1)
B	93 (1)		92 (1)

[a] Molecular weights of carp complement components were determined by SDS-PAGE except that of C1 which was measured by the gel filtration method. Data on human complement components was taken from Law and Reid (1988).
[b] Numbers in parentheses indicate the number of subunits.
[c] Molecular weight of C5b.

and C9 (Stanley and Herz, 1987; Tomlinson *et al.,* 1993) have been deduced by sequence analysis of their cDNA clones and have shown considerable homology to mammalian counterparts. Recently, proteins homologous to mammalian factor I and its cofactor (C4bp or factor H) have been isolated from serum of the sand bass *Parablax neblifer* (Kaidoh and Gigli, 1987;

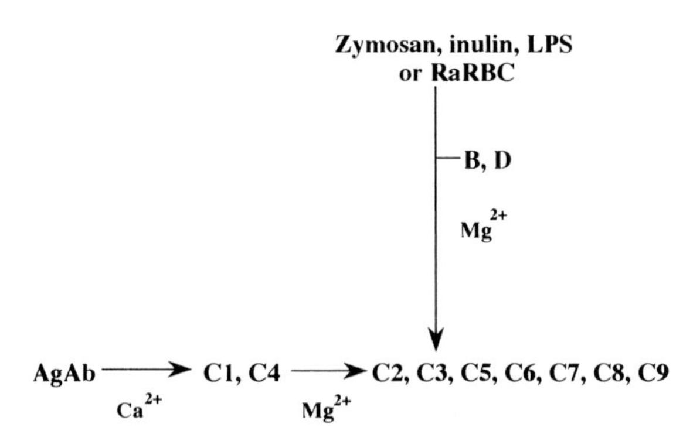

Fig. 2. The classical and alternative pathways of carp complement. AgAb, antigen–antibody complex; LPS, lipopolysaccharide; RaRBC, rabbit erythrocytes; B, factor B; D, factor D.

Dahmen *et al.,* 1994). In contrast to mammalian proteins, the sand bass factor I (155 kDa) and its cofactor (360 kDa) required Ca^{2+} for activation (Kaidoh and Gigli, 1989a,b).

D. Factors Influencing Complement Activity

Seasonal variations in complement activity are observed in carp (Yano *et al.,* 1984) channel catfish (Hayman *et al.,* 1992), and tench (Collazos *et al.,* 1994). The ACP titers (ACH50 values) of tench are high in winter and when CCP activity (Yano *et al.,* 1984; Hayman *et al.,* 1992) as well as the specific immune response (Avtalion, 1981; Bly and Clem, 1991) are depressed.

Sexual maturation of fish also appears to influence complement activity. In rainbow trout, serum bactericidal activity decreased during the spawning period (Iida *et al.,* 1989), while in Atlantic salmon, reduced hemolytic complement activity was observed in fish with signs of sexual maturation (Røed *et al.,* 1992).

E. Genetic Polymorphism of C3

It is known that most of the mammalian complement components exhibit polymorphism (Whitehouse, 1988). Nonaka *et al.* (1985) reported the presence of C3 variants (C3–1 and C3–2) in rainbow trout. While C3-1 was required for hemolysis of sensitized sheep erythrocytes, C3–2 did not promote the hemolytic reaction. Jensen and Koch (1991) have identified three different alleles (S, f1, and f2) of C3 in rainbow trout. Our laboratory has found that the hemolytic activity of carp C3 differs among four variants (unpublished data).

IV. INTERFERON

Interferons (IFNs) are proteins or glycoproteins able to inhibit virus replication. In mammals, three types of IFN (α, β, and γ) are distinguishable on the basis of biological and biochemical properties (Stewart, 1980; Hosoi *et al.,* 1988; Sano *et al.,* 1988). Though IFN-α and IFN-β are different in antigenicity, they have many similarities and are often grouped together as type I IFN. Both IFN-α and -β are induced in cells infected with viruses or incubated with dsRNA (poly I:C), and probably any cell type can produce them. They share 29% amino acid sequence homology and 45% nucleotide homology, are acid stable (pH 2), act via a single receptor, and their genes are located on the same chromosome. It is speculated that IFN-α and IFN-β genes are the products of an ancient gene duplication, perhaps at an early stage of vertebrate evolution (Taniguchi *et al.*, 1980; Miyata *et al.*, 1985). Both IFNs exhibit a broad spectrum of antiviral activity.

In contrast, IFN-γ (type II IFN) is unrelated to IFN-α and IFN-β. It is generally accepted that T-lymphocytes stimulated by antigen or mitogen are the main producers of IFN-γ (Kiener and Spitalny, 1987). It shares less than 10% amino acid sequence homology with IFN-α and IFN-β, and is located on a separate chromosome. IFN-γ is unstable at a low pH (<4) and at a high temperature (>56°C) and acts via a distinct receptor. In addition, it differs from type I IFNs in its spectrum of biological activities which includes cell regulation, cell differentiation, and intercellular communication (Trinchieri and Perussia, 1985). IFN-γ activates cells for immune response such as natural killer cells and macrophages. One of the so-called macrophage-activating factors (MAF) is considered to be IFN-γ, and for this reason, IFN-γ is now included among the lymphokines (cytokines).

IFN-α and IFN-β are glycoproteins with molecular weights of between 16 and 26 kDa, and isoelectric points between pH 6.5 and 7.5. IFN-α is very polymorphic with nine types in humans. IFN-β appears to be of a single type. IFN-γ is reported to be a glycoprotein or protein having a molecular weight (12 to 25 kDa) similar to that of IFN-α and IFN-β. In humans, there are two forms of IFN-γ which are produced by the same gene; the differences in their molecular weights are due to differences in the amount of glycosylation. The role of glycosylated moieties in an IFN molecule is not yet known.

A. Fish Interferon $\alpha\beta$

IFN production has been confirmed in bony fishes but not in cyclostomes or cartilagenous fishes. Fish IFN was demonstrated for the first time in cultured cells derived from fathead minnows by Gravell and Malsberger (1965). They observed that the IPN virus did not replicate if the cultures were maintained at 34°C, however, they did replicate at 23°C. The authors considered that this was not a direct temperature effect on the virus itself but was due to an IFN-like substance. This substance resembled mammalian IFN in that it was nondialyzable, acid stable (pH 2), nonsedimentable (160,000*g* for 14 h), and trypsin labile. Beasley *et al.* (1966) detected an IFN-like substance in the cultures of fin cells (GF) derived from the blue striped grunt. Following a survey of previous work, Beasley and Sigel (1967) concluded that the grant cells produced IFN that was not virus specific but species specific. Oie and Loh (1971) also confirmed IFN production in fathead minnow cells (FHM) infected with mammalian reovirus type II. According to the nomenclature now used for mammals (Stewart, 1980), these IFNs belong to the type I class.

It is now well established that fish cells can secrete IFN-$\alpha\beta$ (IFN-α or IFN-β) molecules in response to virus infection (Kelly and Loh, 1973; Rio *et al.,* 1973; Dorson and de Kinkelin 1974; De Sena and Rio, 1975; de Kinkelin *et al.,* 1977; Baudouy, 1978; de Kinkelin *et al.,* 1982; Sano and Nagakura, 1982; Okamoto *et al.,* 1983; Pintó *et al.,* 1993; Rogel-Gaillard *et al.,* 1993; Snegaroff, 1993) or by exposure to poly I:C, the well-known inducer of IFN-β (MacDonald and Kennedy, 1979; Tengelsen *et al.,* 1989; Eaton, 1990). Most *in vivo* studies have been carried out on rainbow trout. IFN-$\alpha\beta$ was found in the serum of rainbow trout experimentally infected with pathogenic salmonid viruses such as the VHS (viral hemorrhagic septicemia) virus (de Kinkelin and Dorson, 1973), the IHN virus (de Kinkelin and Le Berre, 1974), and the IPN virus (Dorson *et al.,* 1992). When the water temperature is about 15°C, the VHS virus does not cause any overt disease, and at this temperature IFN synthesis was found to occur within 48 h postinfection. As such, rainbow trout IFN-$\alpha\beta$ appears to be involved in the mechanisms leading to the elimination of infection (Dorson and de Kinkelin, 1974; Renault *et al.,* 1991).

B. Fish Interferon γ

Graham and Secombes (1988, 1990) showed that leucocytes from rainbow trout kidney secreted an IFN-γ-like molecule with antiviral and MAF activities when they were stimulated with a mitogen.

C. Physicochemical Properties of Fish Interferons

IFN-$\alpha\beta$ has been isolated twice from rainbow trout. That which was isolated after infection with the VHS virus had a molecular weight of 26 kDa and an isoelectric point of 5.3 (Dorson *et al.,* 1975), whereas that which was isolated following infection with the IPN virus had a molecular weight of 94 kDa and an isoelectric point of 7.1 (De Sena and Rio, 1975). The molecular weight of the latter is extremely high when compared with those reported for other animals (Galabov, 1973). This substance could be a polymeric form of the former IFN.

Recently, Tamai *et al.* (1993a) purified IFN from the culture medium of the flatfish *Paralichthys olivaceus* lymphocyte cell line immortalized by oncogene tranfection. The protein was a glycoprotein of about 16 kDa. cDNA cloning of the molecule showed that it was composed of 138 amino acids. The antiviral activity of the protein was trypsin sensitive and was fairly stable at pH values between 4 and 8. The protein retained about 60% of its activity even at 60°C and showed a broad antiviral activity. These results suggest that the flatfish IFN may be either IFN-α or IFN-β. When the amino acid sequence of the flatfish IFN was compared with human IFN-αN, bovine IFN-αA, horse IFN-α2, rat IFN-α1, human IFN-β, and mouse IFN-β, the sequences were 24, 20, 19, 18, 14, and 12% identical, without counting the gaps, respectively. The amino acid sequence homology between flatfish IFN and human IFN-γ was very low (Tamai *et al.,* 1993b).

V. C-REACTIVE PROTEIN

C-reactive protein (CRP) is the first protein to appear in the plasma of humans and most animals suffering from tissue damage, infection, or inflammation. This acute phase protein has the curious characteristics of recognizing and precipitating C-polysaccharide (CPS), a component of the cell wall of *Streptococcus pneumoniae,* in the presence of Ca^{2+}. CRP was first discovered in human serum (Tillett and Francis, 1930), and since then, CRP has been found in many animal species including invertebrates, the horseshoe crab, and a mollusc *Achatina fulica.*

Mammalian CRP has a very high binding affinity not only for phosphorylcholine (a constituent of CPS), but also for multiple polycations (e.g., histone and protamine) in the presence of Ca^{2+} (Siegel *et al.,* 1974, 1975). In addition, CRP activates the complement system via the classical pathway in combination with its ligands, enhances natural killer– and macrophage-mediated cytotoxicity, and solubilizes endogenous substances derived from damaged cells such as chromatin (Kaplan and Volanakis, 1974; Volanakis and Kaplan, 1974; Osmand *et al.,* 1975; Claus *et al.,* 1977; Mold *et al.,* 1982; Robey *et al.,* 1985; Kilpatrick and Volanakis, 1991). Recently, it has been reported that rat CRP plays an important role in the detoxification of mercury (Agrawal and Bhattacharya, 1989).

It is known that the liver is the site of CRP synthesis, and inflammation causes the hepatocytes that are not producing CRP to commence synthesis (Hurlimann *et al.,* 1966; Kushner and Feldmann, 1978). The CRP concentration in normal serum is reported to be 3–5 μg/ml for humans, 1.5 μg/ml for rabbits, 400–500 μg/ml for rats, and 60 μg/ml for dogs.

Human CRP consists of five identical, nonglycosylated, noncovalently associated polypeptide subunits, each with a single intrachain disulfide bridge (Oliveira *et al.,* 1977, 1979; Volanakis *et al.,* 1978) and two calcium binding sites (Gotschlich and Edelman, 1967). Electron microscopic studies revealed that the subunits are arranged in cyclic pentameric symmetry (in an annular disk-like configuration) (Osmand *et al.,* 1977; Baltz *et al.,* 1982). The native human CRP has a total molecular weight of 110 kDa (187 amino acid residues), an isoelectric point of 6.4, and each subunit has a molecular weight of 24 kDa (Nunomura, 1992).

A. Biological Functions of Fish C-Reactive Protein

In fish, CRP has been isolated from the smooth dogfish *Mustelus canis* (Robey and Liu, 1983; Robey *et al.,* 1983), Japanese eel (Nunomura, 1991), channel catfish (Szalai *et al.,* 1994), rainbow trout (Winkelhake and Chang, 1982; Murai *et al.,* 1990), lumpsucker *Cyclopterus lumpus* (Fletcher and Baldo, 1976; White *et al.,* 1978), tilapia *Tilapia mossambica* (Ramos and Smith, 1978), murrel fish (Mitra and Bhattacharya, 1992), and plaice (Baldo and Fletcher, 1973; Pepys *et al.,* 1978, 1982).

Baldo and Fletcher (1973) first described a CRP-like protein that binds pneumococcal CPS in the presence of Ca^{2+} in plaice serum. They also discovered that subdermal injection of fungal extracts containing CPS-like substances elicited immediate erythematous skin reactions in species of fish with CRP in their sera. Fish, such as the flounder *Platichthys flesus,* that lack CRP failed to react in this way, but reactivity could be transferred with CRP-rich plaice serum or with highly purified plaice CRP itself (Fletcher and Baldo, 1974; Baldo and Fletcher, 1975). Bacterial endotoxin (LPS) caused an increase in the serum concentrations of plaice CRP (White *et al.,* 1981, 1984; White and Fletcher, 1982, 1985). However, since all normal plaice have substantial serum levels (50–70 μg/ml) of CRP, the increase, although significant, was minor compared with the several hundredfold increase of CRP in injured or diseased mammals.

CRP was detected in various tissues of the lumpsucker including the serum, eggs, and sperm (Fletcher and Baldo, 1976; Fletcher *et al.,* 1977; White *et al.,* 1978). In the female, the highest concentration was observed in the eggs, suggesting that most of the CRP synthesized in the female is concentrated in the eggs (Fletcher *et al.,* 1977). In tilapia, a very small amount of CRP was detected in the skin mucus, but none at all was noted in the serum. CRP only became detectable in the serum following the infliction of physical injury (Ramos and Smith, 1978).

CRP isolated from rainbow trout serum reacted with CPS in the presence of Ca^{2+}, and this reaction was inhibited by phosphorylcholine (Winkelhake and Chang, 1982). The CRP–CPS complex suppressed *in vitro* growth of *V. anguillarum* and enhanced the phagocytosis of the bacteria by glass-adherent peritoneal exudate cells, indicating that CRP exhibits a defensive function through the activation of the complement system (Nakanishi *et al.,* 1991). The CRP level in rainbow trout serum increased to threefold that of the control serum after experimental infection with *V. anguillarum* (Murai *et al.,* 1990).

Japanese eel CRP agglutinated rabbit red blood cells and *S. pneumoniae,* and this hemagglutinating activity was inhibited by D-glucosamine and D-mannose (Nunomura, 1991). The eel CRP formed a precipitin line with histone, protamine, poly(L-lysine) and poly(L-arginine) in agarose gel. Channel catfish CRP (100 kDa) precipitated components of the cell wall of *Saprolegnia* sp. as well as CPS in the presence of Ca^{2+} (Szalai *et al.,* 1992a,b, 1994). The serum level of catfish CRP greatly decreased in fish that were injected with *Saprolegnia* sp. (Szalai *et al.,* 1994).

As described above, most fish species appear to possess CRP in their sera. Since phosphorylcholine is a widely occurring determinant in the cell walls or surface structures of invading organisms such as bacteria, fungi, and parasites, it seems highly likely that, like limulin in the horseshoe crab, fish CRP which can recognize phosphorylcholine plays a vital role in the host defense mechanism as a surrogate for immunoglobulin, since fish produce only a single type of immunoglobulin (IgM).

B. Physicochemical Properties of Fish C-Reactive Protein

The plaice CRP molecule (186.8 kDa) consists of 10 noncovalently associated subunits; these are arranged in two pentameric disks interacting face-to-face (Pepys *et al.,* 1982). Upon SDS-PAGE, the CRP molecule yields two distinct subunits. The difference in size between the subunits is due to carbohydrates since the heavier subunit is glycosylated, bearing a single, typical, complex oligosaccharide, while the lighter one is not (Pepys *et al.,* 1982). The amino acid sequence from the N-terminal of plaice CRP shows 42% homology with that of human CRP and 47% homology with that of rabbit CRP (Pepys *et al.,* 1982). The native lumpsucker CRP molecule (125–150 kDa) consists of identical noncovalently bound subunits of 21.5 kDa and its isoelectric point is 5.3 (White *et al.,* 1978).

Winkelhake and Chang (1982) reported that the molecular weight of native rainbow trout CRP is 110 kDa, while that of the subunits is about 20 kDa. However, Murai *et al.* (1990) reported that trout CRP is a trimer glycoprotein (81.4 kDa) which is composed of one monomer subunit (26.6 kDa) and one disulfide-linked dimer (43.7 kDa), but that there exists the possibility that a hexameric disk or a double-stacked hexamer (12 complex subunits), as seen in horseshoe crab CRP (Robey and Liu, 1981), is disrupted and separated into two trimers or four trimers.

Smooth dogfish CRP exists in solution as either a pentameric or hexameric molecule (250 kDa) with dimeric subunits of 50 kDa. Upon the addition of β-mercaptoethanol, these dimeric subunits dissociate into two identical monomeric subunits of 25 kDa (Robey and Liu, 1983; Robey *et al.,* 1983).

Japanese eel CRP is a pentamer (120 kDa) of identical subunits of 24 kDa and, as in the case of human, rabbit, and goat CRPs, has no disulfide bonds among the subunits (Nunomura, 1991). A lectin (123 kDa) has also been purified from eel serum (Matsumoto and Osawa, 1974), but its physico-chemical and biological characteristics are completely different from those of the eel CRP.

Murrel fish CRP (141 kDa) is a doubly stacked pentamer with 10 monomeric subunits (14 kDa), the dimer (28 kDa) being linked by disulfide bridges as in the case of dogfish (Robey *et al.,* 1983). Electron microscopy of channel catfish CRP confirmed that it, like human CRP, is a protein with planar pentagonal symmetry (Szalai *et al.,* 1992a). However, unlike CRPs of other animals, the catfish CRP could not readily be dissociated into subunits.

C. Factors Influencing C-Reactive Protein Activity

1. Season and Sex

The serum level of lumpsucker CRP was greater in males than it was in females, and in both the male and female it showed seasonal fluctuations (Fletcher *et al.,* 1977). In other fish species such as plaice and channel catfish, there was no significant difference in serum CRP levels between the sexes (White *et al.,* 1983; Szalai *et al.,* 1994). In channel catfish, serum levels of CRP were maximal in summer and minimal in winter (Szalai *et al.,* 1994), and in plaice, the highest values for CRP were found between June and September (White *et al.,* 1983).

2. Stress

The CRP level in rainbow trout serum elevated 18-fold over normal serum levels (70 μg/ml) following high temperature shock (Winkelhake and Chang, 1982), while that in channel catfish serum decreased greatly following low temperature shock (Szalai *et al.,* 1994).

3. Administration of Inflammatory Agents

The CRP level in rainbow trout elevated 3- to 20-fold over normal serum levels following an injection of turpentine (Winkelhake and Chang, 1982) or Freund's complete adjuvant (Kodama *et al.,* 1989). The serum level of catfish CRP also elevated rapidly 18-fold over preinjection levels (3.8 mg/ml) following an injection of turpentine (Szalai *et al.,* 1994).

VI. TRANSFERRIN

Transferrin is an iron-binding glycoprotein that plays a central role in the transport of iron between sites of absorption, storage, and utilization in all vertebrate organisms (Putnam, 1975). Transferrin has the electrophoretic mobility of a β-globulin and is present in a high concentration (e.g., 2–3 mg/ml in humans) relative to most other plasma proteins. Iron is an essential element in the establishment of infection by most pathogens (Sussman, 1974), and transferrin limits the amount of endogenous iron available to the pathogens by chelating the metal and making it unavailable for bacterial use (Weinberg, 1974). The amount of transferrin in host blood is therefore an important parameter in deducing the condition of a pathogen-susceptible host.

Human transferrin has a molecular weight of 77 kDa (a single polypeptide chain) and an isoelectric point of 5.5 to 5.6, depending upon the iron content (Aisen, 1980); its iron-binding capacity is estimated to be two atoms per molecule. The complete amino acid sequence of human transferrin has been determined by MacGillivray *et al.* (1982).

Throughout the animal kingdom, transferrin exhibits a high degree of genetic polymorphism. For example, more than 20 electrophoretically detectable variants are known in human transferrin (Putnam, 1975), though this may only reflect the extensive sampling in humans. The transferrin locus has been used for answering zoogeographic and evolutionary questions in closely related vertebrate organisms. In mammals, transferrin phenotypes are reported to be associated with differences in fertility, fetal mortality, and milk production.

A. Biological Functions of Fish Transferrin

Transferrin has been isolated from almost all the species of fish examined. In cyclostomes, transferrin has been isolated from both the Pacific hagfish (Aisen *et al.,* 1972) and lampreys (Boffa *et al.,* 1967; Webster and Pollara, 1969; Macey *et al.,* 1982). In cartilagenous fishes, transferrin has been isolated from the cat shark *Scyllium stellare* (Got *et al.,* 1967) and the lemon shark (Clem and Small, 1967), while in bony fishes, transferrin has been isolated from more than 100 species of fish (Turner and Jamieson, 1987; Jamieson, 1990). Most fish transferrins exhibit polymorphism (Fujino and Kang, 1968; Hershberger, 1970; Utter *et al.,* 1970; Payne, 1971; Menzel, 1976; Valenta *et al.,* 1976; Stratil *et al.,* 1983b; Jamieson, 1990).

In fish, the possible role of transferrin in disease resistance has been examined mostly in coho salmon and the results have been contradictory. Suzumoto *et al.* (1977) reported that there is a differential resistance to bacterial kidney disease (BKD) according to transferrin genotype in coho salmon. The AA genotype was found to be most susceptible, the AC genotype had an intermediate susceptibility, while the CC genotype was the most resistant to the disease. However, the addition of exogenous iron did not appear to increase the pathogenicity of the disease. Pratschner

(1978) found that CC animals showed a greater resistance to vibriosis than did AA animals and that the converse was true for furunculosis. Moreover, Winter *et al.* (1980) examined coho salmon of different stocks, and found that in stocks that showed a differential resistance of genotypes to BKD, the AA type was the most susceptible, but no differences in resistance to vibriosis were observed among the transferrin genotypes. Withler and Evelyn (1990) concluded that the differences they found between coho salmon strains with regard to resistance to BKD were not in fact due to simple possession of the transferrin C allele, because the strain with the greatest resistance had only 4% C allele, while the least resistant strain had 27% C allele.

B. Physicochemical Properties of Fish Transferrin

In cyclostomes such as Pacific hagfish and sea lamprey, transferrins have molecular weights similar to that of human transferrin (77 kDa). Sea lamprey transferrin has an isoelectric point of about 9.1 and has no carbohydrate in its molecule (Boffa *et al.,* 1967; Webster and Pollara, 1969) (Table IV). Six transferrin variants were observed in sea lamprey (Webster and Pollara, 1969). On the other hand, Southern Hemisphere lamprey transferrin has a large molecular weight of 296 kDa and is composed of three identical subunits (78 kDa); it has an isoelectric point of 8.6. In cartilaginous fishes, cat shark transferrin has a molecular weight of 78 kDa and an isoelectric point of 9.2 (Got *et al.,* 1967).

The molecular weights of bony fish transferrins range from 61 to 87 kDa (Bobák *et al.,* 1984). The isoelectric point was estimated to be 5.0 in carp (Valenta *et al.,* 1976) and 5.3 in chum salmon (Hara, 1984). In chum salmon, the iron-binding capacity of transferrin was estimated to be two atoms per molecule (Hara, 1984). Transferrin molecules of silver carp, big head, barbel, pike, brook trout, and Atlantic salmon contain carbohydrate, while those of tench, grass carp, and catfish do not (Hershberger, 1970; Stratil *et al.,* 1983a,b, 1985; Røed *et al.,* 1995). The N-terminal amino acid of grass carp, catfish, pike, and chum salmon transferrin molecules is alanine (Stratil *et al.,* 1983a, 1985; Hara, 1984), while no N-terminal amino acid has been detected in tench, silver carp, big head, or carp (Valenta *et al.,* 1976; Stratil *et al.,* 1983a), indicating that the N-terminal of these transferrin molecules is blocked.

In immunological cross-reactivity tests between the transferrin of chum salmon and that of 12 other fishes, transferrins from the family *Salmonidae* showed cross-reactivity, while those from ayu fish and olive rainbow smelt did not (Hara, 1984). Recently, the complete amino acid sequences of Atlantic salmon (Kvingedal *et al.,* 1993) and medaka *Orizias latipes* (Hirono

Table IV
Physicochemical Properties of Fish Transferrins[a]

Species	MW (kDa)	PI	Carbohydrate	N-terminal amino acid	Source[b]
Eptatretus stoutii (Pacific hagfish)	75–80				(a)
Petromyzon marinus (Sea lamprey)	70–80	9.1–9.2	–	–	(b, c)
Geotria australis (Southern Hemisphere lamprey)	296	8.6			(d)
Scyllium stellare (Cat shark)	78	9.2	+	Alanine (or valine)	(e)
Negaprion brevirostris (Lemon shark)	75				(f)
Anguilla anguilla (European eel)	75				(g)
Cyprinus carpio (Carp)	70	5.0		–	(h)
Ctenopharyngodon idella (Grass carp)	75		–	Alanine	(i)
Hypophthalmichthys molitrix (Silver carp)	79		+	–	(i)
Tinca tinca (Tench)	70		–	–	(i)
Aristichthys nobilis (Big head)	79		+	–	(i)
Barbus meridionalis (Barbel)	76		+		(j)
Silurus glanis (Catfish)	68		–	Alanine	(k)
Esox lucius (Pike)	87		+	Alanine	(k)
Oncorhynchus mykiss (Rainbow trout)	65				(g)
Oncorhynchus keta (Chum salmon)	80	5.3		Alanine	(l)

[a] Abbreviations: MW, molecular weight; PI, isoelectric point.

[b] (a), Aisen *et al.* (1972); (b), Boffa *et al.* (1967); (c), Webster and Pollara (1969); (d), Macey *et al.* (1982); (e), Got *et al.* (1967); (f), Clem and Small (1967); (g), Bobák *et al.* (1984); (h), Valenta *et al.* (1976); (i), Stratil *et al.* (1983a); (j), Stratil *et al.* (1983b); (k), Stratil *et al.* (1985); (l), Hara (1984).

et al., 1995) transferrins have been deduced by sequence analysis of their cDNAs.

VII. LECTIN

Lectins (hemagglutinins) are proteins of nonimmune origin that agglutinate cells and/or precipitate glycoconjugates. Lectins contain at least two sugar-binding sites, and the specificity of lectin is usually defined in terms of the monosaccharide(s) or glycosaccharides that inhibit lectin-induced agglutination or precipitation (Goldstein *et al.,* 1980). Lectins were first discovered in plants, but have since also been found in various tissues and organs of both vertebrates and invertebrates (Sharon and Lis, 1972, 1989a,b; Goldstein *et al.,* 1980). In higher vertebrates, lectins are involved in diverse roles including morphogenesis, polyspermy blocking, serum-screening by the liver, and defense against microorganisms (Yoshizaki, 1990; Kilpatrick *et al.,* 1991).

In mammals, lectins have been placed into several categories, including C-type and S-type. The C-type or calcium-dependent lectins are extracellular or membrane-bound molecules with various carbohydrate specificities, whereas the S-type or thiol-dependent lectins are extracellular or intracellular noncation-dependent molecules with specficity for β-galactosides or more complex β-galactoside-containing oligosaccharides (Drickamer, 1988; Harrison, 1991). There are no structural homologies between C-type and S-type lectins.

A. Biological Functions of Fish Lectin

1. Egg Lectins

The presence of lectins (hemagglutinins) has been recorded in the eggs of many fish species including lamprey, herring, carp, loach, Japanese catfish, smelts, ayu fish, salmonid fishes, sea bass, perch, porgy, and flounder (Table V). The biological role of fish egg lectins has not yet been fully elucidated. However, from the fact that lectins exist in egg cortical granules, it would seem that they are involved in fertilization (prevention of polyspermy), ontogeny (embryo differentiation), and/or defense against microorganisms (Nosek *et al.,* 1983a,b; Krajhanzl *et al.,* 1984; Iwasaki and Inoue, 1985; Sakakibara *et al.,* 1985; Krajhanzl and Kocourek, 1986; Krajhanzl, 1990; Alexander and Ingram, 1992; Hosono and Nitta, 1993).

Chinook salmon egg lectin inhibited the growth of virulent bacteria (*V. anguillarum, Y. ruckeri, A. hydrophila, and Edwardsiella tarda*) that are pathogenic for chinook and other salmonid fishes (Voss *et al.,* 1978). Simi-

Table V
Sugar Specificities, Molecular Weights, and Metal Requirement of Fish Egg Lectins[a]

Species	Erythrocyte	Sugar specificity	MW (kDa) native (subunit)	Metal requirement	Source[b]
Petromyzon marinus (Sea lamprey)	Horse	Sialoglyco-conjugates	(43)	+	(a)
	Horse	Fetuin, mannose, sucrose	>500(30, 65)	C-type	(b)
Clupea harengus (Herring)	Human B	L-Rha (D-Gal)	(17.5)	S-type	(c, d)
Rutilus rutilus (Roach)	Human B	L-Rha (D-Gal)	200 (20, 25, 30)	S-type	(d)
Vimba vimba (Vimba)	Human B	L-Rha (D-Gal)	(24, 35)		(e)
Scardinius erythrophthalmus (Rudd)	Human B	L-Rha (D-Gal)	(23, 36)	–	(e)
Abramis brama (Bream)	Human B	L-Rha (D-Gal)	(19, 30, 42)		(d)
Tinca tinca (Tench)	Human B	L-Rha	(36)		(d)
Hypophthalmichthys molitrix (Silver carp)	Human B, rabbit	L-Rha	90, 170(14, 18)		(d)
Leuciscus hakonensis (Japanese dace)	Human B, rabbit	L-Rha (L-Man, L-Lyx, D-Gal	(25)	–	(f)

Misgurnus anguilli caudatus (Loach)	Human B	L-Rha (L-Man, L-Lyx, D-Gal)	(50)		(g)
Silurus asotus (Japanese catfish)	Human B, rabbit	L-Rha (L-Man, L-Lyx, D-Gal)	38 (33)	–	(h)
Hypomesus transpacificus nipponensis (Pond smelt)	Human B	L-Rha, L-Man, L-Lyx	(68)		(i)
Osmerus eperlanus mordax (Olive rainbow smelt)	Human B, rabbit	L-Rha (L-Man, L-Lyx, D-Gal)	(25, 26, 32)	–	(j)
Plecoglossus altivelis (Ayu fish)	Human B	L-Rha (L-Man, L-Lyx, D-Gal)	(14.8)		(k)
Coregonus peled (White fish)	Human B, rabbit	L-Rha (D-Gal)	100 (20–21.5)		(d)
Coregonus lavaretus maraena (Powan)	Human B, rabbit	L-Rha (D-Gal)	25(27)	+	(l)
Oncorhynchus mykiss (Rainbow trout)	Human B, rabbit	L-Rha	(19, 30)	–	(m)
Oncorhynchus tschwytscha (Chinook salmon)	Human B, rabbit	L-Rha, D-Gal	122		(n)
Oncorhynchus kisutch (Coho salmon)	*A. salmonicida*	L-Rha, D-Gal D-Gal*N*	(24.5)		(o)
Oncorhynchus keta (Chum salmon)	Rabbit	L-Rha, (L-Man, L-Lyx, D-Gal)	(22)	S-type	(p)

(*continues*)

Table V
(continued)

Species	Erythrocyte	Sugar specificity	MW (kDa) native (subunit)	Metal requirement	Source[b]
Oncorhynchus rhodurus (Amago)	Human B, rabbit	L-Rha, (L-Man, L-Lyx, D-Gal)		S-type	(q)
Salmo trutta (Brown trout)	Human B	L-Rha (D-Gal)		S-type	(c)
Dicentrarchus labrax (Sea bass)	Human O	L-Fuc			(r)
Perca fluviatilis (Perch)	Human O	L-Fuc, D-Glc	130–190(13)	+	(s)
Stizostedion lucicoperca (Pikeperch)	Human B	L-Rha	26		(d)
Sponodyliosoma cantharus (Black sea bream)	Human B	L-Rha (D-Gal)	100		(t)
Pleuronectes platessa (Plaice)	Rabbit, rat	D-Man, D-Gal*N*Ac	<232		(u)

[a] Abbreviations: MW, molecular weight; L-Rha, L-rhamnose; D-Gal, D-galactose; L-Man, L-mannose; L-Lyx, D-lyxose; D-Gal*N*, D-galactosamine; L-Fuc, L-fucose; D-Glc, D-glucose; D-Man, D-mannose; D-Gal*N*Ac, *N*-acetyl-D-galactosamine.

[b] (a), Vasta and Marchalonis (1983); (b), Schluter *et al.* (1994); (c), Anstee *et al.* (1973); (d), Krajhanzl (1990); (e), Krajhanzl *et al.* (1978b); (f), Hosono *et al.* (1992); (g), Sakakibara *et al.* (1981); (h), Hosono *et al.* (1993a); (i), Sakakibara *et al.* (1982); (j) Hosono *et al.* (1993b); (k), Sakakibara *et al.* (1985); (l), Krajhanzl *et al.* (1978a); (m), Bildfell *et al.* (1992); (n), Voss *et al.* (1978); (o), Yousif *et al.* (1994b); (p), Kamiya *et al.* (1990); (q) Ozaki *et al.* (1983); (r) Topliss and Rogers (1985); (s) Nosek *et al.* (1983a); (t) Rogers (1978); (u), Bly *et al.* (1986).

larly, roach egg lectin inhibited the growth of *A. hydrophila* and *A. salmonicida* (Zikmundová *et al.,* 1990). Coho salmon egg lectin agglutinated *A. salmonicida,* but showed no such activity with other fish pathogens such as *A. hydrophila, V. anguillarum, Vibrio ordalii,* or *R. salmoninarum* (Yousif *et al.,* 1994b), while chum salmon egg lectin agglutinated *V. anguillarum,* but did not affect the growth of the bacteria (Kamiya *et al.,* 1990). The agglutinating activity of chum salmon eggs varied according to the stage of egg development. In fertilized eggs, the lectin activity showed an apparent decrease after the eyed stage, which is when the vulnerable fertilized egg acquires tolerance against physical and chemical changes within the environment. This activity disappeared completely just before hatching. Rainbow trout egg lectin caused growth inhibition of pathogenic fungi (Balakhnin and Dudka, 1990; Balakhnin *et al.,* 1990). In this case, the levels of the lectin decreased in *Saprolegnia*-infected eggs during development. Fish egg lectins may provide some protection to developing eggs until the immune system reaches a level of sufficient competence (Voss *et al.,* 1978).

2. Skin Mucus Lectins

Fish skin mucus serves as a lubricant in locomotion and a mechanical defense barrier in the prevention of surface colonization by bacteria and fungi. It has been shown that fish skin mucus contains many antipathogenic substances such as immunoglobulin, lysozyme, complement, C-reactive protein, lectin (hemagglutinin), and hemolysin (Ingram, 1980; Ellis, 1981; Fletcher, 1982).

Currently, lectins (hemagglutinins) have been isolated from the skin mucus of hagfish, freshwater eels, moray eel, conger eel, loach, sea catfishes, ayu fish, cusk-eel, dragonet, and flounders (Table VI). It should be noted that most of these fish species have degenerate or no scales on their body surfaces. In the Japanese eel, lectin was seen in club cells within the epidermis, especially in secretory vacuoles (Suzuki and Kaneko, 1986). This fact suggests that in the eel, mucus lectin is derived from club cells. Production of lectin in club cells is also suggested in moray eel (Randall *et al.,* 1981) and Arabian Gulf catfish (Al-Hassan *et al.,* 1986).

At present, the biological function of fish skin mucus lectins is not fully understood, although it has been suggested that they serve in part as a defense mechanism against bacterial infestation on the skin (Ingram, 1980). This possibility may be supported by the fact that skin mucus lectins isolated from windowpane flounder and conger eel agglutinate a marine yeast *Metschnikowia reukaufii* and a marine bacterium *V. anguillarum,* respectively, although their growth is not inhibited (Kamiya and Shimizu, 1980; Kamiya *et al.,* 1988, 1990). In ayu fish, it is reported that skin mucus lectin shows

Table VI
Sugar Specificities, Molecular Weights, and Metal Requirement of Fish Skin Mucus Lectins[a]

Species	Erythrocyte	Sugar specificity	MW (kDa) native (subunit)	Metal requirement	Source[b]
Eptatretus stoutii (Pacific hagfish)	Human O				(a)
Anguilla japonica (Japanese eel)	Human A,B,O; rabbit	Lactose, mucin, asialofetuin	370		(b, c)
Lycodontis nudivomer (Moray eel)	Human A,B,O; rabbit	D-Gal, lactose	<100		(d)
Conger myriaster (Conger eel)	Rabbit	D-Gal, lactose	30(13)	S-type	(e)
	Human A,B,O; rabbit	D-Gal, lactose, asialofetuin	25(12.5)	S-type	(f)
Misgurnus anguillicaudatus (Loach)	Human A,B,O; rabbit	Lactose, mucin, asialofetuin	>300		(c)
	Human A,B; rabbit	Lactose, D-Man, D-GlcNAc, NeuNAc	>300(40, 41)	S-type	(g)

Tachysurus australis (Australian catfish)	Human A,B,O; rabbit		>200		(h)
Arius thalassinus (Arabian Gulf catfish)	Human A,B,O; rabbit	D-Gal, lactose	200	S-type	(i)
Plecoglossus altivelis (Ayu fish)	Rabbit	D-Gal, D-Man, L-Rha, D-Fuc			(j)
Genypterus blacodes (Cusk-eel)	Human A,B,O; rabbit	D-Glc, D-Man, D-Glc*N*Ac	32(8)	C-type	(k)
Repomucenus richardsonii (Dragonet)	Rabbit	D-Gal, lactose, asialofetuin	48(12)	S-type	(l)
Lophopsetta maculata (Windowpane flounder)	Rabbit	Neu*N*Ac	68(16)	S-type	(m)

[a] Abbreviations: MW, molecular weight; D-Gal, D-galactose; D-Man, D-mannose; D-Glc*N*Ac, *N*-acethl-D-glucosamine; Neu*N*Ac, *N*-acetylneuraminic acid; L-Rha, L-rhamnose; D-Fuc, D-fucose; D-Glc, D-glucose.

[b] (a), Spitzer *et al.* (1976); (b), Suzuki (1985); (c), Shiomi *et al.* (1988); (d), Randall *et al.* (1981); (e), Kamiya *et al.* (1988); (f), Shiomi *et al.* (1989); (g), Goto-Nance *et al.* (1995); (h), Di Conza (1970); (i), Al-Hassan *et al.* (1986); (j), Itami *et al.* (1993); (k), Oda *et al.* (1984a); (l), Shiomi *et al.* (1990); (m), Kamiya and Shimizu (1980).

a high affinity to the LPS purified from the *V. anguillarum* cell wall (Itami *et al.*, 1992).

Some plant lectins such as concanavalin A and carnin are known to inhibit the cleavage of starfish (*Asterina pectinifera*) at the blastula stage (Hori *et al.*, 1987). Conger eel mucus lectins (congerins) at a concentration of 25 μg protein/ml inhibited the normal embryonic development of the starfish and lysed the fertilized eggs (Kamiya *et al.*, 1988).

The presence of skin mucus lectins may provide some inhibition of bacterial growth, although the connection between agglutination activity and growth inhibition of bacteria has yet to be clarified. More information regarding chemical and biological properties is necessary in order to determine the role of skin mucus lectins in fishes.

3. Serum Lectins

By the middle of the 1990s, lectins (hemagglutinins) had been isolated from the sera of hagfish, lamprey, freshwater eels, freshwater catfish, salmonid fishes, and murrel fish (Table VII). The biological functions of fish serum lectins are largely unknown. Noteworthy, however, was a mannan-binding lectin which was isolated from Japanese eel serum (Gercken and Renwrantz, 1994). In mammals, it is known that members of this group (mannose- or mannan-binding proteins) can serve as opsonins for the phagocytosis of pathogens (Ezekowitz *et al.*, 1988; Kuhlman *et al.*, 1989), inhibit the infection of H9 lymphoblasts by HIV (Ezekowitz *et al.*, 1989), and be involved in the activation of the classical complement pathway (Ohta *et al.*, 1990). Although the biological functions of the eel mannan-binding lectin have yet to be elucidated, it is tempting to speculate that this lectin is involved in immunological defense reactions similar to those in mammals (Kéry, 1991).

B. Physiochemical Properties of Fish Lectin

In the last two decades, a number of fish lectins have been isolated from various fish species, and their physicochemical properties have been determined, including sugar-binding specificities in most cases (Tables V, VI, and VII). The subunit molecular weights of most fish lectins are within the range of those determined for vertebrate lectins (Harrison, 1991).

The L-rhamnose-specific lectins, which are common among the great majority of fish egg lectins, appear to specifically agglutinate human type B and rabbit erythrocytes (Table V), while L-fucose/D-glucose-specific lectins in fish eggs and serum preferentially agglutinate human type O erythrocytes (Tables V and VII). On the other hand, D-galactose/lactose-specific lectins, which are predominant in the skin mucus, seem to aggluti-

Table VII

Sugar Specificities, Molecular Weights, and Metal Requirement of Fish Serum Lectins[a]

Species	Erythrocyte	Sugar specificity	MW (kDa) native (subunit)	Metal requirement	Source[b]
Eptatretus stoutii (Pacific hagfish)	Sheep		>200		(a)
Petromyzon marinus (Sea lamprey)	Sheep, rabbit		>200	+	(b)
Anguilla anguilla (European eel)	Human O	L-Fuc	50(23)		(c)
	Human O	L-Fuc	80(20)		(d)
	Human O	L-Fuc	(18)	+	(e)
	Human O	L-Fuc, galactan	121 (15)	−	(f)
	Rabbit	Mannan	246(24)		(f)
Anguilla rostrata (American eel)	Human O	L-Fuc	123(10)		(g, h)
Anguilla japonica (Japanese eel)	Human O	L-Fuc	140		(i)
Tandanus tandanus (Freshwater catfish)	Human O		>200		(j)
Oncorhynchus mykiss (Rainbow trout)	Rabbit		>200		(k)
Salmo trutta (Brown trout)	Human A		>200		(l)
Channa punctatus (Murrel fish)	Human A	D-Gal*N*Ac, Forssman-glycolipid	140(68)	C-type	(m, n)
Pleuronectes platessa (Plaice)	Rabbit	D-Man, D-Gal*N*Ac	<232		(o)

[a] Abbreviations: MW, molecular weight; L-Fuc, L-fucose; D-Gal*N*Ac, *N*-acetyl-D-galactosamine.

[b] (a), Linthicum and Hildemann (1970); (b), Gewurz *et al.* (1966); (c), Horejsí and Kocourek (1978); (d), Kelly (1984); (e) Uhlenbruck *et al.* (1982); (f), Gercken and Renwrantz (1994); (g), Springer and Desai (1971); (h), Bezkorovainy *et al.* (1971); (i), Matsumoto and Osawa (1974); (j), Baldo (1973); (k), Hodgins *et al.* (1967; 1973); (l), Holt and Anstee (1975); (m), Manihar and Das (1990); (n), Manihar *et al.* (1990); (o), Bly *et al.* (1986).

nate human ABO and rabbit erythrocytes (Table VII). Fish lectins cannot be classified as to C- or S-type or any other category because only limited work has been undertaken in this area (Alexander and Ingram, 1992), however, certain teleost lectins can be tentatively assigned classes based on those present in higher vertebrates (Tables V, VI, and VII). The S-type molecules appear to be the more dominant type present in teleost eggs and mucus.

The complete amino acid sequence of conger eel skin mucus lectin has been determined by Muramoto and Kamiya (1992). The sequence showed only 33% identity with that of electric eel lectin (Paroutaud *et al.*, 1987). It is difficult to postulate that these amino acid substitutions occurred after the divergence of the two species. However these lectins may be considered isolectins because they are produced in different tissues, that is, the electric eel lectin is produced in the electric organ while the conger eel lectin is produced in the skin (Muramoto and Kamiya, 1992).

C. Pharmacological Action of Fish Egg Lectins

Recently, the pharmacological actions of fish egg lectins have attracted attention. Some egg lectins are mitogens for mammalian lymphocytes (Krajhanzl *et al.*, 1978b; Oda *et al.*, 1984b; Krajhanzl and Kocourek, 1986; Licastro *et al.*, 1991), inhibit protein synthesis (Barbieri *et al.*, 1979; Krajhanzl, 1990), stimulate the release of interleukin 2 (IL-2) from human lymphocytes (Licastro *et al.*, 1991), and enhance the production of prostanoid derivatives such as thromboxane B_2 (TXB_2) and prostaglandin E_2 (PGE_2) by human monocytes (Licastro *et al.*, 1991).

VIII. OTHER SUBSTANCES

The serum and/or mucus of certain fish species contain substances such as hemolysin, proteinase, α_2-macroglobulin, chitinase, α-precipitin, caeruloplasmin, and metallothionein. There is little information concerning these substances, but they may also be involved in the nonspecific defense mechanisms of fish against pathogens.

A. Hemolysin

European eel serum contains a thermo-stable anti-human ABO erythrocyte lysin(s) (Salák *et al.*, 1975). By contrast, the skin mucus of the Japanese eel possesses a thermo-labile hemolysin (290 kDa) with a high degree of specificity toward rabbit erythrocytes (Suzuki, 1985).

B. Proteinase

The skin mucus of rainbow trout, Atlantic salmon, charr, cod, coalfish, plaice, and redfish contains proteases with trypsin-like activity. The mucus protease from rainbow trout was able to destroy Gram-negative bacteria such as *V. anguillarum* (Hjelmeland *et al.,* 1983). In Atlantic salmon, trypsin was detected not only in the epidermal cell layer, but also in the epithelial cell layers of the gills and the intestine (Braun *et al.,* 1990).

C. α_2-Macroglobulin

Fish α_2-macroglobulin (α_2M) has a molecular weight of about 360 kDa which is about half the size of human α_2M (725 kDa) (Starkey and Barrett, 1982). In rainbow trout, α_2M has been shown to inhibit the proteolytic activity of *A. salmonicida* protease (Ellis, 1987), and the difference in α_2M inhibitory activity between two different trout species (rainbow trout and brook trout) has also been found to be correlated with their ability to survive *A. salmonicida* infection (Freedman, 1991), thus suggesting that α_2M may play a role in defense against this infection.

D. Chitinase

Chitinase hydrolyses *N*-acetylglucosamine (Glc*N*Ac) tetramers and higher oligosaccharides including chitin, thus producing Glc*N*Ac dimers (*N,N'*-diacetylchitobiose). The function of chitinase in the serum and other fish tissues is uncertain. It has been suggested that chitinase may function in the defense of fish against chitin-containing pathogens and parasites (Fänge *et al.,* 1976, 1980; Lundblad *et al.,* 1979; Ingram, 1980; Manson *et al.,* 1992), however its distribution pattern in fish tissues does not necessarily support such a hypothesis (Lindsay, 1986, 1987).

E. α-Precipitin

α-Precipitin (a group of 13 proteins) isolated from Atlantic salmon serum reacted with carbohydrates and glycoproteins from several species of fungi and with soluble starch, in particular, amylopectin. The biological function of α-precipitin is unknown, but in the light of its concentration changes during the course of the outbreak of ulcerative dermal necrosis (UDN) that occurred between 1966 and 1975, it would appear to play some defensive role (Alexander and Ingram, 1992).

IX. CONCLUSIONS AND RECOMMENDATIONS FOR FUTURE RESEARCH

It is apparent from the evidence presented that fish have nonspecific humoral defense factors (HDFs) which, for the most part, are not fundamentally different from those present in higher vertebrates. However, it should be realized that a great deal of the currently available information is based upon the study of a relatively small number of fish species, because more attention has been directed toward species that are economically important. It would be appropriate to conduct more studies on other species such as cyclostomes and cartilagenous fishes.

Although fish HDFs resemble their mammalian counterparts with regard to functional and physicochemical properties, they also differ from them in several aspects. For example; (1) the skin mucus and/or eggs of some fish species contain lysozyme, complement, CRP, lectin, and/or hemolysin, which play a role in suppressing bacterial infestation on the body surface and in preventing the mother-to-progeny transmission of some pathogens; (2) lysozyme has substantial antimicrobial activity against Gram-positive and Gram-negative bacteria; (3) fish complement retains its activity even at low temperatures (0–4°C); and (4) the activity of the alternative complement pathway of fish is extremely high compared with that of mammals, indicating that this pathway plays an important role in the fish defense system.

There is a need for more information about the influence of water temperature, season, stress, and sexual maturation of fish HDFs. The ontogeny of fish HDFs and the recognition of complement receptors (CR1, CR2, and C5a receptor) on phagocytic cells still need to be elucidated. It would be interesting to investigate whether fish CRP can activate the complement system, as in higher vertebrates, and whether fish lectins act as opsonins to promote phagocytosis, as in invertebrates. Fish show large intraspecies variations in their susceptibility to microbial infections. The mechanisms underlying such variations should be investigated in relation to the genetic polymorphisms of HDFs. If different degrees of susceptibility to a certain fish disease were shown to be correlated with the different genotypes of HDFs, selective breeding of fish that are resistant to a specific disease would become both feasible and promising. The employment of modern tools, such as the recombinant DNA technique, would also be pertinent either in the bacterial production of HDFs for therapeutic or prophylactic use, or in transgenic experiments aimed at creating enhanced or optimized activity of HDFs in fish to help them to withstand difficult culture conditions. It is hoped that this chapter will provide a convenient starting point for such studies.

REFERENCES

Agrawal, A., and Bhattacharya, S. (1989). Binding property of rat and *Limulus* C-reactive proteins (CRP) to mercury. *Experientia* **45**, 567–570.

Aisen, P. (1980). The transferrins. *In* "Iron in Biochemistry and Medicine. II" (A. Jacobs and M. Worwood, eds.), pp. 87–129. Academic Press, London.

Aisen, P., Leibman, A., and Sia, C.-L. (1972). Molecular weight and subunit structure of hagfish transferrin. *Biochemistry* **11,** 3461–3464.

Alexander, J. B., and Ingram, G. A. (1992). Noncellular nonspecific defence mechanisms of fish. *Annu. Rev. Fish Dis.* **2,** 249–279.

Al-Hassan, J. M., Thomson, M., Summers, B., and Criddle, R. S. (1986). Purification and properties of a hemagglutination factor from Arabian gulf catfish (*Arius thalassinus*) epidermal secretion. *Comp. Biochem. Physiol.* **85B,** 31–39.

Anstee, D. J., Holt, P. D. J., and Pardoe, G. I. (1973). Agglutinins from fish ova defining blood groups B and P. *Vox Sang.* **25,** 347–360.

Avtalion, R. R. (1981). Environmental control of the immune response in fish. *CRC Crit. Rev. Environ. Contr.* **11,** 163–188.

Balakhnin, I. A., and Dudka, I. A. (1990). The interaction of fish mycopathogens with anti-B lectin of *Salmo gardneri* eggs. *Mikol. Fitopatol.* **24,** 224–228.

Balakhnin, I. A., and Dudka, I. A., and Isaeva, N. M. (1990). Testing of fungi on their specific interaction with fish egg lectins. *Mikol. Fitopatol.* **24,** 416–420.

Baldo, B. A. (1973). "Natural" erythrocyte agglutinins in the serum of the Australian freshwater catfish, *Tandanus tandanus* Mitchell. II. Serum fractionation studies. *Immunology* **25,** 813–826.

Baldo, B. A., and Fletcher, T. C. (1973). C-reactive protein-like precipitins in plaice. *Nature* **246,** 145–146.

Baldo, B. A., and Fletcher, T. C. (1975). Phylogenetic aspects of hypersensitivity: immediate hypersensitivity reactions in flatfish. *In* "Immunologic Phylogeny" (W. H. Hildemann and A. A. Benedict, eds.), pp. 365–372. Plenum Press, New York.

Baltz, M., de Beer, F. C., Feinstein, A., Munn, E. A., Milstein, C. P., Fletcher, T. C., March, J. F., Taylor, J., Bruton, C., Clamp, J. R., Davies, A. J. S., and Pepys, M. B. (1982). Phylogenetic aspects of C-reactive protein and related proteins. *Ann. N.Y. Acad. Sci.* **389,** 49–75.

Barbieri, L., Lorenzoni, E., and Stirpe, F. (1979). Inhibition of protein synthesis in vitro by a lectin from *Momordica charantia* and by other hemagglutinins. *Biochem. J.* **182,** 633–635.

Baudouy, A. M. (1978). Relation hôte-virus au cours de la virémie printanière de la carpe. *C. R. Acad. Sci. Paris* **286D,** 1225–1228.

Beasley, A. R., and Sigel, M. M. (1967). Interferon production in cold-blooded vertebrates. *In Vitro* **3,** 154–165.

Beasley, A. R., Sigel, M. M., and Clem, L. W. (1966). Latent infection in marine fish cell tissue cultures. *Proc. Soc. Exp. Biol. Med.* **121,** 1169–1174.

Becherer, J. D., Alsenz, J., and Lambris, J. D. (1989). Molecular aspects of C3 interactions and structural/functional analysis of C3 from different species. *Curr. Top. Microbiol. Immunol.* **153,** 45–72.

Bezkorovainy, A., Springer, G. F., and Desai, P. R. (1971). Physiochemical properties of the eel anti-human blood group H(O) antibody. *Biochemistry* **10,** 3761–3764.

Bildfell, R. J., Markham, R. J. F., and Johnson, G. R. (1992). Purification and partial characterization of a rainbow trout egg lectin. *J. Aquatic Animal Health* **4,** 97–105.

Bly, J. E., and Clem, L. W. (1991). Temperature-mediated processes in teleost immunity: *in vitro* immunosuppression induced by *in vivo* low temperature in channel catfish. *Vet. Immunol. Immunopathol.* **28,** 365–377.

Bly, J. E., Grimm, A. S., and Morris, I. G. (1986). Transfer of passive immunity from mother to young in a teleost fish: haemagglutinating activity in the serum and eggs of plaice. *Pleuronectes platessa* L. *Comp. Biochem. Physiol.* **84A,** 309–313.

Bobák, P., Stratil, A., and Valenta, M. (1984). A comparison of molecular weights of transferrins of various vertebrates. *Comp. Biochem. Physiol.* **79B,** 113–117.
Boffa, G. A., Fine, J. M., Drilhon, A., and Amouch, P. (1967). Immunoglobulins and transferrin in marine lamprey sera. *Nature* **214,** 700–702.
Bower, S. M., and Evelyn, T. P. T. (1988). Acquired and innate resistance to the haemoflagellate *Cryptobia salmositica* in sockeye salmon. (*Oncorhynchus nerka*). *Dev. Comp. Immunol.* **12,** 749–760.
Braun, R., Arnesen, J. A., Rinne, A., and Hjelmeland, K. (1990). Immunohistological localization of trypsin in mucus-secreting cell layers of Atlantic salmon, *Salmo salar* L. *J. Fish Dis.* **13,** 233–238.
Claus, D. R., Siegel, J., Petras, K., Osmand, A. P., and Gewurz, H. (1977). Interactions of C-reactive protein with the first component of human complement. *J. Immunol.* **119,** 187–192.
Clem, L. W., and Small, P. A. (1967). Phylogeny of immunoglobulin structure and function. I. Immunoglobulin of the lemon shark. *J. Exp. Med.* **125,** 893–920.
Collazos, M. E., Barriga, C., and Ortega, E. (1994). Optimum conditions for the activation of the alternative complement pathway of a cyprinid fish (*Tinca tinca* L.). Seasonal variations in the titres. *Fish Shellfish Immunol.* **4,** 499–506.
Culbreath, L., Smith, S. L., and Obenauf, S. D. (1991). Alternative complement pathway activity in nurse shark serum. *Am. Soc. Zool.* **31,** 131A.
Dahmen, A., Kaidoh, T., Zipfel, P. F., and Gigli, I. (1994). Cloning and characterization of a cDNA representing a putative complement-regulatory plasma protein from barred sand bass (*Parablax neblifer*). *Biochem. J.* **301,** 391–397.
Dautigny, A., Prager, E. M., Pham-Dinh, D., Jollés, J., Pakdel, F., Grinde, B., and Jollés, P. (1991). cDNA and amino acid sequences of rainbow trout (*Oncorhynchus mykiss*) lysozymes and their implications for the evolution of lysozyme and lactalbumin. *J. Mol. Evol.* **32,** 187–198.
Day, N. K. B., Good, R. A., Finstad, J., Johannsen, R., Pickering, R. J., and Gewurz, H. (1970). Interactions between endotoxic lipopolysaccharides and the complement system in the sera of lower vertebrates. *Proc. Soc. Exp. Med. Biol.* **133,** 1397–1401.
de Kinkelin, P., and Dorson, M. (1973). Interferon production in rainbow trout (*Salmo gairdneri,* Richardson) experimentally infected with Egtved virus. *J. Gen. Virol.* **19,** 125–127.
de Kinkelin, P., and Le Barre, M. (1974). Nécrose hématopoietique infectieuse des salmonidés: production d'interféron circulant induite après l'infection expérimentale de la truite arc-en-ciel (*Salmo gairdneri* Richardson). *C.R. Acad. Sci. Paris* **279D,** 445–448.
de Kinkelin, P., Baudouy, A. M., and Le Berre, M. (1977). Réaction de la truite fario (*Salmo trutta* L.) et arc-en-ciel (*Salmo gairdneri* Richardson, 1836) à l'infection par un nouveau rhabdovirus. *C. R. Acad. Sci. Paris* **284D,** 401–404.
de Kinkelin, P., Dorson, M., and Hattenberger-Baudouy, A. M. (1982). Interferon synthesis in trout and carp after viral infection. *Dev. Comp. Immunol.* **Suppl. 2,** 167–174.
De Sena, J., and Rio, G. J. (1975). Partial purification and characterization of RTG-2 fish cell interferon. *Infect. Immun.* **11,** 815–822.
Di Conza, J. J. (1970). Some characteristics of natural haemagglutinins found in serum and mucus of the catfish, *Tachysurus australis. Aust. J. Exp. Biol. Med. Sci.* **48,** 515–523.
Dorson, M., and de Kinkelin, P. (1974). Mortalité et production d'interféron circulant chez la truite arc-en-ciel après infection expérimentale avec le virus d'Egtved: influence de la température. *Ann. Rech. Vétér.* **5,** 365–372.
Dorson, M., Barde, A., and de Kinkelin, P. (1975). Egtved virus induced rainbow trout serum interferon: some physicochemical properties. *Ann. Microbiol.* (*Inst. Pasteur*) **126B,** 485–489.

Dorson, M., Torchy, C., and Michel, C. (1979). Rainbow trout complement fixation used for titration of antibodies against several pathogens. *Ann. Vet. Res.* **10,** 529–534.

Dorson, M., De Kinkelin, P., and Torchy, C. (1992). Interferon synthesis in rainbow trout fry following infection with infectious pancreatic necrosis virus. *Fish Shellfish Immunol.* **2,** 311–313.

Drickamer, K. (1988). Two distinct classes of carbohydrate-recognition domains in animal lectins. *J. Biol. Chem.* **263,** 9557–9560.

Eaton, W. D. (1990). Antiviral activity in four species of salmonids following exposure to poly inosinic:cytidylic acid. *Dis. Aquat. Org.* **9,** 193–198.

Ellis, A. E. (1981). Nonspecific defense mechanisms in fish and their role in disease processes. *Dev. Biol. Stand.* **49,** 337–352.

Ellis, A. E. (1987). Inhibition of the *Aeromonas salmonicida* extracellular protease by α_2-macroglobulin in the serum of rainbow trout. *Microb. Pathogen.* **3,** 167–177.

Ezekowitz, R. A. B., Day, L. E., and Herman, G. A. (1988). A human mannose-binding protein is an acute-phase reactant that shares sequence homology with other vertebrate lectins. *J. Exp. Med.* **167,** 1034–1046.

Ezekowitz, R. A. B., Kuhlman, M., Groopman, J. E., and Byrn, R. A. (1989). A human serum mannose-binding protein inhibits *in vitro* infection by the human immunodeficiency virus. *J. Exp. Med.* **169,** 185–196.

Fänge, R., Lundblad, G., and Lind, J. (1976). Lysozyme and chitinase in blood and lymphomyeloid tissues of marine fish. *Mar. Biol.* **36,** 277–282.

Fänge, R., Lundblad, G., Slettengren, K., and Lind, J. (1980). Glycosidases in lymphomyeloid (hematopoietic) tissues of elasmobranch fish. *Comp. Biochem. Physiol.* **67B,** 527–532.

Fletcher, T. C. (1982). Non-specific defence mechanisms of fish. *Dev. Comp. Immunol.* **Suppl. 2,** 123–132.

Fletcher, T. C., and Baldo, B. A. (1974). Immediate hypersensitivity responses in flatfish. *Science* **185,** 360–361.

Fletcher, T. C., and Baldo, B. A. (1976). C-reactive protein-like precipitins in lumpsucker (*Cyclopterus lumpus* L.) gametes *Experientia* **32,** 1199–1200.

Fletcher, T. C., and Grant, P. T. (1968). Glycoproteins in the external mucous secretions of the plaice, *Pleuronectes platessa,* and other fishes. *Biochem. J.* **106,** 12.

Fletcher, T. C., and White, A. (1973). Lysozyme activity in the plaice (*Pleuronectes platessa* L.). *Experientia* **29,** 1283–1285.

Fletcher, T. C., and White, A. (1976). The lysozyme of the plaice *Pleuronectes platessa* L. *Comp. Biochem. Physiol.* **55B,** 207–210.

Fletcher, T. C., White, A., and Baldo, B. A. (1977). C-reactive protein-like precipitin and lysozyme in the lumpsucker *Cyclopterus lumpus.* L. during the breeding season. *Comp. Biochem. Physiol.* **57B,** 353–357.

Freedman, S. J. (1991). The role of alpha$_2$-macroglobulin in furunculosis: a comparison of rainbow trout and brook trout. *Comp. Biochem. Physiol.* **98B,** 549–553.

Fujii, T., and Nakamura, T., Sekizawa, A., and Tomonaga, S. (1992). Isolation and characterization of a protein from hagfish serum that is homologous to the third component of the mammalian complement system. *J. Immunol.* **148,** 117–123.

Fujino, K., and Kang, T. (1968). Transferrin groups of tunas. *Genetics* **59,** 79–91.

Galabov, A. S. (1973). Interferonogenesis and phylogenesis. *Bull. Inst. Pasteur* **71,** 233–247.

Gercken, J., and Renwrantz, L. (1994). A new mannan-binding lectin from the serum of the eel (*Anguilla anguilla* L.): isolation, characterization and comparison with the fucose-specific serum lectin. *Comp. Biochem. Physiol.* **108B,** 449–461.

Gewurz, H., Eugster, G., Muschel, L. H., Finstad, J., and Good, R. A. (1965). Development of the complement system. *Fed. Proc.* **24,** 504.

Gewurz, H., Finstad, J., Muschel, L. H., and Good, R. A. (1966). Phylogenetic inquiry into the origins of the complement system. *In* "Phylogeny of Immunity" (R. T. Smith, P. A. Miescher, and R. A. Good, eds.), pp. 105–117. University of Florida Press, Gainesville.

Giclas, P. C., Morrison, D. C., Curry, B. J., Laurs, R. M., and Ulevitch, R. J. (1981). The complement system of the albacore tuna, *Thunnus alalunga*. *Dev. Comp. Immunol.* **5,** 437–447.

Gigli, I., and Austen, K. F. (1971). Phylogeny and function of the complement system. *Annu. Rev. Microbiol.* **25,** 309–332.

Glynn, A. A. (1969). The complement lysozyme sequence in immune bacteriolysis. *Immunology* **16,** 463–471.

Goldstein, I. J., Hughes, R. C., Monsigny, M., Osawa, T., and Sharon, N. (1980). What should be called a lectin? *Nature* **285,** 66.

Gorski, J. P., Hugli, T. E., and Müller-Eberhard, H. J. (1981). Characterization of human C4a anaphylatoxin. *J. Biol. Chem.* **256,** 2707–2711.

Got, R., Font, J., and Goussault, Y. (1967). Etude sur une transferrine de selacien, la grande roussette (*Scyllium stellare*). *Comp. Biochem. Physiol.* **23,** 317–327.

Goto-Nance, R., Watanabe, Y., Kamiya, H., and Ida, H. (1995). Characterization of lectins from the skin mucus of the loach *Misgurnus anguillicaudatus*. *Fisheries Sci.* **61,** 137–140.

Gotschlich, E. C., and Edelman, G. H. (1967). Binding properties and specificity of C-reactive protein. *Proc. Natl. Acad. Sci. USA* **57,** 706–712.

Graham, S., and Secombes, C. J. (1988). The production of a macrophage-activating factor from rainbow trout *Salmo gairdneri* leucocytes. *Immunology* **65,** 293–297.

Graham, S., and Secombes, C. J. (1990). Do fish lymphocytes secrete interferon-γ? *J. Fish Biol.* **36,** 563–573.

Gravell, M., and Malsberger, R. G. (1965). A permanent cell line from the fathead minnow (*Pimephales promelas*). *Ann. N.Y. Acad. Sci.* **126,** 555–565.

Griffin, B. R. (1984). Random and directed migration of trout (*Salmo gairdneri*) leukocytes: activation by antibody, complement, and normal serum components. *Dev. Comp. Immunol.* **8,** 589–597.

Grinde, B. (1989). Lysozyme from rainbow trout, *Salmo gairdneri* Richardson, as an antibacterial agent against fish pathogens. *J. Fish Dis.* **12,** 95–104.

Grinde, B., Lie, Ø., Poppe, T., and Salte, R. (1988a). Species and individual variation in lysozyme activity in fish of interest in aquaculture. *Aquaculture* **68,** 299–304.

Grinde, B., Jollès, J., and Jollès, P. (1988b). Purification and characterization of two lysozymes from rainbow trout (*Salmo gairdneri*). *Eur. J. Biochem.* **173,** 269–273.

Hanley, P. J., Hook, J. W., Raftos, D. A., Gooley, A. A., Trent, R., and Raison, R. L. (1992). Hagfish humoral defense protein exhibits structural and functional homology with mammalian complement components. *Proc. Natl. Acad. Sci. USA* **89,** 7910–7914.

Hara, A. (1984). Purification and some physicochemical characterization of chum salmon transferrin. *Bull. Japan. Soc. Sci. Fish.* **50,** 713–719.

Harrell, L. W., Etlinger, H. M., and Hodgins, H. O. (1976). Humoral factors important in resistance of salmonid fish to bacterial disease. II. Anti-*Vibrio anguillarum* activity in mucus and observations on complement. *Aquaculture* **7,** 363–370.

Harrison, L. (1991). Soluble β-galactoside-binding lectins in vertebrates. *Lectin Rev.* **1,** 17–39.

Hayman, J. R., Bly, J. E., Levine, R. P., and Lobb, C. J. (1992). Complement deficiencies in channel catfish (*Ictalurus punctatus*) associated with temperature and seasonal mortality. *Fish Shellfish Immunol.* **2,** 183–192.

Hershberger, W. K. (1970). Some physicochemical properties of transferrins in brook trout. *Trans. Am. Fish. Soc.* **99,** 207–218.

Hirono, I., Uchiyama, T., and Aoki, T. (1995). Cloning, nucleotide sequence analysis, and characterization of cDNA for medaka (*Orizias latipes*) transferrin. *J. Mar. Biotechnol.* **2,** 193–198.

Hjelmeland, K., Christie, M., and Raa, J. (1983). Skin mucus protease from rainbow trout, *Salmo gairdneri* Richardson, and its biological significance. *J. Fish Biol.* **23,** 13–22.

Hodgins, H. O., Weiser, R. S., and Ridgway, G. J. (1967). The nature of antibodies and the immune response in rainbow trout (*Salmo gairdneri*). *J. Immunol.* **99,** 534–544.

Hodgins, H. O., Wendling, F. L., Braaten, B. A., and Weiser, R. S. (1973). Two molecular species of agglutinins in rainbow trout (*Salmo gairdneri*) serum and their relation to antigenic exposure. *Comp. Biochem. Physiol.* **45B,** 975–977.

Holloway, Jr., H. L., Shoemaker, C. A., and Ottinger, C. A. (1993). Serum lysozyme levels in paddlefish and walleye. *J. Aquat. Anim. Health* **5,** 324–326.

Holt, P. D. J., and Anstee, D. J. (1975). A natural anti-A agglutinin in the serum of the brown trout (*Salmo trutta*). *Vox Sang.* **29,** 286–291.

Honda, A., Kodama, H., Moustafa, M., Yamada, F., Mikami, T., and Izawa, H. (1985). Response of rainbow trout immunized with formalin-killed *Vibrio anguillarum:* activity of phagocytosis of fish macrophages and opsonizing effect of antibody. *Fish Pathol.* **20,** 395–402.

Honda, A., Kodama, H., Moustafa, M., Yamada, F., Mikami, T., and Izawa, H. (1986). Phagocytic activity of macrophages of rainbow trout against *Vibrio anguillarum* and the opsonising effect of antibody and complement. *Res. Vet. Sci.* **40,** 328–332.

Horejsí, V., and Kocourek, J. (1978). Studies on lectins. XXXVI. Properties of some lectins prepared by affinity chromatography on *O*-glycosyl polyacrylamide gels. *Biochim. Biophys. Acta* **538,** 299–315.

Hori, K., Matsuda, H., Miyazawa, K., and Ito, K. (1987). A mitogenic agglutinin from the red alga *Carpopeltis flabellata. Phytochemistry* **26,** 1335–1338.

Hosoi, K., Utsumi, J., Kitagawa, T., Shimizu, H., and Kobayashi, S. (1988). Structural characterization of fibroblast human interferon-beta1. *J. Interferon Res.* **8,** 375–384.

Hosono, M., and Nitta, K. (1993). Fish roe rhamnose-binding lectins. Possibilities for a new sugar-binding domain structure. *Annu. Rep. Tohoku Coll. Pharm.* **40,** 21–43.

Hosono, M., Kizaki, K., Nitta, K., and Takayanagi, Y. (1992). Partial purification and sugar specificity of Japanese dace (*Leuciscus hakonensis*) roe lectin. *Annu. Rep. Tohoku Coll. Pharm.* **39,** 265–275.

Hosono, M., Kawauchi, H., Nitta, K., Takayanagi, Y., Shiokawa, H., Mineki, R., and Murayama, K. (1993a). Purification and characterization of *Silurus asotus* (catfish) roe lectin. *Biol. Pharm. Bull.* **16,** 1–5.

Hosono, M., Kawauchi, H., Nitta, K., Takayanagi, Y., Shiokawa, H., Mineki, R., and Murayama, K. (1993b). Three rhamnose-binding lectins from *Osmerus eperlanus mordax* (olive rainbow smelt) roe. *Biol. Pharm. Bull.* **16,** 239–243.

Hurlimann, J., Thorbecke, G. J., and Hochwald, G. M. (1966). The liver as the site of C-reactive protein formation. *J. Exp. Med.* **123,** 365–378.

Hyder Smith, S., and Jensen, J. A. (1986). The second component (C2n) of the nurse shark complement system: purification, physicochemical characterization and functional comparison with guinea pig C4. *Dev. Comp. Immunol.* **10,** 191–206.

Iacono, V. J., MacKay, B. J., DiRienzo, S., and Pollock, J. J. (1980). Selective antibacterial properties of lysozyme for oral microorganisms. *Infect. Immun.* **29,** 623–632.

Iida, T., and Wakabayashi, H. (1983). Bactericidal reaction by the alternative pathway of fish complement. *Fish Pathol.* **18,** 77–83.

Iida, T., and Wakabayashi, H. (1988). Chemotactic and leukocytosis-inducing activities of eel complement. *Fish Pathol.* **23,** 55–58.

Iida, T., and Wakabayashi, H. (1993). Resistance of *Edwardsiella tarda* to opsonophagocytosis of eel neutrophils. *Fish Pathol.* **28,** 191–192.

Iida, T., Takahashi, K., and Wakabayashi, H. (1989). Decrease in the bactericidal activity of normal serum during the spawning period of rainbow trout. *Nippon Suisan Gakkaishi* **55,** 463–465.

Ingram, G. A. (1980). Substances involved in the natural resistence of fish to infection—a review. *J. Fish Biol.* **16,** 23–60.

Ingram, G. A. (1987). Haemolytic activity in the serum of brown trout, *Salmo trutta. J. Fish Biol.* **31(Suppl. A),** 9–17.

Ishiguro, H., Kobayashi, K., Suzuki, M., Titani, K., Tomonaga, S., and Kurosawa, Y. (1992). Isolation of a hagfish gene that encodes a complement component. *EMBO J.* **11,** 829–837.

Itami, T., Takehara, A., Nagano, Y., Suetsuna, K., Mitsutani, A., Takesue, K., and Takahashi, Y. (1992). Purification and characterization of lysozyme from a ayu skin mucus. *Nippon Suisan Gakkaishi* **58,** 1937–1944.

Itami, T., Ishida, Y., Endo, F., Kawazoe, N., and Takahashi, Y. (1993). Hemagglutinins in the skin mucus of ayu. *Fish Pathol.* **28,** 41–47.

Iwasaki, M., and Inoue, S. (1985). Structures of the carbohydrate units of polysialoglycoproteins isolated from the eggs of four species of salmonid fishes. *Glycoconjugate J.* **2,** 209–228.

Jamieson, A. (1990). A survey of transferrins in 87 teleostean species. *Anim. Genet.* **21,** 295–301.

Jenkins, J. A., and Ourth, D. D. (1990). Membrane damage to *Escherichia coli* and bactericidal kinetics by the alternative complement pathway of channel catfish. *Comp. Biochem. Physiol.* **97B,** 477–481.

Jenkins, J. A., and Ourth, D. D. (1993). Opsonic effect of the alternative complement pathway on channel catfish peripheral blood phagocytes. *Vet. Immunol. Immunopathol.* **39,** 447–459.

Jenkins, J. A., Rosell, R., Ourth, D. D., and Coons, L. B. (1991). Electron microscopy of bactericidal effects produced by the alternative complement pathway of channel catfish. *J. Aquat. Anim. Health* **3,** 16–22.

Jensen, J. A., Fuller, L., and Iglesias, E. (1973). The terminal components of the nurse shark (*Ginglymostoma cirratum*) C system. *J. Immunol.* **111,** 306–307.

Jensen, J. A., Festa, E., Smith, D. S., and Cayer, M. (1981). The complement system of the nurse shark: hemolytic and comparative characteristics. *Science* **214,** 566–569.

Jensen, L. B., and Koch, C. (1991). Genetic polymorphism of the rainbow trout (*Oncorhynchus mykiss*) complement component C3. *Fish Shellfish Immunol.* **1,** 237–242.

Jensen, L. B., and Koch, C. (1992). Use of monoclonal and polyclonal antibodies to analyse the degradation of rainbow trout C3 in inulin-activated serum. *Fish Shellfish Immunol.* **2,** 241–249.

Johnson, E., and Smith, P. (1984). Attachment and phagocytosis by salmon macrophages of agarose beads coated with human C3b and C3bi. *Dev. Comp. Immunol.* **8,** 623–630.

Jollès, P. (1969). Lysozymes: a chapter of molecular biology. *Angew. Chem. Int. Ed.* **8,** 227–239.

Jollès, P., and Jollès, J. (1984). What's new in lysozyme research? Always a model system, today as yesterday. *Mol. Cell. Biochem.* **63,** 165–189.

Jones, S. R. M., and Woo, P. T. K. (1987). The immune response of rainbow trout, *Salmo gairdneri* Richardson, to the haemoflagellate, *Cryptobia salmositica* Katz, 1951. *J. Fish Dis.* **10,** 395–402.

Kaidoh, T., and Gigli, I. (1987). Phylogeny of C4b-C3b cleaving activity: similar fragmentation patterns of human C4b and C3b produced by lower animals. *J. Immunol.* **139,** 194–201.

Kaidoh, T., and Gigli, I. (1989a). Phylogeny of regulatory proteins of the complement system. Isolation and characterization of a C4b/C3b inhibitor and a cofactor from sand bass plasma. *J. Immunol.* **142,** 1605–1613.

Kaidoh, T., and Gigli, I. (1989b). Phylogeny of the plasma regulatory proteins of the complement system. *Prog. Clin. Biol. Res.* **297,** 199–209.

Kamiya, H., and Shimizu, Y. (1980). Marine biopolymers with cell specificity. II. Purification and characterization of agglutinins from mucus of windowpane flounder *Lophopsetta maculata. Biochim. Biophys. Acta* **622,** 171–178.

Kamiya, H., Muramoto, K., and Goto, R. (1988). Purification and properties of agglutinins from conger eel, *Conger myriaster* (Brevoort), skin mucus. *Dev. Comp. Immunol.* **12,** 309–318.

Kamiya, H., Muramoto, K., Goto, R., Sakai, M., and Ida, H. (1990). Properties of a lectin in chum salmon ova. *Nippon Suisan Gakkaishi* **56,** 1139–1144.

Kaplan, M. H., and Volanakis, J. E. (1974). Interaction of C-reactive protein complexes with the complement system. I. Consumption of human complement associated with the reaction of C-reactive protein with pneumococcal C-polysaccharide and with choline phosphatides, lecithin and sphingomyelin. *J. Immunol.* **112,** 2135–2147.

Kawahara, I., and Kusuda, R. (1988a). Lysozyme activities of staple cultured fishes. *Nippon Suisan Gakkaishi* **54,** 581–584.

Kawahara, I., and Kusuda, R. (1988b). Properties of lysozyme activities in cultured eel. *Nippon Suisan Gakkaishi* **54,** 965–968.

Kelly, C. (1984). Physicochemical properties and N-terminal sequence of eel lectin. *Biochem. J.* **220,** 221–226.

Kelly, R. K., and Loh, P. C. (1973). Some properties of an established fish cell line from *Xiphophorus helleri* (red swordtail). *In Vitro* **9,** 73–80.

Kéry, V. (1991). Lectin-carbohydrate interactions in immunoregulation. *Int. J. Biochem.* **23,** 631–640.

Kiener, P. A., and Spitalny, G. L. (1987). Induction, production and purification of murine gamma interferon. *In* "Lymphokines and Interferons. A Practical Approach" (M. J. Clemens, A. G. Morris, and A. J. H. Gearing, eds.), pp. 15–28. IRL Press, Oxford.

Kilpatrick, D. C., van Driessche, E., and Bøg-Hansen, T. C. (1991). "Lectin Reviews," Vol. 1. Sigma Chemical Co., St. Louis.

Kilpatrick, J. M., and Volanakis, J. E. (1991). Molecular genetics, structure, and function of C-reactive protein. *Immunol. Res.* **10,** 43–53.

Klockars, M., and Roberts, P. (1976). Stimulation of phagocytosis by human lysozyme. *Acta Haemat.* **55,** 289–295.

Kodama, H., Yamada, F., Murai, T., Nakanishi, Y., Mikami, T., and Izawa, H. (1989). Activation of trout macrophages and production of CRP after immunization with *Vibrio anguillarum. Dev. Comp. Immunol.* **13,** 123–132.

Koppenheffer, T. L. (1987). Serum complement systems of ectothermic vertebrates. *Dev. Comp. Immunol.* **11,** 279–286.

Krajhanzl, A. (1990). Egg lectins of invertebrates and lower vertebrates: Properties and biological function. *Adv. Lectin Res.* **3,** 83–131.

Krajhanzl, A., and Kocourek, J. (1986). Fish cortical vesicle lectins—a new group of carbohydrate binding proteins: a review. *In* "Lectins: Biology, Biochemistry, Clinical Biochemistry" (T. C. Bøg-Hansen and E. Van Driessche, eds.), Vol. 5, pp. 257–275. Walter de Gruyter, Berlin.

Krajhanzl, A., Horejsí, V., and Kocourek, J. (1978a). Studies on lectins. XLI. Isolation and characterization of a blood group B specific lectin from the roe of the powan (*Coregonus lavaretus maraena*). *Biochim. Biophys. Acta.* **532,** 209–214.

Krajhanzl, A., Horejsí, V., and Kocourek, J. (1978b). Studies on lectins. XLII. Isolation, partial characterization and comparison of lectins from the roe of five fish species. *Biochim. Biophys. Acta* **532,** 215–224.

Krajhanzl, A., Nosek, J., Habrova, V., and Kocourek, J. (1984). An immunofluorescence study on the occurrence of endogenous lectins in the differentiating oocytes of silver carp (*Hypophthalmichthys molitrix* Valenc.) and tench (*Tinca tinca* L.). *Histochem. J.* **16,** 432–434.

Kuhlman, M., Joiner, K., and Ezekowitz, A. B. (1989). The human mannose-binding protein functions as an opsonin. *J. Exp. Med.* **169,** 1733–1745.

Kushner, I., and Feldmann, G. (1978). Control of the acute phase response. Demonstration of C-reactive protein synthesis and secretion by hepatocytes during acute inflammation in the rabbit. *J. Exp. Med.* **148,** 466–477.

Kusuda, R., and Fukunaga, T. (1987). Characterization of antibody dependent hemolytic activity in eel serum. *Nippon Suisan Gakkaishi* **53,** 2111–2115.

Kusuda, R., and Kitadai, N. (1992). Effects of water temperature on lysozyme activity of eels. *Suisanzoshoku* **40,** 453–456.

Kusuda, R., and Tanaka, T. (1988). Opsonic effect of antibody and complement on phagocytosis of *Streptococcus* sp. by macrophage-like cells of yellowtail. *Nippon Suisan Gakkaishi* **54,** 2065–2069.

Kusuda, R., Kawahara, I., and Hamaguchi, M. (1987). Activities and characterization of lysozyme in skin mucus extract, serum and kidney extract of yellowtail. *Nippon Suisan Gakkaishi* **53,** 211–214.

Kvingedal, A. M., Rørvik, K.-A., and Alestrøm, P. (1993). Cloning and characterization of Atlantic salmon (*Salmo salar*) serum transferrin cDNA. *Mol. Mar. Biol. Biotechnol.* **2,** 233–238.

Lambris, J. D. (1988). The multifunctional role of C3, the third component of complement. *Immunol. Today* **9,** 387–393.

Lambris, J. D., Lao, Z., Pang, J., and Alsenz, J. (1993). Third component of trout complement. cDNA cloning and conservation of functional sites. *J. Immunol.* **151,** 6123–6134.

Law, S. K. A., and Reid, K. B. M. (1988). "Complement," IRL Press, London, UK.

Legler, D. W., and Evans, E. E. (1967). Comparative immunology: hemolytic complement in elasmobranchs. *Proc. Soc. Exp. Biol. Med.* **124,** 30–34.

Legler, D. W., Evans, E. E., and Dupree, H. K. (1967). Comparative immunology: serum complement of fresh-water fishes. *Trans. Am. Fish. Soc.* **96,** 237–242.

Licastro, F., Barbieri, L., Krajhanzl, A., Kocourek, J., and Stripe, F. (1991). A cortical lectin from the oocytes of *Rutilus rutilus* stimulates mitogenic activity and release of soluble factors from human lymphocyte cultures and inhibits protein synthesis in a cell-free system. *Int. J. Biochem.* **23,** 101–105.

Lie, Ø., Evensen, Ø., Sørensen, A., and Frøysadal, E. (1989). Study on lysozyme activity in some fish species. *Dis. Aquat. Org.* **6,** 1–5.

Lindsay, G. J. H. (1986). The significance of chitinolytic enzymes and lysozyme in rainbow trout (*Salmo gairdneri*) defence. *Aquaculture* **51,** 169–173.

Lindsay, G. J. H. (1987). Seasonal activities of chitinase and chitobiase in the digestive tract and serum of cod, *Gadus morhua* (L.). *J. Fish Biol.* **30,** 495–500.

Linthicum, D. S., and Hildemann, W. H. (1970). Immunologic responses of Pacific hagfish. III. Serum antibodies to cellular antigens. *J. Immunol.* **105,** 912–918.

Lobb, C. J., and Hayman, J. R. (1989). Activation of complement by different immunoglobulin heavy chain isotypes of the channel catfish (*Ictalurus punctatus*). *Mol. Immunol.* **26,** 457–465.

Lundblad, G., Fänge, R., Slettengren, K., and Lind, J. (1979). Lysozyme, chitinase and exo-N-acetyl-β-D-glucosaminidase (NAGase) in lymphomyeloid tissue of marine fishes. *Mar. Biol.* **53,** 311–315.

MacDonald, R. D., and Kennedy, J. C. (1979). Infectious pancreatic necrosis virus persistently infects chinook salmon embryo cells independent of interferon. *Virology* **95,** 260–264.

Macey, D. J., Webb, J., and Potter, I. C. (1982). Iron levels and major iron binding proteins in the plasma of ammocoetes and adults of the Southern Hemisphere lamprey *Geotria australis.* Gray. *Comp. Biochem. Physiol.* **72A,** 307–312.

MacGillivray, R. T. A., Mendez, E., Sinha, S. K., Sutton, M. R., Lineback-Zins, J., and Brew, K. (1982). The complete amino acid sequence of human serum transferrin. *Proc. Natl. Acad. Sci. USA* **79,** 2504–2508.

Manihar, S. R., and Das, H. R. (1990). Isolation and characterization of a new lectin from plasma of fish *Channa punctatus. Biochim. Biophys. Acta* **1036,** 162–165.

Manihar, S. R., Varma, K. L., and Das, H. R. (1990). Studies on hemagglutinins (lectins) from plasma of murrel fish, family *Channidae. Indian J. Biochem. Biophys.* **27,** 464–470.

Manson, F. D. C., Fletcher, T. C., and Gooday, G. W. (1992). Localization of chitinolytic enzymes in blood of turbot, *Scophthalmus maximus,* and their possible roles in defense. *J. Fish Biol.* **40,** 919–927.

Matsumoto, I., and Osawa, T. (1974). Specific purification of eel serum and *Cytisus sessilifolius* anti-H hemagglutinins by affinity chromatography and their binding to human erythrocytes. *Biochemistry* **13,** 582–588.

Matsumaya, H., Tanaka, K., Nakao, M., and Yano, T. (1988a). Characterization of the alternative complement pathway of carp. *Dev. Comp. Immunol.* **12,** 403–408.

Matsuyama, H., Nakao, M., and Yano, T. (1988b). Compatibilities of antibody and complement among different fish species. *Nippon Suisan Gakkaishi* **54,** 1993–1996.

Matsuyama, H., Yano, T., Yamakawa, T., and Nakao, M. (1992). Opsonic effect of the third complement component (C3) of carp (*Cyprinus carpio*) on phagocytosis by neutrophils. *Fish Shellfish Immunol.* **2,** 69–78.

McKenzie, H. A., and White, F. H. Jr. (1991). Lysozyme and α-lactalbumin: structure, function, and interrelationships. *Adv. Protein Chem.* **41,** 173–315.

Menzel, B. W. (1976). Biochemical systematics and evolutionary genetics of the common shiner species group. *Biochem. Syst. Ecol.* **4,** 281–293.

Mitra, S., and Bhattacharya, S. (1992). Purification of C-reactive protein from *Channa punctatus* (Bloch). *Indian. J. Biochem. Biophys.* **29,** 508–511.

Miyata, T., Hayashida, H., Kikuno, R., Toh, H., and Kawade, Y. (1985). Evolution of interferon genes. *Interferon* **6,** 1–30.

Mochizuki, A., and Matsumiya, M. (1981). Lysozyme activity in organs of marine fishes. *Bull. Japan. Soc. Sci. Fish.* **47,** 1065–1068.

Mold, C., Du Clos, T. W., Nakayama, S., Edwards, K. M., and Gewurz, H. (1982). C-reactive protein reactivity with complement and effects on phagocytosis. *Ann. N. Y. Acad. Sci.* **389,** 251–262.

Morimoto, T., Iida, T., and Wakabayashi, W. (1988). Chemiluminescence of neutrophils isolated from peripheral blood of eel. *Fish Pathol.* **23,** 49–53.

Möck, A., and Peters, G. (1990). Lysozyme activity in rainbow trout, *Oncorhynchus mykiss* (Walbaum), stressed by handling, transport, and water pollution. *J. Fish Biol.* **37,** 873–885.

Møyner, K., Røed, K. H., Sevatdal, S., and Heum, M. (1993). Changes in nonspecific immune parameters in Atlantic salmon, *Salmo salar* L., induced by *Aeromonas salmonicida* infection. *Fish Shellfish Immunol.* **3,** 253–265.

Munn, C. B., Ishiguro, E. E., Kay, W. W., and Trust, T. J. (1982). Role of surface components in serum resistance of virulent *Aeromonas salmonicida. Infect. Immun.* **36,** 1069–1075.

Muona, M., and Soivio, A. (1992). Changes in plasma lysozyme and blood leucocyte levels of hatchery-reared Atlantic salmon (*Salmo salar* L.) and sea trout (*Salmo trutta* L.) during parr-smolt transformation. *Aquaculture* **106,** 75–87.

Murai, T., Kodama, H., Nakai, M., Mikami, T., and Izawa, H. (1990). Isolation and characterization of rainbow trout C-reactive protein. *Dev. Comp. Immunol.* **14,** 49–58.

Muramoto, K., and Kamiya, H. (1992). The amino acid sequence of a lectin from conger eel, *Conger myriaster,* skin mucus. *Biochim. Biophys. Acta* **1116,** 129–136.

Murray, C. K., and Fletcher, T. C. (1976). The immunohistochemical localization of lysozyme in plaice (*Pleuronectes platessa* L.) tissues. *J. Fish Biol.* **9,** 329–334.

Nakanishi, Y., Kodama, H., Murai, T., Mikami, T., and Izawa, H. (1991). Activation of rainbow trout complement by C-reactive protein. *Am. J. Vet. Res.* **52,** 397–401.

Nakao, M., Yano, T., and Matsuyama, H. (1988). Partial purification and characterization of the fourth component (C4) of carp complement. *Fish Pathol.* **23,** 243–250.

Nakao, M., Yano, T., and Matsuyama, H. (1989). Isolation of the third component of complement (C3) from carp serum. *Nippon Suisan Gakkaishi* **55,** 2021–2027.

Neeman, N., Lahav, M., and Ginsburg, I. (1974). The effect of leukocyte hydrolases on bacteria. II. The synergistic action of lysozyme and extracts of PMN, macrophages, lymphocytes, and platelets in bacteriolysis. *Proc. Soc. Exp. Biol. Med.* **146,** 1137–1145.

Newton, R. A., Raftos, D. A., Raison, R. L., and Geczy, C. L. (1994). Chemotactic responses of hagfish (vertebrata, agnatha) leucocytes. *Dev. Comp. Immunol.* **18,** 295–303.

Nonaka, M. (1994). Molecular analysis of the lamprey complement system. *Fish Shellfish Immunol.* **4,** 437–446.

Nonaka, M., and Takahashi, M. (1992). Complete complementary DNA sequence of the third component of complement of lamprey. *J. Immunol.* **148,** 3290–3295.

Nonaka, M., Yamaguchi, N., Natsuume-Sakai, S., and Takahashi, M. (1981a). The complement system of rainbow trout (*Salmo gairdneri*). I. Identification of the serum lytic system homologous to mammalian complement. *J. Immunol.* **126,** 1489–1494.

Nonaka, M., Natsuume-Sakai, S., and Takahashi, M. (1981b). The complement system of rainbow trout (*Salmo gairdneri*). II. Purification and characterization of the fifth component (C5). *J. Immunol.* **126,** 1495–1498.

Nonaka, M., Fujii, T., Kaidoh, T., Nonaka, M., Natsuume-Sakai, S., and Takahashi, M. (1984a). Purification of a lamprey complement protein homologous to the third component of the mammalian complement system. *J. Immunol.* **133,** 3242–3249.

Nonaka, M., Iwaki, M., Nakai, C., Nozaki, M., Kaidoh, T., Nonaka, M., Natsuume-Sakai, S., and Takahashi, M. (1984b). Purification of a major serum protein of rainbow trout (*Salmo gairdneri*) homologous to the third component of mammalian complement. *J. Biol. Chem.* **259,** 6327–6333.

Nonaka, M., Nonaka, M., Irie, M., Tanabe, K., Kaidoh, T., Natsuume-Sakai, S., and Takahashi, M. (1985). Identification and characterization of a variant of the third component of complement (C3) in rainbow trout (*Salmo gairdneri*). *J. Biol. Chem.* **260,** 809–815.

Nonaka, M., Takahashi, M., and Sasaki, M. (1994). Molecular cloning of a lamprey homologue of the mammalian MHC class III gene, complement factor B. *J. Immunol.* **152,** 2263–2269.

Nosek, J., Krajhanzl, A., and Kocourek, J. (1983a). Studies on lectins. LV. Subcellular localization of an endogenous lectin in fish oocyte. *In* "Lectins: Biology, Biochemistry, and Clinical Biochemistry" (T. C. Bøg-Hansen and A. Spengler, eds.), Vol. 3, pp. 453–459. Walter de Gruyter, Berlin.

Nosek, J., Krajhanzl, A., and Kocourek, J. (1983b). Studies on lectins. LVII. Immunofluorescence localization of lectin present in fish ovaries. *Histochemistry* **79,** 131–139.

Nunomura, W. (1991). C-reactive protein in eel: purification and agglutinating activity. *Biochim. Biophys. Acta* **1076,** 191–196.

Nunomura, W. (1992). C-reactive protein (CRP) in animals: its chemical properties and biological functions. *Zool. Sci.* **9,** 499–513.

Obenauf, S. D., and Hyder Smith, S. (1985). Chemotaxis of nures shark leukocytes. *Dev. Comp. Immunol.* **9,** 221–230.

Obenauf, S. D., and Hyder Smith, S. (1992). Migratory response of nurse shark leucocytes to activated mammalian sera and porcine C5a. *Fish Shellfish Immunol.* **2,** 173–181.

Oda, Y., Ichida, S., Mimura, T., Maeda, K., Tsujikawa, K., and Aonuma, S. (1984a). Purification and characterization of a fish lectin from the external mucus of Ophidiidae. *Genypterus blacodes. J. Pharm. Dyn.* **7,** 614–623.

Oda, Y., Ichida, S., Miura, T., Tsujikawa, K., Maeda, K., and Aonuma, S. (1984b). Mitogenic activity and *in vitro* fertilization inhibitory activity of *Genypterus blacodes* lectin. *J. Pharm. Dyn.* **7,** 849–855.

Ohta, M., Okada, M., Yamashita, I., and Kawasaki, I. (1990). The mechanism of carbohydrate-mediated complement activation by the serum mannose-binding protein. *J. Biol. Chem.* **265,** 1980–1984.

Oie, H. K., and Loh, P. C. (1971). Reovirus type 2: induction of viral resistance and interferon production in fathead minnow cells. *Proc. Soc. Exp. Biol. Med.* **136,** 369–373.

Okamato, N., Shirakura, T., Nagakura, Y., and Sano, T. (1983). The mechanism of interference with fish viral infection in the RTG-2 cell line. *Fish Pathol.* **18,** 7–12.

Oliveira, E. B., Gotschlich, E. C., and Liu, T.-Y. (1977). Primary structure of human C-reactive protein. *Proc. Natl. Acad. Sci. USA* **74,** 3148–3151.

Oliveira, E. B., Gotschlich, E. C., and Liu, T.-Y. (1979). Primary structure of human C-reactive protein. *J. Biol. Chem.* **254,** 489–502.

Oohara, I., Akiyama, T., Aono, H., and Mori, K. (1991). Distribution and interorganic correlations of lysozyme activity in the juvenile bluefin tuna, *Thunnus thynnus. Bull. Natl. Res. Inst. Aquaculture* **19,** 17–26.

Osmand, A. P., Mortensen, R. F., Siegel, J., and Gewurz, H. (1975). Interactions of C-reactive protein with the complement system. III. Complement-dependent passive hemolysis initiated by CRP. *J. Exp. Med.* **142,** 1065–1077.

Osmand, A. P., Friedenson, B., Gewurz, H., Painter, R. H., Hofmann, T., and Shelton, E. (1977). Characterization of C-reactive protein and the complement subcomponent C1t as homologous proteins displaying cyclic pentameric symmetry (pentraxins). *Proc. Natl. Acad. Sci. USA* **74,** 739–743.

Osserman, E. F., Canfield, R. E., Beychok, S. (1974). "Lysozyme." Academic Press, New York.

Ourth, D. D. (1980). Secretory IgM, lysozyme and lymphocytes in the skin mucus of the channel catfish, *Ictalurus punctatus. Dev. Comp. Immunol.* **4,** 65–74.

Ourth, D. D., and Bachinski, L. M. (1987a). Bactericidal response of channel catfish (*Ictalurus punctatus*) by the classical and alternative complement pathways against bacterial pathogens. *J. Appl. Ichthyol.* **3,** 42–45.

Ourth, D. D., and Bachinski, L. M. (1987b). Bacterial sialic acid modulates activation of the alternative complement pathway of channel catfish (*Ictalurus punctatus*). *Dev. Comp. Immunol.* **11,** 551–564.

Ourth, D. D., and Wilson, E. A. (1981). Agglutination and bactericidal responses of the channel catfish to *Salmonella paratyphi. Dev. Comp. Immunol.* **5,** 261–270.

Ourth, D. D., and Wilson, E. A. (1982a). Alternate pathway of complement and bactericidal response of the channel catfish to *Salmonella paratyphi. Dev. Comp. Immunol.* **6,** 75–85.

Ourth, D. D., and Wilson, E. A. (1982b). Bactericidal serum response of the channel catfish against Gram-negative bacteria. *Dev. Comp. Immunol.* **6,** 579–583.

Ozaki, H., Ohwaki, M., Fukada, T. (1983). Studies on lectins of amago *Oncorhynchus rhodurus.* I. Amago ova lectin and its receptor on homologous macrophages. *Dev. Comp. Immunol.* **7,** 77–87.

Paroutaud, P., Levi, G., Teichberg, V. I., and Strosberg, A. D. (1987). Extensive amino acid sequence homologies between animal lectins. *Proc. Natl. Acad. Sci. USA* **84,** 6345–6348.

Payne, R. H., Child, A. R., and Forrest, A. (1971). Geographical variation in the Atlantic salmon. *Nature* **231,** 250–252.

Pepys, M. B., Dash, A. C., Fletcher, T. C., Richardson, N., Munn, E. A., and Feinstein, A. (1978). Analogues in other mammals and in fish of human plasma proteins, C-reactive protein and amyloid P component. *Nature* **273,** 168–170.

Pepys, M. B., De Beer, F. C., Milstein, C. P., March, J. F., Feinstein, A., Butress, N., Clamp, J. R., Taylor, J., Bruton, C., and Fletcher, T. C. (1982). C-reactive protein and serum amyloid P component in the plaice (*Pleuronectes platessa* L.), a marine teleost, are homologous with their human counterparts. *Biochim. Biphys. Acta* **704,** 123–133.

Pintó, R. M., Jofre, J., and Bosch, A. (1993). Interferon-like activity in sea bass affected by viral erythrocytic infection. *Fish Shellfish Immunol.* **3,** 89–96.

Pratschner, G. A. (1978). The relative resistance of six transferrin phenotypes of coho salmon (*Oncorhynchus kisutch*) to cytophagosis, furunculosis, and vibriosis. M.S. Thesis, University of Washington, pp. 1–71. Seattle, Washington.

Putnam, F. W. (1975). Transferrin. *In* "The Plasma Protein" (F. W. Putnam, ed.), Vol. 1, pp. 265–316. Academic Press, New York.

Raftos, D. A., Hook, J. W., and Raison, R. L. (1992). Complement-like protein from the phylogenetically primitive vertebrate. *Eptatretus stouti,* is a humoral opsonin. *Comp. Biochem. Physiol.* **103B,** 379–384.

Raison, R. L., Coverley, J., Hook, J. W., Towns, P., Weston, K. M., and Raftos, D. A. (1994). A cell-surface opsonic receptor on leucocytes from the phylogenetically primitive vertebrate, *Eptatretus stouti. Immunol. Cell Biol.* **72,** 326–332.

Ramos, F., and Smith, A. C. (1978). The C-reactive protein (CRP) test for the detection of early disease in fishes. *Aquaculture* **14,** 261–266.

Randall, J. E., Aida, K., Oshima, Y., Hori, K., and Hashimoto, Y. (1981). Occurrence of a crinotoxin and hemagglutinin in the skin mucus of the moray eel *Lycodontis nudivomer. Mar. Biol.* **62,** 179–184.

Renault, T., Torchy, C., and de Kinkelin, P. (1991). Spectrophotometric method for titration of trout interferon, and its application to rainbow trout fry experimentally infected with viral haemorrhagic septicaemia virus. *Dis Aquat. Org.* **10,** 23–29.

Rijkers, G. T. (1982). Non-lymphoid defence mechanisms in fish. *Dev. Comp. Immunol.* **6,** 1–13.

Rio, G. J., Magnavita, F. J., Rubin, J. A., and Beckert, W. H. (1973). Characteristics of an established goldfish *Carassius auratus* cell line. *J. Fish Biol.* **5,** 315–321.

Robey, F. A., and Liu, T.-Y. (1981). Limulin: a C-reactive protein from *Limulus polyphemus. J. Biol. Chem.* **256,** 969–975.

Robey, F. A., and Liu, T.-Y. (1983). Synthesis and use of new spin-labeled derivatives of phosphorylcholine in a comparative study of human, dogfish, and *Limulus* C-reactive proteins. *J. Biol. Chem.* **258,** 3895–3900.

Robey, F. A., Tanaka, T., and Liu, T.-Y. (1983). Isolation and characterization of two major serum proteins from the dogfish, *Mustelus canis,* C-reactive protein and amyloid P component. *J. Biol. Chem.* **258,** 3889–3894.

Robey, F. A., Jones, K. D., and Steinberg, A. D. (1985). C-reactive protein mediates the solubilization of nuclear DNA by complement *in vitro. J. Exp. Med.* **161,** 1344–1356.

Rogel-Gaillard, C., Chilmonczyk, S., and de Kinkelin, P. (1993). *In vitro* induction of interferon-like activity from rainbow trout leucocytes stimulated by Egtved virus. *Fish. Shellfish Immunol.* **3,** 383–394.

Rogers, D. J. (1978). Lectin-type agglutinins, with anti-BI and anti-HI activity, from the ova of the black sea-bream, *Spondyliosoma cantharus. Med. Lab. Sci.* **35,** 239–245.

Ross, G. D., and Jensen, J. A. (1973a). The first component (C1n) of the complement system of the nurse shark (*Ginglymostoma cirratum*). I. Hemolytic characteristics of partially purified C1n. *J. Immunol.* **110,** 175–182.

Ross, G. D., and Jensen, J. A. (1973b). The first component (C1n) of the complement system of the nurse shark. II. Purification of the first component by ultracentrifugation and studies of its physicochemical properties. *J. Immunol.* **110,** 911–918.

Røed, K. H., Brun, E., Larsen, H. J., and Refstie, T. (1990). The genetic influence on serum haemolytic activity in rainbow trout. *Aquaculture* **85,** 109–117.

Røed, K. H., Fjalestad, K., Larsen, H. J., and Midthjel, L. (1992). Genetic variation in haemolytic activity in Atlantic salmon (*Salmo salar* L.). *J. Fish Biol.* **40,** 739–750.

Røed, K. H., Dehli, A. K., Flengsrud, R., Midthjel, L., and Rørvik, K. A. (1995). Immunoassay and partial characterisation of serum transferrin from Atlantic salmon (*Salmo salar* L.). *Fish. Shellfish Immunol.* **5,** 71–80.

Saggers, B. A., and Gould, M. L. (1989). The attachment of microorganisms to macrophages isolated from the tilapia *Oreochromis spilurus* Gunther. *J. Fish Biol.* **35,** 287–294.

Saha, K., Dash, K., and Sahu, A. (1993). Antibody dependent haemolysin, complement and opsonin in sera of a major carp, *Cirrhina mrigala,* and catfish, *Clarias batrachus,* and *Heteropneustes fossilis. Comp. Immun. Microbiol. Infect. Dis.* **16,** 323–330.

Sakai, D. K. (1981). Heat inactivation of complements and immune hemolysis reactions in rainbow trout, masu salmon, coho salmon, goldfish, and tilapia. *Bull. Japan. Soc. Sci. Fish.* **47,** 565–571.

Sakai, D. K. (1983). Lytic and bactericidal properties of salmonid sera. *J. Fish Biol.* **23,** 457–466.

Sakai, D. K. (1984a). Opsonization by fish antibody and complement in the immune phagocytosis by peritoneal exudate cells isolated from salmonid fishes. *J. Fish Dis.* **7,** 29–38.

Sakai, D. K. (1984b). The non-specific activation of rainbow trout, *Salmo gairdneri* Richardson, complement by *Aeromonas salmonicida* extracellular products and the correlation of complement activity with the inactivation of lethal toxicity products. *J. Fish Dis.* **7,** 329–338.

Sakai, D. K., Suzuki, K., and Awakura, T. (1994). Ontogenesis of salmonid complement and its nonspecific defense to viral infections. *Sci. Rep. Hokkaido Fish Hatchery* **48,** 25–31.

Sakakibara, F., Takayanagi, G., and Kawauchi, H. (1981). An L-rhamnose-binding lectin in the eggs of *Misgurnus anguillicaudatus. Yakugaku Zasshi* **101,** 918–925.

Sakakibara, F., Takayanagi, G., Mukai, C., Sue, H., Terasaki, Y., Nitta, K., and Kawauchi, H. (1982). Isolation and characterization of the blood group B specific lectin from pond smelt eggs (*Hypomesus transpacificus nipponensis*). *Annu. Rep. Tohoku Coll. Pharm.* **29,** 115–124.

Sakakibara, F., Kawauchi, H., and Takayanagi, G. (1985). Blood group B-specific lectin of *Plecoglossus altivelis* (ayu fish) eggs. *Biochim. Biophys. Acta* **841,** 103–111.

Salák, J., Roch, P., and Palousová, Z. (1975). The gel-filtration pattern of eel serum. Separation of hemolysin(s) from haemagglutinins. *J. Chromatogr.* **107,** 234–238.

Salton, M. R. J. (1957). The properties of lysozyme and its action on microorganisms. *Bact. Rev.* **21,** 82–99.

Salton, M. R. J., and Ghuysen, J. M. (1959). The structure of di- and tetra-saccharides released from cell walls by lysozyme and *Streptomyces* F_1 enzyme and the $\beta(1 \rightarrow 4)$ *N*-acetylhexosaminidase activity of these enzymes. *Biochim. Biophys. Acta* **36,** 552–554.

Sankaran, K., and Gurnani, S. (1972). On the variation in the catalytic activity of lysozyme in fishes. *Indian J. Biochem. Biophys.* **9,** 162–165.

Sano, T., and Nagakura, Y. (1982). Studies on viral diseases of Japanese fishes. VIII. Interferon induced by RTG-2 cell infected with IHN virus. *Fish Pathol.* **17,** 179–185.

Sano, E., Okano, K., Sawada, R., Naruto, M., Sudo, T., Kamata, K., Iizuka, M., and Kobayashi, S. (1988). Constitutive long-term production and characterization of recombinant human interferon-gamma from two different mammalian cells. *Cell Struct. Funct.* **13,** 143–160.

Schluter, S. F., Schroeder, J., Wang, E., and Marchalonis, J. J. (1994). Recognition molecules and immunoglobulin domains in invertebrates. *Ann. N.Y. Acad. Sci.* **712,** 74–81.

Sharon, N., and Lis, H. (1972). Lectins: cell agglutinating and sugar-specific proteins. *Science* **177,** 949–959.

Sharon, N., and Lis, H. (1989a). Lectins as cell recognition molecules. *Science* **246,** 227–246.

Sharon, N., and Lis, H. (1989b). "Lectins." Chapman and Hall Ltd., London.

Shiomi, K., Uematsu, H., Yamanaka, H., and Kikuchi, T. (1988). Screening of lectins in fish skin mucus. *J. Tokyo Univ. Fish.* **75,** 145–152.

Shiomi, K., Uematsu, H., Yamanaka, H., and Kikuchi, T. (1989). Purification and characterization of a galactose-binding lectin from the skin mucus of the conger eel *Conger myriaster. Comp. Biochem. Physiol.* **92B,** 255–261.

Shiomi, K., Uematsu, H., Ito, H., Yamanaka, H., and Kikuchi, T. (1990). Purification and properties of a lectin in the skin mucus of the dragonet *Repomucenus richardsonii. Nippon Suisan Gakkaishi* **56,** 119–123.

Siegel, J., Rent, R., and Gewurz, H. (1974). Interactions of C-reactive protein with the complement system. I. Protamine-induced consumption of complement in acute phase sera. *J. Exp. Med.* **140,** 631–647.

Siegel, J., Osmand, A. P., Wilson, M. F., and Gewuz, H. (1975). Interactions of C-reactive protein with the complement system. II. C-reactive protein-mediated consumption of complement by poly(L-lysine) polymers and other polycations. *J. Exp. Med.* **142,** 709–721.

Siwicki, A., and Studnicka, M. (1987). The phagocytic ability of neutrophils and serum lysozyme activity in experimentally infected carp. *Cyprinus carpio* L. *J. Fish Biol.* **31(Suppl A),** 57–60.

Snegaroff, J. (1993). Induction of interferon synthesis in rainbow trout leucocytes by various homeotherm viruses. *Fish. Shellfish Immunol.* **3,** 191–198.

Spitzer, R. H., Downing, S. W., Koch, E. A., and Kaplan, M. A. (1976). Hemagglutinins in the mucus of Pacific hagfish, *Eptatretus stoutii. Comp. Biochem. Physiol.* **54B,** 409–411.

Springer, G. F., and Desai, P. R. (1971). Monosaccharides as specific precipitinogens of eel anti-human blood-group H(O) antibody. *Biochemistry* **10,** 3749–3761.

Stanley, K. K., and Herz, J. (1987). Topological mapping of complement component C9 by recombinant DNA technique suggests a novel mechanism for its insertion into target membranes. *EMBO J.* **6,** 1951–1957.

Starkey, P. M., and Barrett, A. J. (1982). Evolution of α_2-macroglobulin. The demonstration in a variety of vertebrate species of a protein resembling human α_2-macroglobulin. *Biochem. J.* **205,** 91–95.

Stewart, W. E. (1980). Interferon nomenclature. *Nature* **286,** 110.

Stratil, A., Bobák, P., and Valenta, M. (1983a). Partial characterization of transferrins of some species of the family *Cyprinidae. Comp. Biochem. Physiol.* **74B,** 603–610.

Stratil, A., Bobák, P., Tomásek, V., and Valenta, M. (1983b). Transferrins of *Barbus barbus, Barbus meridionalis petenyi,* and their hybrids. Genetic polymorphism, heterogeneity and partial characterization. *Comp. Biochem. Physiol.* **76B,** 845–850.

Stratil, A., Tomásek, V., Clamp, J. R., and Williams, J. (1985). Partial characterization of transferrins of catfish (*Silurus glanis* L.) and pike (*Esox lucius* L.). *Comp. Biochem. Physiol.* **80B,** 909–911.

Studnicka, M., Siwicki, A., and Ryka, B. (1986). Lysozyme level in carp (*Cyprinus carpio* L.). *Bamidgeh* **38,** 22–25.

Sugita, H., Ishii, S., Hajji, N., Karasawa, A., Ohtake, Y., Sayama, Y., and Deguchi, Y. (1989). Bactericidal activity in sera of carp (*Cyprinus carpio*) and crucian carp (*Carassius carassius*). *Bull. Coll. Agr. Vet. Med. Nihon Univ.* **46,** 22–27.

Sunyer, O., and Tort, L. (1994). The complement system of the teleost fish *Sparus aurata. Ann. N.Y. Acad. Sci.* **712,** 371–373.

Sunyer, O., and Tort, L. (1995). Natural hemolytic and bactericidal activities of sea bream *Sparus aurata* serum are affected by the alternative complement pathway. *Vet. Immunol. Immunopathol.* **45,** 333–345.

Sussman, M. (1974). Iron and infection. *In* "Iron in Biochemistry and Medicine" (A. Jacobs and M. Worwood, eds.), pp. 649–699. Academic Press, New York.

Suzuki, Y. (1985). Hemolysin and hemagglutinin in skin mucus of the Japanese eel *Anguilla japonica. Bull. Japan. Soc. Sci. Fish.* **51,** 2083.

Suzuki, Y. (1986). Neutrophil chemotactic factor in eel blood plasma. *Bull. Japan. Soc. Sci. Fish.* **52,** 811–816.

Suzuki, Y., and Kaneko, T. (1986). Demonstration of the mucous hemagglutinin in the club cells of eel skin. *Dev. Comp. Immunol.* **10,** 509–518.

Suzumoto, B. K., Schreck, C. B., and McIntyre, J. D. (1977). Relative resistances of three transferrin genotypes of coho salmon (*Oncorhynchus kisutch*) and their hematological responses to bacterial kidney disease. *J. Fish. Res. Board Can.* **34,** 1–8.

Szalai, A. J., Norcum, M. T., Bly, J. E., and Clem, L. W. (1992a). Isolation of an acute-phase phosphorylcholine-reactive pentraxin from channel catfish (*Ictalurus punctatus*). *Comp. Biochem. Physiol.* **102B,** 535–543.

Szalai, A. J., Bly, J. E., and Clem, L. W. (1992b). Chelation affects the conformation, lability, and aggregation of channel catfish (*Ictalurus punctatus*) phosphorylcholine-reactive protein (PRP). *Comp. Biochem. Physiol.* **102B,** 545–550.

Szalai, A. J., Bly, J. E., and Clem, L. W. (1994). Changes in serum concentrations of channel catfish (*Ictalurus punctatus* Rafinesque) phosphorylcholine-reactive protein (PRP) in response to inflammatory agents, low temperature shock and infection by the fungus *Saprolegnia* sp. *Fish Shellfish Immunol.* **4,** 323–336.

Takahashi, Y., Itami, T., and Konegawa, K. (1986). Enzymatic properties of partially purified lysozyme from the skin mucus of carp. *Bull. Japan. Soc. Sci. Fish.* **52,** 1209–1214.

Tamai, T., Shirahata, S., Sato, N., Kimura, S., Nonaka, M., and Murakami, H. (1993a). Purification and characterization of interferon-like antiviral protein derived from flatfish (*Paralichthys olivaceus*) lymphocytes immortalized by oncogenes. *Cytotechnology* **11,** 121–131.

Tamai, T., Shirahata, S., Noguchi, T., Sato, N., Kimura, S., and Murakami, H. (1993b). Cloning and expression of flatfish (*Paralichthys olivaceus*) interferon cDNA. *Biochim. Biophys. Acta* **1174,** 182–186.

Taniguchi, T., Mantei, N., Schwarzstein, M., Nagata, S., Muramatsu, M., and Weissmann, C. (1980). Human leukocyte and fibroblast interferons are structurally related. *Nature* **285,** 547–549.

Tengelsen, L. A., Anderson, E., and Leong, J. (1989). Variation in fish interferon-like activity: cell line production and IHN virus isolate sensitivity. *FHS/AFS Newsletter* **17,** 4.

Tillett, W. S., and Francis, T. (1930). Serological reactions in pneumonia with a nonprotein somatic fraction of pneumococcus. *J. Exp. Med.* **52,** 561–571.

Tomlinson, S., Stanley, K. K., and Esser, A. F. (1993). Domain structure, functional activity, and polymerization of trout complement protein C9. *Dev. Comp. Immunol.* **17,** 67–76.

Topliss, J. A., and Rogers, D. J. (1985). An anti-fucose agglutinin in the ova of *Dicentrarchus labrax. Med. Lab. Sci.* **42,** 199–200.

Trinchieri, G., and Perussia, B. (1985). Immune interferon: a pleiotropic lymphokine with multiple effects. *Immunol. Today* **6,** 131–136.

Trust, T. J., Courtice, I. D., Khouri, A. G., Crosa, J. H., and Schiewe, M. H. (1981). Serum resistance and hemagglutination ability of marine vibrios pathogenic for fish. *Infect. Immun.* **34,** 702–707.

Turner, R. J., and Jamieson, A. (1987). Transferrins in fishes. *Anim. Genet* **18(suppl. 1),** 70–71.

Uemura, T. (1993). Studies on the membrane attack complex (MAC) of carp complement and on the constituents of the MAC. Ph.D. Thesis, Kyushu University, pp. 1–118. Fukuoka, Japan.

Uemura, T., Nakao, M., and Yano, T. (1992). Isolation of the second component of complement (C2) from carp serum. *Nippon Suisan Gakkaishi* **58,** 727–733.

Uhlenbruck, G., Janssen, E., and Javeri, S. (1982). Two different antigalactan lectins in eel serum. *Immunobiology* **163,** 36–47.

Utter, F. M., Ames, W. E., and Hodgins, H. O. (1970). Transferrin polymorphism in coho salmon (*Oncorhynchus kisutch*). *J. Fish. Res. Board Can.* **27,** 2371–2373.

Valenta, M., Stratil, A., Slechtová, V., Kálal, L., and Slechta, V. (1976). Polymorphism of transferrin in carp (*Cyprinus carpio* L.): genetic determination, isolation, and partial characterization. *Biochem. Genet.* **14,** 27–45.

Vasta, G. R., and Marchalonis, J. J. (1983). Lectins from tunicates and cyclostomes: a biochemical characterization. *In* "Lectins: Biology, Biochemistry, and Clinical Biochemistry" (T. C. Bøg-Hansen and G. A. Spengler, eds.), Vol. 3, pp. 461–468. Walter de Gruyter, Berlin.

Vladimirov, V. L. (1968). Immunity in fish. *Bull. Off. Int. Epiz.* **69,** 1365–1372.

Volanakis, J. E., and Kaplan, M. H. (1974). Interaction of C-reactive protein complexes with the complement system. II. Consumption of guinea pig complement by CRP complexes: requirement for human Clq. *J. Immunol.* **113,** 9–17.

Volanakis, J. E., Clements, W. L., and Schrohenloher, R. E. (1978). C-reactive protein: purification by affinity chromatography and physicochemical characterization. *J. Immunol. Meth.* **23,** 285–295.

Voss, Jr., E. W., Fryer, J. L., and Banowetz, G. M. (1978). Isolation, purification, and partial characterization of a lectin from chinook salmon ova. *Arch. Biochem. Biophys.* **186,** 25–34.

Webster, R. O., and Pollara, B. (1969). Isolation and partial characterization of transferrin in the sea lamprey, *Petromyzon marinus. Comp. Biochem. Physiol.* **30,** 509–527.

Wehnert, S. D., and Woo, P. T. K. (1980). *In vivo* and *in vitro* studies on the host specificity of *Trypanoplasma salmositica. J. Wildlife Dis.* **16,** 183–187.

Weinberg, E. D. (1974). Iron and susceptibility to infectious disease. *Science* **184,** 952–956.

White, A., and Fletcher, T. C. (1982). The effects of adrenal hormones, endotoxin and turpentine on serum components of the plaice (*Pleuronectes platessa* L.). *Comp. Biochem. Physiol.* **73C,** 195–200.

White, A., and Fletcher, T. C. (1985). The influence of hormones and inflammatory agents on C-reactive protein, cortisol, and alanine aminotransferase in the plaice (*Pleuronectes platessa* L.). *Comp. Biochem. Physiol.* **80C,** 99–104.

White, A., Fletcher, T. C., Towler, C. M., and Baldo, B. A. (1978). Isolation of a C-reactive protein-like precipitin from the eggs of the lumpsucker (*Cyclopterus lumpus* L.). *Comp. Biochem. Physiol.* **61C,** 331–336.

White, A., Fletcher, T. C., Pepys, M. B., and Baldo, B. A. (1981). The effect of inflammatory agents on C-reactive protein and serum amyloid P-component levels in plaice (*Pleuronectes platessa* L.) serum. *Comp. Biochem. Physiol.* **69C,** 325–329.

White,A., Fletcher, T. C., and Pepys, M. B. (1983). Serum concentrations of C-reactive protein and serum amyloid P component in plaice (*Pleuronectes platessa* L.) in relation to season and injected lipopolysaccharide. *Comp. Biochem. Physiol.* **74B,** 453–458.

White, A., MacArthur, J. I., and Fletcher, T. C. (1984). Distribution of endotoxin and its effect on serum concentrations of C-reactive protein and cortisol in the plaice (*Pleuronectes platessa* L.). *Comp. Biochem. Physiol.* **79C,** 97–101.

Whitehouse, D. B. (1988). Genetic polymorphisms of animal complement components. *Exp. Clin. Immunogenet.* **5,** 143–164.

Winkelhake, J. L., and Chang, R. J. (1982). Acute phase (C-reactive) protein-like macromolecules from rainbow trout (*Salmo gairdneri*). *Dev. Comp. Immunol.* **6,** 481–489.

Winter, G. W., Schreck, C. B., McIntyre, J. D. (1980). Resistance of different stocks and transferrin genotypes of coho salmon, *Oncorhynchus kisutch,* and steelhead trout, *Salmo gairdneri,* to bacterial kidney disease and vibriosis. *Fish. Bull.* **77,** 795–802.

Withler, R. E., and Evelyn, T. P. T. (1990). Genetic variation in resistance to bacterial kidney disease within and between two strains of coho salmon from British Columbia. *Trans. Am. Fish. Soc.* **119,** 1003–1009.

Woo, P. T. K. (1992). Immunological responses of fish to parasitic organisms. *Annu. Rev. Fish Dis.* **2,** 339–366.

Woo, P. T. K., and Thomas, P. T. (1991). *In vitro* activation of salmonid complement by mammalian antibodies. *Fish Shellfish Immunol.* **1,** 313–315.

Yano, T. (1992). Assays of hemolytic complement activity. *In* "Techniques in Fish Immunology-2" (J. S. Stolen, T. C. Fletcher, D. P. Anderson, S. L. Kaattari, and A. F. Rowley, eds.), pp. 131–141. SOS Publications, Fair Haven, N.J.

Yano, T. (1995). The complement system of fish. *Fish Pathol.* **30,** 151–158.

Yano, T., and Nakao, M. (1994). Isolation of a carp complement protein homologous to mammalian factor D. *Mol. Immunol.* **31,** 337–342.

Yano, T., Ando, H., and Nakao, M. (1984). Optimum conditions for the assay of hemolytic complement titer of carp and seasonal variation of the titers. *J. Fac. Agr. Kyushu Univ.* **29,** 91–101.

Yano, T., Ando, H., and Nakao, M. (1985). Two activation steps of carp complement requiring Ca^{2+} and Mg^{2+} and an intermediate product in immune hemolysis. *Bull. Japan. Soc. Sci. Fish.* **51,** 841–846.

Yano, T., Matsuyama, H., and Nakao, M. (1986). An intermediate complex in immune hemolysis by carp complement homologous to mammalian EAC1,4. *Bull. Japan. Soc. Sci. Fish.* **52,** 281–286.

Yano, T., Matsuyama, H., and Nakao, M. (1988a). Isolation of the first component of complement (C1) from carp serum. *Nippon Suisan Gakkaishi* **54,** 851–859.

Yano, T., Hatayama, Y., Matsuyama, H., and Nakao, M. (1988b). Titration of the alternative complement pathway activity of representative cultured fishes. *Nippon Suisan Gakkaishi* **54,** 1049–1054.

Yano, T., Matsuyama, H., and Nakao, M. (1988c). Formation and characteristics of an intermediate complex EAC1,4,2 in immune hemolysis by carp complement. *Nippon Suisan Gakkaishi* **54,** 1997–2000.

Yoshizaki, N. (1990). Functions and properties of animal lectins. *Zool. Sci.* **7,** 581–591.

Yousif, A. N., Albright, L. J., and Evelyn, T. P. T. (1991). Occurrence of lysozyme in the eggs of coho salmon *Oncorhynchus kisutch. Dis. Aquat. Org.* **10,** 45–49.

Yousif, A. N., Albright, L. J., and Evelyn, T. P. T. (1994a). *In vitro* evidence for the antibacterial role of lysozyme in salmonid eggs. *Dis Aquat. Org.* **19,** 15–19.

Yousif, A. N., Albright, L. J., and Evelyn, T. P. T. (1994b). Purification and characterization of a galactose-specific lectin from the eggs of coho salmon *Oncorhynchus kisutch* and its interaction with bacterial fish pathogens. *Dis. Aquat. Org.* **20,** 127–136.

Zikmundová, J., Svobodová, J., and Krajhanzl, A. (1990). Fish cortical lectins: the bacteriostatic effect of the roach (*Rutilus rutilus* L.) ova lectin. *In* "Lectins: Biology, Biochemistry, and Clinical Biochemistry" (J. Kocourek and D. L. J. Freed, eds.), Vol. 7, pp. 265–271. Sigma, St. Louis.

4

THE SPECIFIC IMMUNE SYSTEM: CELLULAR DEFENSES

MARGARET J. MANNING AND TERUYUKI NAKANISHI

THE FISH IMMUNE SYSTEM:
ORGANISM, PATHOGEN, AND ENVIRONMENT

I. INTRODUCTION

Specific immune responses that are independent of antibody are collectively termed cell-mediated immunity. The term originates from the ability to transfer the antigen-specific response from one individual to another by means of live cells. By the early 1950s it had been shown that the cells responsible for specific cell-mediated immunity are lymphocytes (Mitchison, 1953), now identified as T (thymus-derived) cells. Much of the initial understanding of these cellular defences comes from mammalian transplantation experiments, particularly the rejection of allogeneic grafts. (An allogeneic graft [allograft] is from a member of the same species but of different genetic constitution; cf. a syngeneic graft from an individual of the same genetic constitution, or autograft [self graft] from the same individual.) Allogeneic graft rejection manifests antigen specificity and immunological memory (altered, usually enhanced, reactivity on subsequent exposure to the same antigen). These features are hallmarks of the vertebrate specific immune system.

In mammals, the cell believed to be mainly responsible for the cell-mediated destruction of allografts is the cytotoxic T cell. This cell belongs to a subpopulation of T lymphocytes that bear the CD8 cell surface (membrane) marker and recognize antigen in association with the class I major histocompatibility complex (MHC) antigen (see Section VI). Its biological function is to recognize self class I MHC antigens altered by some fragment of nonself antigen, as seen with the destruction of self cells infected with a virus. The experimental transplantation of foreign tissues is an unnatural situation. However, for the host T cell, the allogeneic class I molecule probably mimics an altered self class I molecule.

In lower vertebrate groups, transplantation reactions have also proved a useful model for the study of cell-mediated immunity. Indeed, the basic phenomena seem to be similar in all vertebrates. There is recognition and rejection of grafts involving cellular infiltration, with an accelerated response on secondary exposure to the same donor antigens. Mononuclear cells (lymphocytes and macrophages) appear to predominate in the cellular reaction but the precise mechanism of destruction of the donor cells is less well understood than for mammals. Allograft recognition with a memory component has also been recorded in invertebrates (Millar and Ratcliffe, 1994) but homologies have yet to be established.

An important event that characterizes the vertebrate lymphocyte and occurs in both humoral and cell-mediated responses, is that the first response to an encounter with an appropriate antigen is the trigger for proliferation of the specifically reactive cell. This important amplification stage has not been described in invertebrates, and may be a significant evolutionary innovation in the vertebrate subphylum. The clonal expansion of receptors has obvious advantages, especially when linked to a high level of specificity. It is therefore not surprising that this advance coincides with the appearance, in vertebrates, of variable region genes for specific antigen receptors.

II. SPECIFIC CELL-MEDIATED IMMUNITY IN FISH

A. Transplantation Immunity: Graft Rejection

Most studies on the immune response to foreign tissue in fish have involved skin grafting using either full thickness skin or scale transplants. The surgical techniques are relatively simple and the rejection phenomena easy to observe. A scale (consisting of a covering layer of epidermis containing melanophores and a capillary network, along with a noncellular scale plate) can be removed from its socket and inserted into a vacant socket from which the scale has been removed (Mori, 1931). The survival time of the scale allograft can be measured from the time of transplantation to the time when the stellate appearances of all melanophores within the scale are no longer visible. Alternatively, clearance of the hyperplastic host tissue which overgrows the graft can be taken as the rejection end point (Rijkers and Van Muiswinkel, 1977).

Full-thickness skin grafts can be made on suitable fish, such as those not too extensively covered with scales, such as the mirror carp (*Cyprinus carpio*). Typically, an area of flank skin of about 0.5 cm^2, containing dermis and epidermis, is transferred to the cleared graft bed of a second fish. Once in place, it is gently dried with a jet of air to aid adherence. The fish is then transferred to a dish of water, with the gills underwater but the graft exposed to the air, until it has recovered from the anesthesia. Alternatively, for young or delicate fish, a slit may be made in the skin of the host fish. The host skin is then lifted from the underlying muscle and the donor graft posted under it through the slit (Tatner, 1990).

The events following transplantation of a scale or skin allograft are basically similar in all species of fish so far examined. Externally, allografts placed on a host that has not previously encountered the donor's tissue (first-set grafts) remain indistinguishable from autografts for the first few days after grafting. Both types of graft appear healthy, with revascularization taking place across the graft bed. Autografts then become completely healed in, whereas allografts begin to lose their pigmentation, with destruction and loss of melanophores. Finally, there is invasive replacement by host tissue accompanied by contracture of the graft bed and scar formation. There is no sloughing off process as such. Unlike mammals, skin grafts in fish become overgrown with host tissue soon after transplantation, and healing involves a gradual return to normal host tissue, leaving just the scar of the rejected allograft (Botham *et al.,* 1980; see also Fig. 1 on the color insert).

Histologically, the host response to an allograft differs markedly from that to autografts. The initial response is an invasion of the allograft by lymphocytes and macrophages. This does not occur in the case of autografts and appears to be common to all fish groups studied (Finstad and Good, 1966; Borysenko and Hildemann, 1969; McKinney *et al.,* 1981). The wave of cellular infiltration begins early in the response. Its peak precedes the cytotoxicity events that result in melanophore breakdown and pigment dispersal (Botham *et al.,* 1980).

The second-set response (repeat grafting from the same donor) shows the specificity and immunological memory characteristic of a cell-mediated immune response. Rejection of a second-set allograft follows a similar pattern to that of the first-set graft but the rate of destruction is accelerated (Hildemann and Haas, 1960). For example, in carp held at 22°C, the mean survival time of the first graft was 14 days; for the second graft it was 7 days (Botham and Manning, 1981; see also Fig. 2).

B. Ontogenetic Aspects of Graft Rejection

It is known from work on a number of species that the immune system of fish is very immature at the time of hatching (Ellis, 1977; van Loon *et al.,* 1981; Nakanishi, 1986a). It is also apparent that once lymphocytic histogenesis is underway in the young fish, the cell-mediated arm of the animal's specific defence mechanisms, as monitored by histoincompatibility reactions, matures rapidly (Kallman and Gordon, 1957; Triplett and Barrymore, 1960; Tatner and Manning, 1983). Furthermore, fry can be primed to yield second-set cell-mediated responses from an early age (Botham *et al.,* 1980; Botham and Manning, 1981). The allograft response is, however, less efficient in these immature fish than in adults. Lymphocytes invade allografts but in much smaller numbers than in older fish and it is uncertain whether full rejection eventually ensues. These aspects of the development of immunocompetence in young fish are discussed in Chapter 6.

C. Allograft Rejection Times

Exact comparisons of mean allograft survival times are not always easy to make, especially when considering differences between fish of widely different groups. There may be considerable dissimilarities in skin type and in the chosen end points for graft survival. For example, in holostean fish with ganoid scales, allograft transplantation presents a different situation from that in teleosts (McKinney *et al.,* 1981). Even within the teleosts, longer mean survival times are recorded for skin grafts relative to scale grafts (Fig. 2).

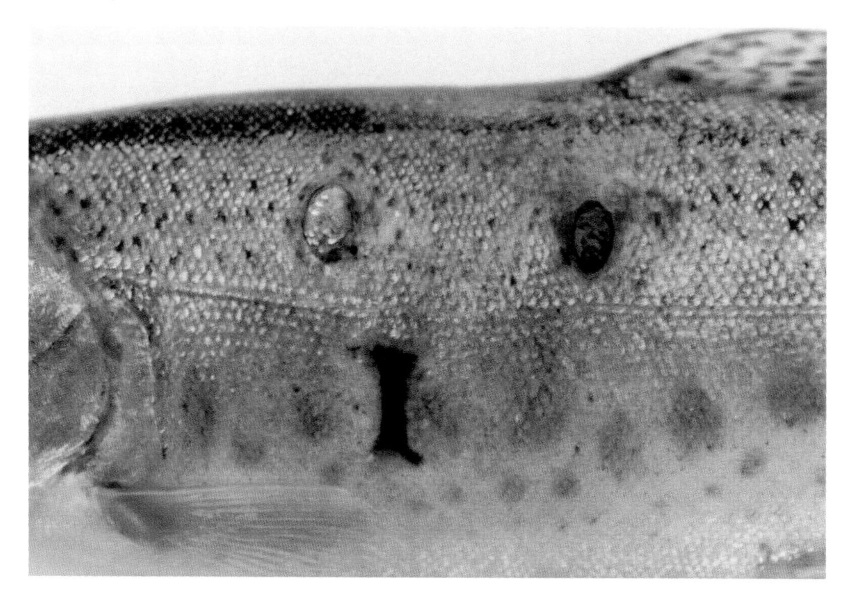

Fig. 1. Autograft and allograft in rainbow trout 1 month post grafting. Autograft (upper left): skin was transplanted from the ventral region (lower left) of the same fish. Note healthy appearance with shining scales. Allograft (upper right): skin was transplanted from a different fish. Note sloughing off of the graft leaving the scar.

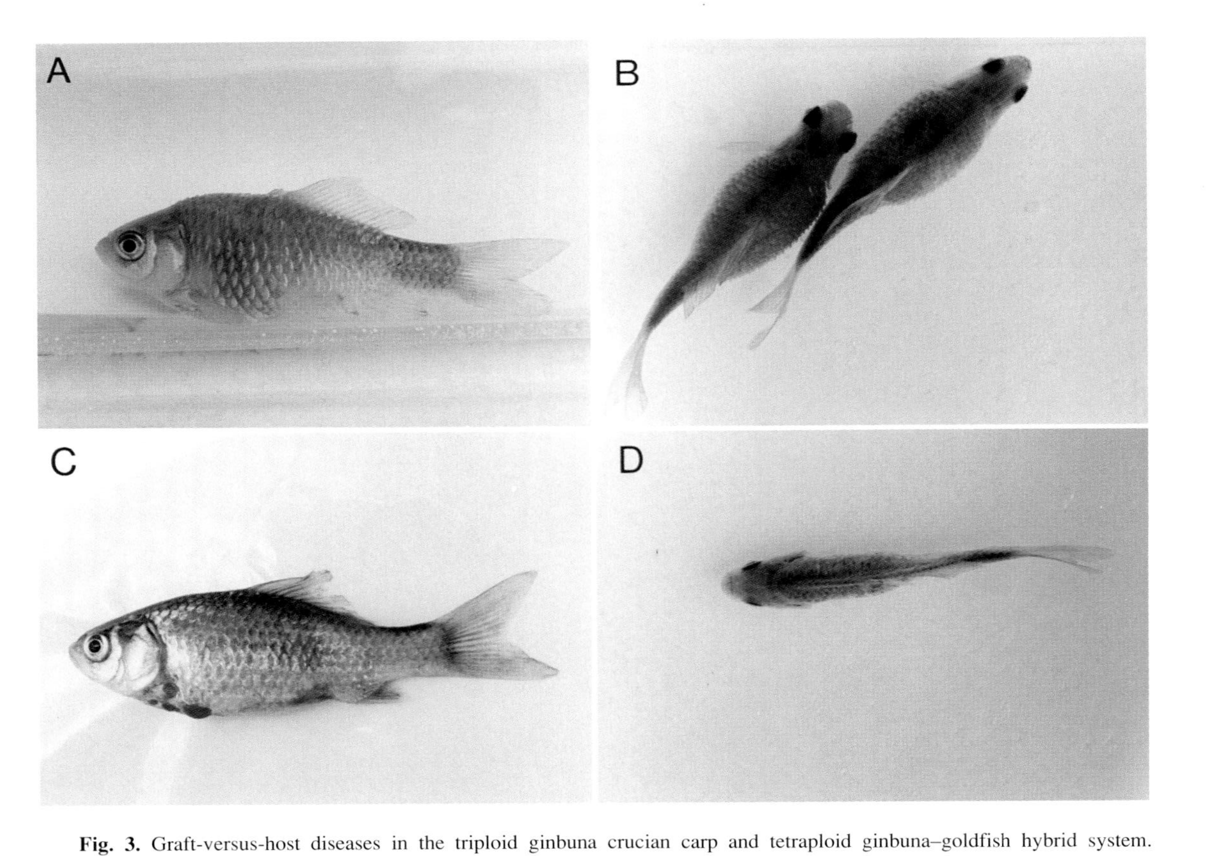

Fig. 3. Graft-versus-host diseases in the triploid ginbuna crucian carp and tetraploid ginbuna–goldfish hybrid system. (A) Side view of fish suffering from graft-versus-host disease to show scale protrusion and hemorrhage at the ventral region. (B) View from the top to show the scale protrusion. (C) Side view to show the severe hemorrhage and the local destruction of ventral skin. (D) Fish becoming very thin and feeble due to a loss of appetite and constipation. Photographs were taken 2 weeks after donor cell injection for (A) and (B), and 3 weeks after for (C) and (D).

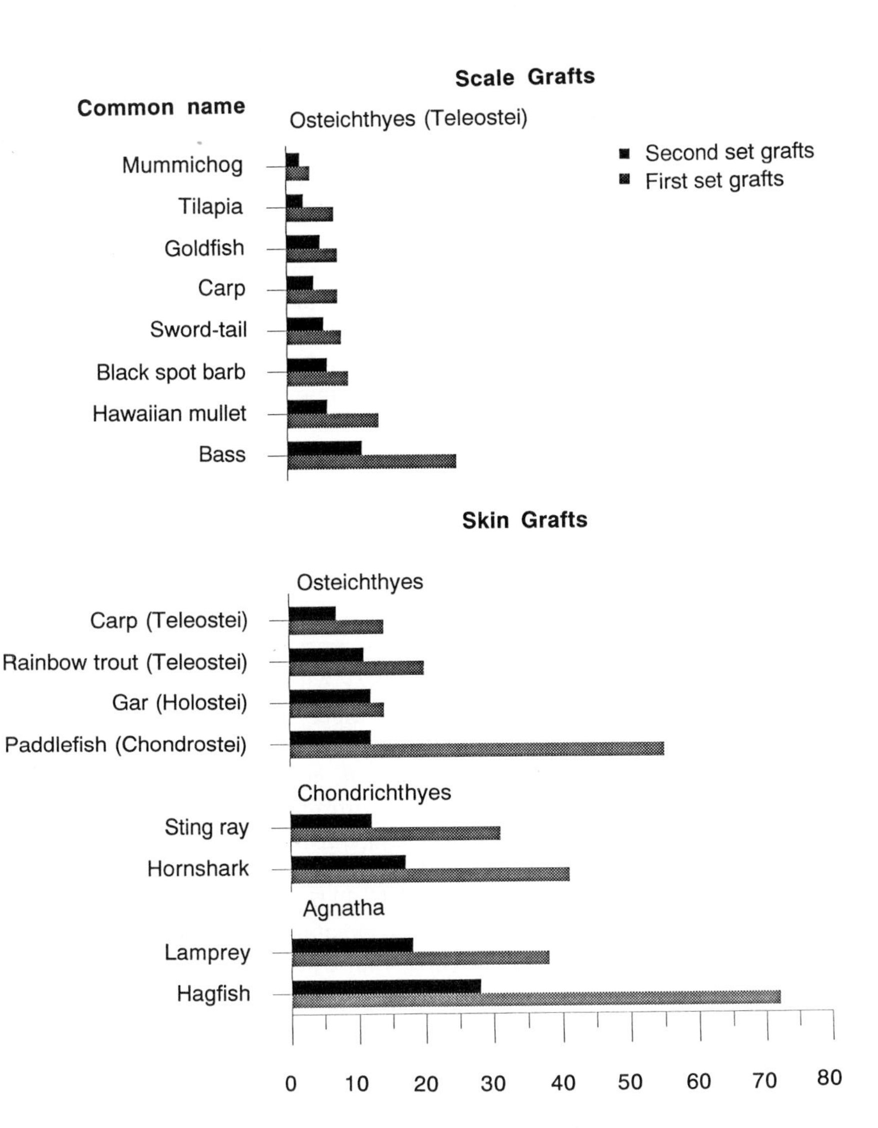

Fig. 2. Survival times of scale and skin allografts in different groups of fish (approximate values). Common name, species, and authors. Scales: Mummichog, *Fundulus heteroclitus* (Cooper, 1964); Tilapia, *Tilapia mossambica* (Sailendri, 1973); Goldfish, *Carassius aureus* (Hildemann and Haas, 1960); Carp, *Cyprinus carpio* (Hildemann and Hass, 1960); Sword-tail, *Xiphophorus maculatus* (Hogarth, 1968); Black spot barb, *Barbus* sp. (Hildemann and Haas, 1960); Hawaiian mullet, *Mughil cephalus* (Leeloy and Hildemann, 1977); Bass, *Micropterus salmoides* (Reid and Triplett, 1968); Skin: Carp, *Cyprinus carpio* (Botham *et al.*, 1980); Rainbow trout, *Salmo gairdneri* (Botham *et al.*, 1980); Gar, *Lepisosteus platyrhincus* (McKinney *et al.*, 1981); Paddlefish, *Polyodon spathula* (Perey *et al.*, 1968); Stingray, *Dasyatis americana* (Perey *et al.*, 1968); Hornshark, *Heterodontus francisci* (Borysenko and Hildemann, 1970); Lamprey, *Petromyzon marinus* (Perey *et al.*, 1968); Hagfish, *Eptatretus stoutii* (Hildemann and Thoenes, 1969).

Physiological factors such as temperature also play a part (Hildemann, 1957). There are even day–night rhythms of immune reactivity (Nevid and Meier, 1993). Also, genetic variability within a species may be restricted if, for example, animals taken from local inbred populations are used. Allowing for such factors, however, it remains clear that allograft rejection in the more primitive groups of fishes is slow (Hildemann, 1970). Figure 2 shows that in agnathans, elasmobranchs, and primitive actinopterygian fish, grafts are rejected in a chronic manner, with first-set grafts surviving for several weeks (>30 days). In higher vertebrates these slow rejection times are believed to be due to cumulative minor histocompatibility reactions. In contrast, teleost fish reject allografts more rapidly in acute responses usually associated with the presence of an MHC, a genetic region whose products are primarily responsible for the rapid rejection of grafts (see Section VI).

D. Mechanisms of Rejection

There is little or no evidence for antibody involvement in the graft rejection mechanisms of fish (Rijkers, 1982). Allograft destruction is elicited only by living tissue, but the nature of the cytotoxic mechanisms of the responding host cells remains to be elucidated. In mammals, histoincompatibility reactions are brought about by specific cytotoxic T cells. Host macrophages also arrive at the reaction site, attracted there and activated by cytokines produced by the host. In fish, on the other hand, although a specific component in the transplantation response has been clearly demonstrated, a specific (CD8-like) cytotoxic T cell has yet to be firmly identified, although there is some evidence for the existence of alloreactive cytotoxic cells (see Section III.E). It is possible that cytokines such as macrophage activating factor (MAF) could be involved with the cellular reactions in the graft since we know that T-like cells producing these factors occur in fish (Graham and Secombes, 1990).

E. Graft-versus-Host Reactions

Graft-versus-host reactions have been well studied in mammals and birds (Clark, 1991), and in amphibians (Clark and Newth, 1972). They have also recently been demonstrated in fish (Nakanishi, 1994). A graft-versus-host reaction can occur when the transplanted donor tissue contains immunologically competent cells that the host is unable to recognize and destroy. This may be because of immaturity or immunosuppression of the host, or its genetic constitution. The grafted cells can then react against the host in a manner that is essentially the reverse of a transplantation (host-versus-graft) reaction. This can lead to extensive tissue damage and death. In mammals, donor T cells play an important role in graft-versus-host reactions through the secretion of cytokines by $CD4^{+ve}$ T cells. Also, $CD8^{+ve}$ donor T cells become highly cytotoxic toward the host cells.

In fish, ploidy manipulation has been used to induce a graft-versus-host reaction (Nakanishi, 1994). Clonal triploid ginbuna crucian carp (*Carassius auratus langsdorfii*) were crossed with goldfish (*Carassius auratus*) to produce tetraploid fish. These tetraploids share all the elements of the triploid genotype but the reverse is not the case. Thus, when immunologically competent lymphoid cells obtained from a triploid donor were injected into the tetraploid host, a typical graft-versus-host reaction ensued (Fig. 3 on the color insert). This commenced about two weeks after injection and, as in mammals, it involved damage to the skin, liver, and lymphoid organs with prominent enlargement of the spleen leading to death of the host. The graft-versus-host reaction was dependent on the number of triploid donor cells injected. It was most effectively induced by head kidney cells and peripheral blood leukocytes (PBL), followed by spleen and thymus cells. Donors had to be sensitized at least twice by grafting the scales of the proposed host to induce the reaction. This suggests that the recognition and/or effector function is a kind of specific event. However, a considerable number of recipients injected with cells from donors that had been sensitized by allogenetically different tetraploids died, suggesting a limited polymorphism of the MHC. Most of the features of acute graft-versus-host reactions observed in the system are quite similar to those found in mammals and birds, thereby providing evidence for the presence of alloreactive T cells in teleosts.

F. Delayed (Cell-Mediated) Hypersensitivity

Delayed (type IV) hypersensitivity reactions (DTH) are usually detected by injecting antigen into the skin of a sensitized individual. A classic reaction is that produced in an animal that has been previously sensitized to tubercle bacilli when either the bacilli or antigenic proteins derived therefrom are injected intradermally. A day or two later, lymphocytes and macrophages begin to infiltrate the injection site which becomes indurated and swollen. In mammals, the responding cells are known to be T lymphocytes and to include both $CD4^{+ve}$ and $CD8^{+ve}$ cells. Cytokines, including IL-2 and IFN-γ, play an important part in mammalian DTH inflammatory reactions.

Experiments using the cellular antigens of mycobacteria to demonstrate DTH reactions in fish include those of Bartos and Sommer (1981) who immunized rainbow trout (*Oncorhynchus mykiss*) with complete Freund's adjuvant containing *Mycobacterium tuberculosis,* then skin-tested the sensitized fish. They found that killed mycobacteria of *M. tuberculosis* or the related fish pathogen *M. salmoniphilum* produced typical DTH skin reactions. Activated macrophages were demonstrated at the site using silicone skin windows. The DTH reactions to mycobacterial antigens are not confined to teleosts; they have also been observed in lamprey, *Petromyzon marinus* (Finstad and Good, 1964), elasmobranchs (guitar fish, *Rhinobatos productus,* and horned shark, *Heterodontus franciscii*), and chondrosteans (paddlefish, *Polyodon spathula*) (Tam *et al.,* 1976).

Attempts to use a DTH skin reaction to screen trout for their cellular immune responses to vaccination against *Yersinia ruckeri* (Gram-negative bacterium that causes enteric redmouth disease) have encountered problems such as possible cross reactivity (Stevenson and Raymond, 1990). In the channel catfish (*Ictalurus punctatus*), immunization with the bacterium, *Flexibacter columnaris,* or with channel catfish virus, *Herpesvirus ictaluri,* failed to result in DTH responses on challenge (Pauley and Heartwell, 1983). In fish it seems that these responses are best elicited by strongly immunogenic and persistent components such as those of mycobacteria and parasites. Delayed type hypersensitivity reactions against parasitic antigens have been demonstrated experimentally in both primitive bony fish and teleosts: against the metazoan parasite *Ascaris* in the bowfin, *Amia calva* (Papermaster *et al.,* 1964), and against the protozoan hemoflagellate *Cryptobia* in rainbow trout (Stevenson and Raymond, 1990; Thomas and Woo, 1990). In addition, histological findings consonant with a DTH reaction have been described in fungal infections of fish, as seen against the spores of *Ichthyophonus* (McVicar and McLay, 1985) and against injected viable *Saprolegnia* hyphae (Bly and Clem, 1992). Similar reactions leading to granulomata formation were reported in an autoimmune response against spermatozoa in experimental autoimmune orchitis in rainbow trout (Secombes *et al.,* 1985).

G. Function of the Thymus: Effects of Thymectomy

1. Thymectomy Operations

In mammals, profound effects have been demonstrated following removal of the thymus from neonatal animals. Thymectomy in the adult, however, has less effect owing to the presence of T lymphocytes that are already established in the circulation. Nevertheless, in the long term (after several months) even adult thymectomy can reveal immune deficiencies as existing peripheral T cells begin to disappear (Metcalf, 1965).

In fish, complete removal of the thymus early in life is technically more difficult than in higher vertebrates or amphibians (Horton, 1994), largely because of its anatomical situation. In fish, in contrast to tetrapod vertebrates, the thymus retains its contact with the pharyngeal epithelium. Also, in many species it lies quite close to the ear and wraps around respiratory muscles that pass from the otic region of the skull to the operculum (Manning, 1994). The problem is greater for some species of fish than for others (Tatner, 1990); for example, the thymus is more accessible in trout than in carp. It can be seen from Fig. 4 that in carp fry, although most of the thymus could be removed, it would be difficult to achieve complete thymic ablation without damaging the muscles used for respiration, making regeneration of thymic tissue from the adjacent pharyngeal epithelium likely.

2. Effects on Allograft Rejection

The effects of thymectomy on allograft rejection has been investigated in both adult and juvenile fish (see Chapter 6). In adult rainbow trout, thymectomy had no effect on the mean survival time of allografts transplanted one month after the operation nor did it affect second-set rejection times (Botham *et al.,* 1980). In younger fish (2- to 4-month-old tilapia) allograft rejection was somewhat delayed (Sailendri, 1973). A similar result was found in rainbow trout thymectomized at 2 months of age and grafted 1 month later (Manning *et al.,* 1982a). Lymphocytic invasion of the allogeneic skin commenced later than in control fish and rejection was still incomplete at the end of the experiment. It should be noted here that thymectomy in these experiments was relatively late in relation to maturation of the lymphoid organs. Even the spleen (which is slow to develop its lymphoid component) can react to antigenic stimulation in 2-month-old fish. The finding that these thymectomized fish were able to recognize and respond to the allogeneic tissue, albeit somewhat less efficiently, was therefore not unexpected.

3. Helper and Suppressor Functions

In mammals it is well established that the thymus is a source of helper T cells. Helper cells are a functional subclass of T cells that are $CD4^{+ve}$ and help to generate cytotoxic T cells and cooperate with B cells in the production of antibody. Antigens that require T-cell help to elicit a response are termed thymus-dependent antigens. The thymus is also a source of suppressor T cells that can down regulate the immune reaction, although the status of this activity remains a matter of debate.

A number of thymectomy experiments have been performed to determine whether the fish thymus is a source of helper T cells and/or suppressor cells. Most have used adult animals and care has been taken to examine the fish to ensure the absence of thymic remnants. Among the putative thymus-dependent antigens that have been used are sheep erythrocytes (SRBC) and the soluble proteins, human gamma globulin (HGG), and dinitrophenylated keyhole limpet hemocyanin (DNP-KLH). Those believed to be thymus-independent (capable of directly stimulating B cells)

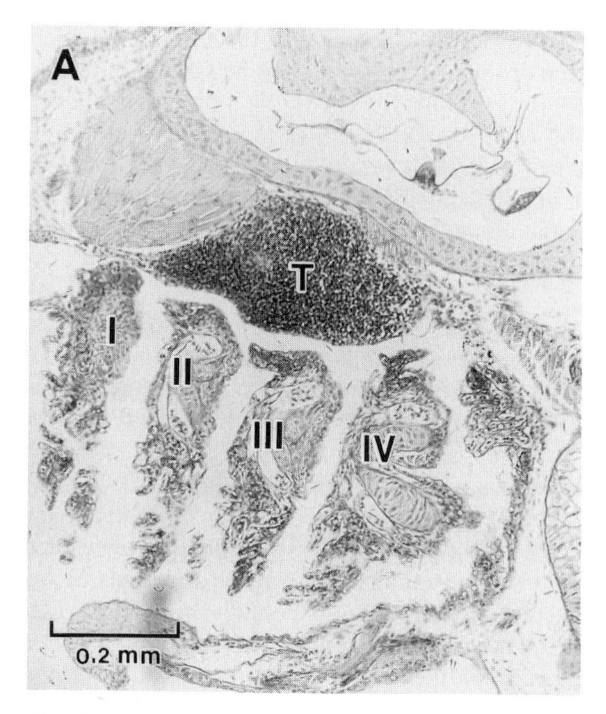
A
T
I
II
III
IV
0.2 mm

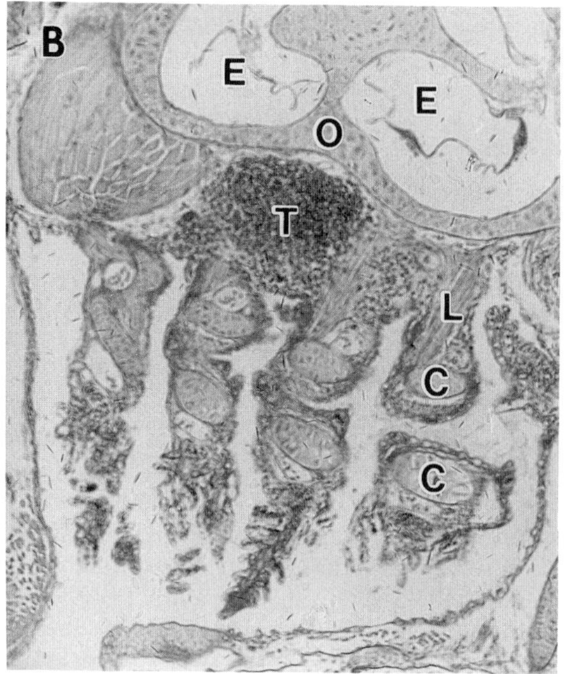
B
E
E
O
T
L
C
C

include polyvinylpyrrolidone (PVP) and killed cells of the bacterium *Aeromonas salmonicida.* These experiments are discussed in Chapter 6 in relation to ontogenetic development and aging.

Adult thymectomy failed to affect antibody production to thymus-independent antigens in rainbow trout (*A. salmonicida,* Manning *et al.,* 1982a; Tatner *et al.,* 1987) or in tilapia, *Oreochromis nilotica* (PVP, Jayaraman *et al.,* 1979). On the other hand, when thymus-dependent antigens are administered, reduced antibody levels have been recorded, but usually only in long-term experiments. Thus in rainbow trout, immunization with HGG at 9 months after adult thymectomy led to lowered antibody production, whereas at 5 months antibody production was normal (Tatner *et al.,* 1987). Similarly, Findlay (1994) obtained normal antibody levels against SRBC and DNP-KLH after short-term (2-month) thymectomy whereas in long-term thymectomized adults, antibody production was significantly reduced. In tilapia, a reduced response to SRBC was found even after short-term adult thymectomy (Jayaraman *et al.,* 1979). A reduced response to SRBC also occurred in the rockfish, *Sebastiscus marmoratus* (Nakanishi, 1986b).

There have been very few attempts at early thymic ablation in fish fry, largely because of the difficulty of the operation. However, microcautery has been used to destroy the thymus of young rainbow trout. Fish that were thymectomized at 1 month posthatch and tested for primary and secondary responses at 2 and 3 months showed normal antibody responses against *A. salmonicida* vaccine, while against HGG the primary response was unaltered but the secondary response was decreased (Tatner, 1986). In contrast, in young rockfish, thymectomy at 1.5 months of age followed by immunization with SRBC 2 weeks later, resulted in a reduced primary antibody response to SRBC accompanied by a marked decrease in the numbers of lymphocytes in the spleen and pronephros (Nakanishi, 1991). Using even younger fish, trout thymectomized at day 14 posthatch were able to produce a proliferative reaction *in vivo* to HGG in Freund's complete

---

Fig. 4. Longitudinal sections through young carp fry, *Cyprinus carpio,* of 9-mm total length (snout to tip of tail): Photographs to illustrate the position of the thymus in relation to early thymectomy experiments. (A) Section through the middle of the thymus. (B) Section through the medial site of the thymus indicating the proximity of the thymus to muscles of respiration. The longitudinal muscle fibers of the levators of the branchial arches run from the otic capsules on to the upper elements of each branchial arch. At the level of this section, note that the thymus lies just above branchial arches II and III, and muscle fibers from these arches intermingle with the thymus at its anterior and posterior edges. Key: I, II, III, and IV, first, second, third, and fourth branchial arches; C, cartilage of the fourth branchial arch; L, levator muscle of the fourth branchial arch; E, ear; O, cartilage of the otic capsule; T, thymus.

adjuvant (measured by tritiated thymidine autoradiography) when immunized 1 month after thymectomy and killed 4–6 weeks later. The response was, however, less intense than in control sham-operated fish (Manning *et al.,* 1982a).

In addition to this tentative *in vivo* evidence for the presence of thymus-derived helper cells in fish, adult thymectomy experiments suggest that in some circumstances the thymus can have a suppressive role. Thus, when adult carp were thymectomized then immunized 4 weeks later, the fish showed elevated serum antibody titers at day 7 to *A. salmonicida* vaccine (Manning *et al.,* 1982b). In rainbow trout, short-term thymectomy in one-year-old fish led to a marked increase in the antibody response to HGG (Manning *et al.,* 1982a,b), while in the adult thymectomy experiments of Findlay (1994), short-term (2 months after thymectomy) immunization with *A. salmonicida* vaccine failed to affect antibody production, whereas in longer-term (8-month) thymectomy experiments the secondary antibody response was elevated.

4. Use of Irradiation

Thymic ablation has also been attempted employing irradiation, either alone (Desvaux and Charlemagne, 1983) or in combination with adult thymectomy (Nakanishi, 1986b). X-irradiation of one-year-old goldfish caused acute depletion of thymocytes at day 1 after irradiation with no sign of thymic repopulation until day 11. The irradiated fish showed a normal initial inflammatory reaction to scale allografts but subsequently (from day 4 onward), the rejection process was impaired and in some cases grafted scales remained healthy for weeks (Desvaux and Charlemagne, 1983). Nakanishi (1986b) working with two- to three-year-old rockfish also found that x-irradiation retarded scale allograft rejection but that rejection times were not further influenced by thymectomy. On the other hand, short-term thymectomy coupled with irradiation enhanced antibody levels to SRBC when compared with the low antibody titers found in nonthymectomized/irradiated fish. It was suggested that suppressor cells resistant to X-rays exist in the fish thymus.

5. Reconstitution Experiments

An important confirmation of the validity of thymectomy findings can come from experiments in which the deficits are reversed by replacing the lost cells. Ideally such experiments require the use of histocompatible lines of fish. Because these were not available at the time, however, restoration after thymectomy has been restricted to the use of autologous thymus or thymic-derived cells. In the experiments of Nakanishi (1986b), the thymus was removed prior to irradiation and autotransplanted back into the fish after irradiation. An essentially similar outcome was obtained by Desvaux and Charlemagne (1983) by shielding the intact thymus during irradiation. These strategies were effective in counteracting the results of thymic ablation. Findlay (1994) reconstituted thymectomized rainbow trout using autologous thymocytes cryopreserved by the method of Tatner (1988) and Tatner and Findlay (1992). The cells were thawed and returned via the caudal vein at the desired time, namely 1 week before secondary immunization. The effects of adult thymectomy already described (Findlay, 1994), i.e., the reduced secondary response to SRBC and DNP-KLH and the elevated secondary response to *A. salmonicida,* were partly restored to normal when the fish were reconstituted with their own cryopreserved cells.

Overall, the outcome of removal of the thymus in fish appears to depend on the completeness of the operation, the stage of development of the fish, the interval between thymectomy and immunization, and doubtless many other factors, including the type and dose of antigen used. The results are compatible with the premise that the thymus is a source of T cells but several matters remain to be clarified; for example, do fish possess distinct subpopulations of thymus-derived cells similar to those of mammals? New possibilities for reconstitution experiments based on the availability of histocompatible animals and, hopefully, on the eventual emergence of well-defined markers for fish T cells may soon provide an impetus for these problems to be revisited.

III. *IN VITRO* STUDIES ON LYMPHOCYTE POPULATIONS

A. T-Cell/B-Cell Heterogeneity

It is now well established that teleost fish possess lymphocyte populations analogous in many respects to the T cells and B cells of mammals (Clem *et al.,* 1991). Early evidence for this heterogeneity in fish was circumstantial, including various functional studies performed *in vitro* on lymphocytes obtained from different organ sources (Etlinger *et al.,* 1976), and demonstrations *in vivo* of a hapten-carrier effect, indicating that production of antibody against a hapten by B cells required carrier-specific (putative T-cell) cooperation (Avtalion *et al.,* 1976; Stolen and Makela, 1976; Ruben *et al.,* 1977). This view of a T-cell/B-cell dichotomy in fish has recently been strengthened by cell separation techniques based on the use of monoclonal antibodies (Sizemore *et al.,* 1984; Graham and Secombes, 1990).

In higher vertebrates, B lymphocytes display immunoglobulin (Ig) on their surface whereas T lymphocytes are characterized by the presence of a different type of antigen-specific receptor, the T-cell receptor (TCR). In fish, monoclonal antibodies raised against homologous serum immunoglobulin have been used to label and separate surface (s)Ig^{+ve} B lymphocytes from the sIg^{-ve} population (which comprises putative T cells together with some other leukocyte types). In these experiments the sIg^{+ve} cells were found to be quite numerous in the lymphocyte populations of the blood, spleen, and kidney, comprising up to 40%, whereas in the thymus only about 2–5% of cells were positively labeled (Lobb and Clem, 1982; Sizemore *et al.,* 1984 [in the channel catfish]; De Luca *et al.,* 1983; Thuvander *et al.,* 1990 [in rainbow trout]; Secombes *et al.,* 1983; Van Diepen *et al.,* 1991 [in carp]; Navarro *et al.,* 1993 [in the gilthead seabream, *Sparus aurata*]). Ultrastructural studies using immunogold-labeled anti-fish Ig revealed that the surface Ig occurs mainly in clusters on the B-cell membrane (Van Diepen *et al.,* 1991; Navarro *et al.,* 1993). When isolated sIg^{+ve} and sIg^{-ve} populations from rainbow trout were radiolabeled *in vitro,* then returned to the fish, their migratory pathways differed. For example, the sIg^{-ve} cells from the blood moved preferentially into the kidney when reinjected into their autologous donor. This phenomenon resembles the homing (ecotaxis) of lymphocytes into specific regions of the lymphoid organs that occurs in higher vertebrates (Tatner and Findlay, 1991).

A method frequently used to separate fish Ig^{+ve} cells from Ig^{-ve} cells for *in vitro* studies is that of panning (Sizemore *et al.,* 1984); a technique in which antibody is used to coat the wells of plastic cell culture plates. When using anti-fish Ig, the antibody binds sIg^{+ve} cells and leaves the nonadherent Ig^{-ve} cells free in suspension. A third cell type important to the immune response, the monocyte/macrophages, can be separated on the basis of its adherent properties. For *in vitro* studies monocytes can be made to adhere to surfaces such as the wells of a microtiter plate, especially when these have been coated with fibronectin (an adhesive glycoprotein). In the experiments of Sizemore *et al.* (1984), baby hamster cell microexudate ("fibronectin")-coated culture wells were employed to isolate monocytes and to demonstrate their role as accessory cells. The ability to separate these different populations of cells has greatly clarified their roles in the immune response of fish and established the B-cell and T-cell nature of their interactions *in vitro.* Enrichment for B cells using anti-Ig monoclonal antibody techniques is quite good; for example, electron microscopy of the sIg^{+ve} fraction isolated from rainbow trout peripheral blood showed that approximately 90% of the cells were lymphocytes, the rest macrophages and neutropils, whereas the sIg^{-ve} fraction contained approximately 30% lymphocytes with the remainder of the sIg^{-ve} cells being predominantly thrombocytes (Marsden *et al.,* 1995).

Attempts to make monoclonal antibodies directed against T cells have produced clones that reacted with most homologous thymocytes and with peripheral sIg^{-ve} lymphocytes. These antibodies appear to be pan-T-cell reagents which also react with other cell types such as neutrophils and some brain cells (Secombes *et al.,* 1983; Miller *et al.,* 1987). T cells have yet to be formally identified in fish, however, as there are no available markers for possible teleost equivalents to the mammalian CD4, CD8, or CD3 molecules. Nevertheless, the rapid advances over the past five years in research into the fish MHC (see Section VI) has included interest in the T-cell antigen receptor that operates in the context of the MHC. T-cell receptor protein sequences which are believed to interact with MHC class I and class II have recently been described in rainbow trout (Partula *et al.,* 1994, 1995a). This research, although advancing rapidly, is still in its early stages. It has not yet been demonstrated in fish, for example, that any of the gene products so far identified are expressed on the surface of a cell (Dixon *et al.,* 1995).

B. Proliferative Responses to Mitogens

In mammals, a number of substances, principally plant lectins such as phytohemagglutinin (PHA) and concanavalin A (Con A), are specific for particular arrays of carbohydrate moieties that occur on T lymphocytes. When present in appropriate doses these T-cell mitogens induce proliferation of T cells but not of B cells. Other mitogens such as bacterial lipopolysaccharides (LPS) activate B cells but not T cells. The majority of studies on mitogen-induced proliferation in fish have been done on teleosts, this being the group for which optimal culture conditions (DeKoning and Kaattari, 1992) and good separation techniques are well established.

In the more primitive lampreys (agnathans), Cooper (1971) showed that lymphocytes from ammocoete larvae respond to PHA *in vitro* by the production of blast cells, suggesting the existence of T-like cells, while similar studies in cartilaginous fish (the nurse shark, *Ginglymostoma cirratum*) using Con A and PHA again yielded a T-cell like response (Lopez *et al.,* 1974; Sigel *et al.,* 1978; Pettey and McKinney, 1981). In these experiments on nurse sharks, leucocyte populations were separated on Percoll density gradients and by their adherence to glass beads. Cells cytotoxic to xenogeneic targets were present among the adherent population while cells of the nonadherent component showed suppressor activity. These regulatory suppressor cells are sIg^{-ve} but do not behave in an MHC-restricted manner (Haynes and McKinney, 1991). Thus, although cells activated by T-cell mitogens are present in subpopulations of shark leukocytes, a T-cell/B-cell-like basis for the heterogeneity has yet to be established in chondrichthian fish.

In teleosts, the ability to respond to T-cell and B-cell mitogens was established some twenty years ago (Etlinger *et al.,* 1976), and a number of studies followed (reviewed by Rowley *et al.,* 1988). These include reports on the channel catfish (Sizemore *et al.,* 1984), carp (Caspi and Avtalion, 1984a), rainbow trout (Warr and Simon, 1983; Tillit *et al.,* 1988; Reitan and Thuvander, 1991), and Atlantic salmon, *Salmo salar,* (Smith and Braun-Nesje, 1982; Reitan and Thuvander, 1991). Documentation of the *in vitro* culture requirements for each species has been important (Rosenberg-Wiser and Avtalion, 1982; Faulmann *et al.,* 1983; DeKoning and Kaattari, 1991). For example, the addition of fetal bovine serum to the culture medium as the serum supplement may not suffice; in salmonids the cells performed better if homologous plasma was used (DeKoning and Kaattari, 1992).

The requirement for accessory cells for the activation of T lymphocytes by mitogens was confirmed for channel catfish peripheral blood lymphocytes by Sizemore *et al.* (1984). It now seems likely that these accessory cells (monocytes) were acting by the secretion of cytokines, possibly interleukin 1 (IL-1) (Miller *et al.,* 1985; Clem *et al.,* 1985). Sizemore *et al.* (1984) separated the sIg^{+ve} cells from sIg^{-ve} cells by panning, and demonstrated that the sIg^{+ve} cells responded to LPS regardless of the presence or absence of monocytes, whereas the sIg^{-ve} population remained unresponsive to either LPS or Con A unless the accessory cells were present, in which case they responded to both mitogens. The reduced but significant response to LPS in the sIg^{-ve} population remains problematic. A similar response of sIg^{-ve} cells to LPS was reported by DeLuca *et al.* (1983) using pronephric leukocytes of rainbow trout. On the other hand, Marsden *et al.* (1995) observed a marked dichotomy in the proliferative responses of peripheral blood leukocytes of rainbow trout. The sIg^{-ve} cells were activated by PHA with very little response to LPS and vice versa for the sIg^{+ve} cell population. Whether this reflects the efficiency of the panning techniques or whether some LPS responsive cells are present in genuinely sIg^{-ve} populations remains uncertain. Koumans–Van Diepen *et al.* (1994) recently noted that, among the peripheral blood lymphocytes of carp, some sIg^{+ve} cells responding to LPS showed only dull immunofluorescent staining. These may have escaped sorting in some experiments.

In rainbow trout, thymocytes proliferate in response to PHA but not to LPS (Reitan and Thuvander, 1991). Essentially similar results were obtained for channel catfish using the thymus from fish about 14 months old and stimulating with Con A (Ellsaesser *et al.,* 1988). Accessory cells (monocytes) were required for the T-cell activation and either autologous or allogeneic cells could be used. Again there was a small response in some cases to LPS. However, this does not detract from the general conclusion that thymocytes are Ig^{-ve} cells that display a T-cell-like response to mitogens. Any discrepancies may be accounted for by the suggestion that the thymus can play a minor role as a residence for a small population of B cells (see Manning, 1994).

C. Mixed Leukocyte Reactions

The mixed leukocyte reaction (MLR) is an *in vitro* test of allogeneic differences. In mammals an MLR typically occurs when a $CD4^{+ve}$ T lymphocyte responds to noncompatible MHC class II molecules on the surface of a stimulator cell (which is usually a monocyte/macrophage or a B lymphocyte). The T cell is activated to become a blast cell and to undergo clonal proliferation. This proliferation can be measured, for example, by the incorporation of tritium-labeled thymidine. The reaction involves the secretion of interleukin 2 (IL-2), a cytokine that is produced by activated $CD4^{+ve}$ T cells and promotes proliferation of any T cells that have the appropriate IL-2 receptors (see later). When cells from two histoincompatible animals are mixed, the T lymphocytes from each animal respond to the stimulator cells of the other. This is known as a two-way MLR. The reaction can be made one way by preventing the proliferation of one set of cells, for example, by irradiation. In the untreated population the T cells can still respond by dividing but in the treated population the cells are only able to act as stimulators.

In fish, MLRs have been described in an agnathan, the Pacific hagfish, *Eptatretus stoutii* (Raison *et al.*, 1987), in a primitive bony fish, the holostean spotted gar, *Lepisosteus platyrhincus* (Luft *et al.*, 1994), and various teleost species: bluegill, *Lepomis macrochirus* (Cuchens and Clem, 1977); carp (Caspi and Avtalion, 1984a,b; Grondel and Harmsen, 1984; Gloudemans *et al.*, 1987); channel catfish (Miller *et al.*, 1986); rainbow trout (Etlinger *et al.*, 1977; Kaastrup *et al.*, 1988); and immature Atlantic salmon (Ellis, 1977). These studies included one-way reactions in the experiments of Caspi and Avtalion (1984b), Miller *et al.* (1986), and Gloudemans *et al.* (1987). Kaattari and Holland (1990) point out the considerable variation in levels of irradiation apparently needed for successful one-way MLRs in different teleost species.

The cellular requirements for the MLR by channel catfish peripheral blood leukocytes has been defined by Miller *et al.* (1986) who separated sIg^{+ve} (B) cells from sIg^{-ve} cells. They also either isolated the monocytes on microtiter plates or depleted them by adherence to Sephadex G10. The results indicated that the sIg^{-ve} (T) cells were those responding by proliferation in the MLR, with monocytes being required as accessory cells. In contrast, monocytes, sIg^{+ve} (B) cells, and sIg^{-ve} could all act as stimulators. These findings for fish are very similar to those for mammals except that mammalian T cells do not normally act as stimulators in MLR. However, this could be explained by the possibility that not all the sIg^{-ve} cells were T cells. There may be some powerful MLR stimulators in the residual leukocyte sIg^{-ve} population.

The production of factors with IL-2–like activity in an MLR of fish has been demonstrated for carp (Caspi and Avtalion, 1984a; Grondel and Harmsen, 1984) with peripheral blood and pronephric leukocytes. Cytokine-like factors were generated into the culture supernatants that had an IL-2–like effect in promoting target blast cells to proliferate. The target cells in these experiments were leukocytes that had been activated with PHA, i.e., putative T-cell lymphoblasts. Interleukin 2–like supernatants could also be found in carp leukocyte cultures activated by mitogens (Caspi and Avtalion, 1984a; Grondel and Harmsen, 1984), and possibly by antigens (Pourreau *et al.,* 1987). This suggests an amplifying or regulatory role of fish T cells similar to that in higher vertebrates.

D. Antigen Presentation and Cellular Cooperation

The role of accessory and antigen-presenting cells in antibody production is described in Chapter 5. In mammals, in addition to the population of $CD4^{+ve}$ helper cells which are effective in eliciting B-cell responses, other $CD4^{+ve}$ cells mediate functions associated with specific cytotoxicity and local inflammatory (cytokine-induced) reactions. These latter aspects (of a possible T-cell/T-cell cooperation) have yet to be defined in fish. Nor is it known whether MHC-restricted antigen-presenting cells of the mammalian dendritic type occur at this stage in phylogeny.

An impetus to studies on fish T-cell reactions comes from the recent availability of functionally active long-term lines of monocyte/macrophages (Vallejo *et al.,* 1991a; Wang *et al.,* 1995). In the channel catfish, these accessory cell lines, as well as lines of T-like cells and of B cells, can be readily established and cloned from normal peripheral blood leukocytes. Unlike with mammals, no special steps were required to transform the leukocytes of catfish. Access to these long-term cell lines makes it possible to keep alive the fish that donated the original cell line for use as a source of autologous cells in future experiments, using the cell line for accessory cells or as target cells (Vallejo *et al.,* 1991a, 1992a; Miller, N. W. *et al.,* 1994). When cells from long-term monocyte-like lines were antigen-pulsed and cocultured with autologous peripheral blood leukocytes from an antigen-primed fish, specific, secondary, *in vitro* proliferative and antibody responses to various thymus-dependent antigens were elicited. These responses failed to occur when the peripheral blood leukocytes were obtained from an allogeneic antigen-primed channel catfish, i.e., not from the original cell line donor. These results have been interpreted as providing evidence that MHC molecules govern antigen presentation and that putative MHC restriction of teleost immune responses ocurs, reminiscent of that found in mammals. In other words, T cells are restricted to interacting with antigen-presenting cells that bear self MHC molecules (Vallejo *et al.*, 1991b,c). The use of such long-term culture lines has recently been applied to studies on specific T-cell cytotoxicity as described in the next section (Yoshida *et al.*, 1995).

E. Specific Cytotoxicity

Cells that can recognize and lyse foreign cells are present in fish. The nonspecific cytotoxic cell (NCC), which is active against xenogeneic targets but not against allogeneic or autologous cells, bears some resemblance to mammalian natural killer (NK) cells and is well documented (see Chapter 2). Other cytotoxic cells with activities rather more akin to mammalian cytotoxic T cells have been described but these have yet to be formally identified and their nature remains problematic. Such cells, which react against modified self cells or against allogeneic cells, have been reported recently in different model systems.

In the study of Verlhac *et al.* (1990), cytotoxicity against hapten-modified autologous cells was detected *in vitro.* Carp kidney leukocytes were sensitized in culture to trinitrophenyl (TNP)-modified spleen cells derived from the same fish. Lysis of the target cells by the sensitized kidney cells was demonstrated by chromium (^{51}Cr) release. It was further demonstrated that the response against autologous TNP-modified target cells was considerably greater than that against allogeneic TNP-modified targets, suggesting that MHC restriction was involved. This is similar to the phenomenon in mammals in which cytotoxic T cells recognize antigen in association with self MHC (class I) surface molecules. In fish, however, the kidney leukocyte that acts as the effector cell has not been identified and the antigen specificity of the reaction has yet to be established.

More recently, channel catfish long-term culture lines (already described in Section D) have been used as a source of either allogeneic or autologous target cells (Yoshida *et al.*, 1995). sIg^{-ve} cells were obtained by depleting peripheral blood leukocyte samples of their B cells and monocytes. These sIg^{-ve} cells were T-like in their ability to kill allogeneic but not autologous leukocytes as shown by the release of ^{51}Cr from labeled long-term culture line targets. They differed from nonspecific cytotoxic cells in their lack of cytotoxicity for xenogeneic (mammalian) lymphoma cell lines. The killing of allogeneic cells was not, however, well defined since cells from an individual fish did not discriminate between different allogeneic cell line targets. This lack of allogeneic discrimination was substantiated by the use of competitive inhibition assays in which inhibition of lysis of the labeled allogeneic targets could be brought about by adding unlabeled cells from different allogeneic lines as well as from the same target line.

These studies on long-term lymphoid cell cultures have also been extended to target cells that were infected with channel catfish virus. Here the cytotoxic peripheral blood leukocytes were able to lyse both allogeneic and autologous virus-infected cells; unlike mammalian T cells, the catfish cells were not MHC-restricted in their cytotoxic activity (Chinchar *et al.*, 1994). These experiments, therefore, demonstrate further populations of fish cytotoxic cells in addition to the previously described NCC cells. How these relate to mammalian T-cell and/or NK-cell lineages has yet to be elucidated but advances in the identification of fish MHC and T-cell receptor molecules may soon help to clarify the situation.

In higher vertebrates the generation of specific $CD8^{+ve}$ cytotoxic T cells occurs when the resting T cell recognizes the MHC class I antigen on an allogeneic cell and is activated to express IL-2 receptors. An IL-2 signal (normally provided by activated $CD4^{+ve}$ T helper cells) then induces its proliferation and maturation into a functional cytotoxic T cell. This reaction can be demonstrated *in vitro* for mammalian cytotoxic T cells using prospective target cells as the stimulators in a one-way MLR and adding ^{51}Cr-labeled blast cells (obtained by mitogenic stimulation) as the target cells. Such assays have not yet been recorded for fish, however (Stet and Egberts, 1991). Nor, to our knowledge, have the modifications adopted by Bernard *et al.* (1979) for the amphibian *Xenopus laevis* been attempted in fish. These authors found that in *Xenopus, in vivo* priming of the responder with the prospective target cells was required before restimulation was carried out *in vitro.*

IV. CYTOKINES IN FISH

A. Identification of Fish Cytokine Activity

Interactions between cells of the immune system are mediated not only by direct cell-to-cell contact but also through the release of soluble factors (cytokines). Cytokines play a regulatory or enhancing role within the immune system, and their range is usually limited to cells in the immediate vicinity of the cytokine-producing cell. Fish produce a number of cytokine-like soluble products that act to orchestrate the events in an immune response. Most of these have been identified in biological assays on the basis of their functional similarity to mammalian cytokine activities. Some have been detected through their cross-reactivity with mammalian cytokines. Tumour necrosis factor α, for example, is a macrophage-derived cytokine with high amino acid homology between species and little species specificity (Hardie *et al.,* 1994). Only recently have molecular techniques been employed (see Secombes, 1994a,b). The polymerase chain reaction (PCR) approach using genomic DNA from a flatfish cell line has given a sequence for IL-2 (Tamai *et al.,* 1992). These authors have also cloned cytokine cDNA and identified an interferon-like antiviral protein sequence in flatfish, *Paralichthys olivaceus* (Tamai *et al.,* 1993). The current status of cytokine research in fish is summarized in Table I.

B. Cytokines Detected in Fish

1. Interleukin 1 (IL-1)

Interleukin-1 is a conserved molecule that appears to have evolved early in the phylogeny of the vertebrates and can cross species barriers in its effects. It manifests a wide range of activities including important functions within the immune system, and is produced mainly by macrophages but also by a variety of other cell types. In mammals, IL-1 serves as a starting point for a number of cascade reactions including its role in the release of

Table I
Identification of Cytokine Activities in Fish[a]

	IL-1	IL-2	IL-3	IL-4	IL-6	TNF	IFN (α, β)	MAF (IFN-γ)	CF	MIF	TGFβ	CSF
Nurse shark										+▲		
Rainbow trout	+△		+△		+△	+△,■	+▲	+▲		+▲	+■,●*	+▲
Carp	+▲,△	+▲			+△	+△			+▲	+▲		
Catfish	+▲,■	+▲		+▲								
Flounder		+●					+●					
Tilapia										+▲		

[a] ●, Identified at cDNA level; ▲, suggested by the function; △, suggested by the cross reaction with antibodies against mammalian cytokines; ■, suggested by the biological cross reactivity with mammalian cytokines; ●*, C. J. Secombes, pers. commun. For references see text.

IL-2 from stimulated T cells and the enhancement of T-cell expression of the IL-2 receptor.

In fish, it has been shown that channel catfish peripheral blood lymphocytes can recognize and respond to human IL-1; also, fish cells can themselves produce an IL-1-like substance. Interleukin 1, which stimulated proliferation in catfish peripheral blood lymphocytes, could be obtained from carp epithelial cell lines (Sigel *et al.*, 1986). An IL-1-like agent was also demonstrated in catfish monocyte culture supernatants. This could substitute for the requirement for monocytes as accessory cells in the *in vitro* immune responses described in Section III (Clem *et al.*, 1985). Interleukin 1 is one of several putative cytokines reported by Ahne (1993, 1994) to be present in the serum of carp infected with spring viremia carp virus and of rainbow trout infected with hemorrhagic septicemia virus or with the virus of infectious hematopoietic necrosis. These serum factors were identified on the basis of their cross reactivity with antibodies against mammalian cytokines using commercially available ELISA systems. They were detected only in the sera of fish that showed neutralizing antibodies to the virus under investigation.

2. INTERLEUKIN 2 (IL-2)

Interleukin-2 was formerly known as T-cell growth factor in recognition of its ability to induce the proliferation of T-cells. In mammals, IL-2 is produced primarily by helper ($CD4^{+ve}$ T) cells although other subsets of T cells may also be involved. Additionally, IL-2 stimulates other cells of the immune system such as NK cells. The receptor for IL-2 is transiently expressed on the T-cell surface following appropriate activation. In fish, soluble factors with IL-2-like activity have been detected following T-cell activation *in vitro* (Caspi and Avtalion, 1984a; see Section III). In the channel catfish, long-term culture lines of T cells immortalized by stimulation with phorbol ester and ionophore (Lin *et al.*, 1992) constitutively produce IL-2-like cytokines (Miller, N. W. *et al.*, 1994). An IL-2 gene has recently been characterized in cultured flatfish leukocytes (Tamai *et al.*, 1992).

3. INTERLEUKIN 4 (IL-4)

In mammals, T cells also release IL-4, a cytokine that stimulates B cells, acting early in the response to trigger the normal B-cell activation pathway. In the channel catfish, the long-term T-cell cultures described above not only produced IL-2 but also a factor that stimulated B cells, i.e., a factor with IL-4-like activity (Miller, N. W. *et al.*, 1994). To date, however, there is only tentative evidence for the occurrence of IL-4 in fish.

4. Interleukin 3 (IL-3) and Interleukin 6 (IL-6)

Mammalian IL-3 is a multipotential hemopoietic cell growth fa[illegible] stimulates growth in the early stages of both lymphoid and hemo[illegible] cell lineages. Interleukin 6, like IL-3 and tumour necrosis factor α ([illegible] is produced early in the mammalian inflammatory response. Interleu[illegible] is also known as interferon (IFN) β_2 because of its antiviral activity. It [illegible] has an effect in promoting B-cell proliferation. Together with IL-1 a[illegible] TNF α, IL-3 and IL-6 have been detected in the serum of virus-infecte[illegible] carp and rainbow trout on the basis of their cross-reactivity with antibodie[illegible] against mammalian cytokines (Ahne, 1993, 1994) but any biological activity of IL-3 and IL-6 in fish has yet to be investigated.

5. Interferons and Macrophage-Activating Factors

The interferons (IFNs) form a heterogenous family of proteins that confer protection against viral infections. They fall into three groups, IFN-α, IFN-β and IFN-γ. IFN-γ has a higher molecular weight than IFN-α and -β, and is termed *immune interferon* in recognition if its role in immune reactions. In mammals, IFN-γ is produced by T cells that have been stimulated by antigens or by mitogens.

It has been known for some time that interferon synthesis can be triggered in fish, either *in vivo* or in cell culture, following infection by various pathogenic viruses (Gravell and Malsberger, 1965; De Sena and Rio, 1975; De Kinkelin *et al.,* 1982). These fish interferons appear to be of the IFN-α or IFN-β types. In young rainbow trout, they have been shown to appear in the serum in response to infectious pancreatic necrosis virus (IPNV) early in ontogeny (Dorson *et al.,* 1992).

Immune interferon (IFN-γ)-like activity has been demonstrated more recently in supernatants from rainbow trout leukocytes stimulated by mitogens. This trout IFN-γ is a macrophage-activating factor (MAF) that induces an elevated respiratory burst in trout macrophages, with increased bacterial killing (Graham and Secombes, 1988). These MAF-containing supernatants conferred viral resistance on a rainbow trout epithelial cell line challenged with IPN virus. They were shown to be produced by sIg^{-ve} cells (Graham and Secombes, 1990; see Chapter 2). Fish, like mammals, may therefore gain protection not only from the broader interferons (α and β) which are induced in cells infected with virus, but also from IFN-γ-like cytokines produced by stimulated T cells.

6. Tumour Necrosis Factor (TNF)

Mammalian TNF is involved in cell-mediated cytotoxic reactions. It is produced by macrophages and there appears to be a synergy between TNF and IFN-γ in the killing of target cells. Like IL-1, TNF is a conserved molecule that manifests a wide range of activities, some of which overlap with those of other cytokines, including IL-1. Tumor necrosis factor α is an important mediator of the host response to Gram-negative bacteria and also plays a role in viral and parasitic infections (Secombes 1994c). In fish, Zelikoff *et al.* (1990) reported the presence of TNFα in rainbow trout macrophages. It was also one of the cytokines detected by Ahne (1993) in the serum of virus-infected fish. Furthermore, it has recently been shown that rainbow trout lymphocytes and macrophages can respond functionally to human recombinant TNFα which suggests that fish leukocytes may possess a specific TNFα receptor. Such cross-reactivity implies an evolutionary conservation of the TNFα receptor—a hypothesis that is supported by the ability of monoclonal antibodies against the human TNF receptor to ablate the fish response to TNF (Jang *et al.,* 1995). It has been found that the response of the trout cells to TNF requires additional signals, from mitogens in the case of lymphocytes or from IFN-γ for macrophages, to enable the TNF to elicit its effect (Hardie *et al.,* 1994; Jang *et al.,* 1995; see also Chapter 2).

7. Transforming Growth Factor β_1 (TGFβ_1)

Transforming growth factor β_1 is a cytokine that deactivates macrophages in mammals, down-regulating the heightened reactions initiated by factors such as macrophage-activating factor (MAF). There is recent evidence that fish macrophages possess receptors that are cross-receptive to mammalian TGBβ_1. Thus bovine TGFβ_1 can down regulate rainbow trout macrophages following their activation by MAF (Jang *et al.,* 1994). A trout cDNA sequence for TGF has recently been obtained with 68% predicted amino acid homology to mammalian TGFβ_1 (Secombes, pers. comm.).

8. Chemotactic Factor (GF) and Macrophage Migration Inhibition Factor (MIF)

Cytokines that affect the movements of leucocytes play a role in the inflammatory responses of mammals. Cytokine-like factors have been described in fish that possibly attract leucocytes to the site of a reaction (CF) or prevent them moving away (MIF). Such factors have been detected by their biological activities in several fish species.

A MIF-like activity has been reported in elasmobranchs and holosteans (McKinney *et al.,* 1976; Manning *et al.,* 1982b) as well as in teleosts (in tilapia [Jayaraman *et al.,* 1979], carp [Manning *et al.,* 1982b], rainbow trout [Blazer *et al.,* 1984; Thomas and Woo, 1990], and eel [Song *et al.,* 1989]). In carp, a chemotactic factor (CF) was detected following stimulation with mitogens or with antigens (Bridges and Manning, 1991), while in rainbow trout, human TNFα had a chemoattractant effect on head kidney neutrophils that could be inhibited by preincubation of the neutrophils with monoclonal antibodies against the TNFα receptors (Jang *et al.,* 1995).

9. Other Mediators of the Inflammatory Response

Other soluble factors that act as mediators of the inflammatory response in fish include those of the complement system (described in Chapter 3) and eicosanoids (leukotrienes and lipoxins) which are produced by fish macrophages (Rowley 1991; see also Chapter 2). In addition, platelet activating factor (PAF) has been identified in carp on the basis of its physicochemical properties, although its biological actions in fish have yet to be verified (Yamaguchi *et al.,* 1990). There is also evidence in fish of a cytokine-like factor that stimulates hemopoietic precursor cells toward the growth of monocyte/macrophage colonies (Kodama *et al.,* 1994). This activity is similar to that of mammalian colony stimulating factor.

V. EFFECTS OF TEMPERATURE

Temperature probably had a major influence on the early evolution of the vertebrate immune system, especially as water temperatures may undergo considerable seasonal and other fluctuations. This could have particular relevance in the phylogeny of the bony fishes, some of whose ancestors in the Palaeozoic era may have dwelt in relatively shallow fresh waters that were subject to temperature changes. As might be expected, ambient temperature has a marked influence on immune reactivity in fish, including the kinetics of specific cell-mediated responses (allograft rejection) (Hildemann, 1970; Rijkers, 1982). In general, the best responses are obtained at the normal summer temperatures of the species concerned. These, of course, vary considerably according to whether the fish is a cold-water, temperate, or warm-water species (O'Neill, 1980). The temperature range at which optimal immune responses are obtained is termed immunologically "permissive." Temperatures below these, but still within the physiological range, tend to be immunosuppressive (nonpermissive temperatures).

Within the complex events that comprise an immune response, some phases are more temperature sensitive than others. This provides a means of analyzing the fish's immune reaction by deliberately raising or lowering the temperature during given stages in the response. This was first attempted *in vivo* by Avtalion and his coworkers studying the humoral antibody response to thymus-dependent antigens in carp. Normal antibody responses could be obtained at nonpermissive temperatures so long as primary immunization was carried out within the permissive temperature range (Avtalion *et al.,* 1976). Also, hapten-carrier responses which are normally inhibited at low temperatures could occur if the carrier preimmunization was performed at the higher temperature (Weiss and Avtalion, 1977). This led to the suggestion that the generation of carrier-specific primary helper (T) cell function is low-temperature sensitive.

More recent studies carried out on the channel catfish using the *in vitro* techniques described in Section III showed that T-cell proliferative responses in MLR and in response to T-cell mitogens (Con A) were suppressed at low temperatures, whereas B-cell reactions to LPS were not (Clem *et al.,* 1984). Also, results using the hapten-carrier effect indicated that virgin T cells, but not memory T cells, are suppressed at the nonpermissive temperatures (Miller and Clem, 1984). Antigen processing and presentation by accessory cells do not appear to be implicated in low-temperature immunosuppression (Vallejo *et al.,* 1992b).

Experiments on catfish in which the temperature was manipulated by moving from permissive temperatures to nonpermissive temperatures (or vice versa) at selected times after stimulation, indicated that low-temperature-induced immunosuppression inhibits the generation/activation of virgin T helper cells. Furthermore, in T-cell proliferation following mitogen stimulation, the events that are blocked at low temperatures occur relatively early in cell activation, in less than 8 h after stimulation (Clem *et al.,* 1984). Cytokine (MAF) production in rainbow trout is also inhibited at nonpermissive temperatures. Again it is the T cell that is the temperature-sensitive cell, the macrophage function itself remaining intact at low temperatures when suitably stimulated with MAF (Hardie *et al.,* 1995).

The mechanism of impairment at nonpermissive temperatures has yet to be elucidated. It was observed that cells obtained from catfish that had been acclimated *in vivo* to a lower temperature over a period of several weeks were able to generate T-cell responses *in vitro* at the formerly nonpermissive temperature (Clem *et al.,* 1984). This led to the suggestion that homeoviscous adaptation may be involved, i.e., a change in the cellular membrane composition to produce more optimal membrane viscosities (Bly and Clem, 1992). However, it now seems unlikely that low-temperature-mediated immunosuppression results from any gross inability of T cells to adapt in this respect, that is, T-cells can increase the ratio of unsaturated:saturated fatty acids in their membrane phospholipids at low temperature (Bly and Clem, 1988). Nevertheless, there may be minor changes in the T-cell plasma membrane that could be investigated in relation to the membrane-associated processes of the immune response (Clem *et al.,* 1991). Because the low-temperature block occurs early in the T-cell activation sequence and because channel catfish T cells can be induced to proliferate *in vitro* at normally nonpermissive temperatures using a combination of exogenously added phorbol ester and calcium ionophore (which together can directly activate protein kinase), the block was proposed to occur before protein kinase C activation (Bly and Clem, 1992).

It is possible that nonspecific immunity may, to some extent, offset any impairment of T-cell function at low temperatures. Low *in vivo* temperatures that inhibited the T-cell mitogenic effect of PHA in carp were found to enhance nonspecific cytotoxic cell (NCC) activity (Le Morvan-Rocher *et al.*, 1995). Also in carp, the granulocyte population was increased and the nonspecific defences were more efficient under temperature conditions unfavorable to antibody production (Rijkers *et al.*, 1981), while in catfish, monocytes produce more IL-1 activity *in vitro* at low temperatures than at higher temperatures (Bly and Clem, 1992). It may even be that, at low temperatures, a diversion of the defence mechanisms away from the more selective and finely tuned T-cell regulated responses back to the nonspecific cellular defences could have some adaptive value. Nevertheless it is known that T-cell-produced cytokines are often required to up regulate the macrophage respiratory burst to levels that will kill virulent strains of bacterial pathogens. Also, in aquaculture the phenomenon known as "winter kill" of fish could perhaps be attributed to outbreaks of infection brought about by immune incompetence at low temperatures (Bly and Clem, 1992).

VI. MAJOR HISTOCOMPATIBILITY COMPLEX (MHC) IN FISH

The highly polymorphic cell surface structures involved in rejection are called MHC antigens. These were initially characterized using alloantibodies produced in one inbred strain of mice immunized with cells of other strains differing only at the MHC. The MHC encodes two classes (I and II) of structurally and functionally distinct glycoproteins that present antigenic peptides to T cells and thus initiate specific immune responses. The MHC is located on the short arm of chromosome 6 in humans (the HLA region) and chromosome 17 in the mouse (the H-2 region) as a single cluster of genes. Though the β_2-microglobulin (*B2m*) gene is not located in the MHC, its gene product is noncovalently associated with MHC class I.

Class I molecules are expressed ubiquitously on all nucleated cell surfaces and present peptides derived from endogenously synthesized proteins to $CD8^{+ve}$ T cells (cytotoxic T cells, Bjorkman and Parham, 1990). The class I molecule comprises a glycosylated heavy chain (45 kDa) and B2m (12 kDa). The class I heavy chain consists of three extracellular domains, designated $\alpha 1$ (*N*-terminal), $\alpha 2$, and $\alpha 3$, a transmembrane region, and a cytoplasmic tail. Among the various class I molecules, some are highly polymorphic (classical class I genes), while others are less polymorphic (nonclassical class I genes).

Distribution of class II molecules is rather specific and expressed mainly in B lymphocytes, macrophages, and activated T cells, which function as the antigen-presenting cell. Class II molecules present peptides derived from exogenously acquired proteins to $CD4^{+ve}$ T cells (helper T cells, Lanzaveichia, 1990). Class II molecules are heterodimers of heavy (α) and light (β) glycoprotein chains. The α chains have molecular weights of 30–34 kDa and the β chains range from 26 to 29 kDa depending on the locus involved.

Although many efforts have been devoted to the identification of MHC-like genes or molecules in fish with conventional approaches, including the cross-hybridization of fish DNA or RNA with probes of higher vertebrates, and the cross-reaction of fish cell-surface molecules with antibodies against mammalian and avian MHC molecules (reviewed by Kaufman *et al.,* 1990; Stet and Egberts, 1991), none of them have succeeded. Hashimoto *et al.* (1990) adopted another strategy for the isolation of MHC genes and first demonstrated the existence of both MHC class I and class II genes in carp using the polymerase chain reaction (PCR). In the isolation of carp MHC genes, they chose two conserved regions surrounding two cysteines in the MHC membrane-proximal domains (class Iα3 and class IIβ2), which belong to the C1 set of the immunoglobulin (Ig) superfamily, as primer sites, since several amino acid residues surrounding the cysteines involved in the intramolecular disulfide bond are well conserved among man, mouse, and chicken (Hashimoto and Kurosawa, 1991).

A. Characteristics and Structure of Fish MHC Genes

Since Hashimoto *et al.* (1990) succeeded in isolating the MHC genes from carp, a number of fish MHC genes have been isolated and sequenced. Now all MHC genes such as class IA, *B2m*, class II*A* and class II*B* genes have been found in both teleosts and elasmobranchs, but not in cyclostomes (Table II). In general, overall organization of fish MHC genes is remarkably similar to that found in their mammalian homologues, although there exists great sequence divergence between fish and mammals. The amino acid sequence similarity between fish and mammalian MHC molecules (class Iα, class IIα, and IIβ chains) is quite low and homology is at most 40%. The position of the glycosylation site and the disulfide forming cysteine residues is well conserved across all species, although the location of the glycosylation site is different among species. In all fish MHC genes studied so far, nearly all the introns are less than 1 kb long and most are less than 100 bp long, therefore short introns seem to be characteristic of fish MHC

Table II
Current Status of the Isolation of MHC Genes in Fishes

Species	Genes[a]	Clones	Sequence type	Reference[b]
Teleosts				
Carp	Class I*A*	*Cyca*-ZA	genomic	1
		Cyca-ZA, ZB, ZC	cDNA	2
		Cyca-UA, -TC	cDNA	3
		Cyca-Z	genomic	4
	Class II*B*	*Cyca*-TLAII(YB)	genomic	1
		Cyca-DAB	cDNA, genomic	5
		Cyca-YB	genomic	4
	B2m	*Cyca*-B2m	cDNA	6
Ginbuna	Class I*A*	*Caau*-ZA, -ZD	cDNA, genomic	2
Zebrafish	Class I*A*	*Brre*-UA(~C)A	cDNA, genomic	7
	Class II*A*	*Brre*-DXA	genomic	8
	Class II*B*	*Brre*-DAB1~4	cDNA	9
		Brre-DA(~F)B	genomic	10
	B2m	*Brre*-B2m, -B2m-G	cDNA, genomic	11
Tilapia	*B2m*	*Orni*-B2m	genomic	6
Cichlid	Class II*B*	*Auha*-M, *Cyfr*-T	cDNA, genomic	12
		Necy-T, *Nili*-M, *Nive*-M, *Pepu*-N, *Pslo*-M, *Psze,* *Thsp*-V	genomic	13
		Meau-, *Mech*-M	genomic	14
Atlantic salmon	Class I*A*	*Sasa*-p23, -p30	cDNA	15
		Sasa-A1-4, -B1	genomic	16
	Class II*B*	*Sasa*-DC, -DB	cDNA	17
		Sasa-DB	genomic	16
Rainbow trout	Class II*B*	*Onmy*-55	cDNA	18
	B2m	*Onmy*		19
Striped bass	Class II*A*	*Mosa*-A	cDNA	20
	Class II*B*	*Mosa*-C, -R, -S	cDNA	21
Perch-like fish	Class II*B*	*Pefl, Gyce*	genomic	22
Crossopterygian				
Coelacanth	Class I*A*	*Lach*-UA, UB, UC, UD	cDNA, genomic	23
Elasmobranchs				
Banded dogfish	Class I*A*	*Trsc*-DS	genomic	24
Nurse shark	Class II*A*	*Gici*-DAA, -DBA	genomic	25
				26
	Class II*B*	*Gici*-8, -11	cDNA	27

[a] The gene designation was shown according to the proposal by Klein *et al.* (1990).

[b] 1. Hashimoto *et al.* (1990). 2. Okamura *et al.* (1993). 3. Van Erp *et al.* (1994). 4. Stet *et al.* (1993). 5. Ono *et al.* (1993a). 6. Dixon *et al.* (1993). 7. Takeuchi *et al.* (1995). 8. Sültmann *et al.* (1993). 9. Ono *et al.* (1992). 10. Sültmann *et al.* (1994). 11. Ono *et al.* (1993b). 12. Ono *et al.* (1993c). 13. Klein *et al.* (1993). 14. Ono *et al.* (1993d). 15. Grimholt *et al.* (1993). 16. Grimholt *et al.* (1994). 17. Hordvik *et al.* (1993). 18. Juul-Madsen *et al.* (1992). 19. Shum *et al.* (1994). 20. Hardee *et al.* (1995). 21. Walker and McConnell (1994). 22. Figueroa *et al.* (1995). 23. Betz *et al.* (1994). 24. Hashimoto *et al.* (1992). 25. Kasahara *et al.* (1992). 26. Kasahara *et al.* (1993). 27. Bartl and Weissman (1994).

genes (Ono *et al.*, 1992, 1993c; Klein *et al.*, 1993; Sültmann *et al.*, 1993). The sequences of the class Iα3 domain, the membrane-proximal domain, are relatively well conserved, whereas the sequences of the respective α1 and α2 domains, the membrane-distal domains, have diverged from one another in cyprinid fishes (Okamura *et al.*, 1993). Several amino acids in the peptide-binding region (PBR), including those that have been proposed to interact with antigenic peptide termini, are highly conserved (Okamura *et al.*, 1993; Grimholt *et al.*, 1993). There are also reports that several amino acids critical for the association with CD4 are conserved within the class IIβ2 in carp (Hashimoto *et al.*, 1990), and those for CD8 within the class Iα3 in banded dogfish, *Triakis scyllia* (Hashimoto *et al.*, 1992). Thus, it is possible that well-conserved amino acids are those important either for peptide binding or in association with CD4 and CD8 molecules. These findings along with the data on the polymorphism at the putative PBR (see later), suggest that fish MHC molecules function in a way similar to those in mammals.

The location of the fish MHC genes in the chromosome or the cosegregation of class I and class II genes in the same locus has yet to be elucidated. However, class I and class II genes may not necessarily be clustered in one locus in fish because more than one cluster of MHC genes is present in chickens (Miller, M. M. *et al.*, 1994) and *Xenopus* (Flajnik *et al.*, 1993), and the existence of several MHC loci in fish has been reported (Ono *et al.*, 1992; Stet *et al.*, 1993; Sültmann *et al.*, 1993). In any case, one should determine whether the genes are related to *bona fide* MHC genes that function in antigen presentation or to other MHC genes that map outside the MHC, such as CD-1 by studying their function, expression patterns, and polymorphism.

B. Polymorphism of MHC Genes

Extensive variability was detected in zebrafish, *Brachydanio rerio* (Ono *et al.*, 1992), cichlid fishes (Ono *et al.*, 1993d), Atlantic salmon (Hordvik *et al.*, 1993), and striped bass, *Morone saxatilis* (Walker and McConnell, 1994) at variable regions of the β1 domain of class II*B* genes and the α1 domain of zebrafish class II*A* genes (Sültmann *et al.*, 1993) which are associated with residues of the putative PBR. At least five or six alleles with numerous amino acid substitutions at the PBR of class II*A* genes (Kasahara *et al.*, 1993) and two alleles at class II*B* genes (Bartl and Weissman, 1994) have been reported from the nurse shark. These findings are similar to those found in functional mammalian class II genes, thereby implying that the function of these genes is probably the same as those of the mammalian MHC. Polymorphism of carp MHC genes has been investigated by using restriction fragment length polymorphism (RFLP) analyses of a limited number of carp lines with different geographical origins. The level of polymorphism of the *Cyca* genes of the strains calculated as the percentage of polymorphic fragments among the total number of fragments observed was 70% for class I and 40–66% for class II genes (Stet *et al.,* 1993).

Extensive MHC variability in cichlid fishes of Lake Malawi has been reported (Klein *et al.,* 1993). They found high sequence variability of class II*B* genes in a sample of species, and that different MHC alleles apparently distributed into different species during adaptive radiation. They suggested that the variability provides a set of molecular markers for studying the manner of speciation during adaptive radiation. Polymorphisms of the MHC can be also utilized for stock identification of fish, as has been reported in mammals. Some studies on this direction are in progress in salmonid species (K. M. Miller, pers. comm.). The MHC and disease association in farm animals has been reported (Van der Zijpp and Egbert, 1989). It would be of great help to breed disease resistant strains in fish, if we could find a polymorphism and its association with disease. In fact, it has been reported that several strains differ in resistance or susceptibility (reviewed by Chevassus and Dorson, 1990). Finally, it should be mentioned that polymorphism can be discussed only with respect to alleles of the same gene and not alleles of different genes at different loci. From this point of view, polymorphism based on true alleles has yet to be demonstrated in fish.

C. Expansion of MHC Genes

In mammals there exists a number of class I genes, including a few polymorphic classical class I genes and many nonclassical class I genes, with low polymorphism. Some strains of mice have more than 40 class I genes, though different species of mammals seem to possess different numbers of distinct class I genes (Stroynowski, 1990). Okamura *et al.,* (1993) demonstrated a dynamic expansion of the genes that encode class I molecules in cyprinid fish. They analyzed the genomic DNA of a carp and a ginbuna crucian carp by Southern hybridization with four probes and detected 2–10 positive fragments with each probe in the blots of the carp and ginbuna DNA. Multiple class I and class II genes in carp are also reported by Stet *et al.* (1993); the class I probe hybridized to 9–12 fragments, whereas the class II probe hybridized to 3–5 fragments in homozygous gynogenetic strains. It therefore appears that the expansion of the MHC occurred more than 400 million years ago when fish first evolved. Thus, a greater number and variety of MHC genes with various functions may exist in fish than expected.

D. Existence of MHC Genes at the Level of Cartilaginous Fish

In sharks and rays, no T-cell-mediated immune responses have been demonstrated (reviewed by McKinney, 1992), and an arguement for the lack of an MHC has been based on the chronic nature of graft rejection. However, Hashimoto *et al.* (1992) demonstrated the existence of a sequence resembling the MHC class Iα3 domains from banded dogfish (*Triakis scyllia*) which do not exhibit acute allograft rejection. Kasahara *et al.* (1992, 1993) also isolated cDNA clones encoding typical MHC class IIα chains from nurse sharks and demonstrated the polymorphism of the genes. These findings suggest that the absence of acute graft rejection in the cartilaginous fish is not a consequence of the lack of MHC polymorphism, and the existence of the MHC and T-cell receptor genes (see later) is not always correlated with acute graft rejection. This encourages efforts to isolate the genes from cyclostomes or other lower vertebrates which also show chronic graft rejection.

E. The Expression of MHC Genes

Since most cDNA has been prepared from the spleen and kidney of fish, it is possible that the MHC genes can be transcribed to mRNA at least in these two organs which are the main lymphoid organs in fish. In northern blot analysis, the zebrafish class II*A* probe hybridized with carp RNA isolated from spleen, pronephros, hepatopancreas and intestine (Sültmann *et al.*, 1993). Similarly, Ono *et al.* (1993b) found the expression of carp class II*B* gene (*Cyca*-DAB) in the same organs as mentioned above, but not in the heart, skeletal muscle, brain, and ovary. Rodrigues *et al.* (1994) also examined the expression of carp class II*B* gene at the level of cells and reported that the expression is high in peripheral blood leucocytes, thymocytes, and head kidney leucocytes, and is present but very low in gut leucocytes and splenocytes. Interestingly, Grimholt *et al.* (1995) reported that a northern blot analysis of numerous salmon tissues showed a high MHC class II*B* gene expression in the gills. The gill, which is highly vascularized, is one of the most important organs for antigen uptake after bath immunization, and many antigen-presenting cells may be present in this organ. Since the initial identification of the MHC in fish resulted from genes but not from molecules, neither the molecules nor thier function have been elucidated. Some trials have been performed to produce polyclonal antibodies against synthetic peptides based on the sequence of isolated genes of carp (Ototake *et al.*, 1995) or expressed proteins of carp (Van Lierop *et al.*, 1995) and ginbuna (Ototake *et al.*, 1995) by using bacterial expression systems. However, these authors found that the molecules differed from the putative MHC molecules in their molecular size and tissue distribution. They may be expressing not putative MHC genes but the genes related to nonclassical MHC genes, if not pseudogenes.

F. The Origin of the MHC

There have been several speculations about the origin of the MHC. The two main theories regarding the origin of the MHC differ in regard to which class of MHC molecules emerged first. One of them proposes that class I MHC was produced by an immunoglobulin domain and the peptide-binding region (PBR) of the 70-kDa heat shock protein (HSP70) gene, based on the structural similarities of the PBR between the MHC and the HSP70 family (Flajnik *et al.,* 1991). The other suggests that the class II MHC genes evolved first and the primordial MHC-like molecules might have been a class IIβ-like homodimer. The α and β chains of class II have the same overall structures and are believed to be produced by gene duplication. Thereafter, the class I and *B2m* genes were produced via a chromosomal inversion event (Kaufman *et al.,* 1990). As for the origin of the PBR of the MHC, one possibility is that the membrane-distal domains might have been derived from a protein family such as HSP70, as already mentioned. Another possibility is that the domain has been derived from the Ig superfamily, since both the MHC class Iα2 and class IIβ1 domains contain a disulfide bond reminiscent of that found in Ig superfamily members (Hashimoto and Kurosawa, 1991). There are a number of arguments on the above theories (Hasimoto and Kurosawa, 1991; Hughes and Nei, 1993), but our knowledge on the evolution of fish MHC genes is too limited to answer these questions at present. In any case, it is certain that the prototype of class I*A* and class II*A* and -*B* genes existed at the time of the appearance of jawed fishes, since all fish MHC genes isolated to date belong to these three major clusters of MHC genes. The intron–exon organization of fish MHC genes, in which only a few complete genomic sequences have been published to date, may help our understanding about its evolutionary origin.

G. T-Cell Receptor (TCR)

Although T cells have not been formally identified in fish with specific molecules like CD4, CD8, or CD3, the genes encoding polypeptides homologous to the TCR molecules have been recently reported in teleosts and elasmobranchs. Partula *et al.* (1994, 1995a,b) have reported cDNA clones from rainbow trout that have sequences very similar to TCR α and β chains

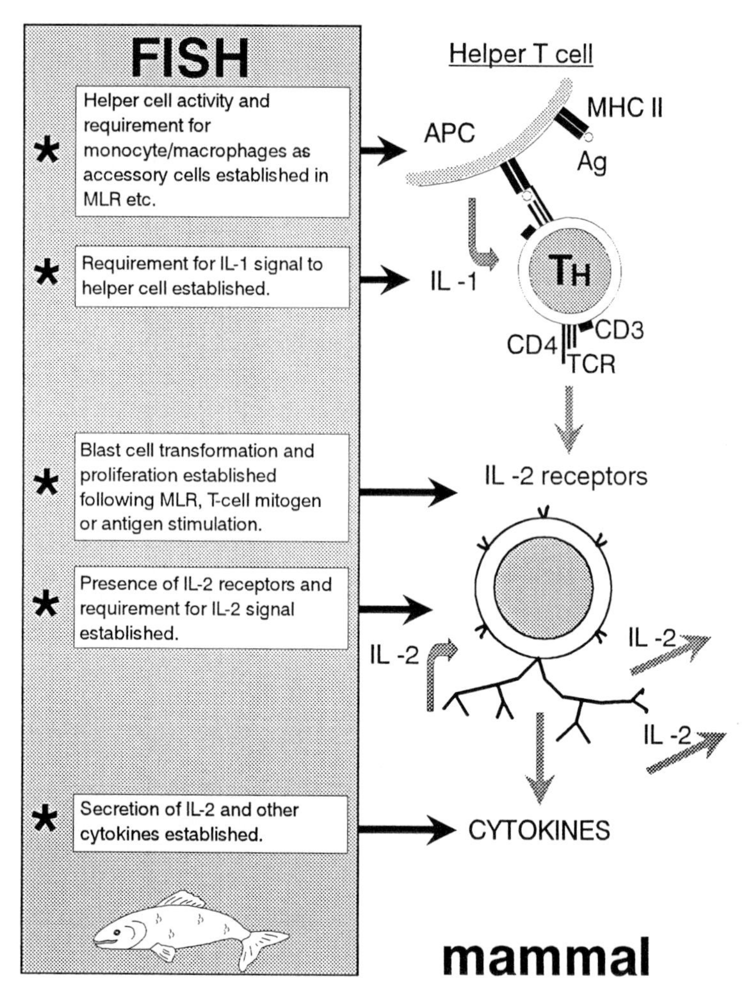

Fig. 5. Present status of research into specific cell-mediated immunity in fish (cf. known responses of mammals). Ag, Antigen; APC, antigen-presenting cell; CD3, CD4, CD8, cluster of differentiation (T-cell surface markers, T3, T4, and T8); IL-1, Interleukin 1; IL-2, Interleukin 2; MHC I, major histocompatibility complex, class I; MHC II, major histocompatibility complex, class II; MLR, mixed leukocyte reaction; TCR, T-cell receptor.

of higher vertebrates. Trout TCR β-chains include Vβ segments belonging to three different families, with Dβ, 10 Jβ segments, and a common Cβ segment (Partula *et al.,* 1995a). Considerable diversity in the complementarity-determining region 3β (CDR3β), which is located at the Vβ–Dβ and Dβ–Jβ junctions, suggests that the repertoire of the TCR β-chain could be considerable even at the level of teleosts. They were also able to find the spe-

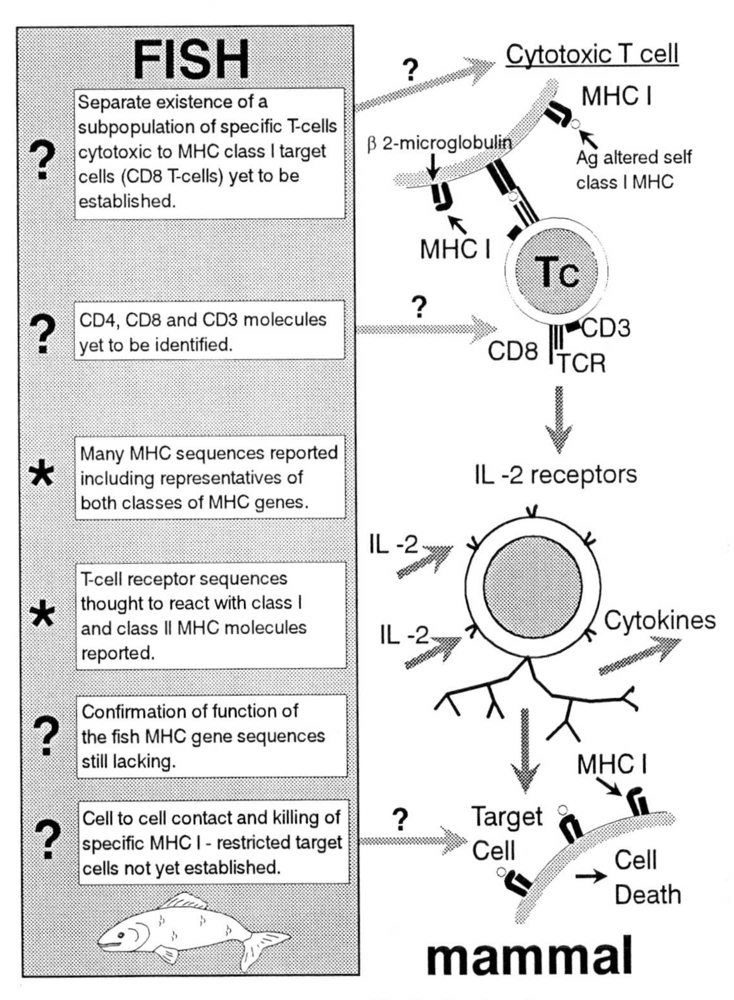

Fig. 5. Continued

cific expression of TCR β-chain mRNA in trout thymocytes and splenocytes. T-cell receptor genes have also been isolated from Atlantic salmon (Hordvik *et al.*, 1995). In elasmobranchs, Rast and Litman (1994) reported the isolation of TCR homologues from the horned shark. By analyzing genomic clones, they found that the germ-line genes are unrearranged and appear to be organized in multiple clusters in the same manner as shark immunoglobulin genes. Interestingly, Greenberg *et al.* (1995) reported new (nurse shark) antigen receptor (NAR) genes, which differed from immunoglobulin and the T-cell receptor. Analyses of the cDNA clones show extensive sequence diversity within the variable domains, which is generated both by rearrangement and by somatic diversification mechanisms. The NAR gene does not cluster with the recently reported shark TCR sequence and is a secreted molecule that undergoes somatic mutation. The finding suggests that sharks use other antigen receptors besides immunoglobulin and TCR during immune responses. The existence of homologues of the antigen receptor in other vertebrates, however, is of great interest as, presumably, these play a specialized role in their immune systems.

VII. DISCUSSION AND FUTURE DIRECTIONS

Figure 5 summarizes some of the resolved and unresolved issues concerning specific cell-mediated immunity in fish.

A variety of cytokines are secreted by fish leucocytes, among which are several factors that in mammals are produced by $CD4^{+ve}$ T cells. These include factors functionally similar to IL-2, IL-3, IL-4, IFN-γ, TGFβ, and TNF. There is also ample evidence for the role of fish helper T cells in cellular cooperation. However, we do not, as yet, possess reagents that could identify CD4 and CD8 molecules, if present on fish lymphocytes. It therefore remains unknown as to whether a separate subpopulation of $CD4^{-ve}/CD8^{+ve}$ cytotoxic cells, responsible for direct cell-to-cell killing of MHC class I target cells, is present at this level in phylogeny. It is theoretically possible that some of the specificity of fish cell-mediated reactions may be endowed by the enhancing effects of specifically induced cytokines. The recent upsurge of studies on the fish MHC and T-cell receptor molecules has, however, created hopes that some of these questions will soon be answered.

As yet, there have been few advances in our knowledge of the stem cell and progenitor sources of fish T cells. Fish have fallen behind their amphibian counterparts as models for studying early T-cell ontogeny in externally developing, free-living embryonic and larval stages. This is partly due to the eggs being more difficult to manipulate than those of amphibians, also to higher requirements for gill respiration which affect the morphology of the fish pharynx and the feasibility of destroying the thymic anlage. Although most of the thymus can be removed early in fish development, it seems unlikely that a virtually complete ablation of the T-cell population will be as easy to achieve in fish as it is in amphibians (*Xenopus*) (Horton *et al.,* 1994).

Nevertheless, ploidy markers and inbred histocompatible lines are now available for fish, while reagents such as monoclonal antibodies directed against relevant surface molecules are a likely future development. We can therefore anticipate an awakened interest in studies, both *in vivo* and *in vitro,* aimed at investigating the progenitor sources of immunologically competent T cells and the tissue arrangements and soluble factors that drive their subsequent differentiation pathways.

REFERENCES

Ahne, W. (1993). Presence of interleukins (IL-1, IL-3, IL-6) and the tumour necrosis factor (TNF alpha) in fish sera. *Bull. Eur. Assoc. Fish Pathol.* **13,** 106–107.

Ahne, W. (1994). Evidence for the early appearance of interleukins and tumor necrosis factor in the phylogenesis of vertebrates. *Immunol. Today* **15,** 137.

Avtalion, R. R., Weiss, E., and Moalem, T. (1976). Regulatory effects of temperature in ectothermic vertebrates. *In* "Comparative Immunology" (J. J. Marchalonis, ed.), pp. 227–238. Blackwell Scientific Publications, Oxford.

Bartl, S., and Weissman, I. L. (1994). Isolation and characterization of major histocompatibility complex class IIB genes from the nurse shark. *Proc. Natl. Acad. Sci. USA* **91,** 262–266.

Bartos, J. M., and Sommer, C. V. (1981). *In vivo* cell-mediated immune response to *M. tuberculosis* and *M. salmoniphilum* in rainbow trout, *Salmo gairdneri. Dev. Comp. Immunol.* **5,** 75–83.

Bernard, C. C. A., Boardman, G., Blomberg, B., and Du Pasquier, L. (1979). Immunogenetic studies on the cell-mediated cytotoxicity in the clawed toad, *Xenopus laevis. Immunogenetics* **9,** 443–454.

Betz, U. A. K., Mayer, W. E., and Klein, J. (1994). Major histocompatibility complex genes of the coelacanth, *Latimeria chalumnae. Proc. Natl. Acad. Sci. USA* **91,** 11065–11069.

Bjorkman, P. J., and Parham, P. (1990). Structure, function, and diversity of class I major histocompatibility complex molecules. *Annu. Rev. Biochem.* **59,** 253–288.

Blazer, V. S., Bennett, R. O., and Wolfe, R. E. (1984). The cellular immune response of rainbow trout, *Salmo gairdneri* Richardson, to sheep red blood cells. *Dev. Comp. Immunol.* **8,** 81–87.

Bly, J. E., and Clem, L. W. (1988). Temperature-mediated processes in teleost immunity: homeoviscous adaptation by channel catfish peripheral blood cells. *Comp. Biochem. Physiol.* **91A,** 481–485.

Bly, J. E., and Clem, L. W. (1992). Temperature and teleost immune functions. *Fish Shellfish Immunol.* **2,** 159–171.

Borysenko, M., and Hildemann, W. H. (1969). Scale (skin) allograft rejection in the primitive teleost, *Osteoglossum bicirrhosum. Transplantation* **8,** 403–412.

Borysenko, M., and Hildemann, W. H. (1970). Reactions to skin allografts in the horn shark, *Heterodontus francisci. Transplantation* **10,** 545–551.

Botham, J. W., and Manning, M. J. (1981). The histogenesis of the lymphoid organs in the carp, *Cyprinus carpio* L, and the ontogenetic development of allograft reactivity. *J. Fish Biol.* **19,** 403–414.

Botham, J. W., Grace, M. F., and Manning, M. J. (1980). Ontogeny of first-set and second-set alloimmune reactivity in fishes. *In* "Phylogeny of Immunological Memory" (M. J. Manning, ed.), pp. 83–92. Elsevier/North Holland Biomedical Press, Amsterdam.

Bridges, A. F., and Manning, M. J. (1991). the effects of priming immersions in various human gamma globulin (HGG) vaccines on humoral and cell-mediated immune responses after intraperitoneal HGG challenge in the carp, *Cyprinus carpio* L. *Fish Shellfish Immunol.* **1,** 119–129.

Caspi, R. R., and Avtalion, R. R. (1984a). Evidence for the existence of an IL-2–like lymphocyte growth promoting factor in a bony fish, *Cyprinus carpio. Dev. Comp. Immunol.* **8,** 51–60.

Caspi, R. R., and Avtalion, R. R. (1984b). The mixed leucocyte reaction (MLR) in carp: Bidirectional and unidirectional MLR responses. *Dev. Comp. Immunol.* **8,** 631–636.

Chevassus, B., and Dorson, M. (1990). Genetics of resistance to disease in fishes. *Aquaculture* **85,** 83–107.

Chinchar, V. G., Stuge, T., Hogan, R. J., Yoshida, S., Miller, N. W., and Clem, L. W. (1994). Recognition and killing of allogeneic and virus-infected lymphoid cells by channel catfish peripheral blood lymphocytes. *Dev. Comp. Immunol.* **18,** Suppl. 1, S81.

Clark, J. C., and Newth, D. R. (1972). Immunological activity of transplanted spleens in *Xenopus laevis. Experientia* **28,** 951–953.

Clark, W. R. (1991). "The Experimental Foundations of Modern Immunology," 4th Edition, John Wiley and Sons, New York.

Clem, L. W., Faulmann, E., Miller, N. W., Ellsaesser, C. F., Lobb, C. J., and Cuchens, M. A. (1984). Temperature-mediated processes in teleost immunity: Differential effects of *in vitro* and *in vivo* temperatures on mitogenic responses of channel catfish lymphocytes. *Dev. Comp. Immunol.* **8,** 313–322.

Clem, L. W., Sizemore, R. C., Ellasaesser, C. F., and Miller, N. W. (1985). Monocytes as accessory cells in fish immune responses. *Dev. Comp. Immunol.* **9,** 803–809.

Clem, L. W., Miller, N. W., and Bly, J. E. (1991). Evolution of lymphocyte populations, their interactions, and temperature sensitivities. *In* "Phylogenesis of the Immune System" (G. W. Warr and N. Cohen, eds.), pp. 191–213. CRC Press, Boca Raton, FL.

Cooper, A. J. (1971). Ammocoete lymphoid populations *in vitro. In* "Fourth Annual Leucocyte Culture Conference" (O. R. McIntyre, ed.), pp. 137–147. Appleton-Century-Crofts, New York.

Cooper, E. L. (1964). The effects of antibiotics and x-irradiation on the survival of scale homografts in *Fundulus heteroclitus. Transplantation* **2,** 2–20.

Cuchens, M. A., and Clem, L. W. (1977). Phylogeny of lymphocyte heterogeneity. II. Differential effects of temperature on fish T-like and B-like cells. *Cell. Immunol.* **34,** 219–230.

De Kinkelin, P., Dorson, M., and Hattenberger-Baudouy, A. M. (1982). Interferon synthesis in trout and carp after viral infection. *Dev. Comp. Immunol. Suppl.* **2,** 167–174.

DeKoning, J., and Kaattari, S. L. (1991). Mitogenesis of rainbow trout peripheral blood lymphocytes requires homologous plasma for optimal responsiveness. *In vitro Cell Dev. Biol.* **27A,** 381–386.

DeKoning, J., and Kaattari, S. L. (1992). Use of homologous salmonid plasma for the improved responsiveness of salmonid leukocyte cultures. *In* "Techniques in Fish Immunology. II" (J. S. Stolen *et al.,* eds.), pp. 61–65. SOS Publications, Fair Haven, NJ.

DeLuca, D., Wilson, M., and Warr, G. (1983). Lymphocyte heterogeneity in the trout, *Salmo gairdneri,* defined with monoclonal antibodies to IgM. *Eur. J. Immunol.* **13,** 546–551.

De Sena, J., and Rio, G. (1975). Partial purification and characterization of RTG-2 fish cell interferon. *Infect. Immun.* **11,** 815–822.

Desvaux, F. X., and Charlemagne, J. (1983). The goldfish immune response. II. Thymic influence on allograft rejection. *Dev. Comp Immunol.* **7,** 563–567.

Dixon, B., Stet, R. J. M., Van Erp, S. H. M., and Pohajdak, B. (1993). Characterization of β_2-microglobulin transcripts from two teleost species. *Immunogenetics* **38,** 27–34.

Dixon, B., Van Erp, S. H. M., Rodrigues, P. N. S., Egberts, E., and Stet, R. J. M. (1995). Fish major histocompatibility genes: An expansion. *Dev. Comp Immunol.* **19,** 109–133.

Dorson, M., De Kinkelin, P., and Torchy, C. (1992). Interferon synthesis in rainbow trout fry following infection with infectious pancreatic necrosis virus. *Fish Shellfish Immunol.* **2,** 311–313.

Ellis, A. E. (1977). Ontogeny of the immune response in *Salmo salar.* Histogenesis of the lymphoid organs and appearance of membrane immunoglobulin and mixed leucocyte reactivity. *In* "Developmental Immunobiology" (J. B. Solomon and J. D. Horton, eds.), pp. 225–231. Elsevier/North Holland Biomedical Press, Amsterdam.

Ellsaesser, C. F., Bly, J. E., and Clem, L. W. (1988). Phylogeny of lymphocyte heterogeneity: The thymus of the channel catfish. *Dev. Comp. Immunol.* **12,** 787–799.

Etlinger, H. M., Hodgins, H. O., and Chiller, J. M. (1976). Evolution of the lymphoid system. 1. Evidence for lymphocyte heterogeneity in rainbow trout revealed by the organ distribution of mitogenic responses. *J. Immunol.* **116,** 1547–1553.

Etlinger, H. M., Hodgins, H. O., and Chiller, J. M. (1977). Evolution of the lymphoid system. III. Evidence for immunoglobulin determinants on rainbow trout lymphocytes and demonstration of mixed leukocyte reaction. *Eur. J. Immunol.* **7,** 881–886.

Faulmann, E., Cuchens, M. A., Lobb, C. J., Miller, N. W., and Clem, L. W. (1983). An effective culture system for studying *in vitro* mitogenic responses of channel catfish lymphocytes. *Trans. Amer. Fish. Soc.* **112,** 673–679.

Figueroa, F., Ono, H., Tichy, H., O'hUigin, C., and Klein, J. (1995). Evidence for insertion of a new intron into an MHC of perch-like fish. *Proc. R. Soc. Lon. B* **259,** 325–330.

Findlay, C. (1994). A study of lymphocyte heterogeneity in the rainbow trout, *Oncorhynchus mykiss.* PhD Thesis, University of Stirling, Scotland.

Finstad, J., and Good, R. A. (1964). The evolution of the immune response. III. Immunologic responses in the lamprey. *J. Exp. Med.* **120,** 1151–1167.

Finstad, J., and Good, R. A. (1966). Phylogenetic studies of adaptive immune responses in the lower vertebrates. *In* "Phylogeny of Immunity" (R. T. Smith, P. A. Miescher, and R. A. Good, eds.), pp. 173–186. University of Florida Press, Gainesville.

Flajnik, M. F., Canel, C., Kramer, J., and Kasahara, M. (1991). Which came first, MHC class I or class II? *Immunogenetics* **33,** 295–300.

Flajnik, M. F., Kasahara, M., Shum, B. P., Salter-Cid, L., Taylor, E., and Du Pasquier, L. (1993). A novel type of class I gene organization in vertebrates: A large family of non-MHC-linked class I genes is expressed at the RNA level in the amphibian *Xenopus. EMBO J.* **12,** 4385–4396.

Gloudemans, A. G. M., Cohen, N., Boersbroek, G. E., Grondel, J. L., Egberts, E., and Van Muiswinkel, W. B. (1987). The major histocompatibility complex (MHC) in fish. Carp family studies using the mixed leucocyte reaction. *In* "Selection, Hybridization, and Genetic Engineering in Aquaculture" (K. Tiews, ed.), pp. 265–268. Heinemann Verlag, Berlin.

Graham, S., and Secombes, C. J. (1988). The production of a macrophage-activating factor for rainbow trout, *Salmo gairdneri,* leucocytes. *Immunology* **65,** 293–297.

Graham, S., and Secombes, C. J. (1990). Do fish lymphocytes secrete interferon γ? *J. Fish Biol.* **36,** 563–573.

Gravell, M., and Malsberger, R. G. (1965). A permanent cell line from the fathead minnow, *Pimephales promelas. Ann. N. Y. Acad. Sci.* **126,** 555–565.

Greenberg, A. S., Avila, D., Hughes, M., Hughes, A., McKinney, E. C., and Flajnik, M. F. (1995). A new antigen receptor gene family that undergoes rearrangement and extensive somatic diversification in sharks. *Nature* **374,** 168–173.

Grimholt, U., Hordvik, I., Fosse, V. M., Olsaker, I., Endresen, C., and Lie, Ø. (1993). Molecular cloning of major histocompatibility complex class I cDNAs from Atlantic salmon (*Salmo salar*). *Immunogenetics* **37,** 469–473.

Grimholt, U., Olsaker, I., De Vries Lindstrøm, C., and Lie, Ø. (1994). A study of variability in the MHC class IIβ 1 and class Iα 2 domain exons of Atlantic salmon, *Salmo salar* L. *Anim. Genet.* **25,** 147–153.

Grondel, J. L., and Harmsen, E. G. M. (1984). Phylogeny of interleukins: Growth factors produced by leucocytes of the cyprinid fish, *Cyprinus carpio* L. *Immunology* **52,** 477–482.

Hardee, J. J., Godwin, U., Benedetto, R., and McConnell, T. J. (1995). Major histocompatibility complex class IIA gene polymorphism in the striped bass. *Immunogenetics* **41,** 229–238.

Hardie, L. J., Chappell, L. H., and Secombes, C. J. (1994). Human tumor necrosis factor α influences rainbow trout, *Onchorhynchus mykiss,* leucocyte responses. *Vet. Immunol. Immunopathol.* **40,** 73–84.

Hardie, L. J., Fletcher, T. C., and Secombes, C. J. (1995). Effect of temperature on macrophage activation and the production of macrophage activating factor by rainbow trout, *Oncorhynchus mykiss,* leucocytes. *Dev. Comp. Immunol.* **18,** 57–66.

Hashimoto, K., and Kurosawa, Y. (1991). Evolution of MHC domains, strategy for isolation of MHC genes from primitive animals. *In* "Molecular Evolution of the Major Histocompatibility Complex" (J. Klein and D. Klein, eds.), NATO ASI Series, Vol. H **59,** pp. 103–109. Springer-Verlag, Berlin.

Hashimoto, K., Nakanishi, T., and Kurosawa, Y. (1990). Isolation of carp genes encoding major histocompatibility complex antigens. *Proc. Natl. Acad. Sci. USA* **87,** 6863–6867.

Hashimoto, K., Nakanishi, T., and Kurosawa, Y. (1992). Identification of a shark sequence resembling the major histocompatibility complex class Iα3 domain.*Proc. Natl. Acad. Sci. USA* **89,** 2209–2212.

Haynes, L., and McKinney, E. C. (1991). Shark spontaneous cytoxicity: Characterisation of the regulatory cell. *Dev. Comp. Immunol.* **15,** 123–134.

Hildemann, W. H. (1957). Scale homotransplantation in the goldfish, *Carassius auratus. Ann. N. Y. Acad. Sci.* **64,** 775–791.

Hildemann, W. H. (1970). Transplantation immunity in fishes: Agnatha, Chondrichthyes, and Osteichthyes. *Transplant Proc.* **11,** 253–259.

Hildemann, W. H., and Haas, R. (1960). Comparative studies of homotransplantation in fishes. *J. Cell. Comp. Physiol.* **55,** 227–233.

Hildemann, W. H., and Thoenes, G. H. (1969). Immunological responses of Pacific hagfish. 1. Skin transplantation immunity. *Transplantation* **7,** 506–521.

Hogarth, P. J. (1968). Immunological aspects of foeto-maternal relations in lower vertebrates. *J. Reprod. Fert. suppl.* **3,** 15–27.

Hordvik, I., Grimholt, U., Fosse, V. M., Lie, Ø., and Endresen, C. (1993). Cloning and sequence analysis of cDNAs encoding the MHC class II β chain in Atlantic salmon (*Salmo salar*). *Immunogenetics* **37,** 437–441.

Hordvik, I., Torvund, J., Samdal, I., and Endresen, C. (1995). Atlantic salmon MHC and T-cell receptor genes. Expression cloning and production of antibody markers. *In* "The Nordic Symposium on Fish Immunology, Reykjavik, Iceland, p. 49.

Horton, J. D. (1994). Amphibians. *In* "Immunology: A comparative approach" (R. J. Turner, ed.), pp. 101–136. John Wiley and Sons, Chichester.

Horton, J., Horton, T., Ritchie, P., Gravenor, I., Gartland, L., and Cooper, M. (1994). Use of monoclonal antibodies to demonstrate absence of T-cell development in early-thymectomized *Xenopus. Dev. Comp. Immunol.* **18,** Suppl. 1, 95.

Hughes, A. L., and Nei, M. (1993). Evolutionary relationships of the classes of major histocompatibility complex genes. *Immunogenetics* **37,** 337–346.

Jang, S. I., Hardie, L. J., and Secombes, C. J. (1994). The effects of transforming growth factor β_1 on rainbow trout, *Orcorhynchus mykiss,* macrophage respiratory burst activity. *Dev. Comp. Immunol.* **18,** 315–323.

Jang, S. I., Mulero, V., Hardie, L. J., and Secombes, C. J. (1995). Inhibition of rainbow trout phagocyte responsiveness to human tumor necrosis factor α (hTNFα) with monoclonal antibodies to the hTNFα 55-Da receptor. *Fish Shellfish Immunol.* **5,** 61–69.

Jayaraman, S., Mohan, R., and Muthukkaruppan, V. R. (1979). Relationship between migration inhibition and plaque forming cell responses to sheep erythrocytes in the teleost *Tilapia mossambicus. Dev. Comp. Immunol.* **3,** 67–75.

Juul-Madsen, H. R., Glamann, J., Madsen, H. O., and Simonsen, M. (1992). MHC class II beta-chain expression in the rainbow trout. *Scand. J. Immunol.* **35,** 687–694.

Kaastrup, P., Nielson, B., Horlyck, V., and Simonsen, M. (1988). Mixed lymphocyte reactions (MLR) in rainbow trout, *Salmo gairdneri,* sibling. *Dev. Comp. Immunol.* **12,** 801–808.

Kaattari, S., and Holland, N. (1990). The one-way mixed lymphocyte reaction. *In* "Techniques in Fish Immunology. I" (J. S. Stolen *et al.,* eds.), pp 165–172. SOS Publications, Fair Haven, NJ.

Kallman, K. D., and Gordon, M. (1957). Transplantation of fins in Xiphophoria fishes. *Ann. N. Y. Acad. Sci.* **71,** 307–318.

Kasahara, M., Vazquez, M., Sato, K., McKinney, E. C., and Flajnik, M. F. (1992). Evolution of the major histocompatibility complex: Isolation of class II A cDNA clones from the cartilaginous fish. *Proc. Natl. Acad. Sci. USA* **89,** 6688–6692.

Kasahara, M., McKinney, E. C., Flajnik, M. F., and Ishibashi, T. (1993). The evolutionary origin of the major histocompatibility complex, Polymorphism of class II α chain genes in the cartilaginous fish. *Eur. J. Immunol.* **23,** 2160–2165.

Kaufman, J., Skjoedt, K., and Salomonsen, J. (1990). The MHC molecules of nonmammalian vertebrates. *Immunol. Rev.* **113,** 83–117.

Klein, D., Ono, H., O'hUigin, C., Vincek, V., Goldschmidt, T., and Klein, J. (1993). Extensive MHC variability in cichlid fishes of Lake Malawi. *Nature* **364,** 330–334.

Klein, J., Bontrop, R. E., Dawkins, R. L., Erlich, H. A., Gyllensten, U. B., Heise, E. R., Jones, P. P., Parham, P., Wakeland, E. K., and Watkins, D. I. (1990). Nomenclature for the major histocompatibility complexes of different species. A proposal. *Immunogenetics.* **31,** 217–219.

Kodama, H., Mukamoto, M., Baba, T., and Mule, D. M. (1994). Macrophage-colony stimulating activity in rainbow trout (*Onchorhynchus mykiss*) serum. *In* "Modulators of Fish Immune Responses" (J. S. Stolen and T. C. Fletcher, eds.), pp. 59–66. SOS Publications, Fair Haven, NJ.

Koumans–Van Diepen, J. C. E., Harmsen, E. G. M., and Rombout, J. H. W. M. (1994). Immunocytochemical analysis of mitogen responses of carp (*Cyprinus carpio* L.) peripheral blood leucocytes. *Vet. Immunol. Immunopathol.* **42,** 209–219.

Lanzavecchia, A. (1990). Receptor-mediated antigen uptake and its effect on antigen presentation to class II-restricted T Lymphocytes. *Annu. Rev. Immunol.* **8,** 773–793.

Leeloy, H. K., and Hildemann, W. H. (1977). Tissue transplantation immunity in Hawaiian mullet (*Mugil cephalus*). *Transplantation* **23,** 109.

Le Morvan-Rocher, C., Troutaud, D., and Deschaux, P. (1995). Effects of temperature on carp leukocyte mitogen-induced proliferation and nonspecific cytotoxic activity. *Dev. Comp. Immunol.* **19,** 87–95

Lin, G. L., Ellsaesser, C. F., Clem, L. W., and Miller, N. W. (1992). Phorbolester/calcium ionophore activate fish leukocytes and induce long-term cultures. *Dev. Comp. Immunol.* **16,** 153–163.

Lobb, C. J., and Clem, L. W. (1982). Fish lymphocytes differ in the expression of surface immunoglobulin. *Dev. Comp. Immunol.* **6,** 473–479.

Lopez, D. M., Sigel, M. M., and Lee, J. C. (1994). Phylogenetic studies on T cells. 1. Lymphocytes of the shark with differential response to PHA and ConA. *Cellular Immunol.* **10,** 287–292.

Luft, J. C., Clem, L. W., and Bly, J. E. (1994). *In vitro* mitogen-induced and MLR-induced responses of leucocytes from the spotted gar, a holostean fish. *Fish Shellfish Immunol.* 4, 153–156.

Manning, M. J. (1994). Fishes. *In* "Immunology: A comparative approach" (R. J. Turner, ed.), pp. 69–100. John Wiley and Sons, Chichester.

Manning, M. J., Grace, M. F., and Secombes, C. J. (1992a). Ontogenetic aspects of tolerance and immunity in carp and rainbow trout: Studies on the role of the thymus. *Dev. Comp. Immunol. Suppl.* **2,** 75–82.

Manning, M. J., Grace, M. F., and Secombes, C. J. (1982b). Developmental aspects of immunity and tolerance in fish. *In* "Microbial Diseases of Fish" (R. J. Roberts, ed.), pp. 31–46. Academic Press, London.

Marsden, M. J., Hamdani, S. H., and Secombes, C. J. (1995). Proliferative responses of rainbow trout, *Aeromonas salmonicida, Fish Shellfish Immunol.* **5,** 199–210.

McKinney, E. C. (1992). Shark lymphocytes: Primitive antigen reactive cells. *Annu. Rev. Fish Dis.* **2,** 43–51.

McKinney, E. C., Ortiz, G., Lee, J. C., Sigel, M. M., Lopez, D. M., Epstein, R. S., and McLeod, T. F. (1976). Lymphocytes of fish: Multipotential or specialized? *In* "Phylogeny of Thymus and Bone Marrow–Bursa Cells." (R. K. Wright and E. L. Cooper, eds.), pp. 73–82. Elsevier/North Holland Biomedical Press. Amsterdam.

McKinney, E. C., McLeod, T. F., and Sigel, M. M. (1981). Allograft rejection in a holostean fish, *Lepisosteus platyrhincus. Dev. Com. Immunol.* **5,** 65–74.

McVicar, A. H., and McLay, H. A. (1985). Tissue response of plaice, haddock, and rainbow trout to the systemic fungus *Ichthyophonus. In* "Fish and Shellfish Pathology" (A. E. Ellis, ed.) pp. 329–346. Academic Press, London.

Metcalf, D. (1995). Delayed effect of thymectomy in adult life on immunological competence. *Nature* (*London*) **208,** 1336.

Millar, D. A., and Ratcliffe, N. A. (1994). Invertebrates. *In* "Immunology: A comparative approach" (R. J. Turner, ed.), p. 29–68. John Wiley and Sons, Chichester.

Miller, M. M., Goto, R., Bernot, A., Zoorob, R., Auffray, C., Bumstead, N., and Briles, W. E. (1994). Two MHC class I and two MHC class II genes map to the chicken *Rfp-Y* system outside the B complex. *Proc. Natl. Acad. Sci. USA* **91,** 4397–4401.

Miller, N. W., and Clem, L. W. (1984). Temperature-mediated processes in teleost immunity: Differential effects of temperature on catfish *in vitro* antibody responses to thymus-dependent and thymus-independent antigens. *J. Immunol.* **133,** 2356–2359.

Miller, N. W., Sizemore, R. C., and Clem, L. W. (1985). Phylogeny of lymphocyte heterogeneity: The cellular requirements for *in vitro* antibody responses of channel catfish leukocytes. *J. Immunol.* **134,** 2884–2888.

Miller N. W., Deuter, A., and Clem, L. W. (1986). Phylogeny of lymphocyte heterogeneity: The cellular requirements for the mixed leukocyte reaction in channel catfish. *Immunology* **59,** 123–128.

Miller, N. W., Bly, J. E., Van Ginkel, F., Ellsaesser, C. F., and Clem, L. W. (1987) Indentification and separation of functionally distinct subpopulations of channel catfish lymphocytes with monoclonal antibodies. *Dev. Comp. Immunol.* **11,** 739–747.

Miller, N. W., Rycyzyn, M. A., Stude, T. B., Luft, J. C., Wilson, M. R., Bly, J. E., and Clem, L. W. (1994). *In Vitro* culture approaches for studying immune cell functions(s) in ectothermic vertebrates. *Dev. Comp. Immunol.* **18,** Suppl. 1, S132.

Mitchinson, N. A. (1953). Passive transfer of transplantation immunity. *Nature* (*London*) **171,** 267.

Mori, Y. (1931). On the transformation of ordinary scales into lateral line scales in the goldfish. *J. Fac. Sci. Imp. Univ. Tokyo* **2,** 185–194.

Nakanishi, T. (1986a). Ontogenetic development of the immune response in the marine teleost *Sebastiscus marmoratus. Bull. Jpn. Soc. Sci. Fish.* **52,** 473–477.

Nakanishi, T. (1986b). Effects of x-irradiation and thymectomy on the immune response of the marine teleost *Sebastiscus marmoratus. Dev. Comp. Immunol.* **10,** 519–527.

Nakanishi, T. (1991). Ontogeny of the immune system in *Sebastiscus marmoratus:* Histogenesis of the lymphoid organs and effects of thymectomy. *Environ. Biol. Fish.* **39,** 135–145.

Nakanishi, T. (1994). The graft-versus-host reaction in teleost. *Dev. Comp. Immunol.* **18,** no. 3, xvi–xvii.

Navarro, V., Quesada, J. A., Abad, M. E., Taverne, N., and Rombout, J. H. W. M. (1993). Immuno(cyto)chemical characterization of monoclonal antibodies to gilthead seabream *Sparus aurata* immunoglobulin. *Fish Shellfish Immunol.* **3,** 167–177.

Nevid, N. J., and Meier, A. H. (1993). A day–night rhythm of immune activity during scale allograft rejection in the gulf killifish, *Fundulus grandis. Dev. Comp. Immunol.* **17,** 221–228.

Okamura, K., Nakanishi, T., Kurosawa, Y., and Hashimoto, K. (1993). Expansion of genes that encode MHC class I molecules in cyprinid fishes. *J. Immunol.* **151,** 188–200.

O'Neill, J. C. (1980). Temperature and the primary and secondary immune responses of three teleosts, *Salmo trutta, Cyprinus carpio,* and *Notothenia rossii* to MS2 bacteriophage. *In* "Phylogeny of Immunological Memory" (M. J. Manning, ed.), pp. 123–130. Elsevier/North Holland Biomedical Press, Amsterdam.

Ono, H., Klein, D., Vincek, V., Figueroa, F., O'hUigin, C., Tichy, H., and Klein, J. (1992). Major histocompatibility complex class II genes of zebrafish. *Proc. Natl. Acad. Sci. USA* **89,** 11886–11890.

Ono, H., Figueroa, F., O'hUigin, C., and Klein, J. (1993a). Cloning of the β_2-microglobulin gene in the zebrafish. *Immunogenetics* **38,** 1–10.

Ono, H., O'hUigin, C., Vincek, V., Stet, R. J. M., Figuero, F., and Klein, J. (1993b). New β chain–encoding MHC class II genes in the carp. *Immunogenetics* **38,** 146–149.

Ono, H., O'hUigin, C., Vincek, V., and Klein, J. (1993c). Exon–intron organization of fish major histocompatibility complex class IIβ genes. *Immunogenetics* **38,** 223–234.

Ono, H., O'hUigin, C., Tichy, H., Klein, J. (1993d). Major histocompatibility complex variation in two species of cichlid fishes from Lake Malawi. *Mol. Biol. Evol.* **10,** 1060–1072.

Ototake, M., Hashimoto, K., Kurosawa, Y., and Nakanishi, T. (1995). Characterization of polyclonal antibodies against carp and ginbuna MHC molecules. *Proc. Jpn. Assoc. Dev. Comp. Immunol.* **7,** D9.

Papermaster, B. W., Condie, R. M., Finstad, J., and Good, R. A. (1964). Evolution of the immune response. I. The phylogenetic development of adaptive immunologic responsiveness in vertebrates. *J. Exp. Med.* **119,** 105–130.

Partula, S., Fellah, J. S., de Guerra, A., and Charlemagne, J. (1994). Identification of cDNA clones encoding the T-cell receptor β-chain in the rainbow trout, *Oncorhynchus mykiss. C.R. Acad. Sci. Paris Life Sci.* **317,** 765–770.

Partula, S., de Guerra, A, Fellah, J. S. and Charlemagne, J. (1995a). Structure and diversity of the T-cell antigen receptor β-chain in a teleost fish. *J. Immunol.* **155,** 699–706.

Partula, S., de Guerra, A, Fellah, J. S., and Charlemagne, J. (1995b). Identification and analysis of genes encoding TcRα chains in the rainbow trout. The Nordic Symposium on Fish Immunology, Reykjavik, Iceland, p. 53.

Pauley, G. B., and Heartwell, C. M. (1983). Immune hypersensitivity in the channel catfish, *Ictalurus punctatus* (Rafinesque) *J. Fish Biol.* **23,** 187–193.

Perey, D. Y. E., Finstad, J. I., Pollara, B., and Good, R. A. (1968). Evolution of the immune response. VI. First- and second-set homograft rejection in primitive fishes. *Lab. Invest.* **19,** 591–597.

Pettey, C. L., and McKinney, E. C. (1981). Mitogen-induced cytotoxicity in the nurse shark. *Dev. Comp. Immunol.* **5,** 53–64.

Pourreau, C. N., Koopman, M. B. H., Hendriks, G. F. R., Evenberg, D., and Van Muiswinkel, W. B. (1987). Modulation of the mitogenic response of carp, *Cyprinus carpio* L., by extracellular products of *Aeromonas salmonicida. J. Fish Biol.* **31,** Suppl. A, 133–143.

Raison, R. L., Gilbertson, P., and Wotherspoon, J. (1987). Cellular requirements for mixed leucocyte reactivity in the cyclostome, *Eptatretus stoutii. Immunol. Cell Biol.* **65,** 183.

Rast, J. P., and Litman, G. W. (1994). T-cell receptor gene homologues are present in the most primitive jawed vertebrates. *Proc. Natl. Acad. Sci. USA* **91,** 9248–9252.

Reid, P., and Triplett, E. L. (1968). Observations on the immune system of *Micropterus salmoides. Transplantation* **6,** 338–341.

Reitan, L. J., and Thuvander, A. (1991). *In vitro* stimulation of salmonid leucocytes with mitogens and with *Aeromonas salmonicida. Fish Shellfish Immunol.* **1,** 297–307.

Rijkers, G. T. (1982). Kinetics of humoral and cellular immune reactions in fish. *Dev. Comp. Immunol. Suppl.* **2,** 93–100.

Rijkers, G. T., and Van Muiswinkel, W. B. (1977). The immune system of of cyprinid fish: The development of cellular and humoral responsiveness in the rosy barb (*Barbus conchonius*). *In* "Developmental Immunobiology" (J. B. Solomon and J. D. Horton, eds.), pp. 233–240. Elsevier/North Holland Biomedical Press, Amsterdam.

Rijkers, G. T., Wiegerinck, J. A. M., Van Oosterom, R., and Van Muiswinkel, W. B. (1981). Temperature dependence of humoral immunity in carp, *Cyprinus carpio. In* "Aspects of Developmental and Comparative Immunology. I" (J. B. Solomon, ed.), pp. 477–482. Pergamon Press, Oxford.

Rodrigues, P. N. S., Hermsen, T. J., Egberts. E., and Stet, R. J. M. (1994). MHC class II expression in lymphoid cells of the carp. *Dev. Comp. Immunol.* **18,** S55.

Rosenberg-Wiser, S., and Avtalion, R. R. (1982). The cells involved in the immune response of fish: III. Culture requirements of PHA-stimulated carp, *Cyprinus carpio,* lymphocytes. *Dev. Comp. Immunol.* **6,** 693–702.

Rowley, A. F. (1991). Eicosanoids: Aspects of their structure, function and evolution. *In* "Phylogenesis of Immune Function" (G. W. Warr, and N. Cohen, eds.), pp. 269–294. CRC Press, Boca Raton, FL.

Rowley, A. F., Hunt, T. C., Page, M., and Mainwaring, G. (1988). Fish. *In* "Vertebrate Blood Cells" (A. F. Rowley, and N. A. Ratcliffe, eds.), pp. 19–127. Cambridge University Press, Cambridge.

Ruben, L. N., Warr, G. W., Decker, J. M., and Marchalonis, J. J. (1977). Phylogenetic origins of immune recognition: Lymphoid heterogeneity and the hapten/carrier effects in the goldfish, *Carassius auratus. Cell. Immunol.* **31,** 266–283.

Sailendri, K. (1973). Studies on the development of lymphoid organs and immune responses in the teleost, *Tilapia mossambicus* (Peters). PhD Thesis, Madurai University, India.

Secombes, C. J. (1994a). The phylogeny of cytokines. *Dev. Comp. Immunol.* **18,** Suppl. 1, S65.

Secombes, C. J. (1994b). The phylogeny of cytokines. *In* "The Cytokine Handbook," 2nd Edition, (A. Thompson, ed.), pp. 567–594. Academic Press, London.

Secombes, C. J. (1994c). Cellular defences of fish: An update. *In* "Parasitic Diseases of Fish" (A. W. Pike, and J. W. Lewis, eds.), pp. 209–224. Samara Publishing Ltd., Tresaith, Wales.

Secombes, C. J., Van Groningen, J. J. M., and Egberts, E. (1983). Separation of lymphocyte subpopulations in carp, *Cyprinus carpio* L., by monoclonal antibodies. Immunohistochemical studies. *Immunology* **48,** 165–175.

Secombes, C. J., Lewis, A. E., Laird, L. M., Needham, E. A., and Priede, I. G. (1985). Experimentally induced immune reactions to gonad in rainbow trout, *Salmo gairdneri. In* "Fish Immunology" (M. J. Manning and M. F. Tatner, eds.), pp. 343–355. Academic Press, London.

Shum, B. P., Azumi, K., and Parham, P. R. (1994). Isolation, genetics, and structural analysis of MHC class I heavy chain and β_2-microglobulin from rainbow trout (*Oncorhynchus mykiss*). *Dev. Comp. Immunol.* **18,** S44.

Sigel, M. M., Lee, J. C., McKinney, E. C., and Lopez, D. M. (1978). Cellular immunity in fish as measured by lymphocyte stimulation. *Marine Fish. Rev.* **40,** 6–11.

Sigel, M. M., Hamby, B. A., and Huggins, E. M. (1986). Phylogenetic studies on lymphokines. Fish lymphocytes respond to human IL-1 and epithelial cells produce an IL-1–like factor. *Vet. Immunol. Immunopathol.* **12,** 47–58.

Sizemore, R. G., Miller, N. W., Cuchens, M. A., Lobb, C. J., and Clem, L. W. (1984). Phylogeny of lymphocyte heterogeneity: The cellular requirements for *in vitro* mitogenic responses of channel catfish leukocytes. *J. Immunol.* **133,** 2920–2924.

Smith, P. D., and Braun-Nesje, R. (1982). Cell-mediated immunity in the salmon: Lymphocyte and macrophage stimulation, lymphocyte/macrophage interactions, and the production of lymphokine-like factors by stimulated lymphocytes. *Dev. Comp Immunol. Suppl.* **2,** 233–238.

Song, Y. L., Lin, T., and Kou, G. H. (1989). Cell-mediated immunity of the eel, *Anguilla japonica* (Temminck and Schlegel). as measured by the migration inhibition test. *J. Fish Dis.* **12,** 117–123.

Stet, R. J. M., and Egberts, E. (1991). The histocompatibility system in teleostean fishes: From multiple histocompatibility loci to a major histocompatibility complex. *Fish Shellfish Immunol.* **1,** 1–16.

Stet, R. J. M., Van Erp, S. H. M., Hermsen, T., Sultmann, H. A., and Egberts, E. (1993). Polymorphism and estimation of the number of MHC Cyca class I and class II genes in laboratory strains of the common carp (*Cyprinus carpio* L). *Dev. Comp. Immunol.* **17,** 141–156.

Stevenson, R. M. W., and Raymond, B. (1990). Delayed-type hypersensitivity skin reactions. *In* "Techniques in Fish Immunology. I" (J. S. Stolen *et al.,* eds.), pp. 173–178. SOS Publications, Fair Haven, NJ.

Stolen, J. S., and Makela, O. (1976). Cell collaboration in marine fish: The effect of carrier preimmunization on the anti-hapten response to NIP and NNP. *In* "Phylogeny of Thymus and Bone Marrow-Bursa Cells" (R. K. Wright and E. L. Cooper, eds.), pp. 93–97. Elsevier/North Holland, Amsterdam.

Stroynowski, I. (1990). Molecules related to class I major histocompatibility complex antigens. *Annu. Rev. Immunol.* **8,** 501.

Sültmann, H., Werner, W. E., Figueroa, F., O'hUigin, C., and Klein, J. (1993). Zebrafish MHC class II α-chain-encoding genes: Polymorphism, expression, and function. *Immunogenetics* **38,** 408–420.

Sültmann, H., Mayer, W. E., Figueroa, F., O'hUigin, C., and Klein, J. (1994). Organization of MHC class IIβ genes in the zebrafish (*Branchydanio rerio*). *Genomics* **23,** 1–14.

Takeuchi, H., Figueroa, F., O'hUigin, C. and Klein, J. (1995). Cloning and characterization of class I MHC genes of the zebrafish, *Brachydanio rerio. Immunogenetics* **42,** 77–84.

Tam, M. R., Reddy, A. L., Karp, R. D., and Hildemann, W. H. (1976). Phylogeny of cellular immunity among vertebrates. *In* "Comparative Immunology" (J. J. Marchalonis, ed.), pp. 98–119. Blackwell Scientific Publications, Oxford.

Tamai, T., Sato, N., Kimura, S., Shirahata, S., and Murakami, H. (1992). Cloning and expression of flatfish interleukin 2 gene. *In* "Animal Cell Technology: Basic and Applied Aspects" (H. Murakami *et al.,* eds.), pp. 509–514. Kluwer Academic Publishers, Netherlands.

Tamai, T., Shirahata, S., Sato, N., Kimura, S., Nonaka, M., and Murakami, H. (1993). Purification and characterization of interferon-like antiviral protein derived from flatfish, *Paralichthys olivaceus,* lymphocytes immortalized by oncogenes. *Cytotechnology* **11,** 121–131.

Tatner, M. F. (1986). The ontogeny of humoral immunity in rainbow trout, *Salmo gairdneri. Vet. Immunol. Immunopathol.* **12,** 93–105.

Tatner, M. F. (1988). The cryopreservation of fish thymocytes and its potential application for immunological reconstitution experiments. *J. Fish Biol.* **33,** 925–929.

Tatner, M. F. (1990). Surgical techniques in fish immunology. *In* "Techniques in Fish Immunology I" (J. S. Stolen *et al.,* eds.). pp. 105–111. SOS Publications, Fair Haven, NJ.

Tatner, M. F., and Findlay, C. (1991). Lymphocyte migration and localization patterns in rainbow trout, *Onchorhynchus mykiss,* studies using the tracer sample method. *Fish Shellfish Immunol.* **1,** 107–117.

Tatner, M. F., and Findlay, C. (1992). Cryopreservation of fish lymphocytes. *In* "Techniques in Fish Immunology II" (J. S. Stolen *et al.,* eds.), pp. 31–34. SOS Publications, Fair Haven, NJ.

Tatner, M. F., and Manning, M. J. (1983). The ontogeny of cellular immunity in the rainbow trout, *Salmo gairdner* Richardson, in relation to the state of development of the lymphoid organs. *Dev. Comp. Immunol.* **7,** 69–75.

Tatner, M. F., Adams, A., and Leschen, W. (1987). An analysis of the primary and secondary antibody responses in intact and thymectomized rainbow trout, *Salmo gairdneri* Richardson, to human gamma globulin and *Aeromonas salmonicida. J. Fish Biol.* **31,** 177–195.

Thomas, P. T., and Woo, P. T. K. (1990). *In vivo* and *in vitro* cell-mediated immune response of rainbow trout, *Oncorhynchus mykiss* (Walbaum), against *Cryptobia salmositica* Katz, 1951 *Sarcomastigophora: Kinetoplastida. J. Fish Dis.* **13,** 423–433.

Thuvander, A., Fossum, C., and Lorenzen, N. (1990). Monoclonal antibodies to salmonid immunoglobulin, characterization, and applicability in immuno-assays. *Dev. Comp. Immunol.* **14,** 415–423.

Tillit, D. E., Giesy, J. P., and Fromm, P. O. (1988). *In vitro* mitogenesis of peripheral lymphocytes from rainbow trout, *Salmo gairdneri. Comp. Biochem. Physiol.* **89A,** 25–35.

Triplett, E. L., and Barrymore, S. (1960). Tissue specificity in embryonic and adult *Cymatogaster aggregata* studied by scale transplantation. *Biol. Bull.* **18,** 463–471.

Vallejo, A. N., Ellsaesser, C. F., Miller, N. W., and Clem, L. W. (1991a). Spontaneous development of functionally active long term monocyte-like cell lines from channel catfish. *In Vitro Cell Dev. Biol.* **27A,** 279–286.

Vallejo, A. N., Miller, N. W., and Clem, L. W. (1991b). Phylogeny of immune recognition: Processing and presentation of structurally-defined proteins in channel catfish immune responses. *Dev. Immunol.* **1,** 137–148.

Vallejo, A. N., Miller, N. W., and Clem, L. W. (1991c). Phylogeny of immune recognition: Role of alloantigens in antigen presentation in channel catfish immune responses. *Immunology* **74,** 165–168.

Vallejo, A. N., Miller, N. W., and Clem, L. W. (1992a). Antigen processing and presentation in teleost immune responses. *Annu. Rev. Fish Dis.* **2,** 73–89.

Vallejo, A. N., Miller, N. W., and Clem, L. W. (1992b). Cellular pathway(s) of antigen processing in fish APC: Effect of varying *in vitro* temperatures on antigen catabolism. *Dev. Comp. Immunol.* **16,** 367–381.

Van der Zijpp, A. J., and Egberts, E. (1989). The major histocompatibility complex and diseases in farm animals. *Immunology Today* **10,** 109–111.

Van Diepen, J. C. E., Wagenaar, G. T. M., and Rombout, J. H. W. M. (1991). Immunocytochemical detection of membrane antigens of carp leucocytes using light and electron microscopy. *Fish Shellfish Immunol.* **1,** 47–57.

Van Erp, S. H. M., Veltman, J. A., Takiuchi, H., Egberts, E., and Stet, R. J. M. (1994). Identification and characterization of a novel MHC class I gene from carp (*Cyprinus carpio* L.). *Dev. Comp. Immunol.* **18,** S 43.

Van Lierop, M. J. C., Roelofs, J., Rodrigues, P. N. S., Van Erp, S. H. M., Egberts, E., and Stet, R. J. M. (1995). Production and characterization of polyclonal antibodies to MHC molecules in fish. Nordic Symposium on Fish Immunology, Reykjavik, p. 48.

Van Loon, J. J. A., Van Oosterom, R., and Van Muiswinkel, W. B. (1981). Development of the immune system in carp. *In* "Aspects of Comparative and Developmental Immunology I" (J. B. Solomon, ed.)., pp. 469–470. Pergamon, New York and Oxford.

Verlhac, V., Sage, M., and Deschaux, P. (1990). Cytotoxicity of carp, *Cyprinus carpio,* leucocytes induced against TNP-modified autologous spleen cells and influence of acclimatization temperature. *Dev. Comp. Immunol.* **14,** 475–480.

Walker, R. A., and McConnell, T. J. (1994). Variability in an MHC Mosa Class II β-chain-encoding gene in striped bass (*Morone saxtilis*). *Dev. Comp. Immunol.* **18,** 325–342.

Wang, R., Neumann, N. F., Shen, Q., and Belosevic, M. (1995). Establishment and characterization of a macrophage cell line from the goldfish. *Fish Shellfish Immunol.* **5,** 329–346.

Warr, G. W., and Simon, R. C. (1983). The mitogen response potential of lymphocytes from the rainbow trout reexamined. *Dev. Comp. Immunol.* **7,** 379–384.

Weiss, E., and Avtalion, R. R. (1977). Regulatory effects of temperature and antigen upon immunity in ectothermic vertebrates. II. Primary enhancement of anti-hapten antibody response in high and low temperatures. *Dev. Comp. Immunol.* **1,** 93–103.

Yamaguchi, T., Sakai, S., and Takeuchi, M. (1990). Identification of platelet-activating factor in the lymphocytes and thrombocytes from carp, *Cyprinus carpio. Nippon Suisan Gakkaishi* **56,** 2119.

Yoshida, S. H., Stuge, T. B., Miller, N. W., and Clem, L. W. (1995). Phylogeny of lymphocyte heterogeneity: Cytotoxic activity of channel catfish peripheral blood leukocytes directed against allogeneic targets. *Dev. Comp. Immunol.* **19,** 71–77.

Zelikoff, J. T., Enane, N. A., Bowser, D., and Squibb, K. S. (1990). Fish macrophage 1: Development of a system for detecting immunomodulating effects of environmental pollutants. Symposium at the Society of Environmental Toxicology and Chemistry, 11th Annual Meeting, Arlington, VA, p. 86.

5

THE SPECIFIC IMMUNE SYSTEM: HUMORAL DEFENSE

STEPHEN L. KAATTARI AND JON D. PIGANELLI

I. INTRODUCTION

The mechanisms involved in specific humoral (antibody) defense have been the most thoroughly explored of all the modes of piscine disease resistance. However, piscine classes such as elasmobranchii and osteichthyes possess extremely wide and varied means of accomplishing physiological goals, including the development of prophylactic immune responses. Unfortunately, even though fish are so diverse, the intensity of effort in analyzing immune function has been focused upon fairly few species, primarily salmonids (*Oncorhynchus* and *Salmo*), catfish (*Ictalurus*), and carp (*Cyprinus*). This may be considered unfortunate; however, the burgeoning importance of aquaculture and the need to increase food production has led

THE FISH IMMUNE SYSTEM:
ORGANISM, PATHOGEN, AND ENVIRONMENT

to an intense research effort with these species. Although specific features of the immune response may vary widely between piscine species, the humoral immune response does share some basic features in form and function with that of mammals. These similarities include basic immunoglobulin structure, the cellular requisites for the induction of antibodies, and the role played by these antibodies in such activities as neutralization, complement fixation, and opsonization. The advances in piscine humoral immunity have been rapid and possess practical applications in the field of vaccinology. Current successes in the analysis of piscine immunoglobulin and cytokine genes have not only revealed fascinating differences in the way physically similar molecules can be constructed, but have provided immunologists with the appropriate tools by which fish immune function may be modulated and more effectively enhanced via vaccination.

It is the scope of this chapter to provide background on the function of the humoral immune system and the application of this knowledge to the immunoprophylaxis of fish. To achieve these goals the reader is introduced to the structure of the piscine antibody molecule and the processes involved in its synthesis (Section I). Following this introduction, four basic effector functions of the antibody moelcule are outlined (Section III). Although not all-inclusive, these functions are those that can be definitely ascribed to fish antibodies. With a working knowledge of the structure and function of the antibody molecule, the cellular requirements for its production are presented (Section IV). The latter sections (V–VII) focus on elements critical to immunoprophylaxis and vaccine development. Section V is dedicated to the process of long-term immunological memory generation, which is essential to vaccine efficacy. In section VI the concept of mucosal immunity is presented. This form of immunity is of particular importance to immunologists dealing with, for most infections, the first line of defense against potential pathogens. Section VII presents an overview of past and current attempts to develop successful vaccines to five salmonid pathogens. Within this section some of the most challenging diseases facing aquaculture are addressed, together with a variety of different approaches to vaccine design and delivery.

II. THE ANTIBODY MOLECULE

Antibodies are among the most structurally complex of biological molecules. This complexity transcends their final protein form and includes the intricate genetic mechanisms required to produce the wide repertoire of antibodies. The need for such complexity is obvious, for within their structure must lie the ability to bind a virtual universe of pathogens and ensure their destruction and removal. The construction and function of such large (~1,000,000 d) and complex molecules has yet to be fully delineated even for the well-studied murine and human systems.

A. Basic Architecture of Fish Immunoglobulins (Igs)

In this description the functional anatomy of the antibody molecule will be approached by dividing the molecule into two regions (Fig. 1): the antigen-binding, amino terminus (Fab) and the carboxy-terminal effector region (Fc). In the teleost this molecule appears primarily as a tetramer, composed of four monomeric subunits, each containing two "heavy" (H) or larger (~72 kDa) protein chains, and two "light" (L) or smaller (~27 kDa) chains (Fig. 1). The same theme is observed in elasmobranchs except that the polymer is a pentamer (Voss and Sigel, 1972). Also there have been reported cases of monomers coexisting with the tetramer in sera (Acton *et al.,* 1971; Clem and McClean, 1975; Lobb and Clem, 1981a,b; Elcombe *et al.,* 1985). Within the tetramer, monomeric subunits have been found to associate either covalently via disulfide bonds, or noncovalently with one another. Each antigen-binding site (Fab) is constructed from the N-terminal portions of each H and L chain, yielding eight binding sites per antibody molecule. Although the H and L chains of an antibody molecule possess different amino acid sequences, all H chains within an antibody are identical to one another and all light chains are identical to one another. Each B cell and its clonal progeny produce a unique antibody molecule. Since all binding sites within an antibody molecule possess the same combination of H and L chains, these binding sites should be identical in their specificity and affinity. One possible exception to this model may have been discovered by Shankey and Clem (1980) in the nurse shark antibody (a pentameter), these investigators determined that the purified antibodies appeared to possess five relatively high-affinity binding sites and five low-affinity binding sites. In comparable studies with mammalian IgM, it also appeared that there may only be five functional binding sites and not ten, as would be expected for a pentameter. In the mammalian studies this phenomenon was attributed to steric hindrance due to occlusion of unbound sites by previously bound antigen (Edberg *et al.,* 1972). However, in Shankey and Clem's studies this possibility was precluded by using small haptenic molecules as the antigens. The authors also isolated distinct isoelectric subpopulations of antibodies from the serum and found that they also demonstrated this dichotomous binding behavior. This latter experiment eliminated the possibility that the authors might be witnessing the affinities possessing two different antibody molecules of distinctly different affinities.

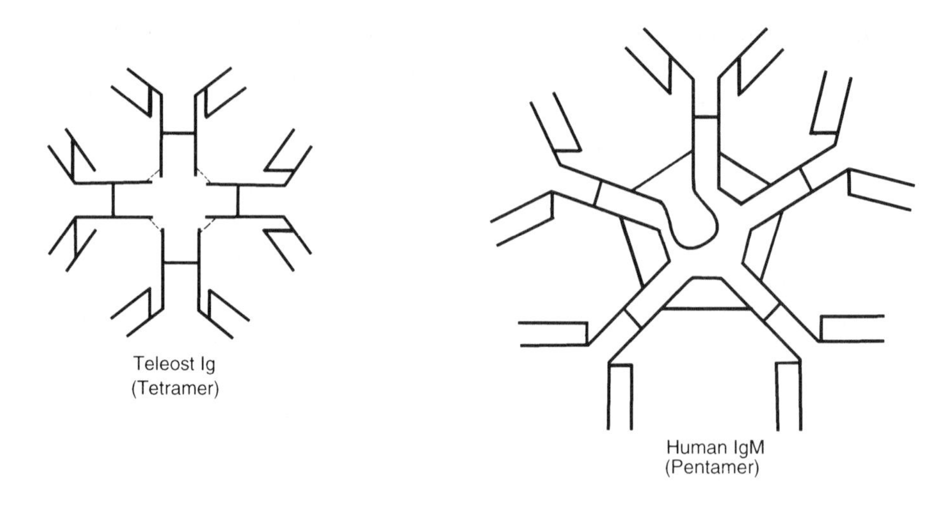
Teleost Ig
(Tetramer)
Human IgM
(Pentamer)

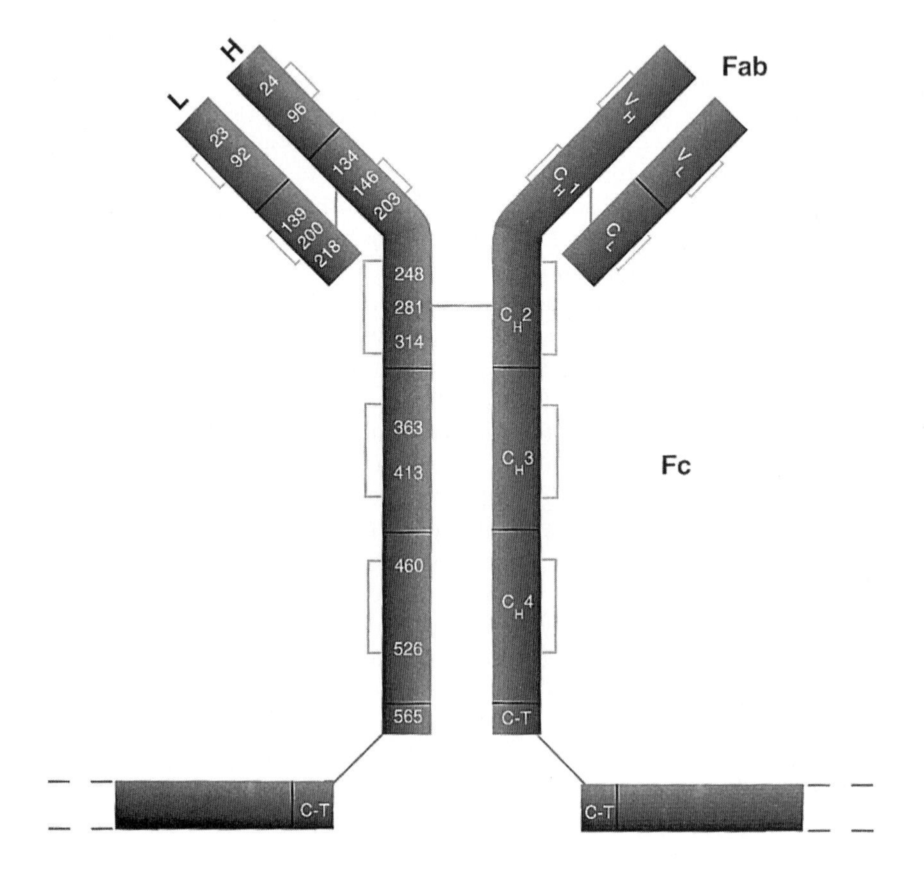
H
L
24
96
23
92
134
146
203
139
200
218
248
281
314
363
413
460
526
565
C-T
Fab
Fc

One possibility then, at least for the nurse shark, would be that each antibody molecule may express two different binding sites.

A molecular inspection of the H chain reveals that it can be divided into five domains and a carboxy-terminal stretch of amino acids, whereas the light chain is composed of two domains (Fig. 1). An immunoglobulin domain is a structural subunit that possesses a certain amount of sequence homology with the other domains and, thus, also physically takes on some conformational similarities with the other domains such as regularly placed intradomain disulfide loops. The N-terminal domain of the heavy chain is termed the V_H (variable-heavy) domain. Within this domain and its light chain counterpart, the V_L (variable-light) domain, resides the antigen-binding site. The variability alluded to by the domain designation refers to the tremendous variability found between amino acid sequences of different antibodies. Four more domains exist in the heavy chain: C_H1, C_H2, C_H3, and C_H4. The "C" in this nomenclature refers to the relative invariance or constancy of the amino acid sequence between different antibodies as compared to that of the V_H domain. There is greater homology among each of these domains than with the V_H, and yet some characteristic differences exist. For example, each domain possesses pairs of similarly placed cysteines, each of which forms a similar intradomain disulfide bond (Fig. 1). These bonds result in a similar loop structure within each domain. However, differences exist, such as the positioning of other specific cysteines in select domains, which result in the bonding of H and L chains into covalent half-mers. In Fig. 1 such a bond is shown to occur between Cys 218 of the light chain and Cys 134 of the heavy chain. In the trout there appears to be two other cysteines available for inter–heavy chain or intersubunit (intermonomer) disulfide bonds (Cys 281 of the C_H2 and Cys 565 of the C-terminal tail). Although the specific residues dedicated to interchain versus intersubunit disulfide bonds have been delineated in mammals (Steiner, 1985), only hypothetical arrangements can be posed at this time for fish Igs.

The assembly of fish antibody molecules via disulfide crosslinking appears to be a means by which structural diversity is generated. This is

Fig. 1. At top, the structure of the secreted forms of teleost immunoglobulin and human IgM are portrayed. The solid black lines represent contiguous heavy or light chains, with the gray lines representing disulfide bonds. Although the intermonomer disulfide positions are confirmed for the human IgM, this has yet to be confirmed in fish. At bottom, the structure of a salmonid monomeric subunit is portrayed, based on available data (Daggfeldt *et al.*, 1993; Lee *et al.*, 1993, Hansen *et al.*, 1994). The numbers represent the positions of available cysteines. The gray brackets represent intradomain disulfides. The interchain disulfides are hypothetical.

distinctly different from what is observed in mammalian species where intersubunit disulfide cross-linking appears to be employed in a more uniform manner. Specifically, mammalian IgM (the primary polymeric Ig) is a pentamer, each monomeric subunit of which is disulfide-linked via a cysteine in the C_H4 (Fig. 1; Davis *et al.,* 1989; Sitia *et al.,* 1990; Wiersma and Shulman, 1995). In fact, it has been determined that complete disulfide cross-linking of all monomeric subunits within mammalian IgM is stringently required. If this process does not occur, the IgM molecule will not be secreted (Sitia *et al.,* 1990), but targeted for digestion (Tartakoff and Vassalli, 1979; Davis *et al.,* 1989; Sitia *et al.,* 1990). The digestion of partially cross-linked IgM is thought to be an essential aspect of the quality control mechanism provided by the endoplasmic reticulum (Hurtley and Helenius, 1989).

In contrast, a variety of teleosts generate considerable structural diversity, by nonuniformly cross-linking monomeric subunits within the tetrameric antibody (Lobb and Clem, 1981a; Warr, 1983; Lobb and Clem, 1983; Lobb, 1985; Ghaffari and Lobb, 1989; Sanchez and Dominguez, 1991). The pattern of this cross-linking diversity varies among fish species. Catfish (*Ictalurus punctatus*) appear to possess tetramers that are either wholly cross-linked, or composed of some combination of $\frac{1}{2}$-, 1-, $1\frac{1}{2}$-, 2-, $2\frac{1}{2}$-, 3- and, $3\frac{1}{2}$-mers (Lobb and Clem, 1983; Lobb, 1985). Sheepshead (*Archosargus probatocephalus*) tetramers are composed of covalently linked tetramers and noncovalently associated dimers (Lobb and Clem, 1981a). Trout (*Oncorhynchus mykiss*) tetramers can be entirely covalently linked, or composed of trimers, dimers, and/or monomers (Sanchez and Dominguez, 1991). The toadfish (*Spheroides glaber*) has also been observed to possess similar structural diversity (Warr, 1983). The precise reason for this structural diversity is unknown at this time, however, it is not unlikely that this may represent a posttranslational mechanism whereby functional diversity may be introduced into immunoglobulin molecules without requiring the dedication of specific structural genes to accomplish the same goals. Obviously, teleosts have developed an alternate mechanism that permits the production and secretion of Ig polymers with varying degrees of disulfide cross-linking between subunits or even a complete absence of such cross-linking, without a resultant dissimilation or deviation from the secretory pathway.

The construction of the membrane receptor form of Ig differs from that of the secreted form, as the heavy chain contains the same V_H (variable-heavy), C_H1, C_H2, and C_H3 domains but a transmembrane coding exon replaces the C_H4 and C-terminal encoding exons (Wilson *et al.,* 1990; Bengten *et al.,* 1991; Hordvik *et al.,* 1992; Hansen *et al.,* 1994). The exchange of the C-terminal secretory tail for the transmembrane sequence is consistent with what is observed in mammals, but the exclusion of the C_H4 domain appears to be a unique feature of fish.

B. Synthesis and Structure of the Antigen-Binding Site

Previous studies in mammals indicate that the number of different binding sites is enormous (Blackwell and Alt, 1988). This great diversity of binding sites can be conceived to be necessary to accommodate the myriad number of possible foreign antigenic structures associated with pathogens. However, this realization also posed a great enigma as to how a genome could encode millions of antibody molecules. The dilemma was resolved upon the characterization of the genes encoding the antigen-binding sites. Through the process of somatic recombination, a limited number of coding sequences (introns) can be randomly spliced together such that millions of possible binding sites can be formed from relatively few gene segments (Schatz *et al.*, 1992). Secondly, each B cell can form an antibody from the association of a light chain and a heavy chain. Random selection of a V_L and V_H then would increase the possible heterogeneity of binding sites. Finally, the resultant binding site may be altered through the process of somatic mutation (Berek and Milstein, 1987).

Variability is particularly great within the short stretches of the V_H and V_L domains called complementarity determining regions (CDRs). There are three such CDRs in the V_H (approximately positions 30–38, 48–64, and 94–103) and three within the V_L (approximately positions 25–40, 50–60, and 90–99). On either side of each CDR is a framework region. These framework regions are relatively invariant and are thought to hold the CDRs in an appropriate position for binding the incoming antigen. Thus, if one aligns the sequences of heavy or light chains from a variety of antibodies, one finds tremendous variability among the amino acids within the CDRs as they represent binding sites that bind structurally unique antigens. There is considerably less variability within the framework regions and virtually none within the constant regions.

Three consistently striking features of fish antibody molecules have been their relative low intrinsic affinity (affinity of the individual binding site), the apparent lack of the ability for serum antibodies to increase in affinity over time after immunization (affinity maturation), and the limited amount of antibody binding site heterogeneity (Clem and Small, 1967; Russell *et al.*, 1970; Voss *et al.*, 1978; Makela and Litman, 1980; Wetzel and Charlemagne, 1985; Cossarini-Dunier *et al.*, 1986). However, the relative overall lack of high-affinity binding sites becomes less of an oddity when compared to the mammalian IgM response (Mochida *et al.*, 1994). In fact, since teleost antibodies are primarily tetrameric, the overall avidity or functional affinity (affinity of the entire molecule) may be comparable or greater than that observed in mammalian IgG antibodies (Fiebig *et al.,* 1977). Indeed, in examining the functional affinity of mammalian IgM versus IgG antibodies, the functional affinity of each is comparable, although the intrinsic affinity is considerably higher for IgG antibodies (Hornick and Karush, 1972).

The diversity of binding sites has been addressed via a number of routes such as the use of scatchard analysis of whole serum antibodies (Shankey and Clem, 1980), relative specificity analysis (Makela and Litman, 1980), isoelectric focusing of immunopurified antibodies (Cossarini-Dunier *et al.,* 1986; Wetzel and Charlemagne, 1985), and anti-idiotypic analysis (Richter and Ambrosius, 1988). In all cases the studies have indicated that little heterogeneity of binding sites exists among the fish species. However, limited heterogeneity is also true for mammalian IgM (Griffiths *et al.,* 1984). These observations lead to the intriguing speculation that the need for the expression of monomeric Ig may have led to the development of mechanisms to produce binding sites with high intrinsic affinity and perhaps a greater repertoire of binding sites. Supportive of this hypothesis is the observation that sharks, which possess a monomeric form of Ig, appear to develop the same high intrinsic affinity in this class of Ig as do the monomeric mammalian counterparts (Voss and Sigel, 1972).

C. Isotypic Diversity among Fish Immunoglobulins

Initial characterizations of fish antibody structure, such as those described above, led to the general supposition that fish Ig was rather comparable to mammalian IgM. This seems to be due primarily to the fact that mammals initially produce a structurally similar polymeric Ig of low intrinsic affinity in response to antigen. Thus, the ability to produce high-affinity binding sites and monomeric forms of Ig was considered an evolutionary advance, and not simply a mammalian adaptation. However, this idea of primitive or perhaps limited functionality is quite likely a mammlian-biased perspective on the function of antibodies. No doubt sentient fish might have adopted a contrary point of view. Whether fish lack isotypic differences remains as an important question, resolution of which will likely have profound implications for our understanding of fish as well as mammalian immune systems. Basically, isotypes are structural differences among Igs that are shared by all members of a species. The molecular basis of these differences lies in the specific sequences encoded by distinct C_H genes (this is, of course, exclusive of allelic variants, allotypes, among individuals). Thus, humans possess specific IgM constant region genes ($C_H\mu$), IgD genes

($C_H\delta$), as well as seven other C_H genes encoding different isotypes (IgG1, IgG2, IgG3, IgG4, IgE, IgA1, IgA2). The isotype can be switched within an antibody-producing cell (B cell) while retaining the original antigen-binding site. The actual switch occurs when the appropriate interleukin signals from an antigen-specific T cell are elicited (Finkelman *et al.,* 1990). Each isotype possesses distinctive C-domain structural differences which lead to distinctly different functional capabilities for that antibody (Winkelhake, 1979). These functional differences include more effective complement activation (IgM, IgG1, IgG3), placental transfer (all IgGs), transfer to mucosal surfaces (IgAs), arming of phagocytes (IgG), and elicitation of allergic responses (IgE). Thus possession of these isotypic differences permits the immune system to exert a more refined control over the humoral response.

Since only tetrameric forms of antibody were originally isolated from fish serum, the supposition was generally made that fish may only possess the one (IgM-like) isotype. However, as with early mammalian studies, once rigorous procedures were used, antibody reagents were developed that could distinguish the isotypic differences in fish. Species in which serologically defined isotypic differences could be observed were, for example, the skate, *Raja kenojei* (Kobayashi *et al.,* 1984; Kobayashi and Tomonaga, 1988), catfish, *I. punctatus* (Lobb and Olson, 1988), trout, *O. mykiss* (Sanchez *et al.,* 1989) and Atlantic salmon, *Salmo salar* (Killie *et al.,* 1991; Hordvik *et al.,* 1992). In the catfish, four distinct heavy chain isotypes can be serologically deduced (Lobb and Olson, 1988). Although the functional relevance of light chain isotypic differences is not quite clear for fish or mammals, they have been observed in both catfish (Lobb *et al.,* 1984) and trout (Sanchez and Dominguez, 1991). Currently in teleosts only a limited number of different C genes has been reported (Ghaffari and Lobb, 1989; Hordvik *et al.,* 1992). However, in elasmobranchs there appear to be minimally 200 C genes (Kobuku *et al.,* 1987; Harding *et al.,* 1990). Obviously if the number of C genes were to be primary factor in determining the number of isotypes, these species would be considered the most isotypically diverse. The C genes sequenced from these species indicate, however, a surprisingly low amount of diversity among sequences.

Isotypic diversity, following the mammalian paradigm, would necessitate that the structural differences required be encoded genetically, i.e., distinct C genes encode for specific isotypes which, in turn, give functionally different antibodies. However, as mentioned previously, structural diversity is not strictly limited to the existence of distinct C-region sequence differences in fish. Structural diversity can be generated from a single gene product, posttranslationally, by such avenues as differential disulfide bond formation between monomeric subunits, or, as in the catfish, between half-mers (Ledford *et al.,* 1993). Although, at present, no functional significance has been ascribed to such structural differences, it is not difficult to hypothesize functional differences. First, the reduction in the number of intersubunit bonds would result in increased flexibility of the molecule. Mammalian IgM, although fairly large, has limited flexibility due to the extensive disulfide cross-linking (Nezlin, 1990). In fish the same antibody molecule would have binding sites of the same specificity and size yet may be capable of accommodating antigenic determinants of different spatial orientations without requiring uniquely specific C gene–encoded isotypes. Furthermore, as of late, it has been determined that the degree of disulfide cross-linking in mammalian Ig also confers varied ability to activate complement (Dorai *et al.,* 1992; Michaelsen *et al.,* 1994; Brekke *et al.,* 1995). If the same applies to fish, then heavily cross-linked molecules may be more effective in fixing complement, whereas more flexible molecules may not need to fix complement, or be used exclusively for some alternate function in antigen elimination.

III. ANTIBODY EFFECTOR MECHANISMS

The purpose of this section is to acquaint the reader with the varied effector functions mediated by the antibody molecule. Such knowledge is essential to understanding the most likely effect an antibody would have on a specific pathogen. Additionally, these reactions can have pathological effects which will be introduced in the last section of the chapter.

A. Neutralization

The most direct effect of the interaction of an antibody with its antigen is the simple physical act of blocking a critical function of the antigen, whether it be that of a receptor, enzymatic active site, or toxigenic determinant. For example, antibodies that are induced in salmonids to the glycoprotein (G protein) of infectious hematopoietic necrosis virus (IHNV) are termed neutralizing antibodies. The G protein is ligand for the cell surface receptor found on the host cell. Binding to the host cell via this receptor permits infection of the cell by the virus. Anti-G protein prevents this from occurring by physically blocking the interaction of the G protein of the virus with its host receptor. Theoretically, if no host receptor can bind to the virus, eventually the virus will be either excreted or otherwise inactivated. Molecular antigens can also be neutralized. A primary example of this is the role of antitoxin in neutralizing toxins. Simple masking of the toxigenic determinant with an antibody will prevent induction of the toxigenic mechanism, often an enzymatic function such as proteolysis or lipolysis.

B. Precipitation and Agglutination

Another effect of the interaction of an antibody molecule with its antigen is a product of its multivalent binding capacity and the subsequent creation of a macromolecular complex of antibodies and antigens. When this process of cross-linking occurs with optimal concentrations of antigen, the cross-linking will result in an extremely large lattice of antigen and antibody. If sufficiently large, this lattice will no longer be soluble in the plasma of the body and will come out of solution. When referring to antigens that are molecular, this process is referred to precipitation, and when the antigen is cellular the process is termed agglutination. The development of such complexes can be a distinct advantage to the host in that particulate matter is more easily phagocytized. Also, the close proximity of many antibodies within the complex also facilitates the activation of complement or adherence of the antibodies to the Fc recpetors (FcR) of phagocytes. Thus these complexes will be scavenged more effectively. This same process can lead to pathological consequences if the concentrations of antibody and antigen are too great (see Section III,D). Precipitation and agglutination are also simple methods that can be employed to monitor the induction of an antibody response, however, since the assay endpoint is the visualization of a precipitate or agglutinate, this method is the least sensitive of all antibody detection methods (Kuby, 1992).

C. Opsonization

Opsonization refers to the process by which either antibody or another immunologically-related molecule promotes the phagocytosis of the antigen by a phagocyte. This is most specifically used with respect to the coating of bacteria, fungi, and parasites with specific antibodies that lead to the interaction and often the phagocytosis of the target (Honda *et al.,* 1986; Whyte *et al.,* 1990). A classical example of how this system works is that of the murine response to *Streptococcus pneumoniae* (Austrian, 1980). *S. pneumoniae* possesses a thick polysaccharide capsule that is antiphagocytic such that as the macrophage approaches the bacterium and makes first contact via its pseudopodia, the immediate recoil of these pseudopodia can be visualized. Therefore, *Streptococcus* is quite capable of avoiding phagocytosis. However, if the bacterium is first opsonized by anticapsular antibodies, the capsular polysaccharide no longer presents a barrier and the macrophages rapidly ingest the coated bacteria. Such is also observed indirectly with salmonid fish and the response to *Diplostomum spathaceum* (Whyte *et al.,* 1990). *In vitro* coincubation of the worms with anti-spathaceum antibodies results in the facilitated digestion of the parasites.

The targeting molecules need not be antibody. The activation of complement leads to the elaboration of cleaved complement components on the target cell surface; in mammals the cleaved component of C3 (C3b) can act as an opsonin (see later). Fish species have demonstrated comparable complement components, thus it is not unlikely that they also can exhibit complement-mediated opsonization. The first step in such complement-mediated opsonization need not be a consequence of an antibody reaction, as certain microbial surfaces of stabilizing spontaneously produced C breakdown products (the alternative pathway).

D. Complement-Mediated Functions

Although our knowledge of the complement system is not complete, it can be asserted with confidence that fish possess a complement system that functions in concert with antibodies, as observed in mammals. This section will not cover complement function in detail as that has been provided by Dr. Yano in this volume. However, as the activation of complement by antibody represents an essential antibody effector mechanism, it will be discussed in this context.

The role of antibody in complement activation can be thought of most simply as a mechanism to focus this potent effector function against specific pathogenic targets. Simply, the binding of antibody to an antigen results in a conformational change of the antibody's Fc region(s) which permits the binding and activation of the first component of the complement system. The binding of this component results in conformational changes which, in turn, facilitates the binding of other complement components. In this manner proteolytic enzymes are produced that cleave (activate) other complement components in a linear fashion, the products of which, in turn, can cleave other specific proenzymes. As activation is a process of enzymatic cleavage, it can be seen that one initial step is tremendously amplified by the subsequent steps, thus resulting in what is termed the complement cascade. At each step of this process proteolytic products are produced that can opsonize pathogens, lyse pathogens, or act as pharmacological mediators of host vascular, muscular, or immune tissue modification.

A brief outline of this process is described here to accompany the scenario depicted in Fig. 2. Again, this description is designed to provide the reader with operational definitions so that the various processes described in this chapter may be appreciated. Figure 2 represents a generalized mammalian scheme over which (shading) the particular piscine homologues have

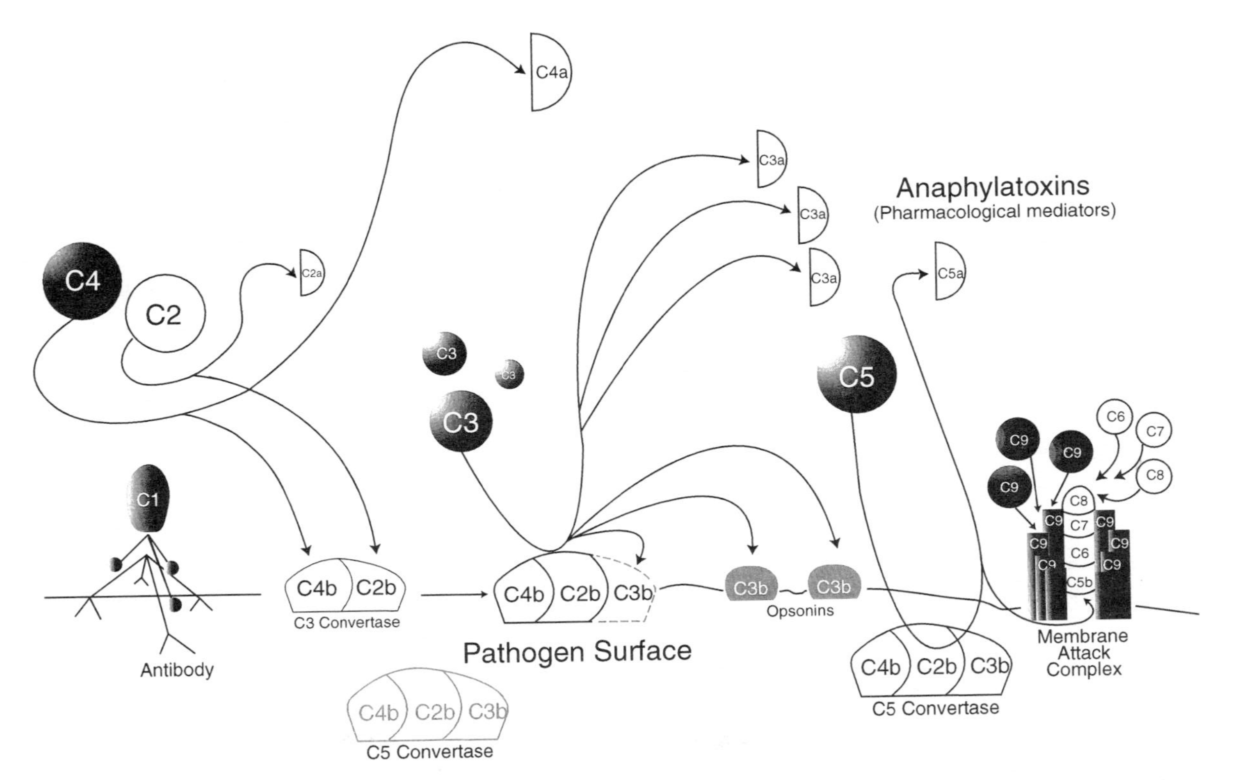

Fig. 2. The process of complement activation. In this figure a tetrameric teleost antibody has reacted with a surface antigen of a pathogen. The general flow of the cascade is from left to right. The shaded components represent complement components that have been observed in fish.

been identified. The isolation and characterization of fish complement components has focused on the most pivotal or critical elements in the cascade (i.e., C3, C9). However, it has become apparent that much of this sequence must be evolutionarily conserved. Molecular evidence, which can be garnered more quickly, may reveal evidence of other components that have not, as yet, been isolated.

Binding the first component of complement (C1) requires the close juxtaposition of more than one monomeric Fc. As mammalian IgM possesses five Fcs, it is the most effective antibody in the activation of complement (Borsos *et al.,* 1981). Indeed, under circumstances where IgM hexamers can be produced, the efficiency of complement activation is enhanced (Davis *et al.,* 1989). The effectiveness of the mammalian monomeric form of Ig (IgG) is also somewhat limited in that, minimally, two IgG molecules must bind to such closely spaced epitopes that their Fcs can be easily crosslinked by a single C1q molecule (Winkelhake, 1979). This C1q has been isolated from carp (Yano *et al.,* 1988) and the nurse shark (Ross and Jensen, 1973a,b).

The relatively high efficiency of complement fixation by polymeric Ig over monomeric Ig was also reported for trout antibodies (Elcombe *et al.,* 1985). Tetrameric anti-TNP antibodies induced in the same fish and possessing the same affinity for the hapten as the monomeric form were found to be 10–100 times more efficient in lysing trinitrophenylated sheep red blood cells than the monomeric form. These two Ig forms also showed some antigenic similarity but were only partially identical. Further, Elcombe *et al.,* (1985) noted that the H chain of the tetrameric form possessed a molecular weight of approximately 60 kDa while the monomeric H chain was 50 kDa. Thus the possibility exists that the effectiveness of complement fixation may not be due simply to the polymeric nature of the tetramer, but also other structural differences that may reside in their unique Fc regions. These authors also posed that the trout monomer may activate complement via a completely different pathway. Application of a treatment that depletes mammalian C4 (NH_4OH) and interrupts the classical pathway completely disrupts only tetramer-induced complement activity, not monomeric activation. It was suggested, therefore, that the monomeric form might act through the alternative pathway.

The operational C1 complex acts to cleave C4 and C2 to generate surface binding components C4b and C2b and soluble components C4a and C2a. The work of Elcombe *et al.* (1985) would seem to indicate that a functional analog of C4 should exist in trout. In mammals, C4a has been recognized as possessing pharmacologic activity and acts to increase vascular permeability and induce smooth muscle contraction (Hugli, 1981). The cell surface complex of C4bC2b is termed C3 convertase. C3 (Nonaka *et al.*, 1984; Nakao *et al.*, 1989) is split into the potent anaphylatoxin C3a, and the opsonin C3b by this convertase. The existence of a C3b-like fragment in trout has been confirmed by protein sequencing (Lambris *et al.*, 1993) and by the generation of functions attributed to the alternative pathway, which requires C3b. Briefly (T. Yano, this volume), the alternative pathway exploits a very simple principle of requiring the presence of the pathogen to help stabilize an unstable complement product (C3b). Once stabilized upon the pathogen surface, and together with a number of serum factors, it behaves as a C3 convertase.

C3b as an opsonin can be recognized by a C3b receptor (CR3) on phagocytic cells. The existence of a possible receptor has been demonstrated in the catfish (Ainsworth, 1994). This was accomplished by using a yeast cell wall material, zymosan, to stabilize C3b. When arrayed with an opsonin such as C3b, phagocytic cells can then recognize and digest the pathogen. Ainsworth (1994) has demonstrated that zymosan can bind neutrophils after exposure to catfish serum. The binding of serum-activated zymosan results in the activation of the neutrophil as evidenced by the generation of chemiluminescent response. The importance of such mechanisms has been demonstrated in fish by exposure of various microbes, microbial cell wall products, or similar synthetic material to fish serum and observing the resultant opsonic behavior in neutrophils (Jenkins and Ourth, 1993).

C3b fragments can also bind to the C3 convertase complex to generate a C5 convertase complex, thereby initiating the final sequence of the complacent cascade, ending in the production of the membrane attack complex (MAC) and lysis of the cell. The MAC is a clustering of C6, C7, C8, and a number of C9 molecules which form a pore within the membrane of the cell. This form of attack is particularly effective on Gram-negative bacterial cells which possess an outer membrane (Fig. 3), or eukaryotic cells which possess no cell wall but only a cellular membrane. Host cells that have been modified, damaged, or virally infected fall into the latter category. Cells that have particularly thick cell walls and/or no surface membrane, (i.e., Gram-positive bacteria or fungi) are resistant to the MAC but are readily affected by the other complement-mediated functions such as opsonization. In fish, C9 has been isolated and the MAC visualized by electron microscopy (Tomlinson *et al.*, 1993). In that study trout anti-erythrocytic serum was employed to form an MAC and lyse rabbit erythrocytes.

The ability of complement to lyse bacterial cells has been demonstrated for Gram-negative bacteria with the alternative complement pathway (Jenkins and Ourth, 1990). It can thus be assumed that Gram-negative bacterial cells can be lysed via complement and serum antibody, as the MAC would be the same whether generated via the alternative or classical pathway.

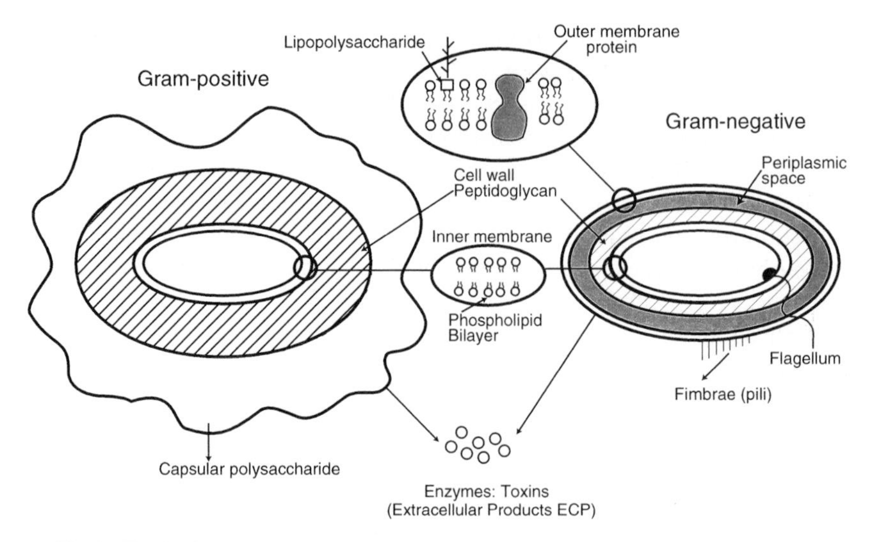

Fig. 3. The basic structures of Gram-positive and Gram-negative bacteria are depicted. Many of the structures listed can serve as protective antigens (i.e., capsular polysaccharides, toxins, outer membrane proteins, and lipopolysaccharides). In the text, *Vibrio anguillarum* and *Aeromonas salmonicida* serve as examples of Gram-negative bacteria, while *Renibacterium salmoninarum* is Gram-positive. *A. salmonicida* also contains an "A layer" which is exterior to the outer membrane and held in place by the O-polysaccharide of the LPS.

Pathological consequences can also ensue from complement activation in that larger complexes will activate greater amounts of complement. High levels of complement activation will lead to the release of a high concentration of pharmacological mediators, which could have potent and detrimental effects on the vasculature and other tissues. Much damage can also ensue with the activation of macrophages and misguided attempts by these cells to phagocytize large amounts of antigen. This can lead to destruction of surrounding normal tissue, or systemic effects when large amounts of circulating complexes are present.

IV. INDUCTION OF B CELLS

From the point of view of B-lymphocyte activation, antigenic structure may be divided into two forms, T-independent (TI, polysaccharides) and T-dependent (TD, proteins). Although exceptions to this rule may be found, most antigens can be classed as one of these two forms. If the antigen is of a different composition (e.g., glycolipid or nucleic acid), work with mammalian systems suggest that it is dealt with as either a T1 or a TD antigen. To date, the most complete elucidation of the cellular pathways employed to induce B cells has been approached only in the catfish and trout. The reasons for this situation are that cell-partitioning and *in vitro* immunization techniques are required to perform these types of cellular analyses. Such technology has been available to mammalian immunologists for approximately the past 30 years, but only for the past 12 years in the field of piscine immunology.

Minimally, the induction of antibody responses to the TI antigen, lipopolysaccharide, requires the auxiliary assistance of an accessory cell such as the monocyte or macrophage (Clem *et al.,* 1985; Tripp, 1988; Vallejo *et al.,* 1990; Kaattari, 1992; Vallejo *et al.,* 1992; Ortega, 1993). The role of the macrophage appears to be primarily the provision of a requisite factor, interleukin 1 (IL-1). Investigators are confident in ascribing this term to the factor as with mammals, in that it performs comparable functions, is induced via the same mechanisms, and appears to possess the same physical characteristics, namely molecular weight (Ortega, 1993). This form of antigen appears to trigger a receptive state in the antigen-specific B lymphocytes of both catfish and trout, while simultaneously inducing the macrophage to produce IL-1. One aspect of this receptive state for the B cell is the ability to receive the IL-1 factor. In trout the IL-1 factor is critical to the differentiation of the B lymphocyte into an antibody-secreting plasma cell, and not simply the enhancement of cellular proliferation.

Protein antigens require a more sophisticated or complex cooperative scheme among other leukocytes. Their need for cellular cooperation was discerned fairly early in the analysis of the induction of antibody responses, however, it relied on the elaboration of antibody responses by separate induction of carrier- and hapten-specific cells. Again, the precise delineation of the leukocytic subpopulations required for the induction of a T-dependent response requires cell-partitioning and *in vitro* culture techniques (Kaattari, 1992). Quite simply, in order to induce an antibody response to a protein antigen, the antigen must be processed by the accessory cell (macrophage) and presented on the cell surface to the T cell (Vallejo *et al.,* 1990; Vallejo *et al.,* 1992), which in turn, elaborates requisite interleukins (Yang *et al.,* 1989). These interleukins then provide the requisite signals and growth factors for B-cell differentiation (Caspi and Avtalion, 1984; Grondel and Harmsen, 1984).

V. MEMORY

Memory has always been considered a hallmark of the specific immune response. However, immunological memory in fish has been defined primarily by employing paradigms based on mammalian responses. These paradigms specifically emphasize a variety of differential responses that occur when mammals respond to protein antigens. These differential responses include logarithmic increases in monomeric Ig concentration and isotype switching (Eisen, 1980), and dramatic increases in the affinity of the monomeric Ig (Griffiths *et al.*, 1984). These specific phenomena either do not occur in fish, or occur to a much lesser degree. It, therefore, has been argued that the development of memory is a recently evolved characteristic. An alternate viewpoint is that the memory response is simply a differentiated response to a secondary exposure to a specific antigen (Hildemann, 1984; Kaattari, 1994). As recently reviewed (Kaattari, 1994), there is a tendency to ascribe a more primitive status to immune functions shared between ectotherms and mammals, particularly when other unique mammalian features are not discernible in the ectotherm. However, each extant species has evolved a molecular and cellular system that provides for essential immune functions. It is possible that many comparable mammalian "nuances" have yet to be recognized in fish, or possibly, fish may have evolved alternate routes to efficaciously execute comparable immune functions.

Close examination of the secondary antibody response in fish reveals a response that is reminiscent of the IgM response observed in mammals. Most commonly, when a differentiated response has been reported, it is the induction of enhanced antibody production. These increases appear to be rather small in comparison to the logarithmic increases observed in the concentration of monomeric Ig (IgG) in mammals (Clem and Sigel, 1965; Finstad and Fichtelius, 1965; Ambrosius and Frenzel, 1972; Tatner, 1986; Arkoosh and Kaattari, 1991), however, they are not insignificant compared to the magnitude of secondary IgM response.

The memory state can also be characterized by the development of an increased sensitivity to antigen. This form of memory is most easily observed in the sensitization of individuals to allergens. The initial exposure to pollens, bee venom, or other allergens usually does not elicit an allergic response (IgE-mediated), however, it can be sufficient to induce immunological memory. A secondary response then results in an allergic response, even if the dose of allergen is less than that of the primary exposure. This phenomenon of increasd antigen sensitivity occurs with antibodies of other isotypes in reponse to nonallergenic antigens in mammals (Celada, 1967), as well as in trout (Arkoosh and Kaattari, 1991). In this latter experiment, trout were challenged with a dose of a TD antigen (TNP-KLH) that was incapable of eliciting a detectable antibody response in naive fish. Fish primed previously with an optimal dose and challenged with a subimmunogenic dose of antigen, however, were able to produce a greater response than that elicited by an optimal primary dose. The author is unaware of similar studies undertaken with other species of fish, however, such a methodology might be a more revealing way to determine if memory exists within a particular animal than simply measuring the elicitation of higher responses to optimal doses of antigen.

Using the technique of limiting dilution analysis (Lefkovits, 1979), the number of antigen-specific B-cell precursors residing in naive and primed (memory) trout have also been determined (Arkoosh and Kaattari, 1991). As the actual number of antibody-secreting cells (plaque-forming cells, PFCs) were determined in this exercise, the contribution of clonal proliferation to the final PFC response could also be assessed. These studies revealed that the enhanced PFC response observed in primed fish was due exclusively to the generation of an enlarged precursor pool. The amount of clonal proliferation induced in native vs. primed lymphocytes was identical, indicating that the memory precursor was not physiologically different from the native precursor. Increased capacity for clonal proliferation has, however, been observed in mammalian species such as the rat (Brooks and Feldbush, 1981).

Isotype switching and affinity maturation are two other common features of mammalian immunological memory that are primarily associated with the monomeric form of immunoglobulin. In both cases fish antibody does not routinely demonstrate either phenomena. As discussed above, the existence of varied isotypic forms has only recently been detected. However, in those cases where serological recognition of isotypes can be made, no switching events have been observed, although shifts in the relative concentration of these isotypes are observed (Lobb and Olson, 1988; Killie *et al.,* 1991).

Often, in mammals, the increase in affinity (affinity maturation) of the monomeric immunoglobulin can be logarithmic, due to the selective induction by antigen of affinity somatic mutants (Griffiths *et al.,* 1984; Berek and Milstein, 1987; Kochs and Rajewsky, 1989). Although dramatic increases in serum antibody affinity have not been observed in fish (Russell *et al.,* 1970; Voss *et al.,* 1978; O'Leary, 1980; Makela and Litman, 1980), evidence has been forthcoming that somatic mutation does occur, at least in elasmobranchs (Hinds-Frey *et al.,* 1993). Thus, it would appear that fish do possess the capacity to generate the variety of binding sites that could lead to affinity maturation, yet they do not exhibit this trait. One possible explanation for this might be the heavy reliance on the polymeric form of Ig. Such a molecule would be functionally quite similar to mammalian IgM and, thus, possess a high functional affinity or avidity, therefore not requiring extensive increases in intrinsic affinity of the individual binding sites (Kaattari, 1994).

VI. MUCOSAL IMMUNITY

The first line of humoral defense, where pathogens can be most effectively blocked or neutralized, is the mucosal surface. Encountering the pathogen at this point can preclude the necessity of the host employing a systemic response. In mammals, secretory IgA and specialized immune organs such the Peyer's patches, tonsils, and appendix are known to participate in this defense. In contrast to such apparently organized structures, Tomonaga *et al.* (1986) demonstrated that elasmobranchs possessed massive intraepithelial lymphocytic aggregations which were located in the central region of the spiral intestine. These aggregations are thought to be analogous to Peyer's patches. Although teleosts also do not appear to possess specialized lymphoid organs or tissues analogous to Peyer's patches, minor subepithelial lymphoid accumulations have been demonstrated in the roach, *Rutilus rutilus,* and perch, *Perca fluviatilis* (Rombout and van den Berg, 1989, Zapata and Solas, 1979).

Pontinus and Ambrosius (1972) have also observed antibody-secreting cells in the lamina propria of the pyloric region in perch after immunization with sheep red blood cells (SRBCs). Fletcher and White (1973) witnessed increased antibody titers within the intestinal mucus of plaice (*Pleuronectes plasteassa* L.) upon oral immunization with heat-killed *Vibiro anguillarum.* Davina *et al.* (1980) observed scattered, lymphoid-like cells in the epithelium or lamina propria of carp, but no true cell clusters were found. Oral administration of *Vibrio* to carp (*Cyprinus carpio*) resulted in an increase in the intraepithelial leukocytes along the entire intestine of the animal (Davina *et al.,* 1982). In most teleosts, the second gut segment appears to have evolved the specialized function of antigen uptake and processing of antigens (Fig. 4; Noaillac-Depreure and Gas, 1973; Strobard and Kroon, 1981; Strobard and Vander Veen, 1981; Nagai and Fujino, 1983; Iida and Yamamoto, 1985; Georgopoulou *et al.,* 1985; Georgopoulou and Vernier, 1986; Fujino *et al.,* 1987; Hart *et al.,* 1988; McLean and Donaldson, 1990). After uptake and processing, the antigens appear in the intraepithelial macrophages located in the second gut segment (Rombout *et al.,* 1985). It is likely that cells called enterocytes in the second gut segment of carp are analogous to the M cells of mammals and may serve a similar function of transferring antigens from the gut epithelium to mucosal lymphoid aggregates (Rombout *et al.,* 1993). These investigators also reported, using monoclonal antibodies and single and double cell-staining techniques, that carp mucosal intraepithelial lymphoid cells are composed of sIg^- (putative T) cells, NK cells, sIg^+ (B) cells, and Ig-binding, antigen-presenting macrophages.

Although an isotypically distinct secretory Ig has yet to be demonstrated in fish, antigen-specific antibody is induced in the skin mucus of orally

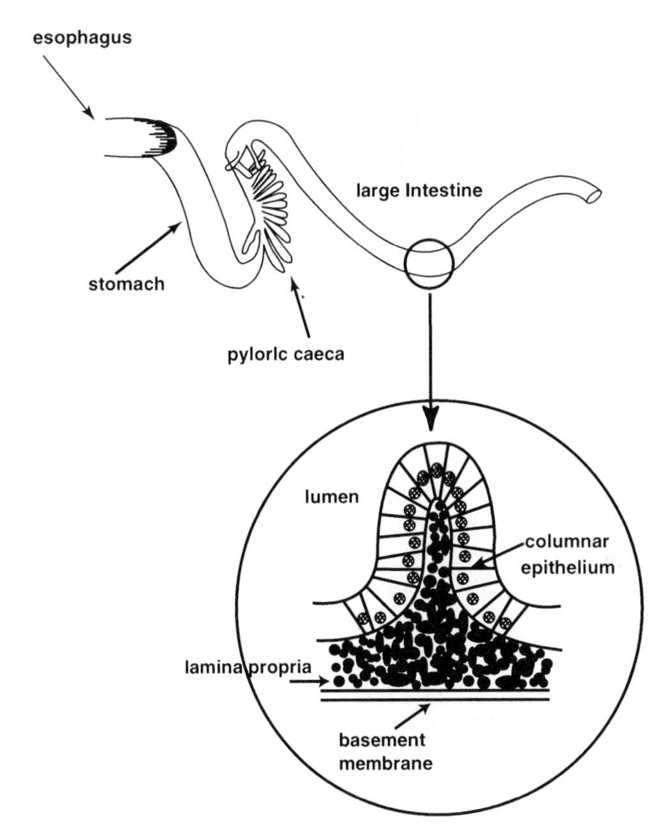

Fig. 4. A simple portrayal of the salmonid gastrointestinal tract is depicted. The large intestine is responsible for the uptake and transport of antigens. The enlargement shows the lumen where antigens are transported to large intraepithelial macrophages in the columnar epithelium. Macrophages process and present antigens to other lymphoid cells. Although obvious Peyer's patch–like structures are not seen in salmonids, many lymphoid-like cells are present, scattered throughout the large intestine (McLean and Donaldson, 1990).

vaccinated fish, while eliciting little serum antibody (Fletcher and White, 1973; Kawai *et al.,* 1981; Lobb, 1986; Rombout *et al.,* 1993). The presence of mucosal antibody without concomitant serum antibody suggests that there may be a regional immune system as observed in mammals. Davidson *et al.* (1993) have reported that the route of immunization determines whether antibody-secreting cells will be generated in the gut of rainbow trout (*Oncorhynchus mykiss*). Specific antibody-secreting cells were detected in both head kidney and intestinal mucosa after either intraperitoneal (ip) injection or oral intubation of *Aeromonas salmonicida* antigen. Antibody was detected in both these tissues but, depending on the route of delivery, differential expression of antibody-secreting cells ensued. Intraperitoneal injection produced antibody-secreting cells in both tissues, with peak antibody secretion at 2 weeks in the kidney, but not until week 7 in the intestinal mucosa. Peak antibody production in orally immunized animals was apparent 3 weeks after immunization in both the intestinal mucosa and head kidney. A second peak of antibody secretion was observed in those animals that were injected, and might have been attributable to adjuvant use. No secondary antibody peak was observed in the oral group but this group did not receive adjuvant with their immunization (Davidson *et al.*, 1993). These data suggest that a separate mucosal or regional immune system exists in rainbow trout. Further evidence for the theory of a separate mucosal immune system has been demonstrated by the differential magnitude of the response seen in the gut, paralleling that seen in the head kidney of ip-immunized animals. The mucosal immune compartment seems to be an active producer of antibody-secreting cells that possesses different kinetics than those of the systemic immune compartment. In conclusion, the evidence presented suggests that all cells necessary for a local or mucosal immune response appear to be present in the gastrointestinal tract of teleost fish.

VII. VACCINATION AND THE ANTIBODY RESPONSE

The study of antibody immunity to disease agents possesses a distinct advantage over the study of cellular immunity in that this mediator of immunity is easily procured from the sera and can be examined for antigenic specificity *in vitro* as well as passively transferred to be examined *in vivo* for biological and prophylactic effects. Although sera can be a very potent tools, critically definitive controls are often forgotten. For example, in viral challenges, the comparison of the protective capabilities of immune vs. normal serum could be due to antibodies, but the induction of other factors such as interferon could play a role. Thus, the determination of the role of specific antibodies requires the use of purified antibodies. Further, the actual specificity of the protective antibody should be determined by selective immunoadsorption and injection (Marquis and Lallier, 1989). Gram-negative bacteria possess an enormous number of molecules that can induce an antibody response, including outer membrane proteins, capsular polysaccharides, proteases, LPS, toxins, and proteases, and within each of these molecules are different epitopes which can selectively affect the prophylactic state. The need for antibodies specifically directed toward specific epitopes can be studied by the selective passive transfer of immunopurified antibodies. Transfer of a high-titered antiserum that recognizes a plethora of antigens may not be as important as the transfer of a low-titered, but purified antibody to a very specific epitope.

A. Conditions Promoting Optimal Immunization

The route, method, and conditions of exposure have direct impacts on the degree of immunity elicited and regionality of the response. These aspects of immunity are essential to the design of delivery systems and the timing of immunization. The ambient water temperature is an essential factor for successful vaccination of fish. Fish vaccinated at less than physiological temperatures display a delayed onset of protective immunity or may not become immune. Bly *et al.* (1992) demonstrated that channel catfish (*Ictalurus punctatus*) could be immunosuppressed by a rapid decrease in environmental water temperature from 22°C to 10°C. Initially this immunosuppression was characterized by the lack of leukocyte infiltrates to saprolegnia-associated skin lesions. Results also indicated that lowering the temperature from 23°C to 11°C over a 24-h period suppressed both B- and T-cell functions for 3–5 weeks, as assessed by *in vitro* responses (Bly and Clem, 1991).

The route of immunization most often used in mammals (injection) proves less practical for fish, and therefore, various other techniques have been developed to mass vaccinate. These include immersion, hyperosmotic immersion, bath, spray, and oral modes of delivery (Ellis, 1988). Although impractical, ip injection has also been used and, in many instances, is the most effective method. Intraperitoneal delivery of adjuvants and antibiotics in concert with antigens also improves the efficacy, yet for large-scale vaccination or vaccination of very small fish (0.5–2.0 g), the ip method is, again, impractical. The handling required for ip injections results in a labor-intensive situation which is also stressful for the fish.

Osmotic shock was one method of delivery that improved uptake of antigenic material from an aqueous environment. This was thought to occur due to a change in gill permeability (Amend and Finder, 1976). This technique was termed hyperosmotic infiltration (HI) and it proved to be of benefit even when used with killed bacterin preparations (Antipa and Amend, 1977, Antipa *et al.*, 1980). Although this method proved to be effective, it was later determined that sufficient levels of antigen exposure and protection could be achieved by simple immersion techniques without subjecting the fish to HI (Gould, 1978). Immersion has become widely used as a primary means of vaccination. Although this method is not a stress-free method of delivery, it is preferred over HI. Studies conducted by Tatner (1987) and Fryer *et al.* (1977) on exposure time revealed that when the concentration of antigen is not a limiting factor, lengthening the immersion time does not result in a greater antigen uptake. The standard method is to expose the animals for a minimum of 20 s to the vaccine in aerated standing water. The disadvantage of immersion vaccination is that it is limited by the mass of fish that can be immunized per unit volume of vaccine. Though immersion vaccination usually provides lower levels of immunity than injection vaccination, the levels of protection are still high enough to justify the use of this method of vaccination (Lillehaug, 1989). The bath method of vaccination is a modification of the immersion method. The fish are exposed to dilute suspensions of the vaccine for times ranging from 30 min to several hours.

Spray vaccination is another modification of the direct immersion method of vaccination. The first attempts at this technology were conducted using a sandblaster that sprayed the fish with a mixture of clay particles and the vaccine (Gould, 1978). It was later determined that the fish could be sufficiently immunized by just spraying the vaccine, as long as the exposure time was adequate.

Oral immunization has a long-established tradition; the first attempts to vaccinate humans and fish was by oral delivery. Immunizing fish orally is an ideal method for mass administration to fish of all sizes without associated handling stress (Fryer *et al.,* 1976, 1977). Early oral vaccination experiments consisted of exposing the animals to antigen in the form of a paste or liquid suspension either coated onto or milled into feed. These early methods resulted in inconsistencies in protection as well as failure to elicit a detectable antibody response (Evelyn, 1984; Hart *et al.,* 1987; McLean and Donaldson, 1990). This failure could be attributed to degradation of pH-sensitive antigens exposed to the gastric portion of the gut. Because of such early failures, oral delivery has remained an underutilized approach to immunization. However, with advent of new technologies, many new oral delivery systems have been developed that are designed to cope with the limitations associated with oral vaccination. Wong *et al.* (1992) delivered *Vibrio* bacterins using an entric coating to protect the antigen from the gastric pH. Such protection was designed to permit intact antigen to reach the lymphoid tissue associated with the second gut segment.

Traditionally, vaccine research has been concerned primarily with the induction of systemic immunity by parenteral immunization. This may be appropriate for diseases in which the infectious agent is introduced parenterally, such as those initiated by trauma to intact skin (i.e., tetanus and malaria), however, most naturally acquired infections occur through mucosal routes: orally, nasally, or genitally (Finlay and Falkow, 1989; McGhee *et al.,* 1991). In these instances, parenterally delivered vaccines may not be the best choice. This is illustrated by the limited efficacy of parenterally administered cholera vaccine (Holmgren *et al.*, 1989). Oral vaccines, however, offer the advantage of stimulating mucosal immunity (Georgopoulou *et al.*, 1985; McGhee *et al.*, 1991; Wong *et al.*, 1992; O'Hagan, 1992). Fish rely heavily on their mucosal coating as a first line of defense to avoid contact with pathogens; therefore, stimulating specific immunity at these mucosal sites should also increase protection (Davidson *et al.*, 1993; see previous). Oral immunization is also safer, easier to administer, and better tolerated. Oral vaccines are also less expensive and require less purity and quality control than parenteral vaccines (O'Hagan, 1992).

Several novel antigen delivery systems are effective in inducing both secretory and systemic immunity following oral administration (Eldridge *et al.*, 1989). These new improvements offer considerable promise for the future and may result in the development of new oral vaccines. Microparticles are one representative of this type of delivery system. Their mechanism of action is based on the reported uptake of particles into Peyer's patches or analogous cells of the gut-associated lymphoid tissue (GALT) (O'Hagan, 1990). After uptake, microparticles are phagocytosed by macrophages where the antigens are processed and presented to lymphoid cells of the gut. Several reports have demonstrated that secretory immunity is stimulated upon oral administration of microparticles (Challacombe *et al.*, 1991; Eldridge *et al.*, 1989; O'Hagan *et al.*, 1989; Wong *et al.*, 1992). It has been demonstrated that these microparticles can also elicit systemic immunity following oral administration (Challacombe *et al.*, 1991; Eldridge *et al.*, 1989). Enteric-coated antigen microspheres (ECAMs) are one type of microparticle delivery system that exploits the use of pH-reactive polymers to protect the antigen being delivered from the harsh environment of the gastric portion of the gastrointestinal tract (Piganelli, 1994; Piganelli *et al.*, 1994). The pH-reactive polymer coating becomes particularly critical when the antigen in question is protein. This protection allows the antigen to be delivered in its native state, which is likely to be essential for immune recognition.

B. Specific Responses to Immunization

The following is a brief summary of the response to five salmonid pathogens wherein significant strides have been made toward vaccination. These are presented to demonstrate the varied humoral responses that can be elicited by vaccination and/or challenge and the roles they may play in protection or pathology.

1. *Vibrio anguillarum*

Vibriosis is a Gram-negative (Fig. 3) bacterial disease of saltwater and migratory fish (Anderson and Conroy, 1970). The severity of the disease can be correlated with stressors such as overcrowding, handling, and water temperature elevation (Fryer *et al.,* 1972). Vibriosis is globally distributed and outbreaks may occur even in immunized fish, with mortalities greater than 70% in high risk areas (Cisar and Fryer, 1969). Antibiotics such as oxytetracycline (terramycin TM50) and sulfonamides have been used to control epizootics but their use can ultimately lead to the development of drug resistant strains of bacteria which may pose an even greater threat, as evidenced in Japan (Aoki *et al.,* 1974). Fortunately, the use and success of the *Vibrio* vaccines has circumvented the need for antibiotics where vibriosis is of concern.

The protective response in fish vaccinated with *Vibrio* vaccines is likely to be primarily antibody-mediated, as passive transfer is sufficient for protection. Harrell *et al.* (1975) demonstrated that passive immunization with immune serum protected rainbow trout from challenge with viable bacteria. Protection was specific, as the effect could be abolished when the sera were absorbed or incubated with *Vibrio* bacterin.

The *Vibrio* vaccines are one of the most effective aquaculture vaccines to date. The data concerning dose, duration, and vaccination times are well established (Fryer *et al.,* 1976). The only limitations that influence the effectiveness of the these vaccines are the size and ontological stage of development of the fish vaccinated. Fish must be between 1.0 and 2.5 g when vaccinated (Johnson and Johnson, 1982; Johnson and Amend, 1983; Johnson *et al.,* 1982).

In contrast to furunculosis and bacterial kidney disease, commercial vaccines for vibriosis required no prior knowledge of virulence factors or protective antigens. The vaccines simply contained inactivated *V. anguillarum* whole cells and extracellular products (Smith, 1988), which were highly immunoprophylactic. These extracellular products contained lipopolysaccharide (LPS) which may either remain cell-associated or be released into the culture supernatant (Chart and Trust, 1984; Evelyn, 1984). LPS (Fig. 3) is considered to be the primary protective antigen due to the fact that heat inactivation, formalin treatment, or even phenol-extracted *V. anguillarum* have all been found to be protective (Evelyn, 1984). *Vibiro* vaccines are delivered by three main methods, including injection, oral, and variations of the immersion method which include, HI, immersion or dip, bath vaccination, spray or shower, and automated immersion (Fryer *et al.,* 1976; Croy and Amend, 1977; Fryer *et al.,* 1977; Gould, 1978; Evelyn, 1984; Austin and Austin, 1987). All methods have been successful, with the ip injection most effective and oral delivery least effective (Agius *et al.,* 1983; Antipa *et al.,* 1980).

In attempts to produce attenuated vaccines, Norqvist *et al.* (1989) employed transposon mutagenesis to develop a live attenuated vaccine for

V. anguillarum. These mutants were screened for rifampin–streptomycin resistance which has been shown to lead to avirulence (Bolin *et al.*, 1985). Resistance to infection was observed for at least 17 weeks, and of particular interest was that this vaccination led to cross-protection for *A. salmonicida*. Protection against both bacteria was observed within 1 week. Immunoblotting with a rabbit antiserum demonstrated recognition of an antigenically similar outer membrane protein in both species. It was not determined if trout also recognized these same antigens. The value of the production of this live avirulent vaccine is that a dose of 10^7 bacteria/ml is sufficient to protect against both vibriosis and furunculosis. Further, no revertants of this bacterial strain were detected in the vaccinated fish.

As a cautionary note, even with this most reliable of vaccines, recent failures of a bivalent vaccine containing *V. ordalii* and *V. anguillarum* have been reported. The vaccine in question was composed of both serotypes 01 and 02, but the offending pathogen was an 02 serotype variant of *V. anguillarum* (Mutharia *et al.*, 1993). Chemical differences between the LPS molecules of the 02 strain used for vaccination and the 02 strain that killed the immunized fish were observed. Of particular interest was the observation that only salmon sera, not rabbit sera, were able to distinguish the antigenic differences of the LPS species of these strains. This confirms the importance of conducting serological analyses with fish species and refraining from reliance on mammalian antisera. Furthermore, the antigenic differences recognized by the salmon were associated with the large molecular weight species of LPS. Thus the protective response appears to be dependent on the appropriate reaction with a somewhat more variable O-polysaccharide component. This scenario highlights the fact that pathogens are constantly evolving and variants can be selected that are resistant to particular antibody specificities. This occurrence should serve as a reminder that the specifics of the immune response should be thoroughly understood, even if apparently efficacious vaccines are available.

2. *Aeromonas salmonicida*

Aeromonas salmonicida, causative agent of furunculosis in salmonids, is a Gram-negative, nonmotile rod (Fig. 3). It is a pathogen of both salmon and trout and is endemic to North America, Europe, and Japan. The disease is named for the raised liquefactive muscular lesions produced by infection with the pathogen (McCarthy and Roberts 1980). Although these lesions appear in chronically infected fish, they are rarely seen during an acute infection, which is characterized simply by the rapid onset of a lethal septicemia. Most if not all species of salmonids ranging from alevin to adult are susceptible to furunculosis. Atlantic salmon and brown trout seem to be the most susceptible and, in a few instances, rainbow trout have been shown to be resistant (Cipriano, 1983).

The lack of a successful vaccine for furunculosis, in contrast to the *Vibrio* vaccines, has been attributed to the difficulty in understanding the virulence factors associated with the disease and what effect they have on the host during the infection (Hastings and Ellis, 1985). Further, Hastings and Ellis (1988) have found that trout appear unable to produce antibodies to any of the known virulence factors that play prominent roles in pathogenesis. In contrast, rabbits immunized to the same mixture of immunogens responded to each of the 16 identifiable protein antigens (including the identifiable virulence factors), whereas trout responded to only four protein antigens (none of which were identifiable virulence factors). The simple bacterin approach that was successful for the previous vaccines has not been successful for furunculosis (Michel, 1985). Efforts to vaccinate fish against *A. salmonicida* began with the efforts of Duff (1942) who demonstrated that long-term oral exposure to killed *A. salmonicida* protected cutthroat trout. However, the results of Duff's experiments could not be consistently repeated. Promising results were also achieved when adjuvants were combined with injectable vaccine preparations. The adjuvants enhanced immunity and may have allowed for a depot effect of the antigen (Patterson and Fryer, 1974).

Udey and Fryer (1978) reported that fresh *A. salmonicida* cells suspended in saline autoagglutinated, but multiple passage of these cells in the laboratory caused this autoagglutination to cease. This loss of autoagglutination correlated with a lack of virulence. Udey and Fryer (1978) determined that this autoagglutination and virulence were dependent on the production of a cell surface layer (A layer) that is external to the outer membrane (Fig. 3). This A layer is composed of a geometric array of the A protein, a 50 kDa protein (Kay *et al.,* 1981). The A layer appears to be held in place by LPS via the O-polysaccharide chains (Belland and Trust, 1985). This A-layer array, when in its correct orientation, permits the adhesion of macrophages. Although a virulence factor, the A-layer protein does not appear to be responsible for observed *A. salmonicida*–mediated macrophage dysfunction (Garduno and Kay, 1992). It is possible that other potent factors may result in this macrophage effect, and that the A layer simply serves to facilitate the uptake of associated toxic factors into the macrophage. These functions can explain why the A layer plays an important role in virulence and has been reported to be a protective antigen (Udey and Fryer, 1978; Munn *et al.,* 1982). The A protein has been shown to aid in the survival of the bacterium in the presence of immune and nonimmune serum as well as to provide resistance to complement-mediated lysis (Munn *et al.,* 1982). Vaccines composed of bacteria possessing the A layer (A-layer$^+$; McCarthy *et al.*, 1983) and those lacking the A layer (A-layer$^-$; Cipriano and Starliper, 1982) have been successful in affording some protection but the protection was generally considered to be marginal. Thornton *et al.* (1993) isolated two mutants of *A. salmonicida,* a slow-growing, amino-glycoside-resistant mutant and a rapidly-growing pseudo-revertant. Both of these mutants continued to exhibit classical virulence factors associated with *A. salmonicida* pathogenesis. However, they differed morphologically from the wild type with respect to the organization of the A layer. Both mutants were avirulent and incapable of sustaining infection. The rapidly-growing, antibiotic-sensitive pseudo-revertant, delivered either intraperitoneally or by immersion, protected fish from challenge with a wild-type virulent strain of *A. salmonicida.* The resistance generated by this live attenuated strain warrants further field trials to determine its potential use as a candidate vaccine. More recently Vaughan *et al.* (1993) have described the development of an aromatic-dependent mutant of *A. salmonicida* that appears to show promise as a live attenuated vaccine. This particular vaccine has been demonstrated to be efficacious in brown trout (*Salmo trutta* L.) upon ip injection. Further, a booster injection also appears to provide for a logarithmic increase in resistance. These live strains might be essential for the elicitation of both the cell-mediated and humoral immune responses, which seem to be critical for full protection against the pathogen (Rockey, 1989; Nikl *et al.*, 1991). The development of cellular immune response assays will be needed in order to determine what potential candidate antigens are involved in eliciting protective cellular immunity. Once these antigens have been identified they can be isolated and used to produce a vaccine.

The work of Kay and colleagues has been instrumental in revealing the need for *in situ* conditions to elicit the antigens normally expressed by a pathogen such as *A. salmonicida* (Thornton *et al.*, 1993). These investigators demonstrated that by growing *A. salmonicida* in chambers within the peritoneal cavity of rainbow trout, completely novel antigens are expressed. Most of these antigens were not observed when the bacteria were grown *in vitro.* These different antigens also included an alternate form of lipopolysaccharide. Among these *in situ*–induced molecules is an extracellular polysaccharide composed of a capsular layer; there is also an apparently unrelated increased resistance to the oxidative killing mechanisms of the host (Garduno *et al.*, 1993). Rabbit antisera prepared against *in situ*–grown cells were much more effective in detecting *A. salmonicida* cells from natural infections (within the fish kidney). Although immunization of fish with *in vitro*–grown cells may lead to an antibody response, it may not necessarily be a protective or appropriate response.

3. *Renibacterium salmoninarum*

Bacterial kidney disease (BKD) is caused by a fastidious, slow-growing, Gram-positive bacterium, *Renibacterium salmoninarum* (Elliott *et al.,* 1989; Fryer and Sanders, 1981). BKD is one of the most prevalent diseases of cultured salmonids (Fryer and Sanders, 1981) and, in spite of its economic importance, there are limited effective methods for controlling BKD (Kaatari *et al.,* 1989).

One reason for this difficulty is that the bacterium is a facultative intracellular parasite, possessing the ability to survive and multiply within the phagocyte. This intracellular nature may allow the bacteria to escape the effects of the humoral arm of the immune response. Also, Bandin *et al.* (1993) has demonstrated that several strains of *R. salmoninarum* can resist killing by rainbow trout macrophages and possibly multiply within the phagocytic cells for up to 4 days.

Current approaches at management include stress reduction, quarantine, chemotherapy, culling, and total destruction of the infected population, with complete sterilization of the facilities harboring these animals. As of yet, no efficacious vaccines exist. Controlling strategies are limited primarily by a lack of understanding the the mechanisms of pathogenesis and how the salmonid responds to infection. Much of the research has been impeded by the technical difficulties related to the culturing of the organism (Fryer and Sanders, 1981; Daly and Stevenson, 1988). These include difficulty in primary pure culture isolation, a slow generation time of 24 h, and the long incubation times (1–4 months) required for experimental challenge of fish (Fryer and Sanders, 1981).

Although favorable vaccination results have been reported by some investigators, (Patterson *et al.,* 1981; McCarthy *et al.,* 1984; Shieh, 1989), others have not confirmed these results (Sakai *et al.,* 1989; Evelyn *et al.,* 1988; Kaatari *et al.,* 1989). McCarthy *et al.* (1984) reported a vaccination trial with two preparations of formalin-inactivated cells of *R. salmoninarum.* The two bacterins used were prepared in four different ways: formalin-killed intact cells, double-strength concentrate of formalin-killed cells, and pH-lysed versions of the previous bacterins. The bacterins were administered without adjuvant by ip injection, immersion, or two-step hyperosmotic infiltration. The fish were challenged by ip injection 36 days after vaccination. No significant protection was afforded by any preparation delivered by HI or immerison. However, those fish vaccinated by ip injection demonstrated protection upon ip challenge. The pH-lysed bacterin was the more efficacious of the modified bacterins. These results were promising, however, McCarthy *et al.* (1984) did not measure antibody titers and used only Gram staining as a method of measuring the dissemination of the infection.

Therefore these results must be interpreted cautiously, as Gram staining lacks the sensitivity of other more sensitive methods of diagnosis. Kaattari *et al.* (1989) prepared a number of cell wall fractions and soluble products and delivered these antigens either alone or with other bacterial antigens. The immunogens were administered ip, orally, and by immersion, with and without Freund's complete adjuvant. None of these early preparations protected fish. In fact, some of the preparations appeared to exacerbate the disease process. This latter effect suggested the possibility that immunosuppression may have been induced by bacterial products such as the p57 cell surface and soluble protein (Wiens and Kaattari, 1991), or that these antigens may have induced hypersensitivity (Kaattari *et al.*, 1989; Sami *et al.*, 1992). Turaga *et al.* (1987) reported that the ability to induce antibodies *in vitro* to an unrelated antigen was suppressed in coho salmon lymphocytes when a soluble antigen fraction derived from a culture of *R. salmoninarum* was incorporated in the culture medium. The degree of reduced antibody production was similar in cultures of lymphocytes from BKD-infected coho salmon and could be correlated with comparable serum soluble antigen levels in the infected fish. These findings suggest that there may be a generalized reduction in the capacity to generate a humoral response in BKD-infected fish.

More recently, the work of Wood (1994) has also demonstrated that the p57 molecule has the ability to occlude immunogenic determinants on the bacterial surface. Proteolytic digestion of this protein resulted in a 20-fold increase in antibody titers. Antibodies generated to this form were not able to bind native *R. salmoninarum* cells but could bind if the p57 molecule was removed. Further, Piganelli (1994) has demonstrated that proteolytic removal of p57 from whole cells, combined with oral administration of this immunogen, can lead to prophylaxis. Comparable levels of protection were not observed if this immunogen was injected rather than orally administered. Also, the levels of serum antibodies (which were quite high in injected fish) did not correlate with protection, indirectly lending support to the concept of an immunopathological disease etiology.

Further vaccine research would gain considerably from the development of a standardized bath challenge procedure that would mimic a more natural route of infection than does ip injection (Murray *et al.*, 1992). Cohabitation might also suffice as a natural challenge method. The ip challenge method bypasses the mucosal immune response totally, relying strictly upon the systemic response which may not be of primary importance in preventing infection (Murray *et al.*, 1992). Monitoring protection under bath or cohabitation challenges becomes difficult, however, as mortalities may take months to accrue. Alternatively, protocols such as the soluble antigen ELISA test that monitor the production of soluble antigen produced by

R. salmoninarum could be employed to accurately monitor the disease throughout the entire challenge period. This procedure would give earlier assessment of infection and could detect carriers amid the survivors (Rockey *et al.,* 1991; Murray *et al.,* 1992).

4. Salmonid Rhabdoviruses (VHSV and IHNV)

Protective antiviral responses are known to be effected by either the induction of neutralizing antibodies or the induction of T cytotoxic (Tc) cells. As the capability of assessing piscine Tc responses is not currently available, the simplest means of evaluating antiviral immune responses, short of disease trials, is by determination of antibody titers. The primary requirement for the induction of an antiviral antibody response to a protein antigen is the processing of the antigen and presentation in the context of the major histocompatibility complex (MHC) II molecules on the macrophage surface (Vallejo *et al.,* 1990; Vallejo *et al.,* 1992). Alternatively, to effect Tc immunity, the susceptible host cell must present the viral antigen in the context of MHC I molecules, which would necessitate an active infection of host cells. For this latter route, a live attenuated vaccine would be required. Induction of antibody response can, however, be easily accomplished with killed vaccines (Fryer *et al.,* 1976; Amend, 1976; White and Fenner, 1986; Bloom, 1989). Care must be taken to ensure that the virus is rendered inactive, yet still retains full immunogenic activity. Killed vaccines are thought to be safer than attenuated vaccines as there is no chance of the virus reverting to a virulent form. The disadvantage associated with killed viral vaccines is the need for repeated immunization. Being particulate antigens, killed viral particles are processed by macrophages and presented in the context of class II MHC molecules, stimulating a strong humoral response but very weak cellular responses.

Attenuated vaccines have the advantage of mimicking a natural infection by the wild-type virus. The attenuated vaccines are constructed to infect the host and stimulate both humoral and cellular arms of the immune system. However, substantial mutations and/or deletions are required so that the virus can no longer replicate satisfactorily and will fail to cause disease (White and Fenner, 1986). This type of vaccination has been successful in mammals (Chanock and Lenner, 1984) and has the advantage of delivering protective antigens to the appropriate site and in the proper context to stimulate the correct immune response. Although the live vaccines stimulate complete immunity, there are some major drawbacks associated with their use. Aside from the reversion to wild type, the possibility also exists that the virus may be avirulent in the species it was specifically designed to vaccinate but in another bystander species it may be highly virulent and cause widespread infection (Fryer *et al.,* 1976).

Subunit vaccines or recombinant vaccines have been derived through the use of cloned DNA that encodes sequences for antigenic determinants. This DNA is isolated and cloned into bacteria, yeast, or insect cells with a baculovirus vector (Gilmore *et al.,* 1988; Koener and Leong, 1990). The first successful recombinant vaccine was developed for the major surface antigens (vp1) of foot-and-mouth disease virus (Kleid, 1981). The advent of cloning these antigens by means of recombinant vectors has also led to the use of synthetically derived peptide vaccines that are constructed from a genetic sequence encoding specific B- and T-cell epitopes. The advantage that peptide vaccines have over live attenuated and killed vaccines is that a specific immune response against the pathogen can be induced without exposing the host to the intact pathogen (Hilleman, 1985). The disadvantages of recombinant synthetic vaccines are that contaminating products such as lipopolysaccharide from the recombinant vector may cause toxic effects when administered, and again, since it is not a live virus, it will not make use of MHC I presentation.

Viral hemorrhagic septicemia virus (VHSV) is a rhabdovirus that infects mainly young trout. This disease presents the most serious problem in continental Europe and is known primarily for its economic impact in the trout industry (de Kinkelin, 1988). Infection usually results in death due to extensive hemorrhaging. The virus initially infects the cells of blood capillaries, hematopoietic tissue, nephrons, and leukocytes (de Kinkelin, 1988). The virus was once thought to cause disease only in rainbow trout but recently it has been shown to infect a wide variety of other species such as lake trout (*Salvelinus namaycush*), brown trout (*Salmo trutta*) (de Kinkelin and LeBerre, 1977), grayling (*Thymallus thymallus*) (Wizigmann *et al.,* 1980), white fish (*Coregonus* sp.) (Ahne and Thosen, 1985), turbot (*Scophthalmus maximus*) (Castric and de Kinkelin, 1984), sea bass (*Dicentrarchus labrax*) (Castric and de Kinkelin, 1984), and pike (*Esox lucius*) (Meier and Vestergaard-Jorgensen, 1980).

Virulent virus is shed in the urine (Neukrich, 1985) and sex fluids. The virus is most abundant in the kidney, spleen, brain, and digestive tract (Wolf, 1988). Transmission is horizontal and may be vector mediated (Wolf, 1988). Asymptomatic carriers can spread the disease to hatcheries via the water. Vaccination would have the greatest impact on fish between 0.5 and 1.5 g. However, fish in this size range cannot be efficiently mass vaccinated by intraperitoneal injection, and immersion methods for delivery of this virus have thus far proven not effective. Early vaccine work on VHSV was conducted by de Kinkelin and LeBerre (1977). They prepared β-propiolactone-inactivated virus and either ip injected or immersed 2-g fish. They determined that one ip injection of 2×10^6 pfu g-fish^{-1} afforded protection upon challenge with 5×10^4 pfu ml^{-1} water. Immersion of 2-g fish in an inactivated virus suspension for 3 h proved to be less effective upon challenge. Neutralizing antibody could not always be detected in those animals demonstrating resistance, which may mean that interference by the vaccinating virus may play a role in protection against challenge. Since a VHS vaccine must be given to very young fish, the requisite ip delivery proved impractical and the research groups concentrated on a live vaccine that could be given by immersions. The live attenuated vaccine strains were produced by successive passage of the virus at progressively increasing temperatures in epithelioma cells (cyprinid EPC). The first live attenuated strain was REVA, which was derived from an F1 serotype produced in RTG (rainbow trout gonad) cells and was attenuated by 240 successive subcultures in these cells at 14°C. This vaccine was intended for use in fish up to 100 g, and protection lasted between 120 and 150 days post vaccination (de Kinkelin and Bearzotti-LeBerre, 1981; Bernard *et al.,* 1983). The second attenuated vaccine (F 25) was produced in EPC cells by multiple passages at elevated temperatures. It induced protection against three VHSV serotypes: 07.71, 23.75, and a wild-type strain belonging to serotype I (de Kinkelin and Bearzotti-Le Berre, 1981; Bernard *et al.,* 1983). The final attenuated strain was a 07.71 variant produced by multiple passage. It afforded protection for 100 days at 10°C. Although the attenuated strains allowed immersion delivery with protection, they were not totally safe; some mortality was recorded as a result of vaccination alone (Bernard *et al.,* 1983).

The use of recombinant subunit vaccine technology has only recently made the use of vaccines an economically feasible method of reducing VHSV infection of trout. Researchers accumulated evidence that antibody directed against the glycoprotein on the surface of the virus could neutralize the effects of the virus and stop the progression of the infective cycle. Therefore, the glycoprotein gene was cloned from VHSV and the neutralizing epitopes identified (Thiry *et al.,* 1990). In a recent trial conducted by N. V. L. Århus and the Laboratory of Gene Expression, University of Århus, the gene for the intermediary protein of the VHSV was expressed in *Escherichia coli* as a fusion protein and induced virus-neutralizing antibodies. These antibodies were detectable by western blots and immunofluorescence of sera from injected trout (Lorenzen *et al.,* 1990). Early results indicated that the recombinant protein also induced protection. It has yet to be demonstrated that these proteins induce protection when delivered by the immersion method. It is likely that immersion would work, as has been demonstrated in a similar recombinant system for infectious hematopoietic necrosis virus (IHNV), which elicits protection when delivered by immersion (Gilmore *et al.,* 1988).

Infectious hematopoietic necrosis virus is a salmonid pathogen that destroys the hemopoietic tissue of salmonid fish (Pilcher and Fryer, 1980). The losses due to hatchery epizootics can reach devastating proportions. For example, at Fraser River Hatchery, British Columbia, Canada, there was a 96% loss attributed to IHNV (Amend *et al.,* 1969). In some instances the total destruction of the infected population was the sole means of control. At present there are no licensed chemotherapeutic drugs available for the control of the disease, which accentuates the need for vaccine.

Although there are five structural proteins, the primary protective antigen appears to be the glycoprotein (G) of the virus, which has been found to induce neutralizing antibodies and is also sufficient for protective immunity (Engelking and Leong, 1989). Although the G protein is glycosylated, deglycoslyated G is capable of inducing a prophylactic response. However, it is quite possible that a conformational epitope must be recognized in order to secure viral neutralization. Such a possibility was postulated by Ristow *et al.* (1993) upon the observation that resistant trout sera, although possessing plaque neutralization titers, appeared negative on western blot. As this assay employs denaturing conditions (SDS and fixation upon nitrocellulose), only linear epitopes would be available upon blotting; all conformational epitopes would have been denatured. Such observations were also made in other experiments (LaPatra *et al.,* 1993). Additionally, in the latter studies the investigators did note a higher number of highly titered sera to the M1 protein. These sera were found to be immunoprophylactic in passive immunization experiments. These observations have prompted the suggestion by LaPatra *et al.* (1993) that the M1 protein may also provide protective epitopes. Several vaccines have been developed for use against IHNV. Amend (1976) and Nishimura *et al.* (1985) reported having success with vaccination of rainbow trout with killed vaccines. Amend used β-propiolactone to inactivate virus, and immunized trout by intraperitoneal injection, which resulted in protection. Nishimura experimented with different methods of formalin inactivation and found several formalin vaccines protective for juvenile trout. The vaccine was most effective when delivered by injection but HI was also capable of stimulating limited immunity. Fryer *et al.* (1976) developed an attenuated strain of INHV by passing the virus multiple times in steelhead trout cell cultures. LD_{50} studies revealed that the attenuation reduced the virulence approximately 100-fold. This vaccine proved to be effective in eliciting protective immunity with only 5% mortality in the vaccinated group as opposed to 90% in the controls. It was also shown that one preparation was capable of giving protection when delivered by immersion; however, some residual virulence was observed (Fryer *et al.,* 1976). These results prompted researchers to halt further experiments because the commercial industry expressed concern about the difficulties of licensing an attenuated live vaccine (Fryer *et al.,* 1976).

Recombinant DNA–derived subunit vaccines have been developed, exploiting the knowledge that the viral glycoprotein purified from one isolate of IHNV would induce protective immunity to a wide variety of IHNV isolates (Engelking and Leong, 1989; Leong *et al.,* 1988; Mourich and Leong, 1991). This information led Leong's group to express, as a fusion protein, an epitope of the glycoprotein of IHNV (Gilmore *et al.,* 1988). Crude lysates of bacteria that expressed this fusion protein were used to immunize fish by immersion; protection was observed in fish after challenge with virulent IHNV (Xu *et al.,* 1991). The mortality values ranged from 0 to 19% for vaccinates versus 64 to 92% for controls, and protection was effected in a number of salmonid species. The cloned product was examined further to reveal what portions of the fusion protein contained the putative B- and T-cell immunodominant epitopes (Xu *et al.,* 1991). The studies of Mourich and Leong (1991) on expressed cloned domains of the G protein revealed a reminiscent theme of piscine immune recognition. As with the reported work of Hastings and Ellis (1988), trout do not appear to be able to recognize as many possible epitopes as can rabbits or mice.

A field trial using this recombinant product demonstrated protection after fish were ponded in water from a source containing resident IHNV-infected fish (Leong, pers. comm.). At 80 days postponding the cumulative percent mortality (cpm) in the control ponds was 27% and the cpm for the vaccinates was 4%. These results demonstrated the effectiveness of the subunit vaccine in the most crucial trial of all, a large-scale field study at a commercial site.

The development of viral vaccines for the aquaculture industry comes under much scrutiny with regard to safety, cost, and effectivness. Prior to the advent of molecular biology techniques, fish farmers had to rely mainly on good animal husbandry as well as careful screening of stocks to avoid viral epizootics. However, the development of recombinant subunit vaccines that avoid the need of attenuated live vaccines for effectiveness, fully meet the criteria for safety and effectiveness. The cost of developing viral vaccines will not be as inexpensive as developing bacterial vaccines because the nature of the technology employed to design and implement viral vaccines is far more difficult.

ACKNOWLEDGMENTS

The authors wish to acknowledge the editorial comments of Dr. C. Ottinger, Mr. D. Shapiro, Ms. I. Khor, Mr. E. H. Van Den Berg, and Ms. I. Kaattari. The authors also wish to acknowledge the financial support derived from USDA CSRS 90-37116-534 and 92-34123-7665 which contributed to the findings of Piganelli and Wood. VIMS contribution number 1958.

REFERENCES

Acton, R. T., Weinhamer, P. F., Hall, S. J., Niedermeier, W., Shelton, E., and Bennett, J. C. (1971). Tetrameric macroglobulins in three orders of bony fish. *Proc. Natl. Acad. Sci.* **68,** 107–111.

Agius, C., Horne, M. T., and Ward, P. D. (1983). Immunization of rainbow trout, *Salmo gairdneri* Richardson, against vibriosis: Comparison of an extract antigen with whole cell bacterins by oral and intraperitoneal routes. *J. Fish Dis.* **6,** 129–134.

Ahne, W., and Thosen, I. (1985). Occurence of VHS in wild white fish (*Coregonus* sp.). *Zentralbl. Vet. Med.* **32,** 73–75.

Ainsworth, J. (1994). A β-glucan inhibitable zymosan receptor on channel catfish neutrophils. *Vet. Immunol. Immunopathol.* **41,** 141–152.

Ambrosius, H., and Frenzel, E. (1972). Anti-DNP antibodies in carps and tortoises. *Immunochemistry* **9,** 65–71.

Amend, D. F. (1976). Prevention and control of viral disease of salmonids. *J. Fish. Res. Bd. Can.* **33,** 1059–1066.

Amend, D. F., and Finder, D. C. (1976). Uptake of bovine serum albumin by rainbow trout from hyperosmotic solutions: A model for vaccinating fish. *Science* **192,** 793–794.

Amend, D. F., Yasutake, W. T., and Mead, A. (1969). A hematopoietic viral disease of rainbow trout (*Salmo gairdneri*). *Trans. Am. Fish. Soc.* **98,** 796–804.

Anderson, J. W., and Conroy, D. A. (1970). Vibriosis diseases in fishes. *In* "A Symposium on Diseases of Fishes and Shellfishes" (S. F. Snieszko, ed.) Special Publication No. 5, pp. 226–272, American Fisheries Society, Washington, DC.

Antipa, R., and Amend, D. F. (1977). Immunization of Pacific salmon: Comparison of intraperitoneal injection and hyperosmotic infiltration of *Vibrio anguillarum* and *Aeromonas salmonicida* bacterins. *J. Fish. Res. Bd. Can.* **34,** 203–208.

Antipa, R., Gould, R., and Amend, D. F. (1980). *Vibrio anguillarum* vaccination of sockeye salmon, *Oncorhynchus nerka* (Walbaum), by direct and hyperosmotic immersion. *J. Fish Dis.* **3,** 161–165.

Aoki, T., Egusa, S., and Arai, T. (1974). Detection of R factor in naturally occurring *Vibrio anguillarum* strains. *Antimicrob. Agents Chemother.* **6,** 534–538.

Arkoosh, M. R., and Kaattari, S. L. (1991). Development of immunological memory in rainbow trout (*Oncorhynchus mykiss*). I. An immunochemical and cellular analysis of the B cell response. *Dev. Comp. Immunol.* **15,** 279–293.

Austin, B., and Austin, D. A. (1987). "Bacterial Fish Pathogens: Diseases in Farmed and Wild Fish." Ellis Norwood Limited, Chichester, UK.

Austrian, R. (1980). Pneumococci. *In* "Microbiology" (B. B. Davis, R. Dulbecco, H. N. Eisen, and H. E. Ginsberg, eds.), pp. 596–606. Harper and Row, Philadelphia.

Bandin, I., Ellis, A. E., Barja, J. L., Secombes, C. J. (1993). Interaction between rainbow trout macrophages and *Renibacterium salmoninarum in vitro. Fish Shellfish Immunol.* **3,** 25–33.

Belland, R. J., and Trust, T. J. (1985). Synthesis, export, and assembly of *Aeromonas salmonicida* A-layer analysed by transposon mutagenesis. *J. Bacteriol.* **163,** 877–881.

Bengten, E., Leandersson, T., and Pilstrom, L. (1991). Immunoglobulin heavy chain cDNA from the teleost Atlantic cod (*Gadus morhus* L.): Nucleotide sequences of secretory and membrane bound forms show an unusual splicing pattern. *Eur. J. Immunol.* **21,** 3027–3033.

Berek, C., and Milstein, C. (1987). Mutation drift and repertoire shift in the maturation of the immune response. *Immunol. Rev.* **96,** 23–41.

Bernard, J., de Kinkelin, P., and Bearzotti, M. (1983). Viral hemorrhagic septicemia of trout: Relation between the G polypeptide, antibody production, and protection of the fish following infection with F25 attenuated variant strain. *Infect. Immun.* **39,** 7–14.

Blackwell, T. K., and Alt, F. W. (1988). Immunoglobulin genes. *In* "Molecular Immunology" (D. B. Hanes and D. M. Glover, eds.), pp. 1–60. IRL, Washington DC.

Bloom, B. (1989). Vaccines for the Third World. *Nature* **342,** 115–120.

Bly, J. E., and Clem, L. W. (1991). Temperature-mediated processes in teleost immunity: *In vitro* immunosuppression induced by *in vivo* low temperature in channel catfish. *Vet. Immunol. Immunopath.* **34,** 365–377.

Bly, J. E., Lawson, L. A., and Clem, L. W. (1992). Temperature effects on channel catfish to a fungal pathogen. *Dis. Aquat. Org.* **13,** 155–164.

Bolin, J., Bortnoy, D. A., and Wolf-Watz, H. (1985). Expression of the temperature-inducible outer membrane proteins of *Yersiniae. Infect. Immun.* **48,** 234–240.

Borsos, T., Chapius, R. M., and Langore, J. J. (1981). Distinction between fixation of C1 and the activation of C by natural IgM anti-hapten antibody: Effect of cell surface hapten density. *Mol. Immunol.* **18,** 863–868.

Brekke, O. H., Michaelsen, T. E., and Sandlie, I. (1995). The structural requirements for complement activation by IgG: Does it hinge on the hinge? *Immunol. Today* **16,** 85–90.

Brooks, K. H., and Feldbush, J. L. (1981). In vitro antigen-mediated clonal expansion of memory B lymphocytes. *J. Immunol.* **127,** 959–967.

Caspi, R. R., and Avtalion, R. R. (1984). Evidence for the existence of an IL-2-like lymphocyte growth promoting factor in a bony fish, *Cyprinus carpio. Dev. Comp. Immunol.* **8,** 51–66.

Castric, J., and de Kinkelin, P. (1984). Experimental study of the susceptibility of two marine fish species, sea bass (*Dicentrachus labrax*) and turbot (*Scophthalums maximus*) to viral hemorrhagic septicemia. *Aquaculture* **41,** 203–212.

Celada, F. (1967). Quantitative studies of the adoptive immunological memory in mice. II. Linear transmission of cellular memory. *J. Exp. Med.* **125,** 199–211.

Challacombe, S. J., Rahman, D., Jeffery, H., Davis, S. S., and O'Hagan, D. T., (1991). Biodegradable microparticles for oral immunization. *Immunology* **2,** 239–242.

Chanock, R. M., and Lerner, R. A., eds. (1984). "Modern Approaches to Vaccines." Cold Spring Harbor, Cold Spring Harbor, NY.

Chart, H., and Trust, T. J. (1984). Characterization of the surface antigen of marine fish pathogens, *Vibrio anguillarum* and *Vibrio ordalli. Can. J. Microbiol.* **30,** 703–710.

Cipriano, R. C. (1983). Resistance of salmonids to *Aeromonas salmonicida:* Relation between agglutinins and neutralizing activities. *Trans. Am. Fish. Soc.* **112,** 95–99.

Cipriano, R. C., and Starliper, C. E. (1982). Immersion and injection vaccination of salmonids against furunculosis with an avirulent strain of *Aeromonas salmonicida. Prog. Fish Cult.* **44,** 167–169.

Cisar, J. O., and Fryer, J. L. (1969). An epizootic of vibriosis in chinook salmon. *Bull. Wild. Dis.* **5,** 73–76.

Clem, L. W., and McClean, W. E. (1975). Phylogeny of immunoglobulin structure and function. VII. Monomeric and tetrameric immunoglobulins of the margate, a marine teleost. *Immunology* **29,** 791–799.

Clem, L. W., and Sigel, M. M. (1965). Antibody responses of lower vertebrates to bovine serum albumin. *Fed. Proc.* **24,** 504.

Clem, L. W., and Small, D. A. (1967). Phylogeny of immunoglobulin structure and function. I. Immunoglobulins of lemon shark. *J. Exp. Med.* **125,** 893–920.

Clem, L. W., Sizemore, R. C., Ellsaesser, C. F., and Miller, N. W. (1985). Monocytes as accessory cells in fish immune responses. *Dev. Comp. Immunol.* **9,** 803–809.

Cossarini-Dunier, M., Desaux, R. S., and Dorson, M. (1986). Variability in humoral responses to DNP-KLH of rainbow trout (*Salmo gairdneri*). Comparison of antibody kinetics and immunoglobulin spectrotypes between normal trouts and trouts obtained by gynogenesis or self-fertilization. *Dev. Comp. Immunol.* **10,** 207–217.

Croy, T. R., and Amend, D. F. (1977). Immunization of sockeye salmon (*Oncorhynchus nerka*) against vibriosis using the hyperosmotic infiltration technique. *Aquaculture* **12,** 317–325.

Daggfeldt, A., Bengten, E., and Pilstrom, L. (1993). A cluster type organization of the loci of the immunoglobulin light chain in Atlantic cod (*Gadus morhus* L.) and rainbow trout (*Oncorhynchus mykiss* Walbaum) indicated by nucleotide sequences of cDNAs and hybridization analysis. *Immunogenetics* **38,** 199–209.

Daly, J. G., and Stevenson, R. M. (1988). Inhibitory effects of salmonid tissue on the growth of *Renibacterium salmoninarum. Dis. Aquat. Org.* **4,** 169–171.

Davidson, G. A., Ellis, A. E., Secombes, C. J. (1993). Route of immunization influences the generation of antibody secreting cells in the gut of rainbow trout (*Oncorhynchus mykiss*). *Dev. Comp. Immunol.* **17,** 373–376.

Davina, J. H. M., Rijkers, G. T., Rombout, J. H. W. M., Timmermans, L. P. M., and Van Muiswinkel, W. B. (1980). Lymphoid and nonlymphoid cells in the intestine of cyprinid fish. *In* "Development of Differentiation of the Vertebrate Lymphocytes." (J. D. Horton, ed.), pp. 129–140. Elsevier/North-Holland Biomedical Press, Amsterdam.

Davina, J. H. M., Parmentier, H. K., and Timmermans, L. P. M. (1982). Effects of oral administration of *Vibrio* bacteria on the intestine of Cyprinid fish. *Dev. Comp. Immunol. Suppl.* **2,** 157–166.

Davis, A. C., Roux, K. H., and Shulman, M. J. (1989). On the structure of polymeric IgM. *Eur. J. Immunol.* **18,** 1001–1008.

de Kinkelin, P. (1988). Vaccination against viral hemorrhagic septicemia virus (VHS). *In* "Fish Vaccination" Ellis, A. E., ed.), pp. 172–192, Academic Press, London.

de Kinkelin, P., and Le Berre, M. (1977). Demonstration de la protection de la truite Arc-en-ciel contre la SHV, par l'administration d'un virus inactive. *Bull. Off. Int. Epizoot.* **83,** 401–402.

de Kinkelin, P., and Bearzotti-Le Berre, M. (1981). Immunization of rainbow trout against viral hemorrhagic septicemia (VHS) with a thermoresistant variant of the virus. *Dev. Biol. Stand.* **49,** 431–439.

Dorai, H., Wesolowski, J. S., and Gillies, S. D. (1992). Role of inter-heavy and light chain disulfide bonds in the effector functions of human immunoglobulin IgG1. *Mol. Immunol.* **29,** 1487–1491.

Duff, D. C. B. (1942). The oral immunization of trout against *Bacterium salmonicida. J. Immunol.* **44,** 87–94.

Earp, B. J., Ellis, C. H., and Ordal, E. J. (1953). Kidney disease in young salmon. Special Reports, Ser. No. 1, State of Washington, Dept. of Fish. 73.

Edberg, S. C., Bronson, P. M., and Van Oss, C. J. (1972). The valency of IgM and IgG rabbit anti-dextran antibody as a function of the size of the dextran molecule. *Immunochemistry* **9,** 273.

Eisen, H. N. (1980). Antibody formation. *In* "Microbiology" (D. B. Davis, R. Dulbecco, H. N. Eisen, and H. E. Ginsberg, eds.), pp 420–450. Harper and Row, Philadelphia.

Elcombe, B. M., Chang, R. J., Taves, C. J., and Winkelhake, J. L. (1985). Evolution of antibody structure and effector functions: Comparative hemolytic activities of monomeric and tetrameric IgM from rainbow trout, *Salmo gairdneri. Comp. Biochem. Physiol.* **80B,** 697–706.

Eldridge, J. H., Meulbroek, J. A., Staas, J. K., Tice, T. R., and Gilley, R. M. (1989). Vaccine-containing biodegradable microspheres specifically enter gut-associated lymphoid tissue following oral administration and induce disseminated mucosal immune response. *In* "Immunobiology of Proteins and Peptides" (V. M. Z. Atassi, ed.), pp. 191–202. Plenum Press, New York.

Elliott, D. G., Pascho, R. J., and Bullock, G. L. (1989). Developments in the control of bacterial kidney disease of Salmonid fishes. *Dis. Aquat. Org.* **6,** 201–215.

Ellis, A. E. (1988). Current aspects of fish vaccination. *Dis. Aquat. Org.* **4,** 159–164.

Engelking, H. M., and Leong, J-A. C. (1989). The glycoprotein of infectious hematopoietic necroses virus elicits neutralizing antibody and protective response. *Virus Res.* **13,** 213–230.

Evelyn, T. P. T. (1984). Immunization against pathogenic vibriosis. *In* "Symposium on Fish Vaccination" (P. de Kinkelin, ed.), pp. 121–150. Office International des Epizooties, Paris.

Evelyn, T. P. T., Ketcheson, J. E., and Prosperi-Porta, L. (1988). Trials with anti-bacterial kidney disease vaccines in two species of Pacific salmon. *In* "International Fish Health Conference," conference handbook, Vancouver, B.C., Fish Health Section, pp. 38. American Fisheries Society, Washington, DC.

Fiebig, H., Gruhn, R., and Ambrosius, H. (1977). Studies on the control of IgM antibody synthesis. III. Preferential formation of anti-DNP antibodies of high functional affinity on the course of the immune response in carp. *Immunochemistry* **14,** 272–276.

Finkelman, F. D., Holmes, J., Katena, I. M., Urban, J. F., Beckman, M. P., Park, L. S., Schooley, K. A., Coffman, R. L., Mosmann, T. R., and Paul, W. E. (1990). Lymphokine control of *in vivo* immunoglobulin isotype selection. *Annu. Rev. Immunol.* **8,** 303–333.

Finlay, B. B., and Falkow, S. (1989). Common themes in microbial pathogenicity. *Microbiol. Rev.* **53,** 210–230.

Finstad, J., and Fichtelius, K. E. (1965). Studies of phylogenetic immunity: Immunologic memory and responsive cellular proliferation in the lamprey. *Fed. Proc.* **24,** 491.

Fletcher, T. C., and White, A. (1973). Antibody production in the plaice, *Pleuronectes platessa* L., after oral and parenteral immunization with *Vibrio anguillarum* antigens. *Aquaculture* **1,** 417–428.

Fryer, J. L., and Sanders, J. E. (1981). Bacterial kidney disease of salmonid fish. *Annu. Rev. Microbiol.* **35,** 273–298.

Fryer, J. L., Nelson, J. S., and Garrison, R. L. (1972). Vibriosis in fish. *In* "Progress in Fishery and Food Science" (R. W. Moore, ed.), Vol. 5., Fisheries Publ., University of Washington, Seattle.

Fryer, J. L., Rohovec, J. S., Tebbit, G. L., McMichael, J. S., and Pilcher, K. S. (1976). Vaccination for control of infectious diseases in Pacific salmon. *Fish Pathol.* **10,** 155–164.

Fryer, J. L., Amend, D. F., Harrel, L. W., Novotony, A. S., Plumb, J. A., Rohovec, J. S., and Tebbit, G. L. (1977). Development of bacterins and vaccines for control of infectious diseases of fish. Sea Grant College Program. Pub. No. Oresu T 77-012. Oregon State University, Corvallis.

Fujino, Y., Ono, S., and Nagai, A. (1987). Studies on the uptake of rabbit's immunoglobulin into the columnar epithelial cells in the gut of the trout, *Salmo gairdneri. Bull. Jpn. Soc. Sci. Fish.* **53,** 367–370.

Garduno, R. A., and Kay, W. W. (1992). Interaction of the fish pathogen *Aeromonas salmonicida* with rainbow trout macrophages. *Infect. Immun.* **60,** 4612–4620.

Garduno, R. A., Thornton, J. C., and Kay, W. W. (1993). *Aeromonas salmonicida* grown *in vivo. Infect. Immun.* **61,** 3854–3862.

Georgopoulou, U., and Vernier, J. M. (1986). Local immunological response in the posterior intestinal segment of the rainbow trout after oral administration of macromolecules. *Dev. Comp. Immunol.* **10,** 529–537.

Georgopoulou, U., Sire, M. F., and Vernier, J. M. (1985). Immunological demonstration of intestinal absorption of proteins by epithelial cells of the posterior intestinal segment and their intracellular digestion in rainbow trout. Ultrastructural and biochemical study. *Biol. of the Cell.* **5,** 269–282.

Ghaffari, S. H., and Lobb, C. J. (1989). Cloning and sequence analysis of channel catfish heavy chain cDNA indicate phylogenetic diversity within the IgM immunoglobulin family. *J. Immunol.* **142,** 1356–1365.

Gilmore, R. D., Engelking, H. M., Manning, D. S., and Leong, J. C. (1988). Expression in *Escherichia coli* of an epitope of the glycoprotein of infectious hematopoietic necrosis virus protects against viral challenge. *Biotechnology* **6,** 295–300.

Gould, R. W. (1978). Development of a new vaccine delivery system for immunizing fish and investigation of the positive antigens in *Vibrio anguillarum.* Ph.D. Thesis, Oregon State University, Corvallis.

Griffiths, G. M., Berek, C., Kaartinen, M., and Milstein, C. (1984). Somatic mutation and the maturation of the immune response to 2-phenyl-oxazolone. *Nature* **312,** 271–275.

Grondel, J. L., and Harmsen, G. M. (1984). Phylogeny of interleukins: Growth factors produced by leucocytes of *Cyprinus carpio* L. *Immunology* **52,** 477–482.

Hansen, J., Leong, J-A., and Kaattari, S. L. (1994). Complete nucleotide sequence of a rainbow trout cDNA encoding a membrane-bound form of immunoglobulin heavy chain. *Mol. Immunol.* **31,** 499–501.

Harding, K. A., Cohen, N., and Litman, G. W. (1990). Immunoglobulin heavy chain gene organization and complexity in the skate, *Raja erinacea. Nucleic Acids Res.* **18,** 1015–1020.

Harrell, L. W., Etlinger, H. M., and Hodgins, H. O. (1975). Humoral factors important in resistance of salmonid fish to bacterial disease. I. Serum antibody protection of rainbow trout (*Salmo gairdneri*) against vibriosis. *Aquaculture* **6,** 211–219.

Hart, S., Wrathmell, A. B., Doggett, T. A., and Harris, J. E. (1987). An investigation of the biliary and intestinal immunoglobulin and the plasma cell distribution in the gall bladder and liver of the common dogfish, *Scyliorhinus canicula* L. *Aquaculture* **67,** 147–155.

Hart, S., Wrathmell, A. B., Harris, J. E., and Grayson, T. H. (1988). Gut immunology in fish: A review. *Dev. Comp. Immunol.* **12,** 453–480.

Hastings, T. S., and Ellis, A. E. (1985). Differences in the production of haemolytic and proteolytic activities by various isolates of *Aeromonas salmonicida. In* "Fish and Shellfish Pathology" (A. E. Ellis, ed.), pp. 69–77. Academic Press, London.

Hastings, T. S., and Ellis, A. E. (1988). The humoral immune response of rainbow trout, *Salmo gairdneri* Richardson, and rabbits to *Aeromonas salmonicida. J. Fish Dis.* **11,** 147–160.

Hildemann, W. H. (1984). A question of memory. *Dev. Comp. Immunol.* **8,** 747–756.

Hilleman, M. R. (1985). New directions in vaccine development and utilization. *J. Infect. Dis.* **151,** 407–414.

Hinds-Frey, K. R., Nishikata, H., Litman, R. R., and Litman, G. W. (1993). Somatic variation precedes extensive diversification of germline sequences and combinatorial joining in the evolution of immunoglobulin heavy chain diversity. *J. Exp. Med.* **178,** 815–824.

Holmgren, J., Clemens, J., Sack, D. A., and Svennerholm, A. M. (1989). New cholera vaccines. *Vaccine* **7,** 94–96, 1989.

Honda, A., Kodama, H., Moustafa, M., Yamada, F., Mikami, T., and Izawa, H. (1986). Phagocytic activity of macrophages of rainbow trout against *Vibrio anguillarum* and the opsonizing effect of antibody and complement. *Res. Vet. Sci.* **40,** 328–332.

Hordvik, I., Voie, A. M., Glette, J., Male, R., and Endresen, C. (1992). Cloning and sequence analysis of two isotypic IgM heavy chain genes from Atlantic salmon, *Salmo salar* L. *Eur. J. Immunol.* **27,** 2957–2967.

Hornick, S. L., and Karush, F. (1972). Antibody affinity. III. The role of multivalence. *Immunochemistry* **9,** 325–342.

Hugli, T. E. (1981). The structural basis for anaphylatoxin and chemotactic functions of C3a, C4a, and C5a. *CRC Crit. Rev. Immunol.* **1,** 321–366.

Hurtley, S. M., and Helenius, A. (1989). Protein oligomerization in the endoplasmic reticulum. *Annu. Rev. Cell Biol.* **5,** 277–307.

Iida, H., and Yamamoto, T. (1985). Intracellular transport of horseradish peroxidase in the absorptive cells of goldfish hindgut *in vitro,* with special reference to the cytoplasmic tubules. *Cell Tissue Res.* **240,** 553–560.

Jenkins, J. A., and Ourth, D. D. (1990). Membrane damage to *Escherichia coli* and bactericidal kinetics by the alternative complement pathway of channel catfish. *Comp. Biochem. Physiol.* **97B,** 477–481.

Jenkins, J. A., and Ourth, D. D. (1993). Opsonic effect of the alternative complement pathway on channel catfish peripheral blood phagocytes. *Vet. Immunol. Immunopath.* **39,** 447–459.

Johnson, D. F., and Johnson, K. A. (1982). Duration of immunity in salmonids vaccinated by direct immersion with *Yersinia ruckeri* and *Vibrio anguillarum* bacterins. *J. Fish Dis.* **5,** 207–213.

Johnson, K. A., and Amend, D. F. (1983). Comparison of efficacy of several delivery methods using *Y. ruckeri* bacterin on rainbow trout, *Salmo gairdneri* Richardson. *J. Fish Dis.* **6,** 337–396.

Johnson, K. A., Flynn, J. K., and Amend, D. F. (1982). Onset of immunity in salmonid fry vaccination by direct immersion in *Vibrio anguillarum* and *Yersinia ruckeri* bacterins. *J. Fish Dis.* **5,** 197–205.

Kaattari, S. L. (1992). Fish B lymphocytes: Defining their form and function. *Annu. Rev. Fish Dis.* **2,** 161–180.

Kaattari, S. L. (1994). Development of a piscine paradigm of immunological memory. *Fish Shellfish Immunol.* **4,** 447–457.

Kaattari, S., Turaga, P., and Wiens, G. (1989). Development of a vaccine for bacterial kidney disease in salmon, pp. 232–253. Bonn. Power Admin. Final Report, Portland.

Kawai, K. K., Kusuda, R., and Itam, T. (1981). Mechanism of protection in ayu orally vaccinated for vibriosis. *Fish Pathol.* **15,** 257–262.

Kay, W. W., Buckley, J. T., Ishiguro, E. E., Phipps, B. M., Monette, J. P. L., and Trust, T. J. (1981). Purification and disposition of a surface protein associated with virulence of *Aeromonas salmonicida. J. Bacteriol.* **147,** 1077–1084.

Killie, J.-K., Espelid, S., and Jorgensen, T. O. (1991). The humoral immune response in Atlantic salmon (*Salmo salar* L.) against the hapten carrier antigen NIP-LPH: The effect of determinant (NIP) density and the isotype profile of anti-NIP antibodies. *Fish Shellfish Immunol.* **1,** 33–46.

Kleid, D. G. (1981). Cloned viral protein vaccine for foot-and-mount disease: Responses in cattle and swine. *Science* **214,** 1125–1129.

Kobayashi, K., and Tomonaga, S. (1988). The second immunoglobulin class is commonly present in cartilaginous fish belonging to the order Rajaformes. *Mol. Immunol.* **25,** 115–120.

Kobayashi, K., Tomonaga, S., and Kajii, T. (1984). A second class of immunoglobulin other than IgM present in the serum of a cartilaginous fish, the skate, *Raja kenojei:* Isolation and characterization. *Mol. Immunol.* **21,** 397–404.

Kobuku, G., Hinds, K., Litman, R., Shamblott, M. J., and Litman, G. W. (1987). Extensive families of constant region genes in a phylogenetically primitive vertebrate indicates an additional level of immunoglobulin complexity. *Proc. Natl. Acad. Sci.* **85,** 5868–5872.

Kochs, C., and Rajewsky, K. (1989). Stable expression and somatic hypermutation of antibody V regions in B-cell differentiation. *Annu. Rev. Immunol.* **7,** 537–559.

Koener, J. F., and Leong, J. (1990). Expression of the glycoprotein gene from a fish rhabdovirus by using baculovirus vectors. *J. Virol.* **1,** 428–430.

Kuby, J. (1992). Antigen-antibody interactions. *In* "Immunology". pp. 121–135, W. H. Freeman and Co., New York.

Lambris, J. D., Lao, A., Pang, J., and Alsenz, J. (1993). Third component of trout complement: cDNA cloning and conservation of functional sites. *J. Immunol.* **151,** 6123–6134.

LaPatra, S. E., Turner, T., Lauda, K. A., Jones, G. R., and Walker, S. (1993). Characterization of the humoral response of rainbow trout to infectious hematopoietic necrosis virus. *J. Aquat. Anim. Health* **5,** 165–171.

Ledford, B. E., Magor, B. G., Middleton, D. L., Miller, R. L., Wilson, M. R., Miller, N. W., Clem, L. W., and Warr, G. W. (1993). Expression of a mouse-channel catfish chimeric IgM molecule in a mouse myeloma cell. *Mol. Immunol.* **30,** 1405–1417.

Lee, M. A., Bengten, E., Daggfeldt, A., Rytting, A-S., and Pilstrom, L. (1993). Characterisation of rainbow trout cDNAs encoding a secreted and membrane-bound Ig heavy chain and the genomic intron upstream of the first constant exon. *Mol. Immunol.* **30,** 641–648.

Lefkovits, I. (1979). Limiting dilution analysis of cells in the immune system. Cambridge University Press, London.

Leong, J. C., Fryer, J. L., and Winton, J. R. (1988). Vaccination against infectious hematopoietic necrosis virus. *In* "Fish Vaccination" (A. E. Ellis, ed.), pp. 193–202. Academic Press, London.

Lillehaug, A. (1989). A cost-effectiveness study of three different methods of vaccination against vibriosis in salmonids. *Aquaculture* **83,** 227–236.

Lobb, C. J. (1985). Covalent structure and affinity of channel catfish anti-dinitrophenyl antibodies. *Mol. Immunol.* **22,** 993–999.

Lobb, C. J. (1986). Structural diversity of channel catfish immunoglobulins. *In* "Fish Immunology" (J. S. Stolen, D. P. Anderson, and W. B. Van Muiswinkel, eds), p. 7–12. Elsevier, Amsterdam.

Lobb, C. J., and Clem, L. W. (1981a). Phylogeny of immunoglobulin structure and function. X. Humoral immunoglobulins of the sheepshead, *Archosargus probatocephalus. Dev. Comp. Immunol.* **5,** 271–282.

Lobb, C. J., and Clem, L. W. (1981b). Phylogeny of immunoglobulin structure and function. XI. Secretory immunoglobulin in the cutaneous mucus of the sheepshead. *Dev. Comp. Immunol.* **5,** 587–596.

Lobb, C. J., and Clem, L. W. (1983). Distinctive subpopulations of catfish serum antibody and immunoglobulin. *Mol. Immunol.* **20,** 811–818.

Lobb, C. J., and Olson, M. O. J. (1988). Immunoglobulin heavy H-chain isotypes in a teleost fish. *J. Immunol.* **141,** 1236–1245.

Lobb, C. J., Olson, M. O., and Clem, L. W. (1984). Immunoglobulin light chain classes in a teleost fish. *J. Immunol.* **132,** 1917–1923.

Lorenzen, N., Olesen, J. F., and Vestergaard-Jørgensen, P. E. (1990). Neutralizing of egtved virus pathogenicity to cell cultures and in fish by monoclonal antibodies to the viral G-protein. *J. Gen. Virol.* **71,** 561–567.

Makela, O., and Litman, G. W. (1980). Lack of heterogeneity in anti-hapten antibodies of a phylogenetically primitive shark. *Nature* **287,** 639–640.

Marquis, H., and Lallier, R. (1989). Efficacy studies of passive immunization against *Aeromonas salmonicida* infection in brook trout. *Salvelinus fontinalis* (Mitchill). *J. Fish Dis.* **12,** 233–240.

McCarthy, D. H., and Roberts, R. J. (1980). Furunculosis of fish and the present state of our knowledge. *In* "Advances in Aquatic Microbiology" (M. R. Droop, and H. W. Jannasch, eds.), pp. 293–341. Academic Press, London.

McCarthy, D. H., Amend, D. F., Johnson, K. A., and Bloom, J. V. (1983). *Aeromonas salmonicida:* Determination of an antigen associated with protective immunity and evaluation of an experimental bacterin. *J. Fish Dis.* **6,** 155–174.

McCarthy, D. H., Croy, T. R., and Amend, D. F. (1984). Immunization of rainbow trout, *Salmo gairdneri* Richardson, against bacterial kidney disease: Preliminary efficacy evaluation. *J. Fish Dis.* **7,** 65–71.

McGhee, J. R., Mestcky, J., Dertzbaugh, M. T., Eldridge, J. H., Hirasawa, M., and Kiyono, H. (1991). The mucosal immune system: From fundamental concepts to vaccine development. *Vaccine* **2,** 75–89.

McLean, E., and Donaldson, E. (1990). Absorption of bioactive peptides by the gastrointestinal tract of fish: A review. *J. Aquat. Anim. Health.* **2,** 1–11.

Meier, W., and Vestergaard-Jorgensen, P. E. (1980). Isolation of VHS virus from pike fry (*Esox lucius*) with haemorrhagic symptoms. *In* "Fish Diseases", Proc. in Life Sciences (W. Ahne, ed.), pp. 8–17. Springer Verlag, Berlin.

Michaelsen, T. E., Brekke, O. H., Aase, A., Sandin, R. H., Bremnes, B., and Sandlie, I. (1994). One disulfide bond in front of the second heavy chain constant region is necessary and sufficient for effector functions of human IgG3 without a genetic hinge. *Proc. Natl. Acad. Sci.* **91,** 9243–9247.

Michel, C. (1985). Failure of anti-furunculosis vaccination of rainbow trout (*Salmo gairdneri*) using extra-cellular products of *Aeromonas salmonicida* as an immunogen. *Fish Pathol.* **20,** 445–451.

Mochida, K., Lou, Y.-H., Hora, A., and Yamauchi, K. (1994). Physical biochemical properties of IgM from a teleost fish. *Immunology* **83,** 675–680.

Mourich, D. V., and Leong, J. C. (1991). Mapping of the immunogenic regions of the IHNV glycoprotein in rainbow trout and mice. pp. 93–100. Proc. of the Second Int. Symp. on Viruses of Lower Vertebrates. (J. L. Fryer and O. A. Lecy, eds.) July 1991, Corvallis, OR.

Munn, C. B., Ishiguro, E. E., Kay, W. W., and Trust, T. J. (1982). Role of surface components in serum resistance of virulent *Aeromonas salmonicida. Infect. Immun.* **36,** 1069–1075.

Murray, C. B., Evelyn, T. P. T., Beacham, T. D., Barner, L. W., Ketcheson, J. E., and Prosperi-Porta, L. (1992). Experimental induction of bacterial kidney disease in Chinook salmon by immersion and cohabitation challenges. *Dis. Aquat. Org.* **12,** 91–96.

Mutharia, L. W., Raymond, B. T., Dekievit, T. R., and Stevenson, R. M. W. (1993). Antibody specificities of polyclonal rabbit and rainbow trout antisera against *Vibrio ordalii* and serotype 0:2 strains of *Vibrio anguillarum. Can J. Microbiol.* **39,** 492–499.

Nakao, M., Yano, T., Matsuyama, H., and Uemura, T. (1989). Isolation of the third component of complement (C3) from carp serum. *Nip. Sui. Gak.* **55,** 2021–2027.

Neukirch, M. (1985). Uptake, multiplication, and excretion of viral hemorrhagic septicemia in rainbow trout (*Salmo gairdneri*). *In* "Fish and Shellfish Pathology" (A. E. Ellis, ed.), pp. 295–300. Academic Press, London.

Nezlin, R. (1990). Internal movements in immunoglobulin molecules. *Adv. Immunol.* **48,** 1–40.

Nikl, L., Albright, L. J., and Evelyn, T. P. T. (1991). Influence of seven immunostimulants on the immune response of coho salmon to *Aeromonas salmonicida. Dis. Aquat. Org.* **12,** 7–12.

Nishimura, T., Sasaki, H., Ushiyama, M., Inoue, K., Suzuki, Y., Ikeya, F., Tanaka, M., Suzuki, H., Kohara, M., Arai, M., Shima, N., and Sano, T. (1985). A trial of vaccination against rainbow trout fry with formalin IHN virus. *Fish Pathol.* **20,** 435–443.

Noaillac-DePeure, J., and Gas, N. (1973). Absorption of protein macromolecules by the enterocytes of the carp (*Cyprinus carpio,* L.) *Z Zell. Mikrobiol. Anat.* **146,** 525–541.

Nonaka, M., Iawaki, M., Nakai, C., Nozaki, M., Kaidoh, J., Nanaka, M., Natsuume-Sakai, S., and Takahashi, M. (1984). Purification of major serum protein of rainbow trout (*Salmo gairdneri*) homologous to the third component of mammalian complement. *J. Biol. Chem.* **259,** 6327–6333.

Norqvist, A., Hagstrom, A., and Wolf-Watz, H. (1989). Protection of rainbow trout against vibriosis and furunculosis by the use of attenuated strains of *Vibrio anguillarum. Appl. Environ. Microbiol.* **55,** 1400–1405.

O'Hagan, D. T. (1990). Intestinal translocation of particulates: Implications for drug and antigen delivery. *Adv. Drug Deliv. Rev.* **5,** 265–295.

O'Hagan, D. T. (1992). Oral Delivery of vaccines: Formulation and clinical pharmacokinetic considerations. *Clin. Pharmacokinet.* **22,** 1–10.

O'Hagan, D. T., Palin, K. J., Davis, S. S., and Artursson, P. (1989). Microparticles as potentially orally active immunological adjuvants. *Vaccine* **7,** 421–424.

O'Leary, D. J. (1980). A partial characterization of high and low molecular weight immunoglobulin in rainbow trout (*Salmo gairdneri*). Ph.D. Thesis, Oregon State Univ., Corvallis.

Ortega, H. O. (1993). Mechanisms of accessory cell function in trout (*Oncorhynchus mykiss*). M.S. Thesis, Oregon State Univ., Corvallis.

Patterson, W. D., and Fryer, J. L. (1974). Immune response of juvenile coho salmon (*Oncorhynchus kisutch*) to *Aeromonas salmonicida:* Cells administered intraperitoneally in Freund's complete adjuvant. *J. Fish. Res. Bd. Can.* **31,** 1751–1755.

Patterson, W. D., Desautels, D., and Weber, J. M. (1981). The immune response of Atlantic salmon, *Salmo salar* L., to the causative agent of bacterial kidney disease, *Renibacterium salmoninarum. J. Fish Dis.* **4,** 99–111.

Piganelli, J. D. (1994). Development of enteric protected vaccines for aquaculture. Ph.D. Thesis. Oregon State Univ., Corvallis.

Piganelli, J. D., Zhang, J. A., Christensen, J. M., and Kaattari, S. L. (1994). Enteric coated microspheres as an oral method for antigen delivery to salmonids. *Fish Shellfish Immunol.* **4,** 179–188.

Pilcher, K. S., and Fryer, J. L. (1980). Viral diseases of fish: A review through 1978, part II. *CRC Crit. Rev. Microbiol.* **7,** 287–364.

Pontinus, H., and Ambrosius, H. (1972). Contribution to the immune biology of poikiothermic vertebrates. IX. Studies on the cellular mechanisms of humoral immune reactions in perch, *Perca fluviatilis* L. *Acta Biol. Med. Ger.* **29,** 319–339.

Richter, R. F., and Ambrosius, H. (1988). The immune response of carp against dextran. An analysis of the idiotypic heterogeneity. *Dev. Comp. Immunol.* **12,** 761–772.

Ristow, S. S., de Avila, J., LaPatra, S. E., and Lauda, K. (1993). Detection and characterization of rainbow trout antibody against infectious hematopoietic necrosis virus. *Dis. Aquat. Org.* **15,** 109–114.

Rockey, D. D. (1989). Virulence factors of *Aeromonas salmonicida* and their interaction with the salmonid host. Ph.D. Thesis, Oregon State Univ., Corvallis.

Rockey, D. D., Gilkey, L. L., Wiens, G. D., and Kaattari, S. L. (1991). Monoclonal antibody analysis of the *Renibacterium salmoninarum* P57 protein in spawning chinook and coho salmon. *J. Aquat. Anim. Health* **3,** 23–30.

Rombout, J. H. W. M., and van den Berg, A. A. (1989). Immunological importance of the second gut segment of carp. I. Uptake and processing of antigens by epithelial cells and macrophages. *J. Fish Biol.* **35,** 13–18.

Rombout, J. H. W. M., Lamers, C. H. J., Hellfrich, M. H., Dekkee, A., and Taverne-Thiele, J. J. (1985). Uptake and transport of intact macromolecules in the intestinal epithelium of the carp (*Cyprinus carpio*) and the possible immunological implications. *Cell Tissue Res.* **239,** 519–530.

Rombout, J. H. W. M., Taverne-Thiele, A. J., and Villena, M. I. (1993). The gut-associated lymphoid tissue (GALT) of carp (*Cyprinus carpio* L.): An immunohistochemical analysis. *Dev. Comp. Immunol.* **17,** 55–66.

Ross, G. D., and Jensen, J. A. (1973a). The first component (C1a) of the complement system of the nurse shark (*Ginglymostoma cirratum*). I. Hemolytic characteristics of partially purified C1a. *J. Immunol.* **110,** 175–182.

Ross, G. D., and Jensen, J. A. (1973b). The first component (C1a) of the complement system of the nurse shark (*Ginglymostoma cirratum*). II. Purification of the first component by ultracentrifugation and studies of its physicochemical properties. *J. Immunol.* **110,** 911–918.

Russell, W. J., Voss, E. W., and Sigel, M. M. (1970). Some characteristics of anti-dinitrophenyl antibody of the gray snapper. *J. Immunol.* **105,** 262–264.

Sakai, M., Atsuta, S., and Kobayashi, M. (1989). Attempted vaccination of rainbow trout, (*Oncorhynchus mykiss*) against bacterial kidney disease. *Bull. Jpn. Soc. Sci. Fish.* **55,** 2105–2109.

Sami, S., Fischer-Scherl, T., Hoffman, R. W., and Pfeil-Putzien. (1992). Immune complex–mediated glomerulonephritis associated with bacterial kidney disease in the rainbow trout (*Oncorhynchus mykiss*). *Vet. Pathol.* **29,** 169–174.

Sanchez, C., and Dominguez, J. (1991). Trout immunoglobulin populations differing in light chains revealed by monoclonal antibodies. *Mol. Immunol.* **11,** 1271–1277.

Sanchez, C., Dominguez, J., and Coll, J. (1989). Immunoglobulin heterogeneity in the rainbow trout, *Salmo gairdneri* Richardson. *J. Fish Dis.* **12,** 459–465.

Schatz, D. G., Oettinger, M. A., and Schlissel, M. S. (1992). V(D)J recombination: Molecular biology and regulation. *Annu. Rev. Immunol.* **10,** 359–383.

Shankey, T. V., and Clem, L. W. (1980). Phylogeny of immunoglobulin structure and function. IX. Intramolecular heterogeneity of shark 19S IgM antibodies to the dinitrophenyl hapten. *J. Immunol.* **125,** 2690–2698.

Shieh, H. S. (1989). Protection of Atlantic salmon against bacterial kidney disease with *Renibacterium Salmoninarum* extracellular toxin. *Microbiol. Lett.* **41,** 69–71.

Sitia, R., Neuberger, M., Alberini, C., Bet, P., Fra, A., Valetti, C., Williams, G., and Milstein, C. (1990). Developmental regulation of IgM secretion: The role of the carboxy-terminal cysteine. *Cell* **60,** 781–790.

Smith, P. D. (1988). Vaccination against vibriosis. *In* "Fish Vaccination" (A. E. Ellis, ed.), 255 pp. Academic Press, San Diego.

Steiner, L. A. (1985). Immunoglobulin disulfide bridges: Theme and variations. *Biosci. Rep.* **5,** 973–989.

Strobard, H. W. J., and Kroon, A. G. (1981). The development of the stomach in *Carias lazera* and the intestinal absorption of protein macromolecules. *Cell Tissue Res.* **215,** 397–425.

Strobard, H. W. J., and Vander Veen, F. H. (1981). The localization of protein adsorption during the transport of food along the intestine of the grass carp, *Ctenopharyngodon idella* (val.) *Histochemistry* **64,** 235–249.

Tartakoff, A., and Vassalli, P. (1979). Plasma cell immunoglobulin M molecules: Their biosynthesis, assembly, and intracellular transport. *J. Cell Biol.* **83,** 284–299.

Tatner, M. R. (1986). The ontogeny of humoral immunity in rainbow trout, *Salmo gairdneri. Vet. Immunol. Immunopath.* **12,** 93–105.

Tatner, M. F. (1987). The quantitative relationship between vaccine dilution, length of immersion time, and antigen uptake, using radiolabeled *Aeromonas salmonicida* bath in direct immersion experiments with rainbow trout, *Salmo gairdneri. Aquaculture* **62,** 173–185.

Thiry, M., Lacocq-Xhonneux, F., Dheur, I., Renard, A., and de Kinkelin, P. (1990). Molecular cloning of the mRNA coding for the G protein of the viral hemorrhagic septicemia (VHS) of salmonids. *Vet. Microbiol.* **23,** 221–226.

Thornton, J. C., Garduno, R. A., Carlos, S. J., and Kay, W. W. (1993). Novel antigens expressed by *Aeromonas salmonicida* grown *in vivo. Infect. Immun.* **61,** 4582–4589.

Tomlinson, S., Stanley, K. K., and Esser, A. F. (1993). Domain structure, functional activity, and polymerization of trout complement protein C9. *Dev. Comp. Immunol.* **17,** 67–76.

Tomonaga, S., Kobayashi, K., Hagiwara, K., Yamaguchi, K., and Awaya, K. (1986). Gut-associated lymphoid tissue (GALT) in elasmobranchs. *Zool. Sci.* **3,** 453–458.

Tripp, R. A. (1988). Glucocorticoid regulation of salmonid B lymphocytes. Ph.D. Thesis, Oregon State Univ., Corvallis.

Turaga, P., Weins, G., and Kaattari, S. (1987). Bacterial kidney disease: The potential role of soluble protein antigen(s). *J. Fish Biol.* **31** (Supp. A), 191–194.

Udey, L. R., and Fryer, J. L. (1978). Immunization of fish with bacterins of *Aeromonas salmonicida. Mar. Fish. Rev.* **40,** 12–17.

Vallejo, A. W., Miller, N. W., Jorgensen, T., and Clem, L. W. (1990). Phylogeny of immune recognition: Antigen processing/presentation in channel catfish immune responses to hemocyanins. *Cell. Immunol.* **130,** 362–377.

Vallejo, A. W., Miller, N. W., and Clem, L. W. (1992). Antigen processing and presentation in teleost immune response. *Annu. Rev. Fish Dis.* **2,** 73–89.

Vaughan, L. M., Smith, P. R., and Foster, T. J. (1993). An aromatic-dependent mutant of the fish pathogen *Aeromonas salmonicida* is attenuated in fish and is effective as a live vaccine against the salmonid disease furunculosis. *Infect. Immun.* **61,** 2172–2181.

Voss, E. W., and Sigel, M. M. (1972). Valence and temporal change in affinity of purified 7S and 18S nurse shark anti-2,4-dinitrophenyl antibodies. *J. Immunol.* **109,** 665–673.

Voss, E. W., Groberg, W. J., and Fryer, J. L. (1978). Binding affinity of tetrameric coho salmon Ig anti-hapten antibodies. *Immunochemistry* **15,** 459–464.

Warr, G. W. (1983). Immunoglobulin of the toadfish, *Spheroides glaber. Comp. Biochem. Physiol.* **76B,** 507–514.

Wetzel, M., and Charlemagne, J. (1985). Antibody diversity in fish. Isoelectrofocalization study of individually purified specific antibodies in three teleost fish species: Tench, carp, and goldfish. *Dev. Comp. Immunol.* **9,** 261–270.

White, D. O., and Fenner, F. J. (1986). "Medical Virology," 3rd edition. Academic Press, Orlando, FL.

Whyte, S. K., Chappell, L. H., and Secombes, C. J. (1990). Protection of rainbow trout, *Oncorhynchus mykiss* (Richardson), against *Diplostomum spathaceum* (Digenea): The role of specific antibody and activated macrophages. *J. Fish Dis.* **13,** 281–291.

Wiens, G. D., and Kaattari, S. L. (1991). Monoclonal antibody characterization of a leukoagglutinin produced by *Renibacterium salmoninarum. Infect. Immun.* **59,** 631–637.

Wiersma, E. J., and Shulman, M. J. (1995). Assembly of IgM. Role of disulfide bonding and noncovalent interactions. *J. Immunol.* **154,** 5265–5272.

Wilson, M. R., Marcus, A., Van Ginkel, F., Miller, N. W., Clem, L. W., Middleton, D., and Warr, G. W. (1990). The immunoglobulin M heavy chain constant region of the channel catfish, *Ictalurus punctatus:* An unusual mRNA splice pattern produces the membrane form of the molecule. *Nucleic Acids Res.* **18,** 5227–5233.

Winkelhake, J. L. (1979). Immunoglobulin structure and effector functions. *Immunochemistry* **15,** 695–714.

Wizigmann, G., Baath, G., and Hoffman, R. (1980). Isolierung des firus der viralen hamorrhagischen septikamie (VHS) aus regenbogen-forellen, Hecht-und Aschenbrut. *Zentralbl. Veterinarmed.* **27,** 79–81.

Wolf, K. (1988). "Fish Viruses and Fish Viral Disease." Cornstock Publishing Associates, Cornell University Press, Ithaca, NY.

Wong, G., Kaattari, S. L., and Christensen, J. M. (1992). Effectiveness of an oral enteric coated vibrio vaccine for use in salmonid fish. *Immunol. Invest.* **21,** 353–364.

Wood, P. (1994). Characterization of the humoral immune response to *Renibacterium salmoninarum* in chinook salmon (*Oncorhynchus tshawytscha*). M.S. Thesis, Oregon State Univ., Corvallis.

Xu, L., Mourich, D. V., Engleking, H. M., Ristow, S., Arnzen, J., and Leong, J. C. (1991). Epitope mapping and characterization of the infectious hematopoietic necrosis virus glycoprotein, using fusion protein synthesized in *E. coli. J. Virol.* **65,** 1611–1615.

Yang, M.-C. W., Miller, N. W., Clem, L. W., and Buttke, T. M. (1989). Unsaturated fatty acids inhibit IL-2 production in thymus-dependent antibody responses *in vitro. Immunology* **68,** 181–184.

Yano, T., Matsuyama, H., and Nakao, M. (1988). Isolation of the first component of complement (C1) from carp serum. *Nip. Sui. Gak.* **54,** 851–859.

Zapata, A., and Solas, M. T. (1979). Gut-associated lymphoid tissue (GALT) in reptilia: Structure of mucosal accumulations. *Dev. Comp. Immunol.* **39,** 477–487.

6

NATURAL CHANGES IN THE IMMUNE SYSTEM OF FISH

MARY F. TATNER

I. INTRODUCTION

Natural factors that have an effect on the immune system of fish include the physiological processes of ontogenetic development, and, at the other end of the life cycle, the progressive decline toward senescence and death. Fish are the largest group of vertebrates and it can be very misleading to generalize about any aspect of their biology. They show a variety of life histories and strategies, and inhabit either fresh, salt, or brackish waters, where they will encounter a range of pathogenic challenges. They are either relatively solitary or live in large social groups, and for widely varying life spans. All these factors will have influenced the selective pressures on the defence system that they have evolved. In addition, the immune system is just one of many physiological processes, and at different times within one individual's life, the requirement for resource (energy) partitioning will

THE FISH IMMUNE SYSTEM:
ORGANISM, PATHOGEN, AND ENVIRONMENT

lead to fluctuations in the resource committed to defence, compared to other demands such as growth, sexual maturation, and breeding. Notwithstanding, there are many common facets of the immune system in the restricted number of fish species that have been studied so far.

The ontogeny of the immune system of other cold-blooded vertebrates, reptiles and snakes, has been well documented (El Deeb and Saad, 1990). Fish lend themselves particularly well to studies of this nature as they produce a large number of free living offspring that are not fully immunocompetent at hatching. In addition, the rate of development of immunity can be experimentally manipulated by controlling the external temperature (Bly and Clem, 1992). On the other hand, very little work has been performed on the immune system of any aged, cold-blooded vertebrate, though this is of interest as the life span of a species would be expected to determine its need for and evolutionary development of an immune memory.

There have been some early reviews on the ontogeny of the immune response in cold-blooded vertebrates (Du Pasquier, 1973), and of fish in particular (Manning *et al.,* 1982a; Manning and Mughal, 1985; Ellis, 1988). This paper will update the knowledge acquired since then and also consider other natural factors that can influence immunity, such as social stress, captivity, smoltification, metamorphosis, and sexual maturation.

Several years ago, Solomon (1978) proposed an age-equivalence model of lymphoid development and immunological function, which suggested that different animals attain the same level of immunological maturity at equivalent physiological ages. This was based on two important milestones of development, the first being the well-documented appearance of small lymphocytes in the thymus, and the second, the determination of peak growth velocity (peak body weight gain per day). It would be of interest and of potential value to apply this theory to what is known of the lymphoid development in various fish species to see if the model applies equally to cold-blooded vertebrates.

II. ONTOGENY OF LYMPHOID ORGAN DEVELOPMENT

A. Histogenesis of the Lymphoid Organs

A complete histological description of the ontogenetic development of the lymphoid organs, the thymus, kidney, spleen, and gut-associated lymphoid tissue, has been performed for relatively few fish species (Table I). From an immunological point of view, the first appearance of lymphocytes within an organ, rather than the first appearance of the organ itself, is a better indication of the development of the lymphoid system. However, it

Table I
Histogenesis of the Lymphoid Organs in Fish

		First appearance of lymphocytes[a]				
Species	Temp°C	Thymus	Head kidney	Spleen	GALT[b]	Author
O. mykiss	14	3 dph	5 dph	6 dph	13 dph	Grace and Manning, 1980
B. conchonius	23	4 dph	4 dph	7 dph	5 dph	Grace, 1981
C. carpio	22	3 dph	6 dph	8 dph		Botham and Manning, 1981
C. carpio		5 mm	6 mm	10 mm		Schneider, 1983
S. salar	4–7	22 dpreh	14 dpreh	42 dph		Ellis, 1977
T. mossambicus	Room temp.	6–8 dlprel	13–16 dmpl	30–80 j		Sailendri, 1973
Pagrus major		13 dph				Kusada *et al.*, 1992
Sparus aurata		47 dph	54 dph	later than 77 days		Josefsson and Tatner, 1993
Zoarces viviparous		2 mnths prep	1 mnth prep	After birth		Bly, 1985
S. canicula	12–14	6–9 mnths prehatch	6–9 mnths prehatch	8 mnths prehatch	8 mnths prehatch	Lloyd-Evans, 1993
S. canicula		1–2 mnth	1–2 mnth	2–5 mnths		Hart *et al.*, 1986
O. mossambicus		9 dpfert	14 dpfert		14 dpfert	Doggett and Harris, 1987
Harpafinger antarticus	4	4 wph	1 hour ph	4 wph		O'Neill, 1989

[a] dpg, Days posthatch; mm, millimeters in length; dpreh, days prehatch; lprel, late prelarval; mprel, mid prelarval; j, juvenile; mnths prepar, months preparturition; dpfert, days postfertilization.
[b] GALT, Gut associated lymphoid tissue

does not necessarily follow that as soon as lymphocytes are first distinguishable, that they are functionally active or mature.

Table I shows that despite difficulties with direct comparisons due to differences in the classification of embryonic stages and the temperatures that animals were raised at, there is a general pattern to the sequential development of the lymphoid organs. In fish, as in all vertebrates, the thymus is the first lymphoid organ to develop and the first to become lymphoid. This is followed by the kidney, with the spleen developing later and remaining predominantly erythroid throughout life, with relatively few lymphocytes. In the dogfish, *Scyliorhinus canicula,* the spleen develops alongside the gut-associated lymphoid tissue (GALT) and epigonal and Leydig organs (Hart *et al.,* 1986).

If we consider one species in detail, the rainbow trout, *Salmo gairdneri* (previously known as *Oncorhynchus mykiss*), the thymus is present as a rudiment at 5 days prehatch at 14°C (Grace and Manning, 1980). It starts its development as a thickening of the epithelial tissue in the dorso-anterior part of the pharynx, without a distinct separation of a thymic bud (Fig. 1)

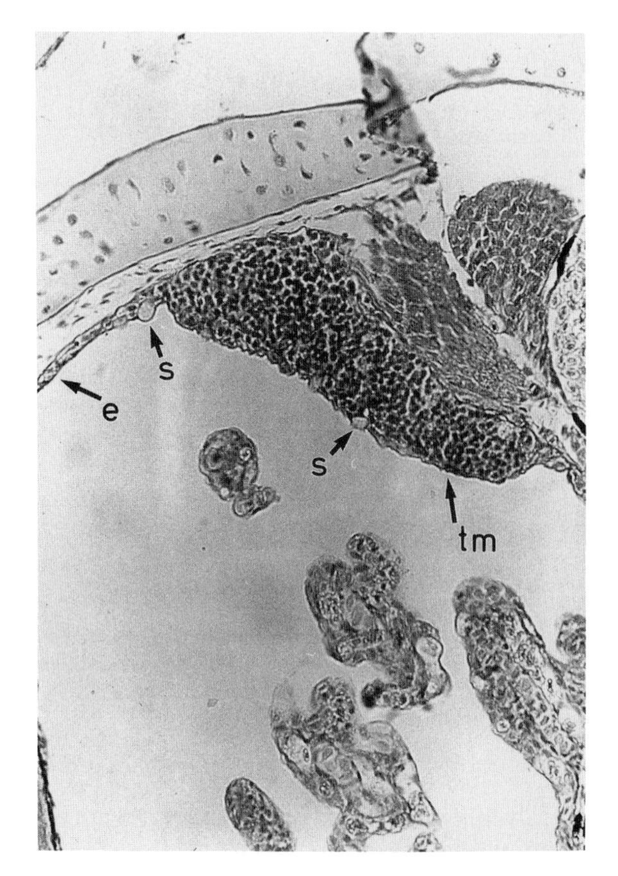

Fig. 1. Thymus gland in rainbow trout fry of 12 days posthatch, at 14°C. Hematoxylin, 8 u, and eosin, T.S., ×250. e, Pharngeal epithelium; s, secretory cell; tm, thymic membrane.

In trout, the thymus remains in a superficial position throughout development and throughout life and is separated from the opercular cavity by only a single layer of cells (Grace and Manning, 1980). In other fish, such as the carp, *Cyprinus carpio,* the thymus becomes more internalized (Botham and Manning, 1981). By day 5 posthatch, following a week of active lymphopoiesis, the trout thymus is clearly a lymphocytic organ, and it is at this time that small lymphocytes are first seen in the kidney (Fig. 2). Several authors have reported the existence of a "bridge" linking the thymus and the kidney (Josefsson and Tatner, 1993, in the seabream, *Sparus aurata;* Sailendri, 1973, in the tilapia, *Tilapia mossambica,* guppy, *Lebistes reticulatus,* and plaice, *Pleuronectes platessa*) but whether there is any movement of cells from one organ to the other, and in which direction, is not known. It has been shown by radiolabeling thymocytes *in situ* in older trout, that cells do migrate from the thymus to the peripheral lymphoid organs (Tatner, 1985), but whether this occurs during ontogeny is not known. It may be that the kidney acquires its lymphoid cell population independently of the thymus, as early thymectomy in fish does not deplete the kidney lymphocyte population to any great extent (Grace, 1981).

As mentioned earlier, the spleen is slow to develop a lymphoid cell population in trout, as in other species studied, and remains predominantly

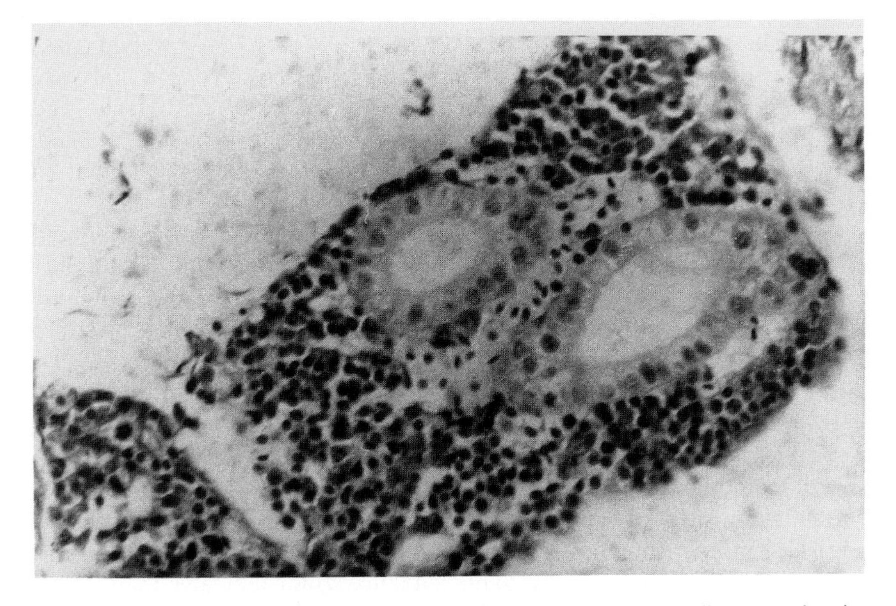

Fig. 2. Kidney of rainbow trout, 5 days posthatch, at 14°C. Hematoxylin, 8 u, and eosin, T.S., ×250.

erythroid throughout life. In studies where the development of the GALT has been related to that of the other lymphoid organs, the GALT acquires its lymphoid cells later than the thymus and kidney (Grace, 1981; Bly, 1985; Hart *et al.,* 1986; Doggett and Harris, 1987).

Few studies have monitored the relative growth of the lymphoid organs. Van Loon *et al.* (1982) showed that in the first few weeks of life in the carp, the thymus contained the largest lymphoid cell pool (70% or about 3×10^5 cells at day 28). At 2 months of age, the total lymphocyte cell pool (5×10^6 cells) was more evenly distributed among the lymphoid organs, with 38% in the mesonephros, 32% in the thymus, 15% in the pronephros, 12% in the peripheral blood, and 3% in the spleen. Bly (1985) showed that the thymus of the viviparous teleost *Zoarces viviparous* reached maximum size 1 month preparturition, when it occupied 0.83% of the total body volume. Thereafter, it decreased in size and by 3 months postparturition occupied only 0.048% of the total volume. Nakanishi (1986a) noted that the thymus of the rock fish, *Sebastiscus marmoratus,* was fully developed and at its maximum ratio of thymus weight to body weight at the time when allograft reactivity was first evident.

Tatner and Manning (1983b) monitored the growth of the lymphoid organs in rainbow trout both in terms of weight and cell numbers, from 1 to 15 months of age. All the lymphoid organs attained their maximum relative weights, expressed as a percentage of body weight, at 2–3 months of age. The organ weights showed a closer correlation to body weight (size) than they did to age. Nakanishi (1986a, 1991) has suggested that body size is more important in determining immunological maturation than age, and that it depends on a certain critical number of lymphocytes being attained. This is certainly true in the ontogeny of the immune response to vaccination (Johnson *et al.,* 1982a,b)

The factors that govern the timing of the development of the lymphoid system in fish are unknown but there are at least two hypotheses. First, natural selection will favor individuals in which development of the immune system coincides with their first exposure to potential pathogens either from their environment or in their food (i.e., hatching in oviparous species, parturition is viviparous species, or at the time of first feeding). It is not known how much antigenic challenge developing eggs and embryos are exposed to. In *Zoarces viviparous,* the embryonic gut contains maternal blood components and tissue debris from resorbed siblings, which would certainly constitute an antigenic challenge, (Bly, 1985) and certain pathogens, for example, infectious pancreatic necrosis virus (IPNV), are transmitted vertically via the eggs. Available evidence so far indicates that even if morphologically mature lymphoid cells are present at these early stages, the animals are not functionally immunocompetent until later, and hence, presumably, rely on other defence mechanisms. A second theory is that development of the lymphoid organs occurs at a preset time during embryogenesis (Solomon, 1978) once a certain "physiological age" has been reached (possibly a critical number of lymphocytes), and is independent of first exposure to antigenic challenge. Further research is needed to determine which of these hypotheses holds true for cold-blooded vertebrates such as fish.

B. Ontogeny of T and B Cells in Lymphoid Organ Development

Antibodies directed against fish immunoglobulin (IgM) have been used to detect the ontogenetic appearance of cells bearing Ig on their surface (sIg^+ B cells) or in their cytoplasm (cIg^+ plasma cells) by a variety of techniques. This can act as an indication of the functional maturity of the cells within the lymphoid organs, as opposed merely to their morphological maturity. Such studies have been performed on the carp, *Cyprinus carpio* (Van Loon *et al.,* 1981, 1982; Koumans van Diepen *et al.,* 1994), rainbow trout, *Oncorhynchus mykiss* (Castillo *et al.,* 1993; Razquin *et al.,* 1990), skate, *Bathyraja aleutica* (Kobayashi *et al.,* 1985), and salmon, *Salmo salar* (Ellis, 1977).

Using a conventional rabbit anti-carp Ig antiserum, Van Loon *et al.,* (1982) detected the first appearance of cells bearing immunoglobulin (sIg^+) or having Ig in their cytoplasms (cIg^+), with cell suspensions of whole animals. The sIg^+ cells could be detected in the thymus and pronephros from day 14 after fertilization (at 21°C), and in the mesonephros and spleen from day 28. Cytoplasmic Ig^+ cells were first seen in the pronephros at day 21. Using monoclonal antibodies, the appearance of sIg^+ and cIg^+ in carp from 2 weeks to 16 months of age was monitored by Koumans van Diepen *et al.* (1994). The sIg^+ cells were first detected in 2-week-old carp, confirming an earlier light microscopic study by Secombes *et al.* (1983a). The age of the fish and the percentage of sIg^+ cells in the lymphoid organs was positively correlated, with the lowest percentage in 2-week-old fish and the highest in the oldest fish. The percentage of sIg^+ cells increased gradually from 3 to 16 months of age, with the values being 15.8–48.1% for blood, 9.7–21.6% for pronephros, 7.2–16.8% for mesonephros, 15.9–21.6% for spleen, and 1.5–3.7% for thymus. For cIg^+ cells, the percentages were very low or absent in spleen, thymus, and blood. Plasma cells were not found in the kidney until the fish were 1 month old (0.17%), and increased to 1% of the total cell population by 8 months of age. This was supported by a steady increase in serum Ig from 3 weeks onward (Van Loon *et al.,* 1981).

In the rainbow trout, sIgM$^+$ cells first appear in the kidney 4–5 days after hatching (Razquin *et al.,* 1990). By one month after hatching, sIg$^+$ cells were also present in the spleen and thymus. Castillo *et al.* (1993) detected lymphocytes showing cytoplasmic IgM in rainbow trout embryos at 12 days prehatch (before any cells positive for sIgM were present). Lymphocytes bearing surface IgM were observed at 8 days prehatch in their study which was conducted at 14°C.

In the salmon, *Salmo salar,* Ellis (1977) demonstrated membrane IgM, using a rabbit anti-salmon IgM, at day 41 posthatch. This coincided with the onset of first feeding. The percentage of positive lymphocytes increased steadily thereafter, reaching 80% by day 48 posthatch.

In the Aleutian skate, *Bathyraja aleutica,* Kobayashi *et al.* (1985) identified two types of Ig-producing cells in the spleen of embryos, one producing a high molecular weight Ig analagous to mammalian IgM, and the other producing a low molecular weight Ig. In the embryos and adults, the ratio of each type of cell was 5:6 and 4:5, respectively. The number of these Ig-producing cells increased with advancing development of the embryos, but was $\frac{1}{20}$th to $\frac{1}{50}$th of that of adults.

The ontogenetic development of T cells has been studied in the carp using mouse anti-carp thymocyte monoclonal antibodies by Secombes *et al.* (1983). One antibody (WC T4) detected determinants present on thymocytes as soon as the thymus was visible histologically (day 4 after fertilization, at 21°C); these determinants were not detected in the pronephros until day 7 postfertilization. Antibodies WCT 8 and WCT 10 recognized determinants on thymocytes from day 7, which were not present on the pronephros until days 16 and 12, respectively. Recently, monoclonal antibodies specific for the thymocytes of the sea bass, *Dicentrarchus labrax,* have been developed by Scapigliati *et al.* (1995).

C. Evidence for Ig in the Eggs and Passive Transfer of Immunity from Mother to Young

It is probable that embryonic fish receive some antigenic challenge before they hatch or are born. It is of interest, therefore, to determine if they possess any defence mechanism at all at these earliest stages of development. Nonspecific humoral factors, such as C-reactive protein and lectin-like agglutinins have been found in fish ova (Ingram, 1980). Immunoglobulin has been detected in the eggs of carp (Van Loon *et al.,* 1981), plaice (Bly *et al.,* 1986), tilapia (Mor and Avtalion, 1989), channel catfish (Hayman and Lobb, 1993), chum salmon (Fuda *et al.,* 1992), and coho salmon (Brown *et al.,* 1994). Castillo *et al.* (1993) quantified the amount of IgM in unfertilized rainbow trout eggs as 11.2 ± 2.6 mg/g egg weight. The levels in the whole fish increased slowly to reach a peak, in terms of milligrams per gram of tissue, at the time of hatching, before slowly declining to initial values by 2 months posthatch. Takemura (1993) studied changes in an immunoglobulin M–like protein during larval changes in the tilapia, *Oreochromis mossambicus.* The levels declined during the prelarval stages, when the yolk was being absorbed, and reached the lowest value (by ELISA) 12 days after hatching, when the yolk was all gone and feeding had started. He concluded that the IgM-like protein was maternally derived and that it was depleted by the time of first feeding, after which the larvae start to produce it for themselves. Bly (1984) studied three species with distinctly different feto–maternal and neonate–maternal relationships, the oviparous plaice, *Pleuronectes platessa,* the viviparous blenny, *Zoarces viviparus,* which displays ovarian gestation, and the viviparous swordtail, *Xiphophorus helleri,* which displays follicular gestation. Using ^{125}I-labeled homologous Ig, Bly (1984) demonstrated uptake of immunoglobulin into the young of the blenny and the swordtail, and into the eggs of the plaice. Bly suggested that the relatively earlier lymphoid development in *Zoarces,* compared to the oviparous species, could result in the embryo being immunologically competent while still within the ovary, and capable of mounting an immune response against both maternal and sibling antigens.

The finding that Ig can be detected in the eggs and larvae of fish raises the possibility that hen (mother) fish could be vaccinated to provide their offspring with protection against various pathogens. This would be particularly beneficial in the case of pathogens that are transmitted vertically and to which the young are particularly susceptible, such as IPN virus. Mor and Avtalion (1989) have demonstrated the specificity of transferred immunogloblin, from mother to egg, when the mother was immunized with various protein antigens, while in the ovoviviparous guppy, immunity was transferred to the newborn fry (Takahashi and Kawahara, 1987). Recently, Sin *et al.* (1994) have demonstrated passive transfer of protective immunity against ichthyopthiriasis in tilapia, *Oreochromis aureus.* Hen fish were vaccinated twice 1 month before spawning with live tomites of *I. multifillis.* The fertilized eggs were collected from the mouths of vaccinated fish and from control nonvaccinated fish and incubated until the fry were free swimming. In addition, other groups were collected without mouthbrooding. On challenge with infective tomites, the control fish with no mouthbrooding all died; control fish with mouth brooding had a 37% survival; fry from vaccinated broodstock without mouth brooding had a 78% survival; whereas fry from vaccinated broodstock that had also been mouth brooded had a survival of 95%. The protective immunity was correlated with the anti-*Ichthyopthiriasis* titers in soluble extracts of fry tissues and in the mothers' plasma. Hence the protective Ig was from two sources: the eggs from the mother and also from the mother's mouth cavity. Subasinghe (1993) had earlier demonstrated the protective effect of mouthbrooding in the same species.

III. AGING OF LYMPHOID TISSUE IN FISH

Very little is known on this aspect of the immune system in fish, though some work has been done on other cold-blooded vertebrates (e.g., in the amphibians, *Bufo viridis,* Saad *et al.,* 1994, and *Rana temporaria,* Plytycz *et al.,* 1994). Although it is generally agreed that it is the size of a fish rather than its age that determines its immunological maturity during ontogeny, it is not known if the same holds true at the other end of the life cycle.

Fish life spans vary considerably, with the Chondrichthyes (sharks) in particular living for a considerable number of years. More commonly, the life span of a particular species is simply not known. In experimental investigations, the age of the fish used is often not reported, and even if the weight is given, it can be inaccurate to use one to predict the other. In fact, fork length shows a better correlation with age than does body weight (Tatner and Manning, 1983b). Tatner and Manning (1983b) monitored the growth and cellular composition of the lymphoid organs in rainbow trout from 1 to 15 months of age. The thymus, spleen, and kidney tissue grew steadily over this period, but all attained their maximum relative weights (expressed as a percentage of body weight) at 2–3 months of age, indicating a time of intense activity and importance in the maturation of the lymphoid system (Fig. 3). The total number of leukocytes in the organs increased with age, but when the size of the organ and the size of the fish were taken into account, the actual numbers decreased with age. An accompanying histological study of the organs showed gradual changes in the structure of the tissues. The thymus showed intense mitotic activity in the first few months, which then slowed down. There were no signs of massive cell death, and it was assumed that emigration of cells to the peripheral lymphoid organs must be occurring. Histological changes indicative of involution were present in the thymus from 9 months onward. The spleen and kidney showed an increase in melanin deposition with age, with the kidney retaining a more varied range of leukocyte types than the spleen. The number of circulating lymphocytes in the blood remained remarkably constant throughout the first year of life.

In the channel catfish, *Ictalurus punctatus,* Ellsaesser *et al.* (1988) found that the number of thymocytes collected from the thymus remained constant from 3 to 10 months of age, increased sharply between 11 and 12 months,

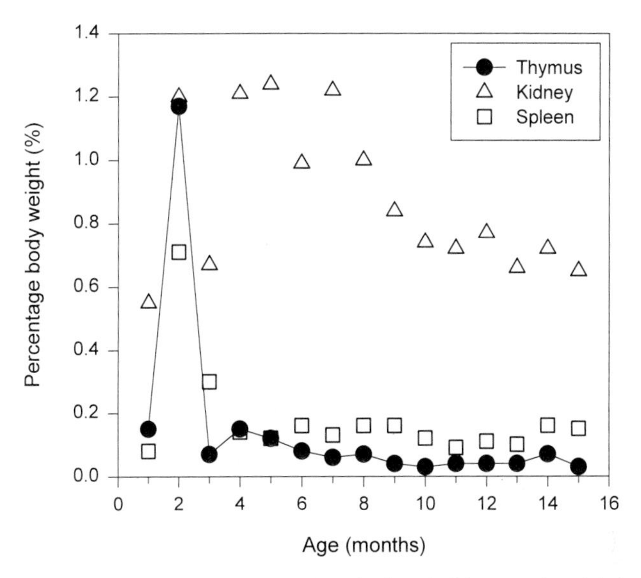

Fig. 3. Graph showing the relative growth of the lymphoid organs in rainbow trout from 1 to 15 months of age. Data from Tatner and Manning (1983b).

and then decreased after 13 months until by 16 months posthatch the thymus had shrunk to a thin epitheloid layer with no detectable lymphocytes.

The most obvious aging effect on the lymphoid tissues is the involution of the thymus gland. The involution of the thymus can be related to age (life span), season, or hormonal changes associated with sexual maturation (Honma and Tamura, 1984; Chilmonczyk, 1992). In some species such as rays, flatfish and cod, the thymus continues to grow after sexual maturity (Lele, 1934), whereas in eels, it has already involuted before sexual maturity (Von Hagen, 1936). In carp and primitive sharks, which are long lived, the thymus does not appear to involute at all (Chilmonczyk, 1992). Significant age-related histological changes occur in the thymus of the Cyprinodont, *Notobranchius guentheri,* with the first signs of senescence appearing at 4 months and complete degeneration by 12 months (Cooper *et al.,* 1983). Early involution has been reported in 3-week-old dogfish, *Scyliorhinus canicula* (Pulsford *et al.,* 1984). Involution is characterized histologically by an increase in connective tissue, a decrease in lymphocyte density, infiltration of adipocytes, the presence of reticular cells and collagenous fibers, and the formation of epithelial cysts (Chilmonczyk, 1992). However, the exact significance of these changes in the structure of the thymus to the immune capabilities of older fish are unclear.

IV. ONTOGENY OF NONSPECIFIC IMMUNITY

In addition to any maternally derived immune protection, newly hatched fry possess a variety of nonspecific defence mechanisms that provide them with protection prior to the specific immune system becoming fully mature. This nonspecific immunity is vitally important during the first few weeks of life when it may be the only, as well as the first, line of defence. However, it contributes to defence throughout the whole of the life cycle, as, being relatively temperature independent, it comes into play at times when the specific immune response is suppressed, for example, at low environmental temperatures.

In addition to the nonspecific lectins and hemagglutinins found in fish eggs and fry mentioned earlier, newly hatched fry possess an efficient phagocytic system. The ontogenetic development of the reticulo-endothelial system has been studied in the rainbow trout by Tatner and Manning (1985) by injecting colloidal carbon into fry at known stages. It was found that fry as early as 4 days posthatch had an efficient phagocytic system, the carbon being engulfed by free-wandering macrophages that accumulated under the skin and in the connective tissue, gut, and gills (Fig. 4). By 14 days posthatch, the pattern of antigen trapping was similar to that found in adults, with the fixed macrophages of the spleen and kidney sequestering the carbon particles from the blood stream. The trapping of carbon in the gills of young fry was interpreted as a specialized mechanism to protect the developing thymus from undesirable and possibly tolerogenic antigen exposure.

Phagocytosis in the juvenile flounder, *Platichthys flesus* L., was studied by Pulsford *et al.*, (1994). Phagocytosis of yeast particles occurred in 35–50% of adherent spleen cells in fish of 4 cm, which was comparable to that of adults, whereas only 2–10% of cells could phagocytose the particles in fish of less than 2.5 cm. The phagocytes of juvenile fish of 6–7 cm in length could produce hydrogen peroxide on stimulation with zymosan.

Interferon is another important component of the immune system in fish (see chapter by Yano, this volume), and its synthesis by rainbow trout fry following infection with IPNV has been demonstrated by Dorson *et al.* (1992) at an age of 1250 degree-days and mean weight of 2.5 g.

V. ONTOGENY OF CELL-MEDIATED IMMUNITY

The ontogenetic development of cellular immunity in fish has been studied mainly by consideration of the allograft rejection response. This is

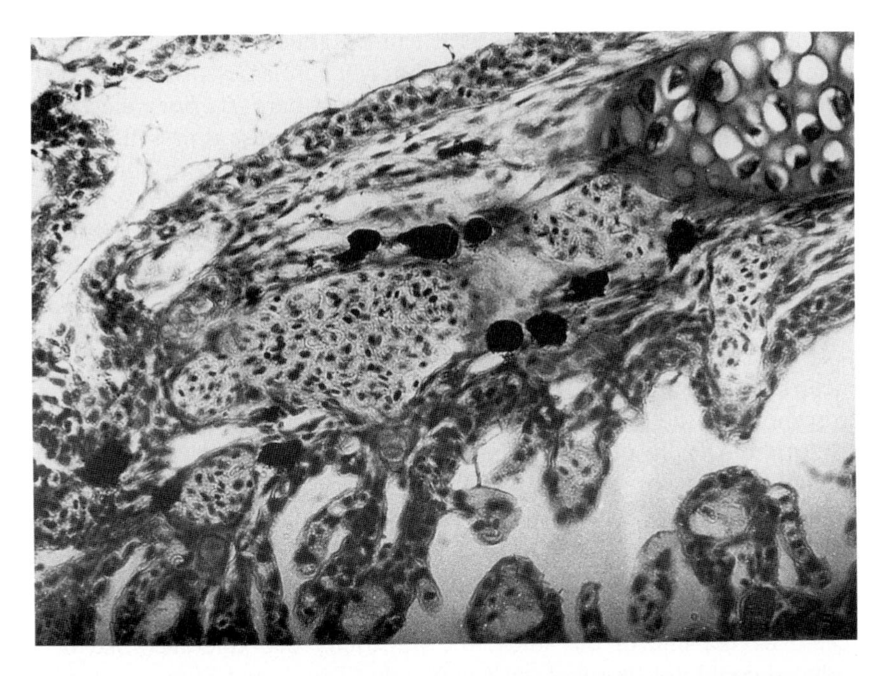

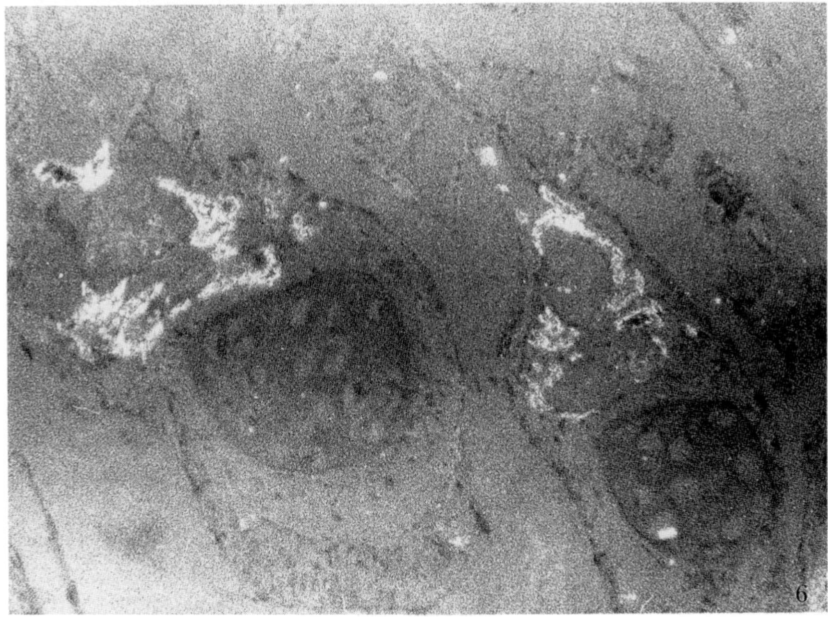

Fig. 4. (a) Trapping of colloidal carbon in the gills of 6-day-old rainbow trout fry. (b) The same viewed under an Ultrapak, which causes the carbon particles to glow.

primarily for technical reasons, as it can be difficult to obtain sufficient numbers of lymphocytes from small fish for *in vitro* assays. Allograft rejection has been studied in the carp, trout, and rosy barb, *Barbus conchonius,* and it is evident that this facet of the immune response is present early in development. It can be assumed, therefore, that the T cells responsible for this response mature functionally soon after their morphological appearance during development. In the rainbow trout, Tatner and Manning (1983a) demonstrated that fry as young as 14 days posthatch (at 14°C) could destroy skin allografts, as assessed by lymphocytic infiltration, pigment breakdown, and phagocytosis of dispersed pigment (Fig. 5). By 26 days posthatch the response was as vigorous as that seen in adult fish, with a mean survival time (MST) of the graft of between 14 and 20 days. The ability to respond to the foreign tissue grafts was correlated with the presence of a morphologically mature lymphoid population in the thymus and the kidney, and the presence of circulating lymphocytes. When grafts were applied at an earlier stage, at 5 days posthatch, they persisted for a while and no lymphocytric infiltration was seen. However, the response was only delayed, for as soon as the lymphoid organs became more mature, the rejection process started. However, as the grafts were not completely rejected by the end of the study, the possibility that tolerance had been induced could not be ruled out completely (Tatner and Manning, 1983a). In amphibians, Horton (1969) had previously noted a delayed rejection process when grafts were applied very early, and suggested that this may be an adaptive process in primitive vertebrates with a free-living larval stage, where antigen exposure could well occur before the immune system was mature enough to deal with it, and where tolerance induction would be inappropriate.

In the carp, Botham and Manning (1981) demonstrated allograft reactivity in fry as early as 16 days posthatch at (22°C), while the rosy barb (*Barbus conchonius*) was fully immunocompetent with respect to the allograft response by 6 months of age, if kept at 24°C (Rijkers and van Muiswinkel, 1977).

In the marine teleost *Sebastiscus marmoratus,* fish as young as 1.5 months were capable of eliciting an allograft response in the same manner as adult fish, as assessed by eye transplantation. By the time the fish were 3 months old, they could reject scale grafts more rapidly than adults (Nakanishi, 1986a). However, direct comparisons between species are problematic, as the temperature the species are maintained at will effect the kinetics of the allograft response (Table II).

The ontogenetic development of an immune memory in allograft reactivity, evidenced by an accelerated rejection time of second-set grafts from the same donor, was studied in carp by Manning *et al.* (1982a,b). An allograft

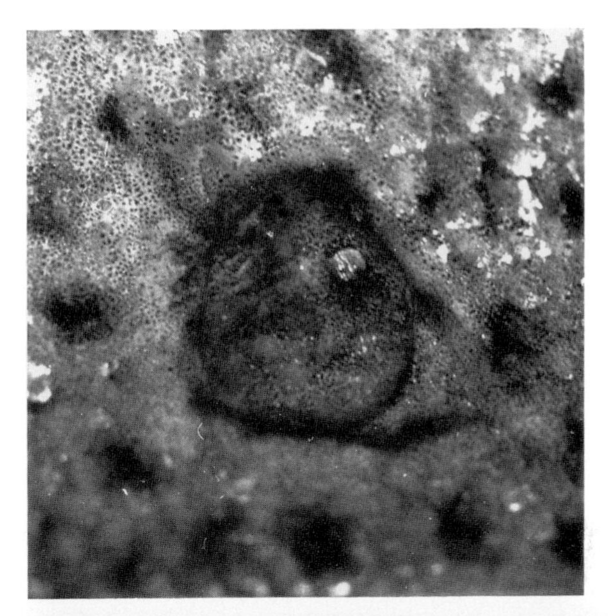

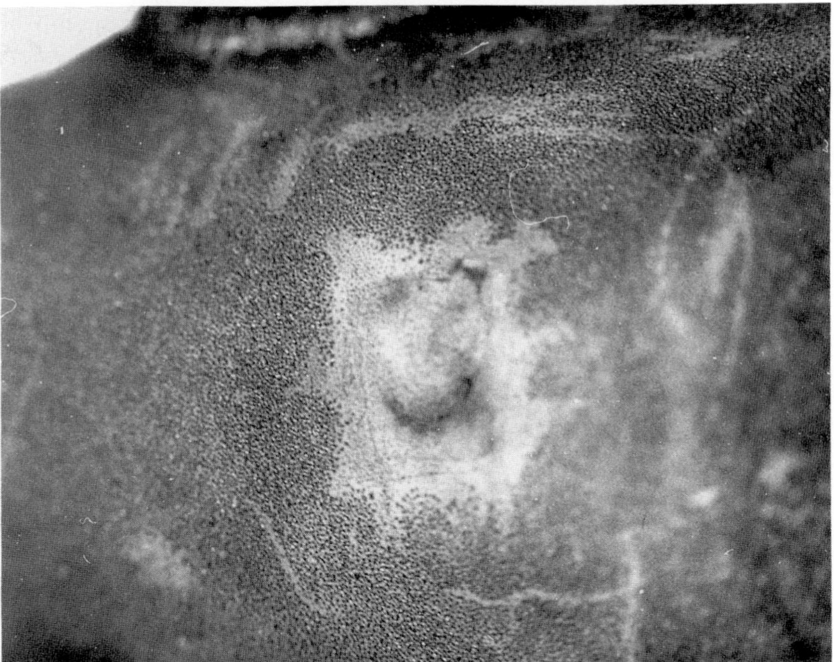

Fig. 5. Allograft rejection in rainbow trout fry of 18 days posthatch. (a) Control autograft has healed in and retains pigment. (b) Allograft has been rejected, with total loss of pigmentation in 21 days at 14°C.

Table II
Ontogeny of Alloimmune Reactivity in Fishes

Species	Age	Response	References
Cyprinus carpio	16 days posthatch	Rejection begins at 16 days postgrafting Anamnestic response	Botham *et al.* (1980)
Oncorhynchus mykiss	5 days 14 days 21 days	No response Rejected by day 30 Rejected by day 20	Tatner and Manning (1983a)
Oreochromis mossambicus	1.5 months	Slow rejection	Sailendri (1973)
Sebastiscus marmoratus	1.5 months	Rejection of eye graft	Nakanishi (1986a)

[a] dpg, Days posthatch; dpreh, days prehatch; lprel, late prelarval; mprel, mid prelarval; j, juvenile; mnths prepar, months preparturition; dpfert, days postfertilization.
[b] GALT, Gut associated lymphoid tissue

applied at 16 days posthatch could induce an anamnestic response on regrafting at 1 month of age, suggesting a rapid maturation of T memory cells in this species.

Another facet of cellular immunity is the proliferation of lymphocytes in response to nonself antigens and mitogens. In salmon, *Salmo salar,* a mixed leukocyte reaction (MLR) was first detected in cell suspensions of whole fry at 45 days posthatch (Ellis, 1977).

A response to the mitogens PHA and Con A (both T-cell mitogens) was detected in isolated spleen and kidney leukocytes from juvenile damselfish (*Pomacentus partitus*) of less than 60 mm total length (McKinney and Schmale, 1993). Lymphocyte proliferation induced by injection of human gamma globulin and Freund's complete adjuvant (FCA) and detected by autoradiography was seen in the spleen of carp by 2 months posthatch (Grace *et al.,* 1981). The results indicate that the cellular immune system develops quickly in the species so far studied (carp, trout, salmon, damselfish, rockfish), though more studies on a greater variety of species in terms of their life spans, reproductive strategies, and phylogenetic positions would be of great value.

VI. ONTOGENETIC DEVELOPMENT OF HUMORAL IMMUNITY

The development of the capacity to produce circulating antibody, which indicates the presence of mature plasma cells, has been studied in the same relatively small number of species as cellular immunity. Generally, it has been found that by the time it is technically feasible to inject and bleed small fish, they are capable of mounting a humoral immune response. This suggests that the actual time when this functional maturity commences could be even earlier than the times reported in the published literature.

An interesting feature in the study of the development of antibody responses has been the discovery that there is a differential capacity to respond to thymus-dependent (TD) and thymus-independent (TI) antigens (see chapter by Kaattari, this volume). This has allowed a more precise analysis of the maturation of the different cell populations involved in these responses, namely the B cells themselves for TI responses, with the addition of the T helper cells and accessory cells for TD responses, on purely functional criteria. There is also the possibility that early exposure to antigen may result in tolerance induction. While this does not appear in cellular immunity, there is a greater probability of it occurring in the humoral immune response based on theoretical grounds.

In carp, van Loon *et al.* (1981) showed that fry of 4 weeks of age were unable to mount a plaque forming cell (PFC) response to sheep red blood cells (SRBC) which are a TD antigen. When reimmunized 3 months later, they were still unable to respond, although control 4-month-old fish receiving their first SRBC injection showed as vigorous a response as adults. However, Manning *et al.* (1982a) showed in older carp fry (8 weeks old, at 22°C) that a primary humoral response could be elicited against human gamma globulin (HGG) in FCA, another thymus-dependent antigen, and to formalin-killed *Aeromonas salmonicida,* which is thymus independent. This suggests that tolerance to TD antigens can be induced only during a certain time period, before 2 months of age in carp; however, the specificity of the suppression was not proven in this case.

In trout fry immunized at 14–17°C with HGG in FCA, or formalin-killed *A. salmonicida* at days 1, 7, 14, and 21 posthatch, and tested 8 weeks later for serum antibody levels, those immunized at day 21 had developed the ability to respond to the bacterial antigen *A. salmonicida* but not to HGG. Following a second injection of the antigens 8 weeks after the first, and further testing 8 weeks later, the fish injected with HGG remained unresponsive to HGG (Secombes, 1981). Again, the specificity of the apparently tolerogenic effect was not established, but it confirmed an earlier study by Etlinger *et al.* (1979) that the ability to respond to thymus-independent antigens precedes the ability to respond to thymus-dependent antigens.

In the rosy barb, *Barbus conchonius,* Rijkers and van Muisiwinkel (1977) demonstrated a PFC response and hemagglutination titers in 4-month-old fish, following injections with SRBC (TD), but the response did not reach adult levels until 9 months. The ontogeny of the primary and secondary humoral antibody responses to SRBC in *Sebastiscus marmoratus* was studied by Nakanishi (1983, 1986a). A detectable hemagglutination titer was first noted 60 days after birth when the fish weighed 0.26 g. Fish injected at 0.02 g (30 days old) showed no response. However, this early injection did not lead to tolerance in this case, as these fish when reinjected at 2.7 g showed a similar antibody response to fish receiving their first injection at this time. Fish receiving a primary injection at 0.3 g showed a good primary response and a heightened secondary response when reinjected 129 days later. The earlier study (Nakanishi, 1983) included fish of the same age but of different sizes, and it was noted that it was the larger fish that showed the better immune response.

Although HGG was tolerogenic when injected into 4-week-old carp (Secombes, 1981), a later study by Mughal and Manning (1984) demonstrated the importance of the route of administration of the antigen to tolerance induction. When the direct immersion method was used to administer the soluble antigen, no effect was seen on subsequent antibody titers following challenge.

Mughal and Manning (1986) investigated the antibody response of juvenile thick-lipped grey mullet (*Chelon labrosus*) to soluble protein antigens. Fish of 6–7 months old, although possessing well-developed lymphoid organs, failed to respond to a single injection of the thymus-dependent antigens HGG or keyhold limpet hemocyanin. However, prior priming of juvenile fish with the antigen was found to potentiate antibody production following challenge with a second dose of the antigens in adjuvant. Priming by the oral route was found to be equally effective as priming by injection.

The ontogeny of antibody production to two antigens delivered by direct immersion was investigated in rainbow trout fry by Tatner (1986). Fry of known ages and weights were immunized with either HGG or *A. salmonicida* at 7 days posthatch and 1, 2, 3, and 4 months posthatch. Half the fry in each group were tested for antibodies 4 weeks after immunization, the remainder were reimmunized and tested again after a further 4 weeks. Appropriate controls to test for tolerance induction and memory responses were included. A primary response to *A. salmonicida* could be detected at 2 months of age, and a memory response (heightened secondary response) at 3 months. A primary response was first detected to HGG at 3 months. There was a period of "unresponsiveness" at earlier ages, which persisted longer for HGG than *A. salmonicida,* but this was not thought to be tolerance as such but a "refractory" state during which antigen was simply not taken up because the immune system was immature and not able to cope with it. Some evidence for this has been provided by immersing rainbow trout fry in radio-labeled bacterial vaccine and measuring antigen uptake directly (Tatner and Horne, 1984), and it has been suggested that the mechanism whereby the antigen is excluded is the phagocytic ability that was noted in the gills of young, but not adult, trout (Tatner and Manning 1985).

Tolerance induction in young fish has been demonstrated only to injected antigens. Where the antigens are presented by direct immersion in a manner similar to which fry would naturally encounter antigens in their environment, no tolerance is seen. The previous reports of tolerance induction may well be experimental artifacts (Table III).

Tatner (1986) noted that the state of immunological maturity in fish may be correlated with the age of the fish, their weight, or simply the naturally rising water temperature (Fig. 6). Of interest, however, is the fact that the period just prior to the first antibody detection (marked with a star in Fig. 6) corresponds to a body weight of approximately 0.24 g. It was at around this weight that, in a previous study on the growth of the lymphoid organs (Tatner and Manning, 1983b), the weight of the thymus reached its maximum percentage body weight, and the number of cells in the thymus reached its peak.

VII. ONTOGENY OF THE RESPONSE TO VACCINATION AND CHALLENGE WITH PATHOGENS

From the practical viewpoint of the increasing use of vaccines in aquaculture, it is of great interest to determine at what age or size fish can respond

Table III
Ontogenetic Development of Humoral Immunity in Fishes

Species	Age	Antigen	Antibody production	Reference
Cyprinus carpio	4 weeks	*A. salmonicida*	+ (Memory)	van Loon *et al.*, (1981)
		HGG	− (Tolerance)	
		SRBC	No PFC	
	2 months	HGG	+	
		A. salmonicida	+	
Oncorhynchus mykiss	2 weeks	*A. salmonicida*	−	Secombs (1981)
		HGG	−	
	3 weeks	*A. salmonicida*	+ (No memory)	
		HGG	− (Tolerance)	
	2 months	*A. salmonicida*	+ (Memory)	
		HGG	+ (Memory)	
Sebastiscus marmoratus	1 month	SRBC	− (No memory)	Nakanishi (1983, 1986a)
	1.5 months	SRBC	−	
	2 months	SRBC	+	

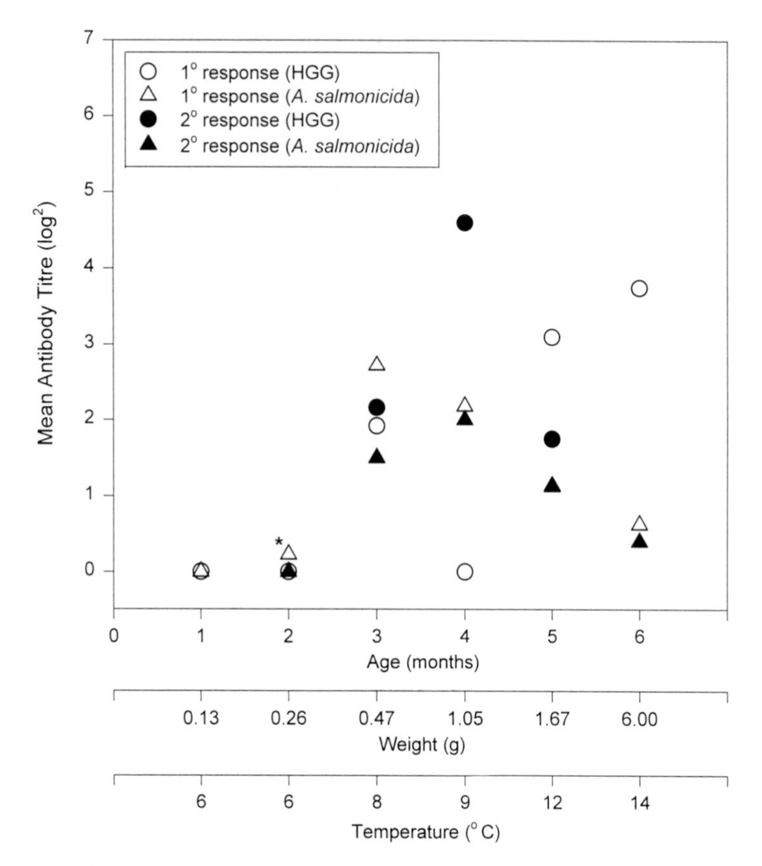

Fig. 6. Graph of the ontogeny of humoral immunity in rainbow trout fry. Data from Tatner (1986).

to vaccination by producing a protective immune response. The vaccination of fry can be fraught with potential problems. There is always the possibility of inducing tolerance, as the thymus dependency or otherwise of the vaccine antigens is often not known. Often, the nature of the protective immunity induced may also not be known (i.e., cellular or humoral), and, as we have seen, these mature at different rates.

The attainment of immunocompetence of any given species in terms of cellular and humoral immunity to particular antigens is known for only a few species; fortuitously, though not surprisingly, these species are those most commonly used in commercial aquaculture at the present time. However, it may be ill advised to extrapolate between one species and another, and, as more species become commercially farmed, a great deal of basic research on the morphological and functional maturation of the immune system of the species in question will need to be performed before attempting to vaccinate their fry.

Wishkovsky *et al.* (1987) studied the effect of fish age on the mortality caused by selected fish pathogens. They compared the mortalities in fingerling carp (70 days, 6–8 g) and adult carp (200–250 g) to a range of injected doses of various pathogenic bacteria. At high doses (greater than 10^4 cells in 0.1 ml), all the fingerlings died, whereas adult carp succumbed only at very high doses of 10^{10} or 10^9. They concluded that the size ratio (1:30) between young and adult fish was not responsible for the variation in mortality rates (two to three orders of magnitude difference) and that the variation was due to the different levels of immunocompetence in the two age groups of fish. Dorson and Torchy (1981) also noted the influence of fish age on mortalities of rainbow trout fry caused by IPNV. The fry showed a decrease in sensitivity with increasing age and ceased to be susceptible to the disease at all when 20 weeks old.

Khalifa and Post (1976) first demonstrated protective immunity in young fish when they successfully immunized 0.3-g rainbow trout fry (which had been feeding for 23 days) against *Aeromonas liquefaciens.* On challenge 3 months later, the vaccinated fish had a mortality of 25% compared with 75% in the controls.

However, it is not practical to routinely immunize fry by injection, even though it provides the best protection levels (see chapter by Kaattari, this volume), and most subsequent work concentrated on the direct immersion method of vaccination. Johnson *et al.* (1982a,b) vaccinated the fry of several salmonid species by direct immersion with either *Yersinia ruckeri* or *Vibrio anguillarum* and monitored the level of protective immunity induced by the survival of the fish after bath challenge with virulent organisms. They found that the minimum size at which maximum protective immunity occurred was between 1.0 and 2.5 g, and that the immunity appeared to be a function of size and not age. Different responses were seen between the different species of salmonids. The duration of the protective immunity varied with the bacterin concentration, and size and species of fish. In fish under 1 g, the duration of protective immunity was longest when the most concentrated bacterin was used. Generally, immunity lasted for 120 days when the fish were vaccinated at 1 g, 180 days for 2-g fish, and in 4-g and heavier fish, the immunity lasted for over a year. This suggested that even though the main lymphocyte populations are functional at an early age, the subpopulations responsible for long-term memory effects may take longer to mature. It is dangerous to conclude that once *any* immunity can be demonstrated, it is fully mature and functional in all its aspects and refinements.

Rainbow trout fry were tested for their susceptibility to experimental infections with *Vibrio anguillarum,* and their ability to mount an immune response against it, from the age of 2 weeks posthatch (0.14 g) onward by Tatner and Horne (1983). Bath challenges were ineffective at inducing vibriosis until 6–8 weeks posthatch. However, at the earliest age tested by intraperitoneal injection (7 weeks posthatch) the fry did succumb to the disease. Protective immunity was evident in fry vaccinated by direct immersion as early as 2 weeks posthatch, when tested by intraperitoneal challenge. By the time the fry reached 0.5 g (10 weeks posthatch), protection levels had reached 50% for direct immersion vaccination and 100% for intraperitoneal vaccination. An oral vaccination, from first feeding onward, proved ineffective at inducing immunity. These same authors showed that no tolerance was induced by preexposure to the *Vibrio* antigen (Tatner and Horne, 1984). Although a period of unresponsiveness to the vaccine was detected in the very earliest groups of fry (vaccinated by necessity by direct immersion), this was found by use of a radiolabeled vaccine preparation to be due to a complete lack of antigen uptake at these stages. It was suggested that the fry "excluded" antigen until their immune systems were able to cope with it. Any conclusions as to tolerance induction or disease susceptibility induced by intraperitoneal injection (a route not naturally encountered by fish fry) must be viewed with caution.

Thorburn and Jansson (1988) investigated the effects of booster vaccination and fish size on protective immunity in trout against *Vibrio anguillarum.* The fish, at 4.1 g and/or 6.3 g, were bath-immunized either once or twice, and bath challenged 1 month after the second vaccination. They found no significant differences in mortalities between the fish that had been boosted at 6.3 g and those receiving a primary vaccination at 6.3 g, that is, the booster did not increase protection levels. However, both 6.3-g groups had lower mortalities than did fish vaccinated at 4.1 g only, indicating some maturation of the protective immune response.

The vaccination of fry against IPNV has received much attention (see chapter by Kaattari, this volume). This disease causes maximum mortalities in fry at first feeding but mortalities become negligible by 6 months of age (Dorson and Torchy, 1981). Early attempts to vaccinate by hyperosmotic shock followed by bath immunization, or by oral vaccination, were unsuccessful, though by 4 weeks fry could be protected by injection immunization and by 5 weeks, shown to produce neutralizing antibodies (Dorson, 1977). This suggests that the fry can be "primed" earlier, without synthesizing antibodies but still protected by a full immune response complete with antibody synthesis later. Direct immersion vaccination of brook trout, *Salvelinus fontinalis,* fry against IPNV (Bootland *et al.,* 1990) indicated that only fry immunized at 2, 3, and 6 weeks posthatch showed protection, when their mean weights were 0.049–0.139 g. An analysis of the growth of the fry suggested that protection against IPNV required immunization during the time of slow weight gain. These authors analyzed the weight and age data for rainbow trout provided by Tatner and Horne (1983, 1984) and found that the fry were protected only if they were immunized at the time of or after the transition to a faster rate of weight gain. Thus, fish weight gain rates appear to be important determinants for immune protection against both IPNV and *V. anguillarum* but in contrasting ways. The minimum weight and age requirement necessary for induction of protection against viruses may be lower than that needed for protection against bacteria, but further studies are required to confirm whether this is a general principle.

It would be very useful if all future studies included data not only on fish age but fish size (weight and length), their growth rate, and the temperature they were maintained at. This would allow valid comparisons to be made between immunization trials using fry, but most importantly, the data could be used in models, such as Solomon's age equivalence model (1978), to enable predictions to be made on the onset of immunocompetence in fry against various pathogens (bacterial, viral, or parasitic), which would be of value in commercial aquaculture.

VIII. AGING EFFECTS ON THE IMMUNE RESPONSE

Very little is known about the effects of aging on the immune response in fish. Most experimental studies use "adult" fish, but their exact age is rarely specified. Although the average body weight is often reported, this correlates poorly with age; the fork length of the fish would serve as a better indicator of their age.

Different fish species have widely varying life spans, from only a year for some tropical species, to several years for salmonids, to many decades as reported for sharks. So it is reasonable to expect that there will be age-related changes in their immune responses.

In mammals, it is generally accepted that with age there is a decline in immune response against exogenous antigens, with an increase in the capacity for immunological self recognition, leading to an increase in autoimmune responses (Wick, 1994). As with the ontogeny of immunity, different aspects of the immune system decline at different rates, for example, in mice the systemic immune functions decline faster than the gut-associated responses (Koyama *et al.*, 1990). In the amphibian *Bufo viridis,* a decline in rosette-forming cells and plaque-forming cell responses, with lower antibody titers was seen in aged toads compared to young ones. This decline was coincident with high levels of thyroxine (T4), prolactin, and cortisol (Saad *et al.,* 1994), reflecting the interaction between the immune and endocrine systems in cold-blooded vertebrates (Maule *et al.,* 1994).

Some indication as to the consequences of aging of the immune responses in fish can be gleaned from adult thymectomy experiments. The thymus does involute in most species, though at various stages in their life cycles, and this could be expected to have an effect on their subsequent immune responses. It was thought that the thymus was important only during the ontogeny of the immune system as the primary source of T cells, but it continues to function in adult life as the replacement source for T cells that die off in the periphery (Metcalf, 1965). Emigration of radiolabeled thymocytes to the peripheral lymphoid organs in rainbow trout has been demonstrated after the age when the initial thymic seeding during ontogeny has occurred (Tatner, 1985), and this was interpreted as a "replenishment" process. However, nothing at all is known about the life span of fish lymphocytes in the peripheral lymphoid organs or in the circulation.

If we consider adult thymectomy to be the equivalent of thymic involution, then what are the consequences of this on the fishes' immune capabilities? The few data available on adult thymectomy in fish appear contradictory at first, but this is mainly due to differences in the time interval between thymectomy and subsequent testing, which is crucial to the outcome observed. Hence, short-term thymectomy will remove only T cells resident in the thymus, whereas long-term thymectomy (i.e., a long time interval between thymectomy and testing) will allow any T cells in the periphery to die off and not be replaced.

In adult carp, which were thymectomized and then immunized four weeks later with *A. salmonicida,* elevated antibody levels were seen at day 7 compared to sham thymectomized fish. These returned to normal levels later in the response. This was taken as tentative evidence for the existence of T suppressor cells in the thymus (Secombes, 1981).

In the marine rockfish, Nakanishi (1986b) used x-irradiation and thymectomy to show that the adult thymus played an important role in the recovery from damage to the immune system caused by the irradiation. Fish that were irradiated only had a complete suppression of the antibody response to sheep red blood cells, and their allograft rejection times were prolonged by a factor of three. Fish that were irradiated 4 days after thymectomy and injected 1 week after irradiation showed a fairly high level of antibody. Animals that were thymectomized, irradiated, but then thymus autoimplanted showed higher levels of antibody when immunized 3 months after irradiation than matched controls. These results strongly suggest a restorative role for the adult thymus in fish, and the combined x-irradiation and thymectomy experiments indicated the presence of X-ray resistant T suppressor cells within the adult thymus gland.

In adult rainbow trout, Tatner *et al.* (1987) found no difference in antibody titer to HGG between control and thymectomized fish 5 months after thymectomy, but after 9 months, the thymectomized fish had lower titers than the controls. This was suggested as being due to the decline in T helper cells in the periphery, required for the response to this T-dependent antigen. No differences were detected in the response to *A. salmonicida* which is T independent.

It is probable that as fish age, there will be a decline in both the nonspecific and specific immune responses, with the thymus-dependent and thymus-independent arms declining differentially. There may be a loss of T-cell suppression especially. There are reports of tumors in older fish (Mulcahy, 1970) and fish can generate autoimmune responses (Laird *et al.*, 1978). Hence, in the future, researchers would be circumspect to look for evidence of these and to determine if the thymus is still present (a histological confirmation is required) when using adult fish in their experiments.

IX. OTHER NATURAL FACTORS AFFECTING THE IMMUNE RESPONSE IN FISH

Apart from ontogeny and aging, there are several other natural factors that affect the immune responses in fish. Stress, induced by social conflict, has been shown to induce both structural and functional alterations in the phagocytes of rainbow trout. The phagocytes showed signs of activation and an increase in phagocytosis rate, though a considerable percentage of the phagocytes showed signs of degeneration in the subordinate fish (Peters *et al.*, 1991). A suppression of natural cytotoxic cell activity, induced by social aggressiveness, has been demonstrated in tilapia (Ghoneum *et al.*, 1988). In species that organize themselves into dominance heirarchies, only the dominant individuals produced antibodies against infection with trypanosomes (Barrow, 1955).

Acute stress (being held out of water for 30 s, hatchery manipulation, or transportation) caused a depression in the number of antibody-producing cells in the kidney of chinook salmon, and an increase in susceptibility to *V. anguillarum* (Maule *et al.,* 1989). In the channel catfish, Ellsaesser and Clem (1986) found a marked lymphopenia and an inability to respond to mitogen stimulation or to undergo primary anti-hapten antibody responses to either T-dependent or T-independent antigens following stress induced by handling and transport. Crowding in the blue gourami (*Trichogaster trichopterus*) markedly reduced the immune response to IPNV (Perlmutter *et al.,* 1973). This was thought to be due to a pheromone-like substance released into the water, because if the water was filtered with methyl chloroform, the suppression was no longer apparent. Miller and Tripp (1982a) noted that the killifish, *Fundulus heteroclitus,* had a lower rosette response to thymus-dependent antigens after having been held in captivity for 4 weeks, compared to newly caught fish. There was no effect on scale allograft rejection or on the response to thymus-independent antigens, suggesting that the suppression was directed specifically to a subpopulation of cells analogous to T helper cells. The immune inhibitory substance was present in the serum of the captive fish (Miller and Tripp, 1982b) and could exert a suppressive effect when injected into control fish 48 h prior to antigen injection, but not if injected simultaneously. The inhibitory substance was a low or very low density lipoprotein, and its production was under the control of the pituitary gland. These reports illustrate the interaction between the immune system and the endocrine system and emphasize the point made at the beginning of the chapter that the immune system cannot be viewed in isolation but as one of the many, interacting physiological processes in the biology of the fish.

Hormones play a major role in the parr-smolt transformation of salmonids, with direct consequences on their immune system. During smoltification in the coho salmon, Maule *et al.* (1987) noted a decrease in the number of plaque-forming cells after injection with *Vibrio anguillarum* and a decrease in the number of leukocytes. These changes were probably due to the direct action of the elevated levels of cortisol at this time. Recently, Meligen *et al.* (1995a) have observed a decrease in total serum protein and IgM levels during the parr-smolt transformation in Atlantic salmon, *Salmo salar,* which returned to the presmolt levels after seawater transfer. Vaccination against *V. salmonicida* during the smolting period (May to July) resulted in lower antibody levels 6 months later compared to fish vaccinated earlier in the year, even though these fish were vaccinated at lower water temperatures (Meligen *et al.,* 1995b). There are other single reports in the literature of natural effects on the immune system of fish. For example, during the metamorphosis of the larval anadromous sea lamprey, *Petromyzon marinus,* there is a complete degeneration of the larval opisthonephros and typhlosole, which lose their hemopoietic capacity (Ardavin and Zapata, 1987).

A circadian rhythm of allograft rejection was discovered in the gulf killifish, *Fundulus gratis,* with rejection occurring predominantly at night (Nevid and Meier, 1993). Seasonal changes have been noted in the antibody response of juvenile summer flounder (*Paralichthys dentatus*) to the hemoflagellate *Trypanoplasma bullocki,* it being greatest in the summer, (Burreson and Frizzell, 1986), and in *Sebastiscus marmoratus,* Nakanishi (1986c) found that the antibody response to SRBC was greatest in the summer. Lopez-Fierro *et al.* (1994) reported a marked depression in a number of immune parameters in farmed rainbow trout during the winter months (lower mitogen responses, lower serum IgM levels, and lower levels of agglutinating antibodies to *Y. ruckeri*). These effects are not simply due to lower environmental temperatures during the winter because Yamaguchi *et al.* (1980) found lower antibody production in rainbow trout against *A. salmonicida* when the fish were immunized in the autumn, compared to those immunized in the spring, even when the temperature was held at a constant 18°C throughout the year.

Sexual maturation and breeding can also have an effect on the immune responses of fish, not only by the action, direct or indirect, of the hormonal effects (Sufi *et al.,* 1980) but also in terms of the partioning of energy resources, as mentioned earlier. Skarstein and Folstad (1994) have suggested that the development and maintenance of carotenoid-based secondary sexual traits in Arctic char, *Salvelinus alpinus,* may have costs with regard to immunity. The immunocompetence handicap hypothesis states that such traits may be used in mate choice for heritable parasite resistance. The carotenoid-based coloration of the char was found to be related to lymphocyte counts. There was a positive relationship between the levels of circulating lymphocytes and the intensities of the pathogenic parasite, *C. trunctatus,* suggesting that only individuals with low intensities of this parasite could afford to pay the immunological costs of ornamentation, signaling their resistance to the parasite.

X. CONCLUSIONS AND SUMMARY

The immune system of fish, as one aspect of their physiology, is affected by the many natural factors they encounter both internally and externally. During ontogeny there is a sequential development of lymphoid organs and immune responses, with nonspecific immunity developing first followed by cell-mediated and then humoral immunity. With humoral immunity, T-independent responses develop earlier than T-dependent ones. This pattern in ontogeny probably reflects the phylogenetic appearance of immunity in vertebrates. During aging the pattern is repeated, with the different facets of immunity declining at different rates. Does the involution of the thymus signal the start of this decline in adult fish? Not enough is known about this aspect as yet, though it is of great interest and fish may well prove to be excellent experimental models for the study of immunogerontology in vertebrates.

Other factors can also affect the immune responses that we seek to measure in the laboratory, such as stress levels, hormonal status, breeding condition, diet, temperature, and the time of year experiments are performed. More complete information on the experimental conditions the fish are maintained in, and on the fish themselves (sex, age, weight, fork length, and growth rate in ontogeny studies) would allow for more valid comparison to be made between species and between studies.

With the increased demand for more applied studies in fish immunology, such as for vaccination trials and for use as bioindicators in ecotoxicology, a standardization of methodology and reporting would be of great value. For example, with more complete information from ontogeny studies, it may be possible to describe a model for fish similar to Solomon's age-equivalence theory (1978), such that all species of fish may be shown to develop immunocompetence at the same physiological age. This would be very useful in predicting the age at which fry would become immunocompetent to various antigens, and hence facilitate their early, safe vaccination. Not enough data is available at the present time to even begin to construct such a model, but hopefully such will become available in the future.

REFERENCES

Ardavin, C. F., and Zapata, A. (1987). Ultrastructure and changes during metamorphosis of the lymphopoeitic tissue of the larval anadromous sea lamphrey, *Petromyzon marinus. Dev. Comp. Immunol.* **11,** 79–93.

Barrow, J. H. (1955). Social behavior in freshwater fish and its effect on resistance to trypanosomes. *Proc. Nat. Acad. Sci.* USA **41,** 676–679.

Bly, J. E. (1984). The ontogeny of immunity in teleost fishes with particular reference to foeto-maternal relationships. Ph.D. Thesis, University College of North Wales, Bangor, UK.

Bly, J. E. (1985). The ontogeny of the immune system in the viviparous teleost *Zoarces viviparous* L. *In* "Fish Immunology" (M. J. Manning and M. F. Tatner, eds.), pp. 327–342. Academic Press, London.

Bly, J. E., and Clem, L. W. (1992). Temperature and teleost immune functions. *Fish Shellfish Immunol.* **2,** 159–172.

Bly, J. E., Grimm, A. S., and Morris, I. G. (1986). Transfer of passive immunity from mother to young in a teleost fish: Hemagglutinating activity in the serum and eggs of plaice, *Pleuronectes platessa* L. *Comp. Biochem. Physiol.* **84A,** 309–313.

Bootland, L. M., Dobos, P., and Stevenson, R. M. W. (1990). Fry age and size effects on immersion immunization of brook trout, *Salvelinus fontinalis* Mitchell, against infectious pancreatic necrosis virus. *J. Fish Dis.* **13,** 113–125.

Botham, J. W., and Manning, M. J. (1981). Histogenesis of the lymphoid organs in the carp, *Cyprinus carpio* L., and the ontogenetic development of allograft reactivity. *J. Fish Biol.* **19,** 403–414.

Brown, L. L., Evelyn, T. P. T., and Iwama, G. K. (1994). On the egg-mediated transfer of passive immunity from coho salmon to their progeny. *6th ISDCI Congress, Dev. Comp. Immunol. Suppl. 1* **18,** 596.

Burreson, E. M., and Frizzell, L. J. (1986). The seasonal antibody response in juvenile summer flounder (*Paralichthys dentatus*) to the hemoflagellate (*Trypanosoma bullocki*). *Vet. Immunol. Immunopathol.* **12,** 395–402.

Castillo, A., Sanchez, C., Domiguez, J., Kaattari, S. L., and Villena, A. J. (1993). Ontogeny of sIgM- and cIgM-bearing cells in rainbow trout. *Dev. Comp. Immunol.* **17,** 419–424.

Chilmonczyk, S. (1992). The thymus in fish: Development and possible function in the immune response. *Annu. Rev. Fish Dis.* **1,** 181–200.

Cooper, E. L., Zapata, A., Garcia Barrutia, M., and Rameriz, J. A. (1983). Aging changes in lymphopoeitic and myelopoeitic organs of the annual cyprinodont fish, *Notobranchius guentheri. Exp. Gerontol.* **18,** 29–38.

Doggett, T. A., and Harris, J. E. (1987). The ontogeny of gut-associated lymphoid tissue in *Oreochromis mossambicus. J. Fish Biol.* **31** (Suppl. A), 23–27.

Dorson, M. (1977). Vaccination trials of rainbow trout against infectious pancreatic necrosis. *Bull. Off. Int. Epizoot.* **87,** 405–406.

Dorson, M., and Torchy, C. (1981). The influence of fish age and water temperature on mortalities of rainbow trout, *Salmo gairdneri* Richardson, caused by a European strain of infectious pancreatic necrosis virus. *J. Fish Dis.* **4,** 213–221.

Dorson, M., de Kinkelin, P., and Torchy, C. (1992). Interferon synthesis in rainbow trout following infection with infectious pancreatic necrosis virus. *Fish Shellfish Immunol.* **2,** 311–314.

Du Pasquier, L. (1973). Ontogeny of the immune response in cold blooded vertebrates. *In* "Current Topics in Microbiology and Immunology," Vol. 61, pp. 38–80, Springer-Verlag, Berlin.

El Deeb, S. O., and Saad, A. H. M. (1990). Ontogenic maturation of the immune system in reptiles. *Dev. Comp. Immunol.* **14,** 151–159.

Ellis, A. E. (1977). Ontogeny of the immune response in *Salmo salar.* Histogenesis of the lymphoid organs and appearance of membrane immunoglobulin and mixed leucocyte reactivity. *In* "Developmental Immunobiology" (J. B. Solomon and J. D. Horton, eds.), pp. 225–231. Elsevier/North Holland Biomedical Press.

Ellis, A. E. (1988). Ontogeny of the immune system in teleost fish. *In*: "Fish Vaccination" (A. E. Ellis, ed.), pp. 20–31, Academic Press, London.

Ellsaesser, C. F., and Clem, L. W. (1986). Hematological and immunological changes in channel catfish stressed by handling and transport. *J. Fish Biol.* **28,** 511–521.

Etlinger, H. M., Chiller, J. M., and Hodgins, H. O. (1979). Evolution of the lymphoid system. IV. Murine T-independent but not T-dependent antigens are very immunogenic in rainbow trout (*Salmo gairdneri*). *Cell. Immunol.* **47,** 400–406.

Fuda, H., Hara, A., Yamazaki, F., and Kobayashi, K. (1992). A peculiar immunoglobulin M (IgM) identified in eggs of chum salmon (*Oncorhynchus keta*). *Dev. Comp. Immunol.* **16,** 415–423.

Ghoneum, M., Faisal, M., Peters, G., Ahmed, I. I., and Cooper, E. L. (1988). Suppression of natural cytotoxic cell activity by social agressiveness in *Tilapia. Dev. Comp Immunol.* **12,** 595–602.

Grace, M. F. (1981). The functional histogenesis of the immune system in rainbow trout, *Salmo gairdneri* Richardson 1836. Ph.D. Thesis, University of Hull, UK.

Grace, M. F., and Manning, J. J. (1980). Histogenesis of the lymphoid organs in rainbow trout, *Salmo gairdneri* Rich. 1836. *Dev. Comp. Immunol.* **4,** 255–264.

Grace, M. F., Botham, J. W., and Manning, M. J. (1981). Ontogeny g lymphoid organ function in fish. *In*: "Aspects of Developmental and Comparative Immunology" (J. B. Solomon, ed.), vol 1, pp. 467–468, Pergamon Press, Oxford.

Hart, S., Wrathmell, A. B., and Harris, J. E. (1986). Ontogeny of gut-associated lymphoid tissue (GALT) in the dogfish, *Scyliorhinus canicula*. *Vet. Immunol. Immunopathol.* **12,** 107–116.

Hayman, J. R., and Lobb, C. J. (1993). Immunoglobulin in the eggs of the channel catfish (*Ictalurus punctatus*). *Dev. Comp. Immunol.* **17,** 241–248.

Honma, Y., and Tamura, E. (1984). Histological changes in the lymphoid system of fish with respect to age, seasonal, and endocrine changes. *Dev. Comp. Immunol. Suppl.* **3,** 239–244.

Horton, J. D. (1969). Ontogeny of the immune response to skin allografts in relation to lymphoid organ development in the amphibian *Xenopus laevis* Daudin. *J. Exp. Zool.* **170,** 449–458.

Ingram, G. A. (1980). Substances involved in the natural resistance of fish to infection: A review. *J. Fish Biol.* **16,** 23–60.

Johnson, K. A., Flynn, J. K., and Amend, D. F. (1982a). Onset of immunity in salmonid fry vaccinated by direct immersion in *Vibrio anguillarum* and *Yersinia ruckeri* bacterins. *J. Fish Dis.* **5,** 197–206.

Johnson, K. A., Flynn, J. K., and Amend, D. F. (1982b). Duration of immunity in salmonids vaccinated by direct immersion with *Yersinia ruckeri* and *Vibrio anguillarum* bacterins. *J. Fish Dis.* **5,** 207–214.

Josefsson, S., and Tatner, M. F. (1993). Histogenesis of the lymphoid organs in sea bream, *Sparus aurata* L. *Fish Shellfish Immunol.* **3,** 35–50.

Khalifa, K. A., and Post, G. (1976). Immune response of advanced rainbow trout to *Aeromonas liquefaciens*. *Prog. Fish Cult.* **38,** 66–68.

Kobayashi, K., Tomonaga, S., Teshima, K., and Kajii, T. (1985). Ontogenic studies on the appearance of two classes of immunoglobulin forming cells in the spleen of the Aleutian skate, *Bathyraja aleutica,* a cartilaginous fish. *Eur. J. Immunol.* **15,** 952–956.

Koumans-van Diepen, J. C. E., Taverne-Thiele, J. C. E., van Rens, B. T. T. M., and Rombout, J. H. W. M. (1994). Immunocytochemical and flow cytometric analysis of B cells and plasma cells in carp, *Cyprinus carpio* L.: An ontogenetic study. *Fish Shellfish Immunol.* **4,** 19–28.

Koyama, K., Hosokawa, T., and Aoike, A. (1990). Aging effect on the immune functions of murine gut-associated lymphoid tissues. *Dev. Comp. Immunol.* **4,** 465–473.

Kusuda, R., Ikemoto, M., and Enzan, H. (1992). Ontogeny of the thymus in red sea bream, *Pagrus major*. Abstracts of the Second Scientific Meeting of the Japanese Association for Developmental and Comparative Immunology. *Dev. Comp. Immunol.* **16,** VI.

Laird, M. L., Ellis, A. E., Wilson, W. R., and Halliday, F. G. T. (1978). The development of the gonadal and immune systems in the Atlantic salmon (*Salmo salar* L.) and a consideration of the possibility of inducing autoimmune destruction of the testis. *Ann. Biol. Anim. Biochem. Biophys.* **18,** 1101–1106.

Lele, S. H. (1934). On the phasical morphology of the thymus gland in some common European fishes and in two cyclostomes. *J. Univ. Bombay* **2,** 33034.

Lloyd-Evans, P. (1993). Development of the lymphomyeloid system in the dogfish, *Scyliorhinus canicula*. *Dev. Comp. Immunol.* **17,** 501–514.

Lopez-Fierro, P., Razquin, B., Alvarez, F., Alonso, J., and Villena, A. (1994). Seasonal changes in the immune system of rainbow trout. *Dev. Comp. Immunol. Suppl.* **18,** S146, Abstract.

Manning, M. J., and Mughal, M. S. (1985). Factors affecting the immune responses of immature fish. *In* "Fish and Shellfish Pathology" (A. E. Ellis, ed.), pp. 27–40. Academic Press, London.

Manning, M. J., Grace, M. F., and Secombes, C. J. (1982a). Developmental aspects of immunity and tolerance in fish. *In* "Microbial Diseases of Fish" R. J. Roberts, ed., pp. 31–46. Academic Press, London.

Manning, M. J., Grace, M. F., and Secombes, C. J. (1982b). Ontogenetic aspects of tolerance and immunity in carp and rainbow trout: Studies on the role of the thymus. *Dev. Comp. Immunol. Suppl.* **2,** 75–82.

Maule, A. G., Schreck, C. B., and Kaattari, S. (1987). Changes in the immune system of coho salmon, *Oncorhynchus kisutch,* during the parr-to-smolt transformation and after implantation of cortisol. *Can. J. Fish. Aquat. Sci.* **44,** 161–166.

Maule, A. G., Tripp, R. A., Kaattari, S. L., and Schreck, C. B. (1989). Stress alters immune function and disease resistance in chinook salmon (*Oncorhynchus tshawytscha*). *J. Endocrinol.* **120,** 135–142.

Maule, A. G., Schrock, R. M., and Fitzpatrick, M. S. (1994). Immune–endocrine interactions during final maturation and senescence of spring chinook salmon. *Dev. Comp. Immunol. Suppl. 1* **18,** 569, Abstract.

McKinney, E. C., and Schmale, M. C. (1993). Immunocompetence of juvenile damselfish. *Fish Shellfish Immunol.* **3,** 395–396.

Meligen, G. O., Stefansson, S. O., Berg, A., and Wergeland, H. I. (1995a). Changes in serum protein and IgM concentration during smolting and early postsmolt period in vaccinated and unvaccinated Atlantic salmon (*Salmo salar* L). *Fish Shellfish Immunol.* **5,** 211–222.

Meligen, G. O., Nilsen, F., and Wergeland, H. I. (1995b). The serum antibody levels in atlantic salmon (*Salmo salar*) after vaccination with *Vibrio salmonicida* at different times during the smolting and early postsmolt period. *Fish Shellfish Immunol.* **5,** 223–236.

Metcalf, D. (1965). Delayed effect of thymectomy in adult life on immunological competence. Nature (London) **208,** 1336–1338.

Miller, N. W., and Tripp, M. R. (1982a). The effect of capitivity on the immune response of the killifish, *Fundulus heteroclitus* L. *J. Fish Biol.* **20,** 301–308.

Miller, N. W., and Tripp, M. R. (1982b). An immunoinhibitory substance in the serum of laboratory held killifish, *Fundulus heteroclitus* L. *J. Fish Biol.* **20,** 309–316.

Mor, A., and Avtalion, R. R. (1989). Transfer of antibody activity from immunised mother to embryo in tilapias. *J. Fish Biol.* **37,** 249–254.

Mughal, M. S., and Manning, M. J. (1984). Antibody responses of young carp, *Cyprinus carpio,* and grey mullet, *Crenimugil labrosus,* immunised with soluble antigens by various routes. *In* "Fish Immunology" (M. J. Manning and M. F. Tatner, eds.), pp. 313–325. Academic Press, London.

Mughal, M. S., and Manning, M. J. (1986). The immune system of juvenile thick lipped grey mullet, *Chelon labrosus* Risso: Antibody responses to soluble protein antigens. *J. Fish Biol.* **29,** 177–186.

Mulcahy, M. F. (1970). The thymus glands and lymphosarcoma in the pike, *Esox lucius* L. (Pisces, Esocidae), in Ireland. *In* "Comparative Leukemia Research 1969" (R. M. Dutcher, ed.), Bibliography of Haematology, 36, pp. 600–609. Karger, Basel.

Nakanishi, T. (1983). Studies on the role of the lymphoid organs in the immune responses of the marine teleost, *Sebastiscus marmoratus.* Ph.D. Thesis, University of Hokkaido, Japan.

Nakanishi, T. (1986a). Ontogenetic development of the immune response in the marine teleost, *Sebastiscus marmoratus. Bull. Jpn. Soc. Sci. Fish.* **52,** 473–477.

Nakanishi, T. (1986b). Effects of x-irradiation and thymectomy on the immune responses of the marine teleost, *Sebasticus marmoratus. Dev. Comp. Immunol.* **10,** 519–528.

Nakanishi, T. (1986c). Seasonal changes in the humoral immune responses and lymphoid tissue of the marine teleost, *Sebastiscus marmoratus. Vet. Immunol. Immunopathol.* **12,** 213–221.

Nakanishi, T. (1991). Ontogeny of the immune system in *Sebastiscus marmoratus:* Histogenesis of the lymphoid organs and effects of thymectomy. *Environ. Biol. Fish.* **30,** 135–145.

Nevid, N. J., and Meier, A. H. (1993). A day–night rhythm of immune activity during scale allograft rejection in the gulf killifish, *Fundulus grandis. Dev. Comp. Immunol.* **17,** 221–228.

O'Neill, J. G. (1989). Ontogeny of the lymphoid organs in an antartic teleost, *Harpagifer antarcticus* (Nototheniodei: Perciformes). *Dev. Comp. Immunol.* **13,** 25–34.

Perlmutter, A., Sarot, D. A., Yu, M. L., Filazzola, R. J., and Seeley, R. J. (1973). The effect of crowding on the immune response of the blue gowami *Trichogaster trichopterus* to infectious pancreatic necrosis (IPN) virus. *Life Sciences.* **13,** 363–375.

Peters, G., Nubgen, A., Roobe, A., and Mock, A. (1991). Social stress induces structural and functional alterations of phagocytes in rainbow trout, *Oncorhynchus mykiss. Fish Shellfish Immunol.* **1,** 17–32.

Plytcycz, B., Mika, J., Jozkowicz, A., and Bigaj, J. (1994). Age-dependent changes of the thymuses of *Rana temporaria. 6th ISDC Congress, Dev. Comp. Immunol. Suppl.* **18,** 597, Abstract.

Pulsford, A., Morrow, W. J. W., and Fange, R. (1984). Structural studies on the thymus of the dogfish, *Scylorhinus canicula. J. Fish Biol.* **25,** 353–360.

Pulsford, A., Tomlinson, M. G., Lemaire-Gony, S., and Glynn, P. J. (1994). Development and immunocompetence of juvenile flounder, *Platichthys flesus* L. *Fish Shellfish Immunol.* **4,** 63–78.

Razquin, B. E., Castillo, A., Lopez-Fierro, P., Alvarez, F., Zapata, A., and Villena, A. (1990). Ontogeny of IgM-producing cells in the lymphoid organs of rainbow trout, *Salmo gairdneri* Richardson: An immuno-and enzyme histochemical study. *J. Fish Biol.* **36,** 159–174.

Rijkers, G. T., and van Muiswinkel, W. B. (1977). The immune system of cyprinid fish. The development of cellular and humoral responsiveness in the rosy barb, *Barbus conchonius. In* "Developmental Immunobiology" (J. B. Solomon and J. D. Horton, eds.), pp. 233–240. Elsevier/North Holland Biomedical Press.

Saad, A. H., Mansour, M. H., Dorgham, V., and Badir, N. (1994). Age-related changes in the immune response of *Bufo viridis. Dev. Comp. Immunol. Suppl.* **18,** 597, Abstract.

Sailendri, K. (1973). Studies on the development of lymphoid organs and immune responses in the teleost, *Tilapia mossambicus* (Peters). Ph.D. Thesis, University of Madurai, India.

Scapigliati, G., Mazzini, M., Mastrolia, Romana, N., and Abelli, L. (1995). Production and characterisation of a monoclonal antibody directed against the thymocytes of the sea bass, *Dicentrarchus labrax* (L) (Teleostea, Percithydae). *Fish Shellfish Immunol.* **5,** 393–406.

Schneider, B. (1983). Ontogeny of fish lymphoid organs. *Dev. Comp. Immunol.* **7,** 739–740.

Secombes, C. J. (1981). Comparative studies on the structure and function of teleost lymphoid organs. Ph.D. Thesis, University of Hull, UK.

Secombes, C. J., van Groningen, J. J. M., van Muiswinkel, W. B., and Egberts, E. (1983). Ontogeny of the immune system in carp, *Cyprinus carpio* L.: The appearance of antigenic determinants on lymphoid cells detected by mouse anti-carp thymocyte monoclonal antibodies. *Dev. Comp. Immunol.* **7,** 455–464.

Sin, Y. M., Ling, K. H., and Lam, T. T. (1994). Passive transfer of protective immunity against Ichthyophthiriasis from vaccinated mother to fry in tilapias, *Oreochromis aureus. Aquaculture* **120,** 229–237.

Skarstein, F., and Folstad, I. (1994). An observational approach to the immunocompetence handicap hypothesis. *Dev. Comp. Immunol. Suppl.* **18,** 147, Abstract.

Solomon, J. B. (1978). Immunological milestones ontogeny. *Dev. Comp. Immunol.* **2,** 409–424.

Subasinghe, R. P. (1993). Effects of immunosuppression on protection of fry from *Ichthyophirius multifiliis* Fouguet 1876 during mouthbrooding of *Oreochromis mossambicus* (Peters). *Fish Shellfish Immunol.* **3,** 97–106.

Sufi, G. B., Mori, K., and Nomura, I. (1980). Involution of the thymus in relation to sexual maturity and steroid hormone treatments in salmonid fish. *Tohoku J. Agricult. Res.* **31,** 97–105.

Takahashi, Y., and Kawahara, E. (1987). Maternal immunity in newborn fry of the ovoviparous guppy. *Nip. Suisan Gakkaishi* **53,** 721–735.

Takemura, A. (1993). Changes in an immunoglobulin (IgM) like protein during larval stages in tilapia, *Oreochromis mossambicus. Aquaculture* **115,** 233–241.

Tatner, M. F. (1985). The migration of labeled thymocytes to the peripheral lymphoid organs in rainbow trout, *Salmo gairdneri* Richardson. *Dev. Comp. Immunol.* **9,** 85–91.

Tatner, M. F. (1986). The ontogeny of humoral immunity in rainbow trout, *Salmo gairdneri. Vet. Immunol. Immunopathol.* **12,** 93–105.

Tatner, M. F., and Horne, M. T. (1983). Susceptibility and immunity to *Vibrio anguillarum* in post-hatching rainbow trout fry, *Salmo gairdneri* Richardson 1836. *Dev. Comp. Immunol.* **7,** 465–472.

Tatner, M. F., and Horne, M. T. (1984). The effects of early exposure to *Vibrio anguillarum* vaccine on the immune response of the fry of the rainbow trout, *Salmo gairdneri* Richardson. *Aquaculture,* **41,** 193–202.

Tatner, M. F., and Manning, M. J. (1983a). The ontogeny of cellular immunity in the rainbow trout, *Salmo gairdneri* Richardson, in relation to the stage of development of the lymphoid organs. *Dev. Comp. Immunol.* **7,** 69–75.

Tatner, M. F., and Manning, M. J. (1983b). Growth of the lymphoid organs in rainbow trout, *Salmo gairdneri,* from one to fifteen months of age. *J. Zool.* (London) **199,** 503–520.

Tatner, M. F., and Manning, M. J. (1985). The ontogenetic development of the reticulo endothelial system in the rainbow trout, *Salmo gairdneri* Richardson. *J. Fish Dis.* **8,** 35–41.

Tatner, M. F., Adams, A., and Leschen, W. (1987). An analysis of the primary and secondary antibody responses in intact and thymectomised rainbow trout, *Salmo gairdneri* Richardson, to human gamma globulin and *Aeromonas salmonicida.* J. Fish Biol. **31,** 177–195.

Thorburn, M. A., and Jansson, E. L. K. (1988). The effects of booster vaccination and fish size on survival and antibody production following *Vibrio* infection of bath vaccinated rainbow trout, *Salmo gairdneri. Aquaculture.* **71,** 285–291.

van Loon, J. J. A., van Oosterom, R., and van Muiswinkel, W. B. (1981). Development of the immune system in carp, *Cyprinus carpio* L. *In* "Aspects of Developmental and Comparative Immunology" (J. B. Solomon, ed.), Vol. 1, pp. 469–470. Pergamon Press, Oxford.

van Loon, J. J. A., Secombes, C. J., Egberts, E., and van Muiswinkel, W. B. (1982). Ontogeny of the immune system in fish: Role of the thymus. *Adv. Exp. Med. and Biol.* **149,** 335–341.

Von Hagen, F. (1936). Die wichtigsten Endokrinen des Fluddals. *Zool. Jab Anat.* **61,** 467–538.

Wick, G. (1994). Aging of the immune response. *Dev. Comp. Immunol. Suppl.* **18,** 591, Abstract.

Wishkovsky, A., Garber, N., and Avtalion, R. R. (1987). The effects of fish age on the mortalitycaused by selected fish pathogens. *J. Fish Biol.* **31 Suppl. A,** 243–244.

Yamaguchi, N., Teshima, C., Kurashige, S., Saito, R., and Mitsuhashi, S. (1980). Seasonal modulation of antibody formation in rainbow trout, *Salmo gairdneri. In* "Aspects of Developmental and Comparative Immunology" (J. B. Solomon, ed.), Vol. 1, pp. 483–484. Pergamon Press, Oxford.

7

ENVIRONMENTAL FACTORS IN FISH HEALTH: IMMUNOLOGICAL ASPECTS

DOUGLAS P. ANDERSON

I. INTRODUCTION

Fish live closely with their aquatic environment; living epidermal cell membranes have direct contact with all materials carried by water. This intimate contact eases the movement of chemicals into and through the mucus, skin, and other external layers and becomes a disadvantage to the fish when nefarious chemicals, pollutants, and contaminants enter the aquatic environment (Adams 1990, Wester *et al.*, 1994). These chemicals can have adverse affects on the fishes' physiological pathways, including those important mechanisms that help protect the fish against diseases: the nonspecific defense mechanisms and the specific immune responses. Without complete immune protection and barriers against viral, bacterial, fungal, and protozoan pathogens that are ubiquitous in the environment, the animal is disadvantaged and becomes susceptible to disease-causing agents. The specific immune system is also important in surveillance and destruction of errant cells such as those involved in tumors or cancer.

THE FISH IMMUNE SYSTEM:
ORGANISM, PATHOGEN, AND ENVIRONMENT

Likewise, the healthy immune system recognizes these as foreign tissues or antigens and destroys them or suppresses further growth.

Immunologists are now studying the effects of immunomodulators, including suppressors and stimulators, on the immune protection in fish, and their putative points of action (Fig. 1). Many recently published scientific research papers show the effects of many different substances on fish, usually acting in suppressive modes (see Zelikoff, 1993; Dunier, 1994; Zelikoff, 1994; Anderson and Zeeman, 1995). The exact mechanisms of action of most of these substances on the defensive systems remain elusive because they are often difficult to trace or pinpoint in the metabolic pathways. Indeed, the contaminating substances often affect other physiological pathways as well as those of the immune system. The effects of doses, temperature, species, and other variables make this a difficult study.

In most cases, we are concerned about the damaging effects of released chemicals on the environment and fish health and immunosuppression. Another side of immunomodulation is immunostimulation, and indeed, with the new techniques of immunostimulation for prevention and treatment of fish diseases, this area of study is also becoming very important in fish culture. In the following chapter we review and comment on the aspects of the nonspecific defense mechanisms and the specific immune response and how they are dependent upon environmental influences.

II. IMMUNOASSAYS

Many of the assays for detecting the changes in the protective mechanisms of fish due to immunomodulations are derived from those used in fish disease diagnostics and immunization programs. Most of the commonly used assays for fish immunomodulation diagnosis are listed in Table I. These important assays are for determining changes in such areas as basic blood characteristics (hematology), the nonspecific defensive mechanisms, and the specific immune response.

Blood monitoring is one of the most popular areas for assays because the animal need not be sacrificed. Simple hematocrit and leukocrit readings give levels or percentages of erythrocytes and leukocytes in the blood, and blood smears on glass microscope slides can be used for determining numbers and/or ratios of leukocytes, or in some cases, subpopulations of these cells. Any aberrations in cell morphology can also be noted during inspection by microscope. While the hemological assays are generally less sensitive as bioindicators of immunomodulation, these assays are easy to do and technical training and laboratory requirements are minimal. Another drawback of hemological assays, especially for fish work, is their wide

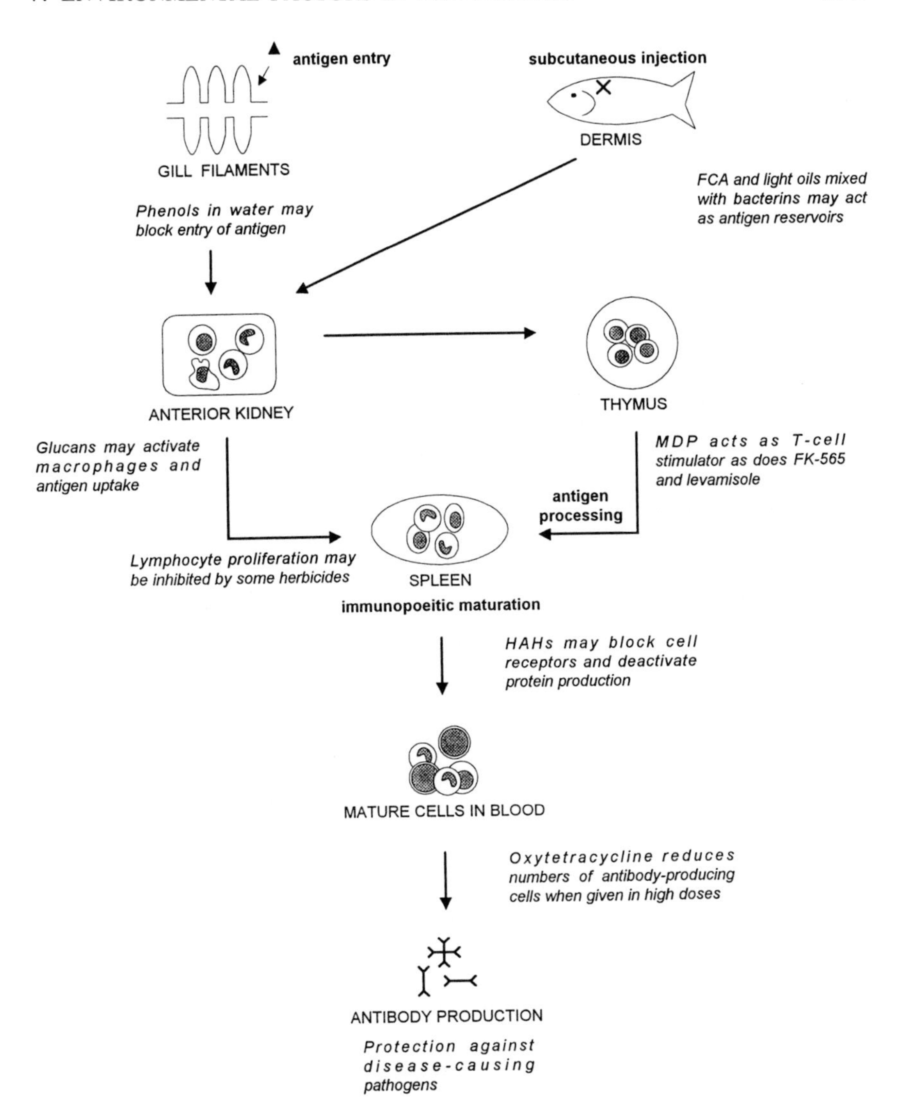

Fig. 1. The schematic shows some of the potential immunomodulators in the aquatic environment that can influence the specific immune response and nonspecific defense mechanisms. Treating fish held in hatcheries with heavy doses of antibiotics to prevent diseases may have adverse effects on the phagocytic cells which are important in the destruction of invasive microorganisms. Heavy metals, common contaminants from industrial processes, in very low doses may actually in some cases have a stimulating effect on the defensive mechanisms, whereas higher doses are known to inhibit the immune response. Much more research needs to be done on putative sites of action of the suppressors or stimulants on the biochemical pathways causing immunological lesions in fish in areas of species sensitivity, timing, and dosages.

Table I
Hematological, Nonspecific Defensive Mechanism, and Specific Immune Response Assays

Hematological and physiological assays—blood samples
- Hematocrit: Percent of red blood cell pack
- Leukocrit: Percent of white blood cell pack
- Cell counts and differentials: Numbers of cells and types
- Lysozyme levels: Enzyme level in blood
- Serum immunoglobulin level: Specific and nonspecific antibody
- Serum protein level: Total protein in serum

Nonspecific defensive mechanism or specific immune response assays
(These assays can be used for either response)
- Phagocytosis: Percents and indexes; engulfment by phagocytic cells
- Rosette-forming cells: Adherence of particles around lymphocytes
- Glass or plastic adherence: Stickiness of phagocytic cells
- Pinocytosis: Engulfment of fluids by phagocytic cells
- Neutrophil activation:
  - Nitroblue tetrazolium dye reduction by oxidative burst
  - Chemiluminescence: light detection from oxidative burst
- Blastogenesis: Mitosis of cells, usually lymphocytes; radioimmunoassay

Specific immune response assays
- Scale rejection: Transplantation indicator
- Delayed hypersensitivity: Allergenic reactions
- Trypan blue: Killer cell activity
- Chromium release: Killer cell activity
- Melanomacrophage centers: Antigen processing cells
  - Antigen accumulation: Concentration in spleen or kidney areas
  - Cell aggregates: Increase in numbers of melanomacrophage cells
- Passive hemolytic plaque assay (Jerne assay): Antibody-producing cells
- Assays measuring serum antibody levels

Agglutination: Clumping of antigen particulates; sometimes antigen-labeled sheep red blood cells (hemagglutination)
- Preciptinin (Ouchterlony gel): Measures soluble antigens in gels
- Immunoelectrophoresis: For defining blood or antigenic components

Counterimmunolectrophoresis
- ELISA: Enzyme-linked immunosorbent assay
- Fluorescent antibody assays: Indirect and direct indicators
- Neutralization: Usually with live viral particles
- Complement fixation: Involves complex serum components as indicators; usually for lysis of blood cells.

Challenge with pathogen: Compare mortalities; determination of LD50s or relative percent survivals among test and control groups of fish

statistical variation, which is often related to the use of outbred feral animals with wide genetic backgrounds.

The assays for testing changes in the nonspecific defense system may require more background training and insight of the technician because they often require more detailed microscope work or sophisticated equipment to note chemical changes. Oxidative burst activity from phagocytic and neutrophil cells taken from the blood or immunopoietic organs is accurately measured by a luminometer. Microscopic and staining methods can also be used for this activity, however, the sensitivity may be compromised. Levels of nonspecific antibody and protein levels in the blood can be easily measured by using differential precipitation techniques. Enzymes with bacteriolytic activity such as lysozyme can also be measured from blood samples with simple microtiter techniques. Other tests for determining cellular immune activation may include macrophage inhibition assays and macrophage aggregation, which may be used for determining either nonspecific or specific changes (Blazer, 1992, Blazer *et al.*, 1994).

Study of the specific immune response may include measuring the levels of humoral antibody with tests such as enzyme immunosorbent assays (ELISA) and microtiter agglutination. These assays are in common use in fish disease diagnosis and can be easily adapted for use in fish toxicology, but usually an immunization regimen is required to induce this specific antibody and the specific cellular responses. The advantage of using serum samples to measure immunomodulation includes the ease of taking blood and storing it for running the assays when convenient. Also, bacterial vaccines are generally good at inducing humoral antibody in healthy fish. *Yersinia ruckeri, Aeromonas salmonicida, Vibrio* sp., and *Edwardsiella ictaluri* bacterins are available through commercial sources. Research projects investigating the immunomodulatory effects caused by contaminating substances can be designed around the use of these standardized bacterins. The specific immune response also involves cellular activation of macrophages, neutrophils, and other phagocytic cells. Again, the techniques of cellular activity can be used. In some cases, T-cell activity might be measured, since these are important controlling cells of the specific immune response and contain specific receptor sites on external membranes.

In most cases, however, the ultimate test for modulation of the immune system by a substance is to show increased suceptibility or resistance to disease by comparing the test animals to control, healthy fish. In the laboratory, this involves challenging the fish with an individual, selected pathogen. Such a protocol was established by Knittel (1981) with the bacterial pathogen *Yersinia ruckeri,* and steelhead trout following an exposure to copper. Other laboratory challenges have been carried out with viruses such as infectious pancreatic necrosis virus and viral hemorrhagic septicemia in salmonids. Susceptibility to pathogens can also be tested by field challenges, but with much less control. Some experiments describe placing test and control fish in cages placed in disease-endemic areas and noting subsequent mortalities. Results from field challenges are often frustrated by lack of control of particular pathogens with respect to dose or virulence. This may result in either the lack of mortalities in either test or control fish, or by both sets of fish succumbing to the disease.

New immunological assays are being added in this rapidly growing field; others are being modified and better adapted to maximize results for different species and environments. A current series of manuals gives some recent updates for this research (Stolen *et al.,* 1990, 1992, 1994, 1995). Perhaps the most important developing area is the use of *in vitro* assays to gather preliminary evidence for immunomodulation before progressing to whole fish studies which include sampling blood or tissues and pathogen challenges. These *in vitro* techniques add another tier to immunological investigations in which the immunopoietic tissues are extracted and whole organs or pieces or cell suspensions are transferred to tissue culture. Experiments on the tissues can be then conducted, in petri dishes for instance, and held under carefully controlled, laboratory environment conditions. Temperature, pH, and length of time are some of the parameters that give the researcher better control of the kinetics of immune reactions. Another distinct advantage is the use of fewer animals in the experiments.

III. IMMUNOSUPPRESSION

Wild fish are exposed to many chemicals in their environment that may possess immunomodulatory potential. Some of these substances affecting fish health are from natural sources, such as the low pH in some lakes resulting from accumulation of tannic acid from forest degradation. This acidity can also be created by humans, as is especially evident in northeastern United States, eastern Canada, and mid-Europe areas located in the fall of industrial air effluents. Leaching from naturally occurring erosion in locations of heavy metal and sulfur deposits can also damage aquatic life, as can effluents from active or inactive mining areas. While pH changes have not been directly linked to immune defense changes, the reduction of food and habitat changes affect the health of fish, leading to increased disease susceptibility.

In another category, the source of most contaminating substances is anthropomorphic, as in chemicals derived from agricultural use. Herbicides, fungicides, and insecticides are certainly a major cause of reduction in fish populations; residual drainage of these substances and their derivatives are found in creeks, rivers, and marine waters and sediments. Fertilizers high in nitrogen and phosphates drain into aquatic environments, causing disruptions in the food chain and, thus, the general health of fish. Pollutants also can be those materials that are purposely placed into the water to affect fish such as those used in the prevention or treatment of diseases. Heavy antibiotic use in salmonid aquaculture for the purpose of treating bacterial, protozoan, and fungal pathogens is a concern. Other chemicals are used to prevent excessive algae and other plant growth in fish pond farms and marine aquaculture facilities.

In many of the following studies in which the effects of contaminants and pollutants on fish were tested, the mechanisms of suppression are not known; however, if available, analogies can be drawn and used from mammalian examples. These problems are complicated in some cases by the presence of several to many potential immunomodulators in combination, such as is known to happen with crude oil and sewage sludge pollutants. These may contain heavy metals, aromatic hydrocarbons, and combinations of other inhibitory substances that can have agonist and antagonistic characteristics.

A. Metals

Many problems in aquatic contamination and its subsequent effect on fish health have arisen from mining, manufacturing, and processing waste moving into waters and sediments. Table II lists some of the reports from the literature concerning immunological tests of immunosuppression caused by the presence of metals. Extrapolating these results to estimate the effects of the metals and the actual concentrations that fish encounter in the environment is a topic open to debate. Some research is now direted toward bringing the water or sediments from the field environment into the laboratory and exposing fish to those "natural" concentrations in order to relate laboratory data to the field. However, the presence of concentrations of other substances from the environment and their combined effects is also a topic for debate.

It has long been recognized by fish culturists that copper piping cannot be used in hatchery systems because low amounts of this metallic ion kill fish. Hetrick *et al.* (1979) showed that salmonids held in low dilutions of copper and exposed to infectious hematopoietic necrosis virus became more susceptible to the disease, demonstrating this probable toxicity and effects on the protective mechanisms of fish. Other experiments indicating reduction in protective mechanisms with copper include those by Knittel (1981) who showed increases in susceptibility of steelhead trout (*Salmo gairdneri*) to *Y. ruckeri,* and Baker *et al.,* (1983) who demonstrated the increased susceptibility of chinook salmon (*Oncorhynchus tshawytscha*) to *Vibrio anguillarum.* For fish, copper seems to be one of the most serious heavy metal contaminants in the environment.

Table II
Nonspecific Defense Mechanisms and Specific Immune Response Parameters of Fish Affected by the Presence of Metals

Metal	Parameter	Fish species	Reference
Aluminum	Reduced chemiluminescence	Rainbow trout	Elsasser *et al.*, 1986
Arsenic	Phagocytosis elevated or lowered	Rainbow trout	Thuvander *et al.*, 1987
Cadmium	Elevated serum antibody	Rainbow trout	Thuvander, 1989
	Lymphocyte number and mitogenic response reduced	Goldfish	Murad and Houston, 1988
	Chemiluminescence reduced	Rainbow trout	Elsasser *et al.*, 1986
	Serum antibody heightened	Cunners	Robohm, 1986
	Lowered serum antibody	Striped bass	Robohm, 1986
Chromium	Serum antibody reduced	Brown trout, carp	O'Neill, 1981a
Copper	Chemiluminescence reduced	Rainbow trout	Elsasser *et al.*, 1986
	Serum antibody reduced	Brown trout	O'Neill, 1981a
	Antibody-producing cells reduced	Rainbow trout	Anderson *et al.*, 1989
	Susceptibility to *Vibrio anguillarum* increased	Eel	Rodsaether *et al.*, 1977
	Susceptibility to IHNV increased	Rainbow trout	Hetrick *et al.*, 1979
Lead	Serum antibody reduced	Brown trout	O'Neill, 1981b
Mercury	Lymphocyte numbers reduced	Barb	Gill and Pant, 1985
	Susceptibility to IPNV increased	Blue gourami	Roales and Perlmutter, 1977
Nickel	Serum antibody reduced	Brown trout	O'Neill, 1981a
Zinc	Serum antibody reduced	Brown trout	O'Neill, 1981a

In vitro assays describing the immunosuppressive effects of copper on the antibody-producing cells in isolated rainbow trout (*Oncorhynchus mykiss*) spleen cells shows the direct effect on the humoral immune response (Anderson *et al.*, 1989). Spleen fragments were incubated in tissue culture media for 14 days after being injected with *Y. ruckeri* bacterins. The higher concentrations of copper in the media inhibited the formation of the antibody-producing cells. Phagocytic uptake and chemiluminescence are also inhibited *in vitro* as shown in short-term assays by Elsasser *et al.,* (1986).

Aluminum, cadmium, mercury, magnesium, arsenic, lead, and selenium are some of the other heavy metals that affect fish health. O'Neill (1981a,b) was among the first to show that lead and cadmium injected at high doses caused a reduced circulatory antibody response in brown trout (*Salmo trutta*). In more refined experiments, Thuvander (1989) showed that doses of cadmium (0.7 or 3.6 μg/liter) in the water caused reduced mitogenic activity in fish immunized with *Vibrio anguillarum.* No clinical or histological changes were observed in the exposed fish, and no differences in protection were noted. Indeed, a higher degree of protection was noted by MacFarlane *et al.* (1986) when fish were held in water with low levels of cadmium and exposed to the pathogen that causes gill diseases in fish, *Cytophaga columnaris.* This example illustrates the difficulty that often arises in working with low levels of the heavy metals when determining toxicity and immunomodulation effects. Sometimes, low levels of heavy metals are immunostimulatory, perhaps because of stress and subsequent hyperactivity. Indeed, in protection tests, the metals may affect the pathogen's physiology as well, generally raising or lowering the numbers of pathogens in the challenge dose.

In many of these methods used to determine the effects of chemicals and pollutants on the immune response in fish, *in vitro* assays are becoming important. Dunier and Siwicki (1993) studied the *in vitro* effect of magnesium on the immune system of carp (*Cyrpinus carpio*). Lymphocyte proliferation was suppressed at all concentrations tested; phagocytosis of *Y. ruckeri* bacteria was suppressed at high concentrations.

Metallothionein, a detoxifying serum protein, binds to some heavy metals. Fish, as most other animals, produce this protein upon exposure, and as such it may be useful as an indicator of metals in the environment. The mechanism of damage to the biochemical pathways in fish by metals may include modifying enzymes, particularly the mixed-function oxidases. The different levels of metallothionein and subsequent metal ions available in various fish species may help explain differences in degrees of protection upon challenge to a pathogen.

B. Aromatic Hydrocarbons, Including Polychlorinated Biphenyls (PCBs)

For many years, PCBs were used in electrical transformers as liquid insulators. These lipophilic chemicals are very long lasting and tend to accumulate as they are passed through the food chain. There has been great concern about the effects of PCBs on aquatic life, however, any direct effects in the natural environment have been difficult to substantiate other than by inference from laboratory experiments using high doses (Table III).

Cleland *et al.* (1988) added a PCB, Aroclor 1254, to the diets of rainbow trout for up to 12 months; no effect on the cellular plaque-forming cell response was observed. Aroclor 1232 was injected intraperitoneally into channel catfish by Jones *et al.* (1979); circulating antibody titers were not effected, but there was a reduction of phagocytic cell activity. A slight delay in humoral antibody production to the bacterium *Escherichia coli* was found by Stolen (1985) in summer flounder exposed to Aroclor 1254.

The mechanisms of the action of halogenated hydrocarbons on the immune response may be through the cytosolic aryl hydrocarbon receptor (Ah-R), resulting in the induction of cytochrome P450AI, a mixed function

Table III
Nonspecific Defense Mechanism and Specific Immune Response Assays in Fish Affected by Aromatic Hydrocarbons

Compound	Parameter	Fish species	Reference
Phenol	Antibody-producing cells reduced	Rainbow trout	Anderson *et al.*, 1984
Benzidine	Nonspecific agglutination rise		Middlebrooks and Meador, 1984
Polychlorinated biphenyls (PCB)			
Aroclor 1254	No effect	Rainbow trout	Cleland *et al.*, 1988
Aroclor 1254	Antibody-producing cells reduced	Coho salmon	Cleland *et al.*, 1989
Aroclor 1254	Serum antibody slightly reduced	Summer flounder	Stolen, 1985
Aroclor 1232	Susceptibility to disease increased	Channel catfish	Jones *et al.*, 1979
Aroclor 1254/1260	Susceptibility to disease increased	Rainbow trout	Mayer and Mayer, 1985
Chlorinated dioxin (TCDD)	Mitogenic response partially suppressed	Rainbow trout	Spitsbergen *et al.*, 1986
	Susceptibility to IHNV	Rainbow trout	Spitsbergen *et al.*, 1988
Polynuclear aromatic hydrocarbons (PAHs)	Macrophage activity reduced	Spot	Weeks and Warinner, 1986
	Macrophage activity reduced	Hogchoker	Weeks and Warinner, 1986
	Melanomacrophage numbers reduced	Flounder	Payne and Fancey, 1989

oxygenase responsible for metabolism of the aromatic hydrocarbons. The Ah-R is constitutive and relatively nonspecific, that is, it is found on most cell membranes and can be blocked by several aromatic compounds.

Fish have a natural avoidance instinct for many of the aromatic hydrocarbons, so if new effluents resulting from industrial release and buildup of these chemicals in ancient migration paths occur, the natural fish migrations may be disrupted. Cases have been presented showing that phenols and other aromatics are often present in the aquatic environment and can influence the immune system of salmonids. Anderson *et al.* (1984) showed the reduction of splenic antibody-producing cells in rainbow trout held in low concentrations (10 ppm) of phenol before bath immunization with *Y. ruckeri* bacterins. These bacterins contain high percentages of lipopolysaccharides and have been shown to access the fish through gill membranes and other epidermal areas. They speculated that the gills may have been temporarily damaged, perhaps by receptor blockage, and thus the effect of the vaccine was reduced. If so, the blockage was temporary because these same fish could be successfully immunized when the time between phenol exposure and immunization was increased. Phenols and complex aromatic hydrocarbons are components of pulp manufacturing; these substances have long been held in suspicion by fish pathologists. 2,4,6-Trichlorophenol along with several other cogenetors from bleached pulp mill effluents were found to inhibit the immune response in rainbow trout *in vitro* (Voccia *et al.*, (1995).

In long-term studies, Arkoosh *et al.* (1994a,b) have shown the effects in coho salmon (*O. kisutch*) of swimming through contaminated environments during seaward migration or while returning to freshwater to spawn. The Duwamish river, a highly industrialized area near Seattle, Washington, has long been known to have high concentrations of PCBs, especially in sediments. It is believed that these substances are damaging to the immune memory responses involving T lymphocytes. It has been shown that fish from the contaminated areas have a lowere secondary immune response, which is dependent upon T-lymphocyte mediation. The work was partially built on earlier research with aflatoxins, which are recognized inducers of liver tumors in rainbow trout. Originally, the aflatoxins were found in trout feed that contained rancid cotton seed oil. Kaattari *et al.* (1994) have also demonstrated the inhibited T-lymphocyte response in fish exposed to aflatoxins. This work is important because it demonstrates an actual step in the immune pathways that is affected by the contaminating substance.

Channel catfish were exposed to 3,3′,4,4′,5-pentachlorobiphenyl (PCB 126) to investigate its effect on hematological, immunological, and enzyme biomarkers (Rice and Schlenk, 1995). This comprehensive study showed that the nonspecific cytotoxic cell activity was reduced by exposure, and the specific immune response as measured by numbers of specific antibody-secreting cells was elevated by contaminant exposures at the lowest doses. However, the hematological indexes, except for neutrophil numbers, remained unaffected. They concluded that fish may not be as sensitive to halogenated aromatic hydrocarbons as mice with regard to their humoral immune compartment.

C. Pesticides

The dangers of pesticides and their breakdown products on the environment and the immune systems of fishes merit thorough investigation. Some of the experiments with fish are given in Table IV. As a consequence of the research findings that bird eggshells weaken when high doses of DDT

Table IV
Nonspecific Defense Mechanisms and Specific Immune Response Parameters of Fish Affected by Pesticides

Compound	Parameter	Fish species	Reference
Endrin	Phagocytic, antibody-producing cell activities reduced	Rainbow trout	Bennett and Wolke, 1987a
Malathion	Lymphocyte number reduced	Channel catfish	Areechon and Plumb, 1990
Methyl bromide	Thymic necrosis	Medaka	Wester *et al.*, 1988
Trichlorphon	No effect	Carp	Cossarini-Dunier *et al.*, 1990
	Phagocytic, neutrophilic, lysozyme activity reduced	Carp	Siwicki *et al.*, 1990
DDT	Antibody-producing cell, serum antibody reduced	Goldfish	Sharma and Zeeman, 1980
Lindane	No effect	Carp	Cossarini-Dunier *et al.*, 1987
Atrazine	No effect	Carp	Cossarini-Dunier *et al.*, 1987
Bayluscide	Serum African antibody reduced	Catfish	Faisal *et al.*, 1988
Tributyltin	Chemiluminescence reduced	Oyster toadfish, hogchoker, croaker	Wishkovsky *et al.*, 1989
	Chemiluminescence	Toadfish	Rice and Weeks, 1990

(1,1,1-trichloro-2,2-bis(*p*-chlorophenyl)ethane) are consumed the present pesticide products on the market and approved for use in the United States have been well tested. One of the first studies showing immunomodulation with fish was done using DDT. It was found that this polychlorinated biphenyl caused leukopenia, reduction in spleen weight, and suppression of the humoral immune response in gold fish (Zeeman and Brindley, 1981). As DDT is very persistent in the environment and often employed against aquatic insects or their larvae, it was important to reduce or stop its use.

Bennett and Wolke (1987a,b) presented studies showing that endrin had little effect on the phagocytic action in rainbow trout after 60 days of exposure; the specific immune response assayed by determining the numbers of antibody-producing cells and humoral antibody was reduced. In a similar study with Lindane (γ-hexachlorocyclohexane), Cossarini-Dunier *et al.* (1987) showed that this herbicide had no effect on hematocrit or antibody production in rainbow trout immunized with a *Y. ruckeri* bacterin.

Tributyltin, an important antifouling agent used in exterior paints for boats, has immunosuppressive properties on the chemiluminescence response of macrophages in toadfish (*Opsanus tau*). Rice and Weeks (1990) proposed that the calcium flux across the macrophage membrane may be suppressed, impairing this function. Their field studies showed the immunosuppressive properties of high concentrations of this pesticide in the natural environment where tributyltin was heavily used on naval ships in the Elizabeth River in Virginia.

Atrazine (2-chloro-4-ethylamino-6-isopropylamino-trazine), an affective herbicide widely used in Europe, is responsible for fish kills when it contaminates aquatic enviornments after field spraying. The herbicide was fed to carp for 84 days and changes were not observed in immunological parameters such as splenic index (weight) (Cossarini-Dunier *et al.,* 1988). This also correlated with their *in vitro* exposures. Using mugilids (*Liza ramada* and *L. Aurata*), however, Biagianti-Risbourg (1990) found that after 7 days of exposure to 0.025–0.28 mg/liter, the livers contained degenerated macrophages and other cellular aberrations.

D. Drugs: Chemicals and Antibiotics Used in Treating Fish Diseases

Extensive use of antibiotics for prevention and treatment of diseases in fish held in aquaculture pens has lead to concern about the accumulation of the drugs in the environment (Table V). Samples of sediments below hanging net pens, for example, have been found to have high levels of residual tetracycline, sulfamerzine, or other drugs. Indeed, a major concern is the transfer of drug-resistant factors from resident bacteria to fish disease-

Table V
Nonspecific Defense Mechanisms and Specific Immune Response Assays in Fish Affected by the Antibiotics, Drugs, and Other Chemicals

Compound	Parameter	Fish species	Reference
Oxytetracycline	Mitogenic response reduced	Carp	Gondel and Boesten, 1982
	Antibody-producing cells reduced	Rainbow trout	van Muiswinkel *et al.*, 1985
	Antibody-producing cells reduced	Rainbow trout	Siwicki *et al.*, 1989
Aflatoxin B-1	B-cell memory loss	Rainbow trout	Arkoosh and Kaattari, 1987
Cortisol/ Kenalog-40	Antibody-producing cells reduced	Rainbow trout	Anderson *et al.*, 1982
Hydrocortisone	Phagocytic activity reduced	Striped bass	Stave and Roberson, 1985

causing bacteria, increasing levels of drug resistance and making therapy more difficult. In some cases, immunization regimens have included the addition of these antibiotics, supposedly to give treatment at the same time as immunization.

Oxytetracycline, the antibiotic most used by fish culturists in treatment of bacterial diseases, has been long known to be immunosuppressive in fish, reducing the numbers of antibody-producing cells (Grondel and Boesten, 1982; van Muiswinkel *et al.*, 1985; Siwicki *et al.*, 1989). Whether the therapeutic drug is injected, fed, or given by bath, an immunosuppressive effect is evident. Oxolinic acid, a more recent addition to the list of drugs used for treating fish bacterial pathogens, was found to not have immunosuppressive characteristics when given at therapeutic levels (Siwicki *et al.*, 1989).

IV. IMMUNOSTIMULATION

Some substances can also benefit or stimulate fish immune protection systems. Aquaculturists are interested in the use of adjuvants and immunostimulants for preventing fish diseases (Anderson, 1992). This subject was first introduced in fish culture with the addition of adjuvants to bacterins and vaccines for inducing a greater respone than if the immunogens are given alone. Some of the adjuvants and immunostimulators are listed in Table VI. In some studies using Freund's complete adjuvant (FCA), it was found that the adjuvant could be given alone to elevate the nonspecific

Table VI
Immunostimulants and Adjuvants and Their Mode of Action[a]

- Complete Freunds Adjuvant (CFA): Antigen reservoir, T-lymphocyte stimulator
- Incomplete Freunds Adjuvant (IFA): Antigen Reservoir
- Muramyl dipeptide: T-lymphocyte stimulator
- Levamisole: T-lymphocyte stimulator
- FK-565 (streptomyces olivaceogriseus derivative): T-lymphocyte stimulator
- Glucans: Macrophage activator
- ISK (fish and product derivative): Unknown
- Quaternary ammonium compounds: Unknown
- Chitin and chitosan: Unknown, maybe similar to glucans
- Ete (tunicate derivative): Phagocytosis
- Vitamin C (ascorbic acid): Antioxidant, multiple cell stimulator
- Vitamin E (α-tocopherol): Antioxidant, B- and T-lymphocyte stimulator
- Bacterial endotoxins: B-lymphocyte stimulators

[a] These compounds affect the nonspecific defensive mechanisms and specific immune response of fish. In many cases the mode of action is of the substance in fish is uncertain or unknown.

defense mechanisms (Cipriano and Pyle, 1985; Dunier, 1985). Although immunostimulants as such rarely occur alone in the natural environment, the subject is worth discussing here because these substances may become useful in preventing fish diseases in aquaculture pens and hatcheries.

Many substances can be immunostimulative; currently, glucans, chitins, bacterial lipopolysaccharides, and light oils and levamisole hold promise for further research development. Other immunostimulants of interest for preventing fish diseases include muramyl dipeptide, quanterary ammonium compounds, and various animal products such as ISK, and immune stimulating complex from fish and tunicate extracts (Kodama *et al.,* 1993; Jeney and Anderson, 1993b; McCumber *et al.,* 1981; Stanley *et al.,* 1995).

The glucans, derived from yeast cell walls and certain higher plants, when injected or fed to fish have excellent immunostimulatory properties. Yano *et al.* (1991) showed the β-1,6-branched β-1,3-glucans were effective in carp (*C. carpio*). Jeney and Anderson (1993a) used a glucan derived from barley that showed increased activity in nonspecific defense mechanisms and in protection against challenges of *Y. ruckeri.* Robertson *et al.* (1994) summarized their studies wtih β-glucans as increasing the nonspecific resistance of fish against infections of bacterial pathogens. Glucan treatment of Atlantic salmon (*Salmo salar*) induced protection against *Vibrio salmonicida.* Chen and Ainsworth (1992) also showed that a yeast glucan induced protection in channel catfish (*Ictalurus* sp.) against the pathogen *Edwardsiella ictaluri.*

The nonspecific immunostimulants are effective on a short-term basis because the specific immune response is not involved. Anderson and Siwicki (1994) administered a chitin derivative, chitosan, to brook trout (*Salvelinus fontinalis*) by injection and immersion and found that high levels of protection occurred 1, 2, and 3 days after, but protection was greatly reduced by day 14. Injection of the chitosan was also more effective than simple immersion.

Several glucan products are marketed commercially. VitaStim (Taito Co., Tokyo, Japan) and Macrogard (KS Biotec-Mackymal, Tromso, Norway) are used in supplementing fish feeds. At present in the United States, the governmental regulating agencies have not determined whether these substances are to be classified as drugs or feed additives.

The importance of vitamin supplements in fish feeds is recognized by manufacturers. Blazer (1992) summarizes fish disease research on vitamins C and E and shows examples of these vitamins as useful antioxidants and stimulators of many facets of the immune system; most studies indicate increased resistance to specific diseases. Lysozyme, complement, and antibody levels, and resistance to *A. salmonicida* were elevated in Atlantic salmon given diets supplemented with vitamin C (Waagbo *et al.*, 1993).

V. IMMUNOREVERSAL

There has been some speculation about undoing the damaging effects of pollutants and contaminants that are immunosuppressive to fish. If the harmful agents in the environment cannot be avoided, scientists may be able to reverse the suppressive effects by administering stimulants. Reconstitution or restoration of the defensive functions may be possible if some of the progenitor cells can be stimulated. Siwicki *et al.* (1995) have begun some research in this area with chitosan, Finnstim, and dimerized lysozyme as the immunostimulants. Oral administration of the immunostimulants restored some functions of the immune response after suppression by the organophosphorus insecticides trichlorfon and dichlorfon. Dunier *et al.* (1995) also present evidence that vitamin C and the pharmacological drug Nitrogranulin were able to offset of prevent decreased phagocytosis, lymphocyte proliferation, and antibody response *in vitro* after the immunopoietic tissues were exposed to the organochlorine insecticide Lindane.

The damaging effects of pollutants on fish populations can be reversed when the environment is cleaned up. An example comes from a study by Baumann and Harshbarger (1995) showing that there was a decline of liver neoplasms in populations of wild brown bullhead catfish after the closing of a coking plant on the Black River in Ohio. Aromatic hydrocarbons were significantly reduced in the river sediments. The populations of these bottom-feeding fish recovered and returned to near normal conditions.

VI. CONCLUSIONS

The many different substances in the environment that can suppress or affect the immune response of fish and resultant protection against disease-causing agents makes life hazardous. When immunological techniques can contribute to recognizing the effects of the agents, then we can make efforts to eliminate the dangerous substances from the environment. For example, if a single chemical leaves a mark of a specific deficiency or lesion in the immune pathway, forensic tracing can reveal the source. Unfortunately, our science is only beginning to approach these levels of sophistication. Keeping fish healthy in the wild environment and in aquaculture situations should be one of the major goals of our society.

REFERENCES

Adams, S. M., ed. (1990). "Biological Indicators of Stress in Fish," American Fisheries Society. Bethesda, Maryland.

Anderson, D. P. (1992). Immunostimulants, adjuvants, and vaccine carriers in fish: Applications to aquaculture. *Annu. Rev. Fish Dis.* **2,** 281–307.

Anderson, D. P., Roberson, B. S., and Dixon, O. W. (1982). Immunosuppression induced by a corticosteroid or an alkylating agent in rainbow trout (*Salmo gairdneri*) administered a *Yersinia ruckeri* bacterin. *Dev. Comp. Immunol. Suppl.* **2,** 197–204.

Anderson, D. P., Dixon, O. W., and van Ginkel, F. W. (1984). Suppression of bath immunization in rainbow trout by contaminant bath pretreatments. *In* "Chemical Regulation of Immunity of Veterinary Medicine" (M. Kende, J. Gainer, and M. Chirigos, eds.), pp. 289–293. Alan R. Liss, New York.

Anderson, D. P., Dixon, O. W., Bodammer, J. E., and Lizzio, E. F. (1989). Suppression of antibody-producing cells in rainbow trout spleen sections exposed to copper in vitro. *J. Aquat. Anim. Health* **1,** 57–61.

Anderson, D. P. and Siwicki, A. K. (1994). Duration of protection against *Aeromonas salmonicida* in brook trout immunostimulated with glucan or chitosan by injection or immersion. *The Progressive Fish-Culturist* **56,** 258–261.

Anderson, D. P., and Zeeman, M. G. (1995). Immunotoxicology in fish. *In* "Aquatic Toxicology" (G. Rand, ed.), pp. 371–404. Taylor and Francis, Washington DC.

Areechon, N., and Plumb, J. A. (1990). Sublethal effects of malathion on channel catfish, *Ictalurus punctatus. Bull. Environ. Contam. Toxicol.* **44,** 435–442.

Arkoosh, M. R., and Kaatari, S. L. (1987). Effect of early aflatoxin B-2 exposure on *in vivo* and *in vitro* antibody response in rainbow trout (*Salmo gairdneri*). *J. Fish Biol. Suppl. A* **31,** 19–22.

Arkoosh, M. R., Stein, J. E., and Casillas, E. (1994a). Immunotoxicology of an anadromous fish: Field and laboratory studies of B cell–mediated immunity. *In* "Modulators of Fish Immune Responses: Models for Environmental Toxicology, Biomarkers, and Immunostimulators" (J. S. Stolen and T. C. Fletcher, eds.), Vol. 1, pp. 33–48. SOS Publications, Fair Haven, NJ.

Arkoosh, M. R., Clemons, E., Myers, M., and Casillas, E. (1994b). Suppression of B cell–mediated immunity in juvenile chinook salmon (*Oncorhynchus tshawytscha*) after exposure to either a polychlorinated aromatic hydrocarbon or to polychlorinated biphenyls. *Immunopharmacol. Immunotoxicol.* **16,** 293–314.

Baker, R. J., Knittel, M. D., and Fryer, J. L. (1983). Susceptibility of chinook salmon (*Oncorhynus tshawytscha*) and rainbow trout (*Salmo gairdneri*) to infection with *Vibrio anguillarum* following sublethal copper exposure. *J. Fish Dis.* **6,** 267–275.

Baumann, P. C., and Harshbarger, J. C. (1995). Decline in liver neoplasms in wild brown bullhead catfish after coking plant closes and environmental PAHs plummet. *Environ. Health Perspect.* **103,** 168–170.

Bennett, R. O., and Wolke, R. E. (1987a). The effect of sublethal endrin exposure on rainbow trout (*Salmo gairdneri* Richardson). I. Evaluation of serum coritsol concentrations and immune responsiveness. *J. Fish Biol.* **31,** 375–385.

Bennett, R. O., and Wolke, R. W. (1987b). The effect of sublethal endrin exposure on rainbow trout (*Salmo garidneri* Richardson). II. The effect of altering serum cortisol concentrations on the immune response. *J. Fish Biol.* **31,** 387–394.

Biagianti-Risbour, S. (1990). Contribution a l'etude du foie juveniles de muges Teleosteens (Mugilides) contamines experimentalement per l'atrazine (s-triazine herbicide): Interet en ecotoxicologie. Ph.D. Thesis, Academie de Montpellier, University of Perpignan, France.

Blazer, V. S. (1992). Nutrition and disease resistance in fish. *Annu. Rev. Fish Dis.* **2,** 309–323.

Blazer, V. S., Facey, D. E., Fournie, J. W., Courtney, L. A., and Summers, J. K. (1994). Macrophage aggregates as indicators of environmental stress. *In* "Modulators of Fish Immune Responses: Models for Environmental Toxicology/biomarkers, Immunostimulators" (J. S. Stolen and T. C. Fletcher, eds.), Vol. 1, pp. 169–185. SOS Publications, Fair Haven, NJ.

Chen, D., and Ainsworth, A. J. (1992). Glucan administration potentiates immune defence mechanisms of channel catfish, *Ictalurus punctuatus* Rafinesque. *J. Fish Dis.* **15,** 295–304.

Cipriano, R. C., and Pyle, S. W. (1985). Adjuvant-dependent immunity and the agglutinin response of fishes against *Aeromonas salmonicida,* cause of furunculosis. *Can. J. Fish. Aquat. Sci.* **42,** 1290–1295.

Cleland, G. B., McElroy, P. J., and Sonstegard, R. A. (1988). The effect of dietary exposure to Aroclor 1254 and/or Mirex on humoral immune expression of rainbow trout (*Salmo gairdneri*). *Aquat. Toxicol.* **12,** 141–146.

Cleland, G. B., McElroy, P. J., and Sonstegard, R. A. (1989). Immunomodulation in C56B1/6 mice following consumption of halogenated aromatic hydrocarbon-contaminated coho salmon (*Oncorhynchus kisutch*) from Lake Ontario, Canada. *J. Toxicol. Environ. Health* **27,** 477–486.

Cossarini-Dunier, M., Monod, G., Damael, A., and Lepot, D. (1987). Effecs of γ-hexchlorocyclohexane (lindane) on carp (*Cyprinus carpio*). 1. Effect of chronic intoxication on humoral immunity in relation to tissue pollutant levels. *Ecotoxicol Environ. Safety* **13,** 339–345.

Cossarini-Dunier, M., Demael, A., Riviere, J. L., and Lepot, D. (1988). Effects of oral doses of the herbicide atrazine on carp (*Cyrpinus carpio*). *Ambio* **17,** 401–405.

Cossarini-Dunier, M., Damael, A., and Siwicki, A. K. (1990). *In vivo* effect of the organophosphorus insecticide trichlorphon on the immune response of carp (*Cyprinus carpio*): Effect of contamination on antibody production in relation to residue level in organs. *Ecotoxicol. Environ. Safety* **19,** 93–98.

Dunier, M. (1985). Effect of different adjuvants on the humoral immune response of rainbow trout. *Dev. Comp. Immunol.* **9,** 141–146.

Dunier, M. B. (1994). Effects on environmental contaminants (pesticides and metal ions) on fish immune systems. *In* "Modulators of Fish Immune Responses: Models for Environmental Toxicology, Biomarkers, and Immunostimulators" (J. S. Stolen and T. C. Fletcher, eds.), pp. 123–139. SOS Publications, Fair Haven, NJ.

Dunier, M., and Siwicki, A. K. (1993). Effects of pesticides and other organic pollutants in the aquatic environment on immunity of fish: A review. *Fish Shellfish Immunol.* **3,** 423–438.

Dunier, M., Siwicki, A. K., Verlhac, V., Vergnet, Ch., and Studnicka, M. (1995). The immunotoxic effect of Lindane on the specific and non specific immune response of rainbow trout and treatment with Nitrogranulin and vitamin C. *In* "Modulators of Immune Responses" (J. S. Stolen, ed.), p. 145. SOS Publications, Fair Haven, NJ, Abstract.

Elsasser, M. S., Roberson, B. S., and Hetrick, F. M. (1986). Effects of metals on the chemiluminescent response of rainbow trout (*Salmo gairdneri*) phagocytes. *Vet. Immunol. Immunopathol.* **12,** 243–250.

Faisal, M., Cooper, E. L., El-Mofty, M., and Sayed, M. A. (1988). Immunosuppression of *Clarias lazera* (Pisces) by a molluscicide. *Dev. Comp. Immunol.* **12,** 85–97.

Gill, T. S., and Pant, J. C. (1985). Mercury-induced blood anomalies in the freshwater teleost *Barbus conchonius. Water Air Soil Pollut.* **24,** 165–171.

Grondel, J. L., and Boesten, H. J. A. M. (1982). The influence of antibiotics on the immune system. I. Inhibition of the mitogenic leukocyte response *in vitro* by oxytetracycline. *Dev. Comp. Immunol.* **2,** 211–216.

Hetrick, F. M., Knittel, M. D., and Fryer, J. L. (1979). Increased susceptibility of rainbow trout to infections hematopoietic necrosis virus after exposure to copper. Applied Environmental Microbiology. 37:198–201.

Jeney, G., and Anderson, D. P. (1993a). Glucan injection or bath exposure given alone or in combination with a bacterin enhance the nonspecific defence mechanisms in rainbow trout (*Oncorhynchus mykiss*). *Aquaculture* **116,** 315–329.

Jeney, G., and Anderson, D. P. (1993b). Enhanced immune response and protection in rainbow trout to *Aeromonas salmonicida* bacterin following prior immersion in immunostimulants. *Fish Shellfish Immunol.* **3,** 51–58.

Jones, D. H., Lewis, D. H., Eurell, T. W., and Cannon, M. S. (1979). Alteration of the immune response of channel catfish (*Ictalurus punctatus*) by polychlorinated biphenyls. Animals as monitors of environmental pollutants. Symposium on Pathobiology of Environmental Pollutants; Animal Models and Wildlife as Monitors. National Academy of Sciences, Washington, DC (Abstract). pp. 385–386.

Kaattari, S. L., Adkison, M., Shapiro, D., and Arkoosh, M. R. (1994). Mechanisms of immunosuppression by aflatoxin B-1. *In* "Modulators of Fish Immune Responses: Models for Environmental Toxicology, Biomarkers, and Immunostimulators," (J. S. Stolen and T. C. Fletcher, eds.), Vol. 1, pp. 151–167. SOS Publications, Fair Haven, NJ.

Knittel, M. D. (1981). Susceptibility of steelhead trout, *Salmo gairdneri* Richardson, to redmouth infection, *Yersinia ruckeri* following exposure to copper. *J. Fish Dis.* **4,** 33–40.

Kodama, H., Yoshikatsu, H., Masafumi, M., Tsuyoshi, B., and Azuma, I. (1993). Activation of rainbow trout (*Oncorhynchus mykiss*) phagocytes by muramyl dipeptide. *Dev. Comp. Immunol.* **17,** 129–130.

MacFarlane, R. D., Bullock, G. L., and McLaughlin, J. J. A. (1986). Effects of five metals on susceptibility of striped bass to *Flexibacter columnaris. Trans. Am. Fish. Soc.* **115,** 227–231.

Mayer, K. S., and Mayer, F. L. (1985). Waste transforms oil and PCB toxicity to rainbow trout. *Trans. Am. Fish Soc.* **114,** 869–886.

McCumber, L. J., Trauger, L. T., and Sigel, M. M. (1981). Modification of the immune system of the American eel, *Anguilla rostrata,* by ETE. *Dev. Biol. Stand.* **49,** 289–294.

Middlebrooks, B. L., and Meador, C. B. (1984). Effects of benzidine exposure on the immune response of an estuarine fish (*Cyprinodon variegatus*). Abstracts of the Annual Meeting of the American Society for Microbiology. p. 82. American Society for Microbiology Publishers, Washington, DC. Abstract.

Murad, A., and Houston, A. H. (1988). Leukocytes and leukopoietic capacity on goldfish, *Carassius auratus,* exposed to sublethal levels of cadmium. *Aquat. Toxicol.* **13,** 141–154.

O'Neill, J. G. (1981a). The humoral immune response of *Salmo trutta* L. and *Cyprinus carpio* L. exposed to heavy metals. *J. Fish Biol.* **19,** 297–306.

O'Neill, J. G. (1981b). Effects of intraperitoneal lead and cadmium on the humoral immune response of *Salmo trutta. Bull. Environ. Contam. Toxicol.* **27,** 42–48.

Payne, J. F., and Fancey, L. F. (1989). Effect of polycyclic aromatic hydrocarbons on immune response in fish: Change in melanomacrophage centers in flounder (*Psseudopleuronectes americanus*) exposed to hydrocarbon-contaminated sediments. *Mar. Environ. Res.* **28,** 431–435.

Rice, C. D., and Schlenk, D. (1995). Immune function and cytochrome P4501A activity after acute exposure to 3,3′,4,4′,5-pentachlorobiphenyl (PCB 126). *J. Aquat. Anim. Health* **7,** 195–204.

Rice, C. D., and Weeks, B. A. (1990). The influence of in vitro exposure to tributylin on reactive oxygen formation in oyster toadfish macrophages. *Arch. Environ. Contam. Toxicol.* **19,** 854–857.

Roales, R. R., and Perlmutter, A. (1977). The effects of sublethal doses of methylmercury and copper, applied singly and jointly, on the immune response of the blue gourami (*Trichogaster trichopterus*) to viral and bacterial antigens. *Arch. Environ. Toxicol.* **5,** 325–331.

Robertsen, B., Engstad, R. E., and Jorensen, J. B. (1994). Beta-glucans as immunostimulants in fish. *In* "Modulators of Fish Immune Responses: Models for Environmental Toxicology, Biomarkers, and Immunostimulators" (J. S. Stolen and T. C. Fletcher, eds.), pp. 83–99. SOS Publications, Fair Haven NJ.

Robohm, R. A. (1986). Paradoxical effects of cadmium exposure on antibacterial antibody responses in two fish species: Inhibition in cunners (*Tautogolabrus adspersus*) and enhancement in striped bass (*Morone saxatilis*). *Vet. Immunol. Immunopathol.* **12,** 251–262.

Rodsaether, M. C., Olafsen, J., Raa, J., Myhre, K., and Steen, J. B. (1977). Copper as an initiating factor in vibriosis (*Vibrio anguillarum*) in eel (*Anguilla*). *J. Fish Biol.* **10,** 17–21.

Sharma, R. P., and Zeeman, M. G. (1980). Immunologic alternation by environmental chemicals: Relevance of studying mechanisms versus effects. *J. Immunopharmacol.* **2,** 285–307.

Siwicki, A. K., Anderson, D. P., and Dixon, O. W. (1989). Comparisons of nonspecific and specific immunomodulation by oxolinic acid oxytetracycline and levamisole in salmonids. *Vet. Immunol. Immunopathol.* **23,** 195–200.

Siwicki, A. K., Cossarini-Dunier, M., Studnicka, M., and Damael, A. (1990). *In vivo* effect of the organophosphorus insecticide trichlorphon on immune response of carp (*Cyprinus carpio*): Effect of high doses of trichlorphon on nonspecific immune response. *Ecotoxicol. Environ. Safety* **19,** 99–105.

Siwicki, A. K., Studnicka, M., and Morand, M. (1995). Restoration of cellular and humoral immunity after suppression induced by organophosphorus insecticides. *In* "Modulators of Immune Responses: The Evolutionary Trail" (J. S. Stolen, ed.), SOS Publications, Fair Haven, NJ. Abstract.

Spitsbergen, J. M., Schat, K. A., Kleeman, J. M., and Peterson, R. E. (1986). Interactions of 2,3,7,8-tetrachlorodibenzo-*p*-dioxin (TCDD) with immune responses of rainbow trout. *Vet. Immunol. Immunopathol.* **12,** 263–280.

Spitsbergen, J. M., Schat, K. A., Kleeman, J. M., and Peterson, R. E. (1988). Effects of 2,3,7,8-tetrachlorodibenzo-*p*-dioxin (TCDD) or Aroclor 1254 on the resistance of rainbow trout, *Salmo gairdneri* Richardson, to infectious hematopoietic necrosis virus. *J. Fish Dis.* **11,** 73–83.

Stanley, L. A., Hayasaka, S. S., and Schwedler, T. E. (1995). Effects of the immunomodulator *Ecteinascidia turbinata* extract on *Edwardsiella ictaluri* infection of channel catfish. *J. Aquat. Anim. Health* **7,** 141–146.

Stave, J. W., and Roberson, B. S. (1985). Hydrocortisone suppresses the chemiluminescent response of striped bass phagocytes. *Dev. Comp. Immunol.* **9,** 77–84.

Stolen, J. S. (1985). The effect of the PCB, Aroclor 1254, and ethanol on the humoral immune response of a marine teleost to a sludge bacterial isolate of E. coli. *In* "Marine Pollution and Physiology. Recent Advances" (F. J. Vernberg, ed.), pp. 419–426. Academic Press, New York.

Stolen, J. S., Fletcher, T. C., Anderson, D. P., Roberson, B. S., and van Muiswinkel, W. B. eds. (1990). "Techniques in Fish Immunology I." SOS Publications, Fair Haven, NJ.

Stolen, J. S., Fletcher, T. C., Anderson, D. P., Kaattari, S. L., and Rowley, A. F. eds. (1992). "Techniques in Fish Immunology II." SOS Publications, Fair Haven, NJ.

Stolen, J. S., Fletcher, T. C., Rowley, A. F., Zelikoff, J. T., Kaattari, S. L., and Smith, S. A., eds. (1994). Fish Immunology III." SOS Publications, Fair Haven, NJ.

Stolen, J. S., Fletcher, T. C., Smith, S. A., Zelikoff, J. T., Kaattari, S. L., Anderson, R. S., Soderhall, K., and Weeks-Perkins, B. A., eds. (1995). Techniques in Fish Immunology III. SOS Publications. Fair Haven, NJ.

Thuvander, A. (1989). Cadmium exposure of rainbow trout (*Salmo gairdneri* Richardson): Effects on immune functions. *J. Fish Biol.* **35,** 521–529.

Thuvander, A., Norrgren, L., and Fossum, C. (1987). Phagocytic cells in blood from rainbow trout, *Salmo gairdneri* (Richardson), characterized by flow cytometry and electron microscopy. *J. Fish Biol.* **35,** 31:197–208.

van Muiswinkel, W. B., Anderson, D. P., Lamers, C. H. J., Egberts, E., van Loon, J. J. A., and Ijssel, J. P. (1985). Fish immunology and fish health. *In* "Fish Immunology" (M. J. Manning and M. F. Tatner, eds.), pp. 1–8. Academic Press, London.

Voccia, I., Sanchez-Dardon, J., Dunier, J. M., and Fournier, M. (1995). Four chemicals found in bleached pulp mill effluents were tested *in vitro* on the immune response of rainbow trout (*Oncorhynchus mykiss*). *In* "Modulators of Immune Responses" (J. S. Stolen, ed.), p. 141. SOS Publications, Fair Haven, NJ. Abstract.

Waagbo, R., Glette, J., Raa-Nilsen, E., and Sandness, K. (1993). Dietary vitamin C and disease resistance in Atlantic salmon (*Salmo gairdneri*). *Fish Physiol. Biochem.* **12,** 61–73.

Weeks, B. A., and Warinner, J. E. (1986). Functional evaluation of macrophages in fish from a polluted estuary. *Vet. Immunol. Immunopathol.* **12,** 313–320.

Wester, P. W., Canton, J. H., and Dormans, J. A. M. A. (1988). Pathological effects in freshwater fish *Poecilia reticulata* (guppy) and *Oryzias latipes* (medaka) following methyl bromide and sodium bromide exposure. *Aquat. Toxicol.* **12,** 323–344.

Wester, P. W., Vethaak, A. D., and van Muiswinkel, W. B. (1994). Fish as biomarkers in immunotoxicology. *Toxicology* **86,** 213–234.

Wishkovsky, A., Mathews, E. S., and Weeks, B. A. (1989). Effect of tributyltin on the chemiluminescent response of phagocytes from three species of estuarine fish. *Arch. Environ. Contam. Toxicol.* **18,** 826–831.

Yano, T., Matsuyama, H., and Mangindaan, R. E. P. (1991). Polysaccharide-induced protection of carp, *Cyprinus carpio,* against bacterial injection. *J. Fish Dis.* **14,** 577–582.

Zeeman, M G., and Brindley, W. A. (1981). Effects of toxic agents upon fish immune systems: A review. *In* "Immunologic Considerations in Toxicology" (R. P. Sharma, ed.), pp. 1–47. CRC Press, Boca Raton, FL.

Zelikoff, J. T. (1993). Metal pollution–induced immunomodulation in fish. *Annu. Rev. Fish Dis.* **3,** 305–325.

Zelikoff, J. T. (1994). Immunological alteration as indicators of environmental metal exposure. *In* "Modulators of Fish Immune Responses: Models for Environmental Toxicology, Biomarkers, and Immunostimulators" (J. S. Stolen and T. C. Fletcher, eds.), pp. 101–110. SOS Publications, Fair Haven, NJ.

8

IMMUNOMODULATION: ENDOGENOUS FACTORS

CARL B. SCHRECK

I. INTRODUCTION

The general health of fish is a function of their environment, the nature of the pathogen(s), and factors intrinsic to the fish themselves. Hans Selye (1936, 1950) first discovered that stress can impair the health of animals, an observation now known to be true for fishes. The status of health depends on the fish's genetic composition, prior history, and the quality of the present environment for both fish and pathogen (Snieszko, 1974; Wedemeyer, 1970, 1974; Wedemeyer, *et al.,* 1976). Acute or chronically stressful situations influence the ability of fish to resist microparasites or other environmental insults. Fish health is a broad topic that transcends well beyond effects of pathogenic organisms, however. Nutrition and water quality are obviously important in determining the health of fish. Poor nutrition can result in poor health even in the absence of pathogens. Poor water quality can likewise impair the general health of fish. In this chapter I concentrate only on health as related to pathogenic organisms and the immune system as influenced by stress. I exclude discussion of toxicants as stressful agents and their potential consequences on health and disease; these are subjects of chapters by Anderson and others in this volume. I also do not consider nutritional aspects of health and immunity; review of how various nutritional deficits could interact with stressors to impair general health and disease resistance are beyond the scope of this chapter.

Modes of action of stress on health operate via a variety of mechanisms and over various temporal scales. Stressors can be selective forces that influence the genetic structure of populations and hence the resistance of subsequent generations to microparasites. Environmental factors that are stressful to fish may actually provide a more optimal environment for pathogenic organisms and consequently increase their virulence. Stress may also influence the ability of fish to resist pathogenic insults by affecting the specific and nonspecific components of the immune system. In this chapter I review what is known about the effects of stress on fish health, particularly from the perspective of the immune system. Because stressful environments may also affect the pathogen directly or have populational genetic consequences for the fish relative to disease resistance, these factors are considered as well. The ontogeny of the immune system per se is the subject of Chapter 6 by Tatner (this volume). I will, however, address developmental aspects of the relationship among stress, immune system, and disease resistance as the physiological stress response changes through development.

Presentation of a clear paradigm of how stress affects a fish's ability to resist pathogens is confounded by the fact that pathogens themselves can be considered part of a fish's environment, and when these microorganisms result in disease they can induce stress. It is important to distinguish between stress resulting from environmental factors other than pathogenic agents and stress resulting from the microparasites directly.

What is meant by "stress" is in itself confounding. Various definitions have been reviewed by Pickering (1981), and they range from the causative stimulus to the physiological response (Pickering, 1981; Barton and Iwama, 1991). Using Pickering's (1981) example, I use the word "stress" when referring to the stimulus (sometimes referred to as the "stressor") and "stress response" when referring to consequences of the stress. It is useful when considering stress to think in terms of those impositions or events that extend physiological mechanisms beyond their normal variation or function during homeostasis (Brett, 1958).

II. FRAMEWORK OF THE STRESS RESPONSE

Stressful situations result in a cascade of events that are transduced centrally and communicated via the nervous and endocrine systems (Mazeaud *et al.,* 1977; Donaldson, 1981; Schreck, 1981; Barton and Iwama,

1991) and predispose fish to disease (Wedemeyer, 1970, 1974) (Fig. 1). The classical physiological stress response typically involves the release of catecholamines and cortisol, the main interrenal steroid of teleosts. These

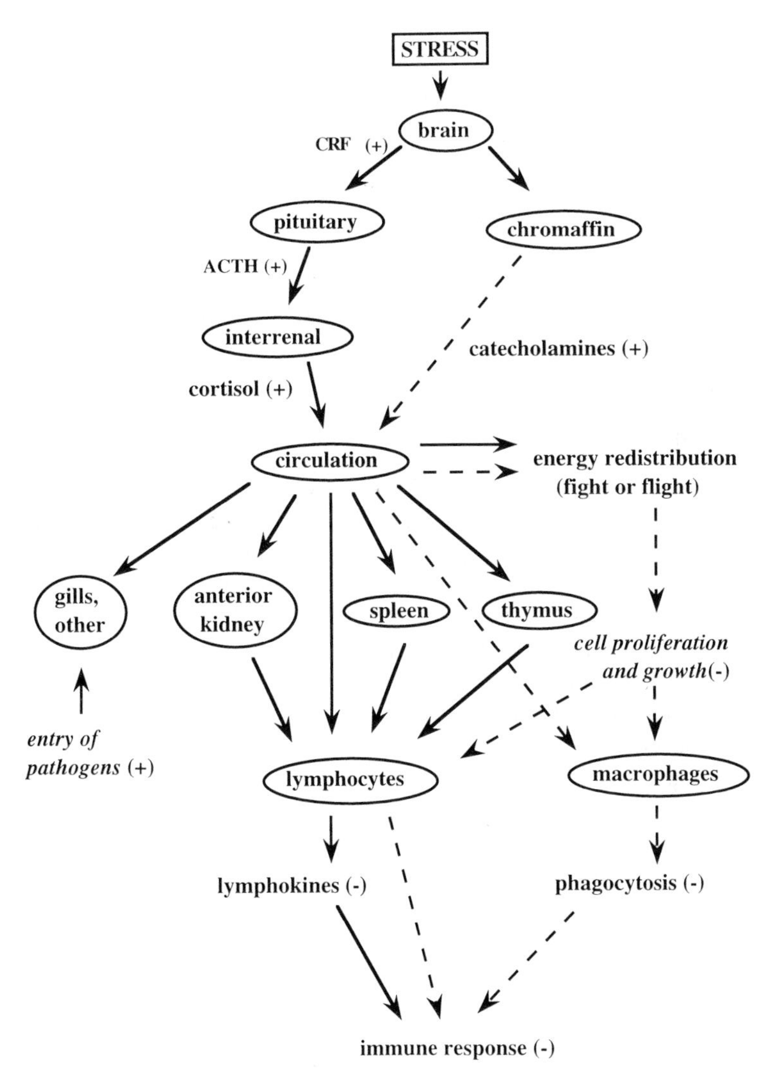

Fig. 1. Schematic of the physiological stress response and mechanisms whereby the immune system is affected. Pathways denoted by solid arrows are those mediated by the hypothalamic-hypophyseal-interrenal axis, and those with dashed arrows are mediated by catecholamines. Italicized words denote modes of action that are inferred but not based on direct data. (+) and (−) indicate the direction of the response.

hormones directly or indirectly result in the secondary and tertiary responses that can be generally classified as those involved with energy mobilization and metabolism, hydromineral balance, and other major physiological functions that may affect disease resistance (Schreck, 1981; Columbo *et al.,* 1990; Schreck and Li, 1991). Energy budgets are adjusted, making shifts such that readily usable carbohydrate sources like glucose are available from the circulation during stress. Adjustments to osmoregulatory systems are also made, apparently in response to water and electrolyte imbalances that accompany stress, and this is an energy consumptive process (Schreck, 1982). Any restructuring of energy sources may have deleterious effects on fish health because the process of cellular and humoral disease resistance is energy demanding. Reproduction can also be disrupted by stress (Ayson, 1989; Pottinger *et al.,* 1995; Contreras-Sanchez *et al.,* 1996), a time in the life cycle when fish can be particularly vulnerable to pathogens as discussed subsequently.

A. Stress and Disease Resistance

Circulating levels of leukocytes change through time when fish are stressed (Weinreb, 1958; Pickford *et al.,* 1971a,b,c; McLeay, 1975; Ellsaesser and Clem, 1987; Schreck *et al.,* 1993). Changes following physical types of stressors include depression in the number of lymphocytes in the circulation (Pickering *et al.,* 1982; Pickering and Pottinger, 1987; del Vale *et al.,* 1995) or the number of these cells relative to the number of erythrocytes (Barton *et al.,* 1987; Barton and Zitzow, 1995). Peters and Schwarzer (1985) provided a review of changes in hemopoietic tissue of rainbow trout, *Oncorhynchus mykiss,* consequent to stress, with the general finding that production of new blood cells was reduced and also that the final destruction of immune-competent cells contributed to susceptibility to disease. Negative effects of stress on disease resistance appear to be due to the depression of antibody-synthesizing capability via effects on lymphocytes. Maule *et al.* (1989) reported that juvenile chinook salmon, *O. tshawytscha,* that had been stressed by handling were more vulnerable to *Vibrio anguillarum* than were unstressed controls. This debilitation in disease resistance was correlated with a depression in the ability of pronephric and splenic lymphocytes to produce antibody.

Even very brief stressful experiences can result in rather long-lasting effects. Chinook salmon juveniles stressed by handling for 30–60 s exhibited severe suppression in numbers of antibody-producing pronephric lymphocytes; there was a rebound in immune capacity by the next day. Disease resistance capacity was also enhanced in the stressed fish over that of controls at 1 day after the brief stress in experiments conducted in the fall, and 1 and 7 days post-stress in tests run in the spring (Maule *et al.,* 1989).

The response of certain elements of the immune system can also be quite rapid after the onset of stress. Four plasma antigens increased in concentration within 5 min of a handling stress in rainbow trout, suggesting that the acute phase response is independent of gene activation (Demers and Bayne, 1994).

Stress affects bacteriocidal activity of leukocytes. Atlantic salmon stressed by confinement had reduced leukocyte respiratory burst and bacteriocidal activities (Thompson *et al.,* 1993). Cells from pronephric or circulatory populations may have different responses to the stressor. The production of antibody following immunization with the bacterin *Aeromonas salmonicida* was reduced in the stressed fish (Thompson *et al.,* 1993).

Little work has been done to assess the effects of stress on phagocytes in fish. However, evidence suggests that stressors such as anesthetic plus injection or noise, and confinement of rainbow trout resulted in reduced phagocytic activity of macrophages of both head kidney and spleen (Narnaware *et al.,* 1994).

B. Endocrine Mediation

Stress effects on the immune system may be mediated by the endocrine system (Clem *et al.,* 1990). Catecholamines and corticosteroids are the major stress hormones of vertebrates. Essentially nothing is known about the adrenergic control systems in fishes (Nilsson, 1984) relative to mediation of physiological factors responsible for health during stress. Cortisol is the major corticosteroid produced by teleostean fishes (Columbo *et al.,* 1971; Henderson and Kime, 1987; Schreck, 1992), and considerable evidence exists that this steroid has a direct effect on the immune system and disease resistance (Chilmonczyk, 1982; Peters and Schwarzer, 1985; Kaattari and Tripp, 1987; Tripp *et al.,* 1987; Thomas and Lewis, 1987; Maule *et al.,* 1987, 1989; Pickering, 1989; Pickering and Pottinger, 1989; Schreck *et al.,* 1993). Cortisol can affect distribution of leukocytes (Rogers and Matossian-Rogers, 1982), result in lymphocytopenia (Pickering, 1984), and suppress the growth of lymphocytes (Grimm, 1985). The stress-related immunosuppression is in large part due to this corticosteroid. Coho salmon, *O. kisutch,* that had their plasma levels of cortisol elevated by implants to concentrations characteristic of stressed fish demonstrated changes in numbers of leukocytes similar to those found in acutely stressed fish (Maule and Schreck, 1990a). In other experiments, coho salmon juveniles receiving cortisol implants had reduced numbers of splenic antibody-producing cells, splenic lymphocytes, and circulating leukocytes. Oral administration of cortisol also resulted in a slight decrease in IgM concentration in masu salmon (*O. masou*) (Nagae *et al.,* 1994b). A clear effect of cortisol on disease resistance was demonstrated in these fish, for they experienced greater mortality than controls when exposed to *Vibrio anguillarum* (Maule *et al.,* 1987). Similarly, exogenous cortisol administered to brown trout, *Salmo trutta,* increased their susceptibility to *Saprolegnia* infection and furunculosis (Pickering and Duston, 1983). Administration of synthetic corticosteroid to juvenile channel catfish, *Ictalurus punctatus,* increased the susceptibility of the fish to *Edwardsiella ictaluri* infection (Antonio and Hedrick, 1994). Confounding the relationship between corticosteroids and health is the fact, as shown by Robertson *et al.* (1987), that disease may also act as a stressor, thereby resulting in elevated cortisol titers.

Some species of teleosts do not fit the general paradigm for the endocrine stress response, i.e., extremely rapid (seconds) secretion of catecholamines and rapid (seconds to minutes) secretion of cortisol following the onset of stress. For example, plasma cortisol required an hour to become elevated after the onset of stress in the sea raven, *Hemitripterus americanus,* and only severe stress resulted in increases in catecholamine concentrations (Vijayan and Moon, 1994).

The mechanism of action of cortisol appears to be through a specific receptor in the leukocytes (Maule and Schreck, 1990b, 1991). Acute stress increased the number of receptors in splenic and pronephric leukocytes, reduced the affinity for the hormone in splenic leukocytes, but did not affect the affinity for the hormone in pronephric leukocytes. However, prolonged stress lowered the affinity for cortisol in receptors from both head kidney and spleen. Leukocytes of coho salmon after prolonged cortisol exposure had responses in cortisol receptors similar to those found in chronically stressed fish (Maule and Schreck, 1990b, 1991). Experiments where leukocytes were incubated in cortisol-enriched medium indicated that cortisol's effects are mediated through some lymphokine (Tripp *et al.,* 1987).

Administration of cortisol to channel catfish caused hematologic and immunologic changes in circulating leukocytes similar to those caused by acute physical stress. Basically, exposure to stress or the hormone for hours resulted in a decline in the number of circulating lymphocytes and an increase in neutrophils; the remaining lymphocytes in the circulation were no longer capable of responding to mitogenic stimuli. However, the effects of cortisol on depression of the mitogenic response may not be direct (Ellsaesser and Clem, 1987).

Corticosteroids are also known to induce apoptosis in homeotherms (Wyllie, 1987; Motyka and Reynolds, 1991). Peripheral leukocytes of channel catfish stressed by confinement in a net for hours did not have as much apoptosis as those from unstressed controls. This was apparently due to a factor in the circulation of the stressed fish, for cells of unstressed fish cultured in plasma from stressed fish exhibited reduced apoptosis. However, the presumptive elevation in cortisol concentration in the stressed fish may not have been the responsible factor, because *in vitro* experiments did not show any effects of the steroid on apoptosis in leukocytes (Alford *et al.,* 1994).

Mediation of phagocytosis via the endocrine system is probable, but *in vitro* experiments with rainbow trout macrophages incubated with cortisol failed to show any effects. However, both alpha- and beta-adrenergic agonists were very depressive (Narnaware *et al.,* 1994). Catecholamine evidently affects the metabolism of reactive oxygen species in salmonids (Bayne and Levy, 1991a,b).

Most consequences of elevated corticosteroids appear to lead to mostly negative effects on disease resistance capacity, and it is difficult to reconcile why such a physiological response to stress would not be maladaptive. Recent work with mammalian systems suggests that while certain aspects of immunity are depressed during stress, those at the periphery of the animal are enhanced, thus facilitating immune challenge at the site of antigen entry into the organism (Dhabhar *et al.,* 1995; Dhabhar and McEwen, 1996; May, 1996).

III. VARIABLES THAT AFFECT HEALTH AND STRESS

A. Genetics and Health/Stress

An emerging literature that suggests that the ability of fish to resist pathogens has a relatively high heritability (Fjalestad *et al.,* 1991). Also, fish that inherit resistance to a pathogen maintain that resistance throughout their lives (Snieszko *et al.,* 1959). For example, individual fish of the same population may have different abilities to resist bacteria such as *Vibrio anguillarum, Aeromonas salmonicida,* and *Renibacterium salmoninarum* (Beacham and Evelyn, 1992a). Different genetic strains of chinook salmon and pink salmon, *O. gorbuscha,* also exhibited variation in resistance to these pathogens in challenge experiments (Beacham and Evelyn, 1992b,c). Atlantic salmon have a high heritability to susceptibility to the causative agents for furunculosis (Gjedrem *et al.,* 1991; Gjedrem and Gjoeen, 1995), bacterial kidney disease, and coldwater vibriosis (Gjedrem and Gjoeen, 1995); Arctic char, *Salvelinus alpinus,* have a high heritability to fungal resistance (Nilsson, 1992). Selective breeding appears successful at reducing the effects of dropsy in carp, *Cyprinus carpio,* and furunculosis and octomytosis in brook trout, *S. fontinalis,* and brown trout (Ilyassov, 1986). The genetic control of *Ceratomyxa shasta* resistance, however, does not fit a simple Mendelian model (Ibarra *et al.,* 1994).

Physiological and/or biochemical mechanisms conferring resistance to microparasites can have a strong genetic basis (Chevassus and Dorson, 1988). For example, Suzumoto *et al.* (1977), Pratschner (1978), and Winter *et al.* (1980) demonstrated a linkage between resistance to the causative agent of bacterial kidney disease and transferrin genotype of coho salmon. Experiments with rainbow trout and Atlantic salmon demonstrated that the antibody response to diphtheria toxoid was under considerable genetic control. The heritability was found to be around 0.2 in the trout and 0.12 in the salmon (Eide *et al.,* 1994). Unfortunately, there is presently no understanding of whether or not selection for traits associated with the physiological stress response and those concerned with resistance to a specific pathogen are in any way linked. However, genes involved in disease resistance mechanisms are beginning to be identified (Trobridge and Leong, 1994). Variation in serum hemolytic activity appears to be under additive genetic control in Atlantic salmon (Roeed *et al.,* 1992). Genetic strains of coho salmon and tilapia, *Oreochromis niloticus,* were experimentally infected with *Vibrio;* the strains having the greatest resistance to the pathogens displayed a more active natural immune system as judged by measurement of phagocyte respiratory burst activity, plasma lysozyme activity, and differential leukocyte counts (Balfry *et al.,* 1994). It also appears that genotypic correlates of resistance to one pathogen may not be generalized to others. For example, while Winter *et al.* (1980) detected a strong correlation between salmon transferrin genotype and *R. salmoninarum* resistance, no such correlation was evident for the causative agent of vibriosis.

There can also be genetic variants of pathogens that differ in their virulence (Engelking and Leong, 1989; Engelking *et al.,* 1991; Drolet *et al.,* 1993; Bootland *et al.,* 1994; Kim *et al.,* 1994a,b; Leong, 1994). It is entirely possible that fish stocks and pathogen strains may have evolved as coadapted gene complexes such that the microorganism does not "kill" (viz., display extreme virulence) its host. Consequently, one could infer that the effects of stress on pathogen resistance might be less severe.

Stressors can also be selective forces. There is a genetic basis for the physiological response of fish to stressors other than pathogenic microorganisms (Schreck, 1981; Schreck and Li, 1991). Individual fish in a population may have differing genetically determined physiological responses to a stressor. For example, while the corticosteroid stress response was surprisingly similar in two different strains of trout (Pottinger and Moran, 1993), individual rainbow trout or Atlantic salmon can have different circulating levels of cortisol following a stressful experience, and this difference has a genetic basis (Fevolden *et al.,* 1991, 1993; Pottinger *et al.,* 1992, 1994). Using selective breeding they were able to create populations of trout with a "low" or a "high" cortisol response when challenged with the same stressor. Genetic lines of rainbow trout selected for high or low cortisol responses to stress had differing but inconsistent resistances to pathogens. While the "low stress" line was more resistant to *A. salmonicida* challenge, it was less resistant when challenged with *V. anguillarum* (Fevolden *et al.,* 1992). Other physiological traits that respond to stress are also heritable and some of these correlate to general health. For example, approximately one half of the response in hematocrit and red blood cell numbers to physical disturbance of the ayu, *Plecoglossus altivelis,* could be explained by genetic control (Del Valle and Taniguchi, 1995).

B. Environment and Health/Stress

Considerable evidence indicates that environmental conditions affect both health and disease resistance in fishes (see Chapters 6 and 7; Pulsford *et al.,* 1995). Numerous environmental variables have been shown to affect the physiological stress response in teleosts (Schreck, 1981; Schreck and Li, 1991). Environmental conditions and seasonality appear to affect circulating lymphocytes (McLeay, 1975). Environmental stress (holding the fish in the laboratory) in killifish, *Fundulus heteroclitus,* produced a factor that depressed the immunocytoadherence response to red blood cells (Miller and Tripp, 1982a). This factor was most likely a low or very low density lipoprotein. Hypophysectomy decreased the amount of the immune inhibition, suggesting that production of the factor is under endocrine control. In this regard, the stressor apparently affects the subpopulation of lymphocytes that are analogous to the mammalian T helper cell (Miller and Tripp, 1982b).

Temperature appears to be a major factor that can modify the response to stress. Barton and Schreck (1987) found that temperature affected the carbohydrate response to stress more than the interrenal response in juvenile chinook salmon. Diet and nutritional status also modified the stress response in chinook salmon (Barton *et al.,* 1988) as did prior exposure to stress; chinook salmon exposed to a series of acute stresses experienced cumulative physiological responses to each subsequent stressor (Barton *et al.,* 1986). Thermal shock, however, did not increase susceptibility of rainbow trout to *Flexibacter columnaris,* and in some tests resistance may actually have been enhanced by the temperature (Poston *et al.,* 1985). It is difficult to discern when temperature becomes stressful to fish; in general, fish reared in cooler water may have lower immune responses than those from warmer conditions (Rijkers *et al.,* 1980; Avtalion, 1981; Bly and Clem, 1992).

Presence of suspended solids in the water may stress fish and lower their ability to resist pathogens. Rainbow trout exposed to suspended volcanic ash were more susceptible to *F. columnaris* (Poston *et al.,* 1985). Similarly, coho salmon and rainbow (steelhead) trout exposed to high but natural levels of suspended topsoil, kaolin clay, and volcanic ash experienced temporary elevations in plasma cortisol. These fish had reduced disease resistance capacity, as demonstrated in trout that had been held in water with suspended topsoil for 2 days and then challenged with *V. anguillarum* (Redding *et al.,* 1987).

Unfortunately, there is little evidence linking the effects of stress on the immune systems or other disease resistance mechanisms under various environmental conditions. General environmental stress associated with hypoxic and hypercarbic conditions in a small, weed-infested pond caused decreases in macrophages in several species including *Labeo rohita, Cirrhinus mrigala,* and *Catla catla.* The main secondary effects were due to *Argulus* sp., *Gyrodactylus* sp., *Trichodina* sp., and *Myxobolus* sp. (Radheyshyam *et al.,* 1993). Salonius and Iwama (1991, 1993) concluded that the differences in plasma cortisol concentration, number of pronephric antibody producing cells, hematocrits, and differential blood cell counts observed among hatchery and wild coho and chinook salmon stocks exposed to acute handling stress can be attributed to rearing environment because the fish were of the same genetic strain.

A fish's social environment may also affect its health and general well-being. Social rank in dominance hierarchies may correlate to stress, with the individuals that are most subordinate being the most stressed. This has been inferred in a number of species of teleosts (Erickson, 1967; Noakes and Leatherland, 1977; Delventhal, 1978; Klinger *et al.,* 1979; Scott and Currie, 1980) and confirmed by Ejike and Schreck (1980) who found an inverse relationship between social rank and plasma cortisol concentrations in coho salmon. Peters *et al.* (1980) found that, relative to dominant individuals, subordinate eels, *Anguilla anguilla,* had decreased spleen weights and total leukocyte counts, but increased leukocrits due to increases in the number of large granulocytes. Subordinate fish also do not appear as capable as fish higher up in the social hierarchy at resisting pathogens. Assessment of rainbow trout that were subordinate showed classical signs of stress and greater infection rates when challenged with *Aeromonas hydrophila.* In addition, the pathogen spread to more organs in these fish (Peters *et al.,* 1988). Hierarchical status apparently affects cell-mediated immunity. Aggressive encounters between two tilapia species resulted in suppression of nonspecific cytotoxicity and mitogen-stimulated proliferation of head kidney leukocytes in subordinates. These effects appeared to be mediated by humoral factor(s) (Faisal *et al.*, 1989). Basically, intraspecific social interactions suppress pathogen-defense mechanisms and contribute to susceptibility to disease (Peters and Schwarzer, 1985).

It is difficult to separate stress due to aggressive encounters from other density-dependent interactions such as those found under normal aquaculture rearing densitites. Evaluation of coho salmon reared under various densities that all resulted in "good production" of fish found that resting plasma cortisol concentration was related to density and metabolic waste loading; these relationships changed over the period of smoltification (Patiño *et al.*, 1986). An inverse relationship was apparent between rearing density and the ability of the fish to resist *V. anguillarum* challenge (Schreck *et al.*, 1985). Juvenile coho salmon that had been reared at a higher fish density for several weeks and then reared for 2 weeks under lower density conditions had plasma cortisol and numbers of antibody-producing cells that were similar to those that had been reared continuously at the lower density in contrast to those raised continuously at the higher density. This suggests that the response to subtle differences in rearing conditions is quite dynamic (Maule *et al.*, 1987).

Handling and crowded rearing conditions also affect resting cortisol levels and the ability of Atlantic salmon, *Salmo salar,* to form antibody-producing cells (Mazur and Iwama, 1993). Nonspecific immune parameters can be affected by crowded rearing conditions (Yin *et al.*, 1995). Serum lysozyme activity and bactericidal complement activity were depressed and cortisol concentrations were elevated in channel catfish at high population density. Phagocytic activity was not affected. These fish also experienced a transient reduced ability to resist *A. hydrophila* infection (Yin *et al.*, 1995).

C. Ontogeny and Health/Stress

The ability of fish to resist pathogens and respond to stress, and the interactive effects of stress and disease resistance capacity of fish, change seasonally and through ontogeny (Fig. 2). The ontogeny of the immune system in fish has been described by Ellis (1988). Passive immunity during early development may be initiated by vertical transfer from maternal sources into the egg (Brown *et al.*, 1994; Yousif *et al.*, 1994a,b, 1995). Antibodies to bacterial kidney disease, if injected into maturing female coho salmon, help confer a degree of immunity in developing embryos (Brown *et al.*, 1990).

1. Early Life History and Juvenile

Unfertilized eggs receive a repertoire of hormones including cortisol and the sex steroids from maternal sources that are present during early

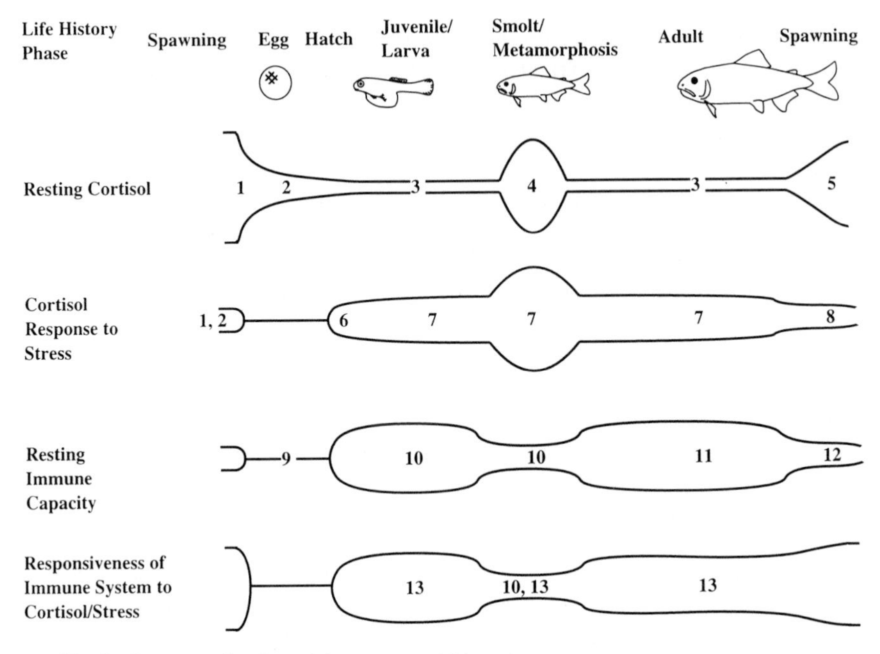

Fig. 2. Conceptualization of the pattern of (1) resting cortisol, (2) stress-stimulated cortisol, (3) antibody production capacity of resting fish, and (4) the ability of cortisol (stress) to suppress the immune system (greater height indicates greater leukocyte suppressibility) at different life history stages in fish with larval or smolt stages. The patterns are based to a large extent on information published on salmonids, the most studied fishes in this context. The series starts with the parent at spawning and concludes with spawning fish; it can, therefore, be considered a continuum. The height of each pattern reflects the magnitude of the concentration or response and is figurative and not precise. Numbers refer to references that support the pattern at a particular stage. Patterns depicted without numbers are based on my inference. References: 1. Caldwell *et al.,* 1991; Contreras-Sánchez, 1996. 2. Yeoh *et al.,* 1996a,b; Contreras-Sánchez, 1996. 3. Barton *et al.,* 1985; Schreck, 1992. 4. Barton *et al.,* 1985; de Jesus *et al.,* 1991; Nagae *et al.,* 1994. 5. Hane and Robertson, 1959; Fagerlund, 1967; Delahunty *et al.,* 1979; Donaldson and Fagerlund, 1968, 1972; Maule, 1996. 6. Pottinger and Musowe, 1994; Barry *et al.,* 1995; Feist and Schreck, 1996, and unpublished data. 7. Specker, 1982; Barton *et al.,* 1985; de Jesus *et al.,* 1991. 8. Sumpter *et al.,* 1987. 9. Brown *et al.,* 1994; Yousif *et al.,* 1994a,b. 10. Maule *et al.,* 1987, 1989; Muona and Sovio, 1992; Nagae *et al.,* 1994a. 11. Maule *et al.,* 1996. 12. Robertson and Wexler, 1960; Robertson *et al.,* 1961a,b; Hane *et al.,* 1966; Richards and Pickering, 1978; Pickering and Christie, 1980; Iida *et al.,* 1989; Ridgway, 1960, 1962; Maule *et al.,* 1996. 13. Maule and Schreck, 1990a; Maule *et al.,* 1993.

development. This has been demonstrated in tilapia (Rothbard *et al.,* 1987), salmonids (Feist *et al.,* 1990; de Jesus and Hirano, 1992; Schreck *et al.,* 1991; Feist and Schreck, 1996; Yeoh *et al.,* 1996a,b), and Japanese flounder, *Paralichthys olivaceus* (de Jesus *et al.,* 1991). Interestingly, it appears that the female may buffer her eggs and consequently protect resulting embryos from debilitatingly high corticosteroid concentrations that could result from her own elevated corticosteroid levels that accompany final maturation and spawning. Contreras-Sànchez (1996) found that plasma of female rainbow trout contained greatly elevated cortisol titers at the time of spawning in both stressed and unstressed individuals, while the ovarian fluid had considerably lower levels, similar to those found in ovulated eggs. Fry that were the progeny of stressed females had the same resistance to vibriosis as did those from unstressed controls, even though eggs and resulting progeny from females that had received physical disturbances for many weeks prior to spawning tended to be more variable in size and quality. The mechanism of this buffering in rainbow trout can be attributed to the elevation in concentration of corticosteroid-binding globulin as spawning approaches (Caldwell *et al.*, 1991). The globulin reduces the amount of free cortisol that could pass into the ovarian fluid and, subsequently, the egg.

The hypothalamo-hypophyseal-interrenal axis becomes responsive to stress rather early. Developing rainbow trout respond to stress by producing cortisol within about a week of hatching (Pottinger and Musowe, 1994; Barry *et al.*, 1995), several weeks before the yolk sac is absorbed. Data from my laboratory (Grant Feist, unpublished) demonstrated a cortisol-stress response in rainbow trout within 1 to 2 days posthatching.

Little is known about dynamics or involvement of endogenous hormones such as cortisol in species that undergo metamorphosis, other than work by de Jesus *et al.* (1991) who showed that concentrations of this corticosteroid in the bodies of Japanese flounder is dynamic during this period. Nothing is known about stress and health of fish at this stage.

More is known about stress, hormonal changes, and risk to infection during the period of parr-smolt transformation in anadromous salmonids. The pituitary-interrenal axis becomes activated during smoltification (McLeay, 1975), and the interrenal tissue undergoes hypertrophy and resting plasma levels of cortisol increase (Specker, 1982; Barton *et al.*, 1985). The metabolic clearance of cortisol also becomes reduced (Patiño *et al.*, 1985). During this period the cortisol stress response becomes elevated. For example, coho salmon given identical acute stresses over the course of smoltification had progressively greater cortisol elevations above resting levels (Barton *et al.*, 1985). Salmonids also appear to be more at risk of infection during this period of their life cycle, most likely due to a reduced number of cells that produce antibody in response to antigen. Maule *et al.* (1987) found that the number of splenic antibody-producing cells decreased during this period; in addition, the total number of circulating leukocytes relative to erythrocytes and in the spleen were reduced. During the parr-smolt transformation period, the capacity to produce antibody declined, but the proportion of splenic leukocytes that produced antibody-producing cells was greater than that of pronephric leukocytes. In addition, *in vitro* studies with splenic leukocytes of parr showed that these cells were generally more sensitive to the suppressive effects of cortisol compared to leukocytes from head kidney that were insensitive; these trends reversed somewhat during smoltification. Glucocorticoid receptor numbers increased and affinity for ligand decreased in pronephric cells at about the time that they became responsive to suppression by cortisol (Maule *et al.*, 1993). Lysozyme activities and leukocyte and lymphocyte numbers also decrease in smolting Atlantic salmon (Muona and Sovio, 1992), and serum concentration of immunoglobulin M changes over the parr-smolt transformation in masu salmon (Nagae *et al.*, 1994a). The interrelationships among stage during parr-smolt transformation, resting cortisol titer, the cortisol response to stressors, the immune system, and disease resistance capacity is obviously complex. As smolts, the fish appear more sensitive to stress and are more vulnerable to infection than are younger or older individuals.

2. Adult

Mature fish appear to be less capable of resisting pathogens than fish at other maturational stages. In mature salmonids the thymus becomes involuted, and pronephric and splenic lymphocytes decline in number (Robertson and Wexler, 1960). There are other immune deficiencies at this time as well including lowered bactericidal serum activity or nonspecific antibody production (Iida *et al.*, 1989; Maule *et al.*, 1996), and the inability to readily form isohemagglutinins (Ridgway, 1960, 1962).

The ability of fish to resist pathogens decreases around the time of spawning, when infection rates are often elevated (Richards and Pickering, 1978; Pickering and Christie, 1980). Immunodeficiencies become evident during this period of the life cycle (Ridgway, 1960, 1962; Iida *et al.*, 1989) when the fish appear more sensitive to stressors (Schreck *et al.*, 1995a). Circulating levels of cortisol become elevated during spawning in goldfish, *Carassius auratus*, (Delahunty *et al.*, 1979) and particularly in semelparous species such as Pacific salmon, *Oncorhynchus* spp. (Hane and Robertson, 1959; Fagerlund, 1967; Donaldson and Fagerlund, 1968, 1972; Maule *et al.*, 1996). Pacific salmon become Cushingoid around the time of spawning and eventually die due to infection (Robertson and Wexler, 1957; Robertson *et al.*, 1961a,b; Hane *et al.*, 1966). They have hyperplasic interrenals and elevated circulating corticosteroid concentrations, and degenerated spleen, thymus, kidney, liver, and thyroid (Robertson *et al.*, 1961a,b). Chinook salmon adults that are in freshwater for many months prior to spawning exhibited elevated cortisol titers and few antibody-producing cells in the peripheral circulation; slight reversals in these trends were evident if the fish were held in cooler, more stable conditions (Maule *et al.*, 1996). In addition, as the fish matured, estradiol, testosterone, 11-ketotestosterone, and androstenedione concentrations correlated with the number of antibody-producing cells in the circulation of females but not males. Lysozyme activity potentially also correlates to these sex hormones (Maule *et al.*, 1996). The common paradigm is that the corticosteroids depress the immune systems of adult Pacific salmon, rendering them vulnerable to pathogenic insult that often becomes the proximate cause of death.

Chronic stress will reduce circulating levels of testosterone (Pickering *et al.*, 1987; Safford and Thomas, 1987; Sumpter *et al.*, 1987; Carragher *et al.*, 1989; Pickering, 1989; Carragher and Sumpter, 1990a,b), and high plasma androgen levels may reduce stress-induced cortisol production (Sumpter *et al.*, 1987; Pottinger *et al.*, 1995). This presents somewhat of an enigma, for testosterone is a major sex hormone in both females and males and becomes elevated around the final stages of maturation in fish (Fitzpatrick *et al.*, 1986, 1987). This is interesting because Slater and Schreck (1993) have shown that this androgen is nearly as immunosuppressive as cortisol when salmonid lymphocytes are incubated *in vitro* with the steroid. When lymphocytes were exposed to physiological levels of testosterone plus cortisol, characteristic of mature salmonid fish, effects of the steroids were nearly additive, producing strong suppression of antibody formation capacity. While both steroids appear to operate via receptor-regulated processes, each has its own receptor (Maule and Schreck, 1990b; Slater *et al.*, 1995a,b) and they apparently operate through different mechanisms. Cortisol appears to interfere with the production of some lymphokine such as an interleukin during early phases of the antibody formation pathway (Tripp *et al.*, 1987). Testosterone on the other hand apparently reduces immunocompetence by killing lymphocytes (Slater *et al.*, 1995a). If stress reduces the ability of mature fish to resist pathogens, the role of sex hormones may not be that important in this process. It is difficult to reconcile the immunosuppressive effect of testosterone and its suppression by stress, as stress should then provide a benefit with regard to disease resistance. Further, cortisol is clearly immunosuppressive and it should be lowered instead of elevated during maturation if androgens regulate corticosteroids in the unstressed animal as suggested. Direct experimentation to evaluate the role of testosterone on disease resistance is limited to the work of Slater and Schreck (unpublished data) who found that chinook salmon juveniles treated with androgen to elevate their circulating titers to near mature-fish levels had the same ability to resist *V. anguillarum* as control fish.

D. Condition and Health

Little work has been done to elucidate potential relations between the physical (and perhaps mental) condition of fish and their ability to resist stress, or to determine how this would affect their general health and disease resistance. Even efforts to relate physical condition of fish with their overall fitness have yielded equivocal results (Burrows, 1969; Wendt and Saunders, 1972; Cresswell and Williams, 1983; Shchurov *et al.*, 1986; Kazakov *et al.*, 1987; Mikhaylenko, 1990; Khovanskij *et al.*, 1992; Khovanskiy *et al.*, 1993; Evenson and Ewing, 1993; Hernandez *et al.*, 1993; Wiley *et al.*, 1993; Young and Cech, 1993). However, Schreck *et al.* (1995b) were able to use positive conditioning to physical disturbance to either enhance or maintain the ability of chinook salmon to resist *A. salmonicida* when challenged subsequent to a rather severe stressful experience. Conditioning not associated with reward also lessened the severity of the furunculosis when fish were exposed to pathogen following the stressful experience, but these fish did not perform as well as those that had been positively conditioned. Circulating levels of cortisol also returned to prestress levels more rapidly following the stress in fish that had been positively conditioned than in either those conditioned in a manner not associated with reward or in unconditioned control fish. Interestingly, the pathogen-resistance benefits of condition ing were apparently not associated with increased numbers of antibody-producing cells as determined by passive hemolytic plaque assay (Schreck *et al.*, 1995b).

IV. SYNTHESIS AND SUMMARY

Stress can have marked effects on the health of fishes. Prolonged exposure to stress is particularly detrimental to the health of fish. However, even very brief stressful experiences can depress certain aspects of the cellular and humoral immune systems, and consequently, lower pathogen-resistance capacity. The physiology of disease prevention and resistance is mediated by the endocrine system, particularly the steroid cortisol. Once a stressor is perceived, a cascade of neuroendocrine events ensues that generally leads to elevation of this steroid hormone in circulation; it then affects lymphocytes and antibody production. Action of this hormone and others appear to operate by activation of specific receptors in the lymphocyte. Stress can also depress macrophage activity and distribution of leukocytes into various body compartments.

The physiological stress response and the immune system are under genetic influence. In addition, the fish's past and present rearing environment affect both the physiological response to stress and the ability to resist pathogens. Different ontological stages have differing magnitudes in the physiological response to similar stressors and differing pathogen tolerances. In addition, the effect of physiological stress factors such as cortisol on disease defense mechanisms change through ontogeny.

ACKNOWLEDGMENTS

I very much appreciate the constructive criticism and advice provided by Drs. B. A. Barton, C. J. Bayne, G. K. Iwama, A. D. Pickering, and C. H. Slater during the preparation of this chapter.

REFERENCES

Alford, P. B., III, Tomasso, J. R., Bodine, A. B., and Kendall, C. (1994). Apoptotic death of peripheral leukocytes in channel catfish: Effect of confinement-induced stress. *J. Aquat. Anim. Health* **6,** 64–69.

Antonio, D. B., and Hedrick, R. P. (1994). Effects of the corticosteroid kenalog on the carrier state of juvenile channel catfish exposed to *Edwardsiella ictaluri. J. Aquat. Anim. Health* **6,** 44-52.

Avtalion, R. (1981). Environmental control of the immune response in fish. *CRC Crit. Rev. Environ. Control* **11,** 163–188.

Ayson, F. G. (1989). The effect of stress on spawning of broodfish and survival of larvae of the rabbitfish, *Siganus guttatus* (Block). *Aquaculture* **80,** 241–246.

Balfry, S. K., Shariff, M., Evelyn, T. P. T., and Iwama, G. K. (1994). The importance of the natural immune system in disease resistance of fishes. *Int. Symp. Aquatic Animal Health,* Seattle, WA. University of California, Davis, p. 110.

Barry, T. P., Parrish, J. J., and Malison, J. A. (1995). Ontogeny of the cortisol stress response in larval rainbow trout. *Gen. Comp. Endocrinol.* **97,** 57–65.

Barton, B. A., and Iwama, G. K. (1991). Physiological changes in fish from stress in aquaculture with emphasis on the response and effects of corticosteroids. *Annu. Rev. Fish. Dis.* **1,** 3–26.

Barton, B. A., and Schreck, C. B. (1987). Influence of acclimation temperature on interrenal and carbohydrate stress responses in juvenile chinook salmon (*Oncorhynchus tshawytscha*). *Aquaculture* **62,** 299–310.

Barton, B. A., and Zitzow, R. E. (1995). Physiological responses of juvenile walleyes to handling stress with recovery in saline water. *Prog. Fish-Cult.* **57,** 267–276.

Barton, B. A., Schreck, C. B., Ewing, R. D., Hemmingsen, A. R., and Patiño, R. (1985). Changes in plasma cortisol during stress and smoltification in coho salmon, *Oncorhynchus kisutch. Gen. Comp. Endocrinol.* **59,** 468–471.

Barton, B. A., Schreck, C. B., and Sigismondi, L. A. (1986). Multiple acute disturbances evoke cumulative physiological stress responses in juvenile chinook salmon. *Trans. Am. Fish. Soc.* **115,** 245–251.

Barton, B. A., Schreck, C. B., and Barton, L. D. (1987). Effects of chronic cortisol administration and daily acute stress on growth, physiological conditions, and stress responses in juvenile rainbow trout. *Dis. Aquat. Org.* **2,** 173–185.

Barton, B. A., Schreck, C. B., and Fowler, L. G. (1988). Fasting and diet content affect stress-induced changes in plasma glucose and cortisol in juvenile chinook salmon. *Prog. Fish-Cult.* **50,** 16–22.

Bayne, C. J., and Levy, S. (1991a). The respiratory burst of rainbow trout, *Oncorhynchus mykiss* (Walbaum), phagocytes is modulated by sympathetic neurotransmitters and the "neuro" peptide ACTH. *J. Fish. Biol.* **38,** 609–619.

Bayne, C. J., and Levy, S. (1991b). Modulation of the oxidative burst in trout myeloid cells by adrenocorticotropic hormone and catecholamines: Mechanisms of action. *J. Leukocy. Biol.* **50,** 554–560.

Beacham, T. D., and Evelyn, T. P. T. (1992a). Genetic variation in disease resistance and growth of chinook, coho, and chum salmon with respect to vibriosis, furunculosis, and bacterial kidney disease. *Trans. Am. Fish. Soc.* **121,** 456–485.

Beacham, T. D., and Evelyn, T. P. T. (1992b). Population and genetic variation in resistance of chinook salmon to vibriosis, furunculosis, and bacterial kidney disease. *J. Aquat. Anim. Health* **4,** 153–167.

Beacham, T. D., and Evelyn, T. P. T. (1992c). Population variation in resistance of pink salmon to vibriosis and furunculosis. *J. Aquat. Anim. Health* **4,** 168–173.

Bly, J. E., and Clem, L. W. (1992). Temperature and teleost immune functions. *Fish Shellfish Immunol.* **2,** 159–171.

Bootland, L. M., Lorz, H. V., Rohovec, J. S., and Leong, J. C. (1994). Experimental infection of brook trout with infectious hematopoietic necrosis virus types 1 and 2. *J. Aquat. Anim. Health* **6,** 14–148.

Brett, J. R. (1958). Implications and assessments of environmental stress. *In* "Investigations of Fish-power Problems" (P. A. Larkin, ed.), H. R. MacMillian Lectures in Fisheries, University of British Columbia, Vancouver, British Columbia.

Brown, L. L., Albright, L. J., and Evelyn, T. P. T. (1990). Control of vertical transmission of renibacterium salmoninarum by injection of antibiotics into maturing female coho salmon *Oncorhynchus kisutch. Dis. Aquat. Org.* **9,** 127–131.

Brown, L. L., Evelyn, T. P. T., and Iwama, G. K. (1994). Vertical transfer of passive immunity from coho salmon via their eggs. Int. Symp. Aquatic Animal Health, Seattle, WA. University of California, Davis, p. W-6.1.

Burrows, R. E. (1969). The influence of fingerling quality on adult salmon survivals. *Trans. Am. Fish. Soc.* **98,** 777–784.

Caldwell, C. A., Kattesh, H. G., and Strange, R. J. (1991). Distribution of cortisol among its free and protein-bound fractions in rainbow trout (*Oncorhynchus mykiss*): Evidence of control by sexual maturation. *Comp. Biochem. Physiol.* **99A,** 593–595.

Carragher, J. F., and Sumpter, J. P. (1990a). The effect of cortisol on the secretion of sex steroids from cultured ovarian follicles of rainbow trout. *Gen. Comp. Endocrinol.* **77,** 403–407.

Carragher, J. F., and Sumpter, J. P. (1990b). Corticosteroid physiology in fish. *Prog. Clin. Biol. Res.* **342,** 487–492.

Carragher, J. F., Sumpter, J. P., Pottinger, T. G., and Pickering, A. D. (1989). The deleterious effects of cortisol implantation on reproductive functions in two species of trout, *Salmo trutta* L. and *Salmo gairdneri* Richardson. *Gen. Comp. Endocrinol.* **76,** 310–321.

Chevassus, B., and Dorson, M. (1988). Genetics of resistance to disease in fishes. 3. Int. Symp. on Genetics in Aquaculture, (T. Gjedrem, ed.), pp. 83–107. Trondheim (Norway).

Chilmonczyk, S. (1982). Rainbow trout lymphoid organs: Cellular effects of corticosteroids and anti-thymocyte serum. *Dev. Comp. Immunol.* **6,** 271–280.

Clem, L. W., Bly, J. E., Ellsaesser, C. F., Lobb, C. J., and Miller, N. W. (1990). Channel catfish as an unconventional model for immunological studies. *J. Exp. Zool.* **4,** 123–125.

Colombo, L., Pesaventi, S., and Johnson, D. W. (1971). Patterns of steroid and metabolism in teleost and ganoid fishes. *Gen. Comp. Endocrin. Supp.* **3,** 245–253.

Colombo, L., Pickering, A. D., Schreck, C. B., and Belvedere, P. (1990). Stress-inducing factors and stress reaction in aquaculture. *In* "European Aquaculture Society Special Publication N. 12" (N. De Pauw, and R. Billard, eds.), pp. 93–112. Bredene, Belgium.

Contreras-Sánchez, W. M. (1996). Effects of stress on the reproductive performance and physiology of rainbow trout (*Oncorhynchus mykiss*). M.S. Thesis, Oregon State University, Corvallis, Oregon.

Cresswell, R. C., and Williams, R. (1983). Post-stocking movements and recapture of hatchery-reared trout released into flowing waters-effect of prior acclimation to flow. *J. Fish Biol.* **23,** 265–276.

de Jesus, E. G., and Hirano, T. (1992). Changes in whole body concentrations of cortisol, thyroid hormones, and sex steroids during early development of the chum salmon, *Oncorhynchus keta. Gen. Comp. Endocrinol.* **85,** 55–61.

de Jesus, E. G., Hirano, T., and Inui, Y. (1991). Changes in cortisol and thyroid hormone concentrations during early development and metamorphosis in the Japanese flounder, *Paralichthys olivaceus. Gen. Comp. Endocrinol.* **82,** 369–376.

Delahunty, G., Schreck, C. B., Specker, J., Olcese, J., Vodicnik, M. J., and deVlaming, V. (1979). The effects of light reception on circulating estrogen levels in female goldfish, *Carassius auratus:* Importance of retinal pathways versus the pineal. *Gen. Comp. Endocrinol.* **38,** 148–152.

del Valle, G., and Taniguchi, N. (1995). Genetic variation of some physiological traits of clonal ayu (*Plecoglosus altivelis*) under stressed and nonstressed conditions. *Aquaculture* **137,** 193–202.

del Valle, G., Taniguchi, N., and Tsujimura, A. (1995). Effects of stress on some hematological traits of chromosome manipulated ayu, *Plecoglosus altivelis. Suisanzoshoku Jpn. Aquacult. Soc.* **43,** 89–95.

Delventhal, H. (1978). Experimentelle Stressuntersuchungen am Europaischen Aal, *Anguilla anguilla* (Linne, 1758)—physiologische und ethologische Aspekte. Diplomarbeit, University of Hamburg, Hamburg.

Demers, N. E., and Bayne, C. J. (1994). Plasma proteins of rainbow trout (*Oncorhynchus mykiss*): immediate responses to acute stress. *In* "Modulators of Fish Immune Responses" (J. S. Stolen and T. C. Fletcher, eds.), Vol. 1, pp 1–10.

Dhabhar, F. S., and McEwen, B. S. (1996). Stress-induced enhancement of antigen-specific cell-mediated immunity. *J. Immunol.* **156,** 2608–2615.

Dhabhar, F. S., Miller, A. H., McEwen, B. S., and Spencer, R. L. (1995). Effects of stress on immune cell distribution dynamics and hormonal mechanisms. *J. Immunol.* **154,** 5511–5527.

Donaldson, E. M. (1981). The pituitary-interrenal axis as an indicator of stress in fish. *In* "Stress and Fish" (A. D. Pickering, ed.), pp 11–47. Academic Press, London.

Donaldson, E. M., and Fagerlund, U. H. M. (1968). Changes in the cortisol dynamics of sockeye salmon (*Oncorhynchus nerka*) resulting from sexual maturation. *Gen. Comp. Endocrinol.* **11,** 552–561.

Donaldson, E. M., and Fagerlund, U. H. M. (1972). Corticosteroid dynamics in Pacific salmon. *Gen. Comp. Endocrinol.* **3,** 254–265.

Drolet, B. S., Rohovec, J. S., and Leong, J. C. (1993). Serological identification of infectious hematopoietic necrosis virus in fixed tissue culture cells by alkaline phosphatase immunocytochemistry. *J. Aquat. Anim. Health* **5,** 265–269.

Eide, D. M., Linder, R. D., Stromsheim, A., Fjalestad, K., Larsen, H. J. S., and Roed, K. H. (1994). Genetic variation in antibody response to diphtheria toxoid in Atlantic salmon and rainbow trout. *Aquaculture* **127,** 103–113.

Ejike, C., and Schreck, C. B. (1980). Stress and social hierarchy rank in coho salmon. *Trans. Am. Fish. Soc.* **104,** 423–426.

Ellis, A. E. (1988). Ontogeny of the immune system in teleost fish. *In* "Fish-vaccination" (A. E. Ellis, ed.), pp. 20–30. Academic Press, London.

Ellsaesser, C. G., and Clem, L. W. (1987). Cortisol-induced hematologic and immunologic changes in channel catfish (*Ictalurus punctatus*). *Comp. Biochem. Physiol.* **88A,** 589–594.
Engelking, H. M., and Leong, J. C. (1989). Glycoprotein from infectious hematopoietic necrosis virus (IHNV) induces protective immunity against five IHNV types. *J. Aquat. Anim. Health* **1,** 291–300.
Engelking, H. M., Harry, J. B., and Leong, J. C. (1991). Comparison of representative strains of infectious hematopoietic necrosis virus by serological neutralization and cross-protection assays. *Appl. Environ. Microbiol.* **57,** 1372–1378.
Erickson, J. G. (1967). Social hierarchy, territoriality, and stress reactions in sunfish. *Physiol. Zool.* **40,** 40–48.
Evenson, M. D., and Ewing, R. D. (1993). Effect of exercise of juvenile winter steelhead on adult returns to Cole Rivers Hatchery, Oregon. *Prog. Fish-Cult.* **55,** 180–183.
Fagerlund, U. H. M. (1967). Plasma cortisol concentration in relation to stress in adult sockeye salmon during the freshwater stage of their life cycle. *Gen. Comp. Endocrinol.* **8,** 197–207.
Faisal, M., Chiappelli, F., Weiner, H., Ahmed, I., and Cooper, E. L. (1989). Role of endogenous opioids in modulating some immune functions in hybrid tilapia. *J. Aquat. Anim. Health* **1,** 301–306.
Feist, G., and Schreck, C. B. (1996). Brain-pituitary-gonadal axis during early development and sexual differentiation in the rainbow trout, *Oncorhynchus mykiss. Gen. Comp. Endocrinol.* **102,** 394–409.
Feist, G., Schreck, C. B., Fitzpatrick, M. S., and Redding, J. M. (1990). Sex steroid profile of coho salmon (*Oncorhynchus kisutch*) during early development and sexual differentiation. *Gen. Comp. Endocrinol.* **80,** 299–313.
Fevolden, S. E., Refstie, T., and Roeed, K. H. (1991). Selection for high and low cortisol stress response in Atlantic salmon (*Salmo salar*) and rainbow trout (*Oncorhynchus mykiss*). *Aquaculture* **95,** 53–65.
Fevolden, S. E., Refstie, T., and Roeed, K. H. (1992). Disease resistance in rainbow trout (*Oncorhynchus mykiss*) selected for stress response. *Aquaculture* **104,** 19–29.
Fevolden, S. E., Refstie, T., and Gjerde, B. (1993). Genetic and phenotypic parameters for cortisol and glucose stress response in Atlantic salmon and rainbow trout. *Aquaculture* **118,** 205–216.
Fitzpatrick, M. S., van Der Kraak, G., and Schreck, C. B. (1986). Profiles of plasma sex steroids and gonadotropin in coho salmon (*Oncorhynchus kisutch*) during the parr-smolt transformation and after implantation of cortisol. *Can. J. Fish. Aquat. Sci.* **44,** 161–166.
Fitzpatrick, M. S., Redding, J. M., Ratti, L. D., and Schreck, C. B. (1987). Plasma testosterone concentration predicts the ovulatory response of coho salmon (*Oncorhynchus kisutch*) to gonadotropin-releasing hormone analog. *Can. J. Fish. Aquat. Sci.* **44,** 1351–1357.
Fjalestad, K. T., Gjedrem, T., and Gjerde, B. (1991). Genetic improvement of disease resistance in fish: An overview. 4. Int. Symp. On Genetics in Aquaculture (G. A. E. Gall and H. Chen, eds.), pp. 65–74. Elsevier, Amsterdam (1993).
Gjedrem, T., and Gjoeen, H. M. (1995). Genetic variation in susceptibility of Atlantic salmon, *Salmo salar* L., to furunculosis, BKD, and cold water vibriosis. *Aquacult, Res.* **26,** 129–134.
Gjedrem, T., Salte, R., and Magnus-Gjoeen, H. (1991). Genetic variation in susceptibility of Atlantic salmon to furunculosis. *Aquaculture* **97,** 1–6.
Grimm, A. S. (1985). Suppression by cortisol of the mitogen-induced proliferation of peripheral blood leukocytes from plaice, *Pleuronectes platessa* L. *In* "Fish Immunology" (M. J. Manning and M. F. Tatner, eds.), pp. 263–271. Academic Press, London.
Hane, S., and Robertson, O. H. (1959). Changes in plasma 17-hydroxycorticosteroids accompanying sexual maturation and spawning of the Pacific salmon (*Oncorhynchus tshawytscha*) and rainbow trout (*Salmo gairdneri. Proc. Natl. Acad. Sci. U.S.A.* **45,** 886–893.

Hane, S., Robertson, O. H., Wexler, B. C., and Krupp, A. (1966). Adrenocortical response to stress and ACTH in Pacific salmon (*Oncorhynchus tshawytscha*) and steelhead trout (*Salmo gairdneri*) at successive stages in the sexual cycle. *Endocrinology* **78,** 791–800.

Henderson, I. W., and Kime, D. E. (1987). The adrenal cortical steroids. *In* "Vertebrate Endocrinology: Fundamentals and Biomedical Implications" (P. K. T. Pang and M. P. Schreibman, eds.), pp 121–142. Academic Press, London.

Hernandez, M. D., De Costa, J., Mendiola, P., and Zamora, S. (1993). Lactacidemia as an index of the response to adaptation to stressing exercise. *In* "4th National Congress on Aquaculture, 21–24 September 1993." Vilanova de Arousa, Galicia, Spain. 743–747.

Ibarra, A. M., Hedrick, R. P., and Gall, G. A. E. (1994). Genetic analysis of rainbow trout susceptibility to the myxosporean, *Ceratomyxa shasta. Aquaculture* **120,** 239–262.

Iida, T., Takahashi, K., and Wakabayashi, H. (1989). Decrease in the bactericidal activity of normal serum during the spawning period of rainbow trout. *Bull. Jpn. Soc. Sci. Fish.* **55,** 463–465.

Ilyassov, Y. I. (1986). Genetic principles of fish selection for disease resistance. World Symposium on Selection, Hybridization and Genetic Engineering in Aquaculture of Fish and Shellfish for Consumption and Stocking, Bordeaux (France). In "Selection, hybridization and genetic engineering in aquaculture (K. Tiews, ed.), pp. 455–469. FAO, Rome (1987).

Kaattari, S. L., and Tripp, R. A. (1987). Cellular mechanisms of glucocorticoid immunosuppression in salmon. *J. Fish Biol. Suppl. A* **31,** 129–132.

Kazakov, R. V., Minina, E. V., Shchurov, I. L., and Shustov, Y. A. (1987). Changes in the biochemical properties of the skin resulting from physical training of the hatchery-produced young freshwater *Salmo salar. J. Ichthyol.* **27,** 500–505.

Khovanskij, I. E., Natochin, Y. V., and Shakhatova, E. I. (1992). Effect of physical load on the osmoregulatory capacity of hatchery-reared young chum salmon, *Oncorhynchus keta. J. Ichthyol.* **32,** 133–139.

Khovanskiy, I. Y., Natochin, Y. V., and Shakhmatova, Y. I. (1993). Effect of physical exercise on osmoregulatory capability in hatchery-reared juvenile chum salmon, *Oncorhynchus keta. J. Ichthol.* **33,** 36–43.

Kim, C. H., Winton, J. R., and Leong, J. C. (1994a). Neutralization-resistant variants of infectious hematopoietic necrosis virus have altered virulence and tissue tropism. *J. Virol.* **68,** 8447–8453.

Kim, C. H., Winton, J. R., and Leong, J. C. (1994b). Characterization of neutralization-resistant IHNV variants. International Symposium on Aquatic Animal Health, Seattle, WA (USA). University of California, Davis, p. W-4.5.

Klinger, H., Peters, G., and Delventhal, H. (1979). Physiologische und morphologische effekte von sozialem stress beim Aal, *Anguilla anguilla* L. Verhandlungen der Deutsche Zoologischen Gesellschaft. 246.

Loeng, J. C. (1994). Molecular approaches to taxonomy, epizootiology, and diagnosis of viral diseases of fish. International Symposium on Aquatic Animal Health, Seattle, WA (USA). University of California, Davis.

Maule, A. G., and Schreck, C. B. (1990a). Changes in numbers of leukocytes in immune organs of juvenile coho salmon (*Oncorhynchus kisutch*) after acute stress or cortisol treatment. *J. Aquat. Anim. Health* **2,** 298–304.

Maule, A. G., and Schreck, C. B. (1990b). The glucocorticoid receptors in leukocytes and gill of juvenile coho salmon (*Oncorhynchus kisutch*). *Gen. Comp. Endocrinol.* **77,** 448–455.

Maule, A. G., and Schreck, C. B. (1991). Stress and cortisol treatment changed affinity and number of glucocorticoid receptors in leukocytes and gill of coho salmon. *Gen. Comp. Endocrinol.* **84,** 83–93.

Maule, A. G., Schreck, C. B., and Kaattari, S. L. (1987). Changes in the immune system of coho salmon (*Oncorhynchus kisutch*) during the parr-to-smolt transformation and after implantation of cortisol. *Can. J. Fish Aquat. Sci.* **44,** 161–166.

Maule, A. G., Tripp, R. A., Kaattari, S. L., and Schreck, C. B. (1989). Stress alters immune function and disease resistance in chinook salmon (*Oncorhynchus tshawytscha*). *J. Endocrinol.* **120,** 135–142.

Maule, A. G., Schreck, C. B., and Sharpe, C. (1993). Seasonal changes in cortisol sensitivity and glucocorticoid receptor affinity and number in leukocytes of coho salmon. *Fish Physiol. Biochem.* **10,** 497–506.

Maule, A. G., Schrock, R., Slater, C., Fitzpatrick, M. S., and Schreck, C. B. (1996). Immune and endocrine responses of adult chinook salmon during freshwater immigration and sexual maturation. *Fish Shellfish Immunol.* **6,** 221–233.

May, M. (1996). Skin-deep stress. *Am. Sci.* **84,** 224–225.

Mazeaud, M. M., Mazeaud, F., and Donaldson, E. M. (1977). Primary and secondary effects of stress in fish: Some new data with a general review. *Trans. Am. Fish. Soc.* **16,** 201–212.

Mazur, C. F., and Iwama, G. K. (1993). Handling and crowding stress reduces number of plaque-forming cells in Atlantic salmon. *J. Aquat. Anim. Health* **5,** 98–101.

McLeay, D. J. (1975). Variation in the pituitary-interrenal axis and the abundance of circulating blood-cell types in juvenile coho salmon, *Oncorhynchus kisutch,* during stream residence. *Can. J. Zool.* **53,** 1882–1891.

Mikhaylenko, V. C. (1990). Impact of training on the survival of lake salmon, *Salmo salar* sebago, juveniles. *J. Ichthyol.* **30,** 13–21.

Miller, N. W., and Tripp, M. R. (1982a). An immunoinhibitory substance in the serum of laboratory held killifish, *Fundulus heteroclitus* L. *J. Fish Biol.* **20,** 309–316.

Miller, N. W., and Tripp, M. R. (1982b). The effect of captivity on the immune response of the killifish, *Fundulus heteroclitus* L. *J. Fish Biol.* **20,** 301–308.

Motyka, B., and Reynolds, J. D. (1991). Apoptosis is associated with the extensive B-cell death in the sheep ileal Peyer's patch and the chicken bursa of Fabricius: A possible role in B-cell selection. *Eur. J. Immunol.* **21,** 1951–1958.

Muona, M., and Soivio, A. (1992). Changes in plasma lysozyme and blood leukocyte levels of hatchery-reared Atlantic salmon (*Salmo salar* L.) and sea trout (*Salmo trutta* L.) during parr-smolt transformation. *Aquaculture* **106,** 75–87.

Nagae, M., Fuda, H., Hara, A., Saneyoshi, M., and Yamauchi, K. (1994a). Changes in serum concentrations of immunoglobulin M (IgM), cortisol, and thyroxine (T sub (4)) during smoltification in the masu salmon, *Oncorhynchus masou. Fish. Sci.* **60,** 241–242.

Nagae, M., Fuda, H., Ura, K., Kawamura, H., Adachi, S., Hara, A., and Yamauchi, K. (1994b). The effect of cortisol administration on blood plasma immunoglobulin M (IgM) concentrations in masu salmon (*Oncorhynchus masou*). *Fish Physiol. Biochem.* **13,** 41–48.

Narnaware, Y. K., Baker, B. I., and Tomlinson, M. G. (1994). The effect of various stresses, corticosteroids, and adrenergic agents on phagocytosis in the rainbow trout, *Oncorhynchus mykiss. Fish Physiol. Biochem.* **13,** 31–40.

Nilsson, J. (1992). Genetic variation in resistance of Arctic char to fungal infection. *J. Aquat. Anim. Health* **4,** 126–128.

Nilsson, S. (1984). Adrenergic control systems in fish. *Mar. Biol. Lett.* 127–144.

Noakes, D. L. G., and Leatherland, J. F. (1977). Social dominance and interrenal cell activity in rainbow trout, *Salmo gairdneri* (Pisces, Salmonidae). *Environ. Biol. Fish.* **2,** 131–136.

Patiño, R., Schreck, C. B., and Redding, J. M. (1985). Clearance of plasma cortisol in coho salmon, *Oncorhynchus kisutch,* at various stages of smoltification. *Comp. Biochem. Physiol.* **82A,** 531–535.

Patiño, R., Schreck, C. B., Banks, J. L., and Zaugg, W. S. (1986). Effects of rearing conditions on the developmental physiology of smolting coho salmon. *Trans. Am. Fish. Soc.* **115,** 828–837.

Peters, G., and Schwarzer, R. (1985). Changes in hemopoietic tissue of rainbow trout under influence of stress. *Dis. Aquat. Org.* **1,** 1–10.

Peters, G., Delventhal, H., and Klinger, H. (1980). Physiological and morphological effects of social stress on the eel, *Anguilla anguilla* L. *Arch. Fisch. Wissen.* **30,** 157–180.

Peters, G., Faisal, M., Lang, T., and Ahmed, I. (1988). Stress caused by social interaction and its effect on susceptibility to Aeromonas hydrophila infection in rainbow trout, *Salmo gairdneri. Dis. Aquat. Org.* **4,** 83–89.

Pickering, A. D. (1981). Introduction. *In* "Stress and Fish" (A. D. Pickering, ed.), pp 1–9. Academic Press, London.

Pickering, A. D. (1984). Cortisol-induced lymphocytopenia in brown trout, *Salmo trutta* L. *Gen. Comp. Endocrinol.* **53,** 252–259.

Pickering, A. D. (1989). Environmental stress and the survival of brown trout, *Salmo trutta. Freshwater Biol.* **21,** 47–55.

Pickering, A. D., and Christie, P. (1980). Sexual differences in the incidence and severity of ectoparasitic infestation of the brown trout, *Salmo trutta* L. *J. Fish Biol.* **6,** 669–683.

Pickering, A. D., and Duston, J. (1983). Administration of cortisol to brown trout, *Salmo trutta* L., and its effects on the susceptibility to Saprolegnia infection and furunculosis. *J. Fish Biol.* **23,** 163–175.

Pickering, A. D., and Pottinger, T. G. (1987). Crowding causes prolonged leucopenia in salmonid fish, despite interrenal acclimation. *J. Fish Biol.* **32,** 701–712.

Pickering, A. D., and Pottinger, T. G. (1989). Stress responses and disease resistance in salmonid fish: Effects of chronic elevation of plasma cortisol. *Fish Physiol. Biochem.* **7,** 253, 258.

Pickering, A. D., Pottinger, T. G., and Christie, P. (1982). Recovery of the brown trout, *Salmo trutta* L., from acute handling stress: A time-course study. *J. Fish Biol.* **20,** 229–244.

Pickering, A. D., Pottinger, T. G., Carragher, J. F., and Sumpter, J. P. (1987). The effects of acute and chronic stress on the levels of reproductive hormones in the plasma of mature male brown trout, *Salmo trutta* L. *Gen. Comp. Endocrinol.* **68,** 249–259.

Pickford, G. E., Srivastava, A. K., Slicher, A. M., and Pang, P. K. T. (1971a). The stress response in the abundance of circulating leukocytes in the killifish, *Fundulus heteroclitus.* I. The cold-shock sequence and the effects of hypophysectomy. *J. Exp. Zool.* **177,** 89–96.

Pickford, G. E., Srivastava, A. K., Slicher, A. M., and Pang, P. K. T. (1971b). The stress response in the abundance of circulating leukocytes in the killifish, *Fundulus heteroclitus.* II. The role of catecholamines. *J. Exp. Zool.* **177,** 97–108.

Pickford, G. E., Srivastava, A. K., Slicher, A. M., and Pang, P. K. T. (1971c). The stress response in the abundance of circulating leukocytes in the killifish, *Fundulus heteroclitus.* III. The role of adrenal cortex and a concluding discussion of the leukocyte-stress syndrome. *J. Exp. Zool.* **177,** 109–118.

Poston, T. M., Neitzel, D. A., Aberneth, C. S., and Carlile, D. W. (1985). Effects of suspended volcanic ash and thermal shock on susceptibility of juvenile salmonids to disease. 8th Symp. On Aquatic Toxicology and Hazard Assessment (R. C. Bahner and D. J. Hansen, eds.), pp. 359–374. Fort Mitchell, KY.

Pottinger, T. G., and Moran, T. A. (1993). Differences in plasma cortisol and cortisone dynamics during stress in two strains of rainbow trout (*Oncorhynchus mykiss*). *J. Fish. Biol.* **43,** 121–130.

Pottinger, T. G., and Musowe, E. M. (1994). The corticosteroidogenic response of brown and rainbow trout alevins and fry during a "critical" period. *Gen. Comp. Endocrinol.* **95,** 350–362.

Pottinger, T. G., Pickering, A. D., and Hurley, M. A. (1992). Consistency in the stress response of individuals of two strains of rainbow trout *Oncorhynchus mykiss. Aquaculture* **103,** 275–298.

Pottinger, T. G., and Moran, T. A., and Morgan, J. A. W. (1994). Primary and secondary indices of stress in the progeny of rainbow trout (*Oncorhynchus mykiss*) selected for high and low responsiveness to stress. *J. Fish. Biol.* **44,** 149–163.

Pottinger, T. G., Balm, P. H. M., and Pickering, A. D. (1995). Sexual maturity modifies the responsiveness of the pituitary–interrenal axis to stress in male rainbow trout. *Gen. Comp. Endocrinol.* **98,** 311–320.

Pratschner, G. A. (1978). The relative resistance of six transferrin phenotypes of coho salmon (*Oncorhynchus kisutch*) to cytophagosis, furunculosis, and vibriosis. M.S. Thesis, University of Washington, Seattle.

Pulsford, A. L., Crampe, M., Langston, A., and Glynn, P. J. (1995). Modulatory effects of disease, stress, copper, TBT, and vitamin E on the immune system of flatfish. *Fish Shellfish Immunol.* **5,** 631–643.

Radheyshyam, Sharma, B. K., Sarkar, S. K., and Chattopadhyay, D. N. (1993). Environmental stress associated with disease outbreak in fish in an undrainable weed infested pond. *Environ. Ecol.* **11,** 118–122.

Redding, J. M., Schreck, C. B., and Everest, F. H. (1987). Physiological effects on coho salmon and steelhead of exposure to suspended solids. *Trans. Am. Fish. Soc.* **116,** 737–744.

Richards, R. H., and Pickering, A. D. (1978). Frequency and distribution of *Saprolegnia* infection in wild and hatchery-reared brown trout, *Salmo trutta* L., and char, *Salvelinus alpinus* (L.). *J. Fish Dis.* **1,** 69–82.

Ridgway, G. J. (1960). Blood types in pacific salmon. Special Sci. Report, Fisheries, no. 324. U.S. Fish and Wildlife Service.

Ridgway, G. J. (1962). Demonstration of blood types in trout and salmon by isoimmunization. *Ann. N.Y. Acad. Sci.* **97,** 111–118.

Rijkers, G. T., Frederix-Wolters, E. M. H., and Van Muiswinkel, W. B. (1980). The immune system of cyprinid fish. Kinetics and temperature dependence on antibody-producing cells in carp (*Cyprinus carpio*). *Immunology* **41,** 91–97.

Robertson, L., Thomas, P., Arnold, C. R., and Trent, J. M. (1987). Plasma cortisol and secondary stress responses of red drum to handling, transport, rearing density, and a disease outbreak. *Prog. Fish-Cult.* **49,** 1–12.

Robertson, O. H., and Wexler, B. C. (1957). Pituitary degeneration and adrenal tissue hyperplasia in spawning Pacific salmon. *Science* **25,** 1295–1296.

Robertson, O. H., and Wexler, B. C. (1960). Histological changes in the organs and tissues of migrating and spawning pacific salmon (genus *Oncorhynchus*). *Endocrinology* **66,** 222–239.

Robertson, O. H., Krupp, M. A., Favour, C. B., Hane, S., and Thomas, S. F. (1961a). Physiological changes occurring in the blood of the Pacific salmon (*Oncorhynchus tshawytscha*) accompanying sexual maturation and spawning. *Endocrinology* **68,** 733–746.

Robertson, O. H., Krupp, M. A., Thomas, S. F., Favour, C. B., Hane, S., and Wexler, B. C. (1961b). Hyperadrenocorticism in spawning, migratory, and nonmigratory rainbow trout (*Salmo gairdnerii*): Comparison with Pacific salmon (Genus *Oncorhynchus*). *Gen. Comp. Endocrinol.* **1,** 473–484.

Roeed, K. H., Fjalestad, K., Larsen, H. J., and Midthjel, L. (1992). Genetic variation in hemolytic activity in Atlantic salmon (*Salmo salar* L.). *J. Fish Biol.* **40,** 739–750.

Rogers, P., and Matossian-Rogers, A. (1982). Differential sensitivity of lymphocyte subsets to corticosteroid treatment. *Immunology* **46,** 841–848.

Rothbard, S., Moav, B., and Yaron, Z. (1987). Changes in steroid concentrations during sexual ontogenesis in tilapia. *Aquaculture* **83,** 153–166.

Safford, S. E., and Thomas, P. (1987). Effects of capture and handling on circulatory levels of gonadal steroids and cortisol in the spotted seatrout, *Cynoscion nebulosus. In* "Reproductive Endocrinology of Fish" (D. W. Idler, L. W. Crim, and J. M. Walsh, eds.), pp 312. Memorial University, St. John's, Canada.

Salonius, K., and Iwama, G. K. (1991). The effect of stress on the immune function of wild and hatchery coho salmon juveniles. 8th. Ann. Meet. Of the Aquaculture Association of Canada, (St. Andrews, N.B.). **91,** 47–49.

Salonius, K., and Iwama, G. K. (1993). Effects of early rearing environment on stress response, immune function, and disease resistance in juvenile coho (*Oncorhynchus kisutch*) and chinook salmon (*O. tshawytscha*). *Can. J. Fish. Aquat. Sci.* **50,** 759–766.

Schreck, C. B. (1981). Stress and compensation in teleostean fishes: Response to social and physical factors. *In* "Stress and Fish" (A. D. Pickering, ed.), pp 295–321. Academic Press, London.

Schreck, C. B. (1982). Stress and rearing of salmonids. *Aquaculture* **28,** 241–249.

Schreck, C. B. (1992). Glucocorticoids: Metabolism, growth, and development. *In* "The Endocrinology of Growth, Development, and Metabolism in Vertebrates." (M. P. Schreibman, C. G. Scanes, and P. K. T. Pang, eds.), pp. 367–392. Academic Press, New York.

Schreck, C. B., and Li, H. W. (1991). Performance capacity of fish: Stress and water quality. *In* "Aquaculture and Water Quality" (D. E. Brune and J. R. Tomasso, eds.), Advances in World Aquaculture, Vol. 3, 21–29. World Aquaculture Society, Baton Rouge, Louisiana.

Schreck, C. B., Patiño, R., Pring, C. K., Winton, J. R., and Holway, J. E. (1985). Effects of rearing density on indices of smoltification and performance of coho salmon, *Oncorhynchus kisutch. Aquaculture* **44,** 253–255.

Schreck, C. B., Fitzpatrick, M. S., Feist, G. W., and Yeoh, C.-G. (1991). Steroids: Developmental continuum between mother and offspring. *In* "Proc. 4th Internal. Symp. Reprod. Physiol. Fish" (A. P. Scott, J. P. Sumpter, D. E. Kime, and M. S. Rolfe, eds.), pp. 256–258. Fish Symp 91. Sheffield, England.

Schreck, C., Maule, A. G., and Kaattari, S. L. (1993). Stress and disease resistance. *In* "Recent Advances In Aquaculture" (J. F. Muir and R. J. Roberts, eds.), pp 170–175. Blackwell Scientific Publications. Oxford, UK.

Schreck, C. B., Fitzpatrick, M. S., and Currens, K. P. (1995a). Pacific Salmon (*Oncorhynchus* spp.). *In* "Broodstock Management and Egg and Larval Quality" (N. R. Bromage and R. J. Roberts, eds.). Blackwell Science Ltd., University Press, Cambridge.

Schreck, C. B., Jonsson, L., Feist, G., and Reno, P. (1995b). Conditioning improves performance of juvenile chinook salmon, *Oncorhynchus tshawytscha,* to transportation stress. *Aquaculture* **135,** 99–100.

Scott, D. B. C., and Currie, C. E. (1980). Social hierarchy in relation to adrenocortical activity in *Xiphophorus helleri* Heckel. *J. Fish Biol.* **16,** 265–277.

Selye, H. (1936). A syndrome produced by diverse nocuous agents. *Nature* **138,** 32.

Selye, H. (1950). Stress and the general adaptation syndrome. *Brit. Med. J.* **1,** 1383–1392.

Shchurov, I. L., Smirnov, Y. A., and Shustov, Y. A., (1986). Adaptation of hatchery-reared young salmon, *Salmo salar* L., to river conditions. 2. Behavior and feeding of trained young salmon in the river. *Vopr. Ikhtio.* **26,** 871–874.

Slater, C. H., and Schreck, C. B. (1993). Testosterone alters the immune response of chinook salmon (*Oncorhynchus tshawytscha*). *Gen. Comp. Endocrinol.* **89,** 291–298.

Slater, C. H., Fitzpatrick, M. S., and Schreck, C. B. (1995a). Androgens and immunocompetence in salmonids: Specific binding in and reduced immunocompetence of salmonid lymphocytes exposed to natural and synthetic androgens. *Aquaculture* **136,** 363–370.

Slater, C. H., Fitzpatrick, M. S., and Schreck, C. B. (1995b). Characterization of an androgen receptor in salmonid lymphocytes: Possible link to androgen induced immunosuppression. *Gen. Comp. Endocrinol.* **100,** 218–225.

Snieszko, S. F. (1974). The effects of environmental stress on outbreaks of infectious diseases of fish. *J. Fish Biol.* **6,** 197–208.

Snieszko, S. F., Dunbar, C. E., and Bullock, G. L. (1959). Resistance to ulcer disease and furunculosis in eastern brook trout (*Salvelinus fontinalis*). *Prog. Fish-Cult.* **21,** 111–116.

Specker, J. L. (1982). Interrenal function and smoltification. *Aquaculture* **28,** 59–66.

Sumpter, J. P., Carragher, J., Pottinger, T. G., and Pickering, A. D. (1987). The interaction of stress and reproduction in trout. *In* "Reproductive Physiology of Fish" (D. R. Idler, L. W. Crim, and J. M. Walsh, eds.), pp 299–302. Memorial Univ. Press, St John's, Newfoundland, Canada.

Suzumoto, B. K., Schreck, C. B., and McIntyre, J. D. (1977). Relative resistance of three transferrin genotypes of coho salmon (*Oncorhynchus kisutch*) and their hematological responses to bacterial kidney disease. *J. Fish. Res. Board Can.* **34,** 1–8.

Thomas, P., and Lewis, D. H. (1987). Effects of cortisol on immunity in red drum, *Sciaenops ocellatus. J. Fish Biol.* **31A,** 123–127.

Thompson, I., White, A., Fletcher, T. C., Houlihan, D. F., and Secombes, C. J. (1993). The effect of stress on the immune response of Atlantic salmon (*Salmo salar* L.) fed diets containing different amounts of vitamin C. *In* "European Aquaculture Soc." 19 (M. Carrillo, L. Dahle, J. Morales, P. Sorgeloos, N. Svennevig, and J. Wyban, eds.), p. 276. World Aquaculture Int. Conf., Torremolinos, Spain.

Tripp, R. A., Maule, A. G., Schreck, C. B., and Kaattair, S. L. (1987). Cortisol-mediated suppression of salmonid lymphocyte responses *in vitro. Develop. Comp. Immunol.* **11,** 565–576.

Trobridge, G. D., and Leong, J. C. (1994). Identification and characterization of the antiviral Mx genes of rainbow trout. International Symposium on Aquatic Animal Health, Seattle, WA (USA). University of California, Davis, p. W-21.4.

Vijayan, M. M., and Moon, T. W. (1994). The stress response and the plasma disappearance of corticosteroid and glucose in a marine teleost, the sea raven. *Can. J. Zool.* **72,** 379–386.

Wedemeyer, G. (1970). The role of stress in the disease resistance of fishes. In "A Symposium on Disease of Fishes and Shellfishes" (S. F. Snieszko, ed.), Spec. Publ. No. 5, pp. 30–35. Am. Fish. Soc., Washington, DC.

Wedemeyer, G. (1974). Stress as a predisposing factor in fish diseases. Fish Disease Leaflet 38. Fish and Wildlife Service, Washington, DC.

Wedemeyer, G. A., Meyer, F. P., and Smith, L. (1976). Environmental stress and fish diseases. *In* "Diseases of Fishes" Book 5 (S. F. Snieszko and H. R. Axelrod, eds.), T. F. H. Publications, Neptune City, NJ.

Weinreb, E. L. (1958). Studies on the histology and histopathology of the rainbow trout, *Salmo gairdneri* irideus. I. Hematology: Under normal and experimental conditions of inflammation. *Zoologica* **43,** 145–155.

Wendt, C. A. G., and Saunders, R. L. (1972). Changes in carbohydrate metabolism in young Atlantic salmon in response to various forms of stress. *Int. Atlantic Salmon Found. Spec. Publ. Ser.,* **4,** 55–82.

Wiley, R. W., Whaley, R. A., Satake, J. B., and Fowden, M. (1993). An evaluation of the potential for training trout in hatcheries to increase poststocking survival in streams. *N. Am. J. Fish. Manage.* **13,** 171–177.

Winter, G. W., Schreck, C. B., and McIntyre, J. D. (1980). Resistance of different stocks and transferrin genotypes of coho salmon, *Oncorhynchus kisutch,* and steelhead trout, *Salmo gairdneri,* to bacterial kidney disease and vibriosis. *Fish. Bull.* **77,** 795–802.

Wyllie, A. H. (1987). Apoptosis: Cell death in tissue regulation. *J. Pathol.* **153,** 313–316.

Yeoh, C.-G., Schreck, C. B., Feist, G. W., and Fitzpatrick, M. S. (1996a). Endogenous steroid metabolism is indicated by fluctuations of endogenous steroid and steroid glucuroinide levels in early development of the steelhead trout (*Oncorhynchus mykiss*). *Gen. Comp. Endocrinol.* **103,** 107–114.

Yeoh, C.-G., and Schreck, C. B. (1996b). Effects of hormone on early development of salmonids and profiles of cortisol and cortisol glucuronide during embryogenesis and beyond. *Aquaculture (In press)*

Yin, Z., Lam, T. J., and Sin, Y. M. (1995). The effects of crowding stress on the nonspecific immune response in fancy carp (Cyprinus carpio L.). *Fish Shellfish Immunol.* **5,** 519–529.

Young, P. S., and Cech, Jr., J. J. (1993). Improved growth, swimming performance, and muscular development in exercise-conditioned young-of-the-year striped bass (*Morone saxatilis*). *Can. J. Fish. Aquat. Sci.* **50,** 703–707.

Yousif, A. N., Albright, L. J., and Evelyn, T. P. T. (1994a). *In vitro* evidence for the antibacterial role of lysozyme in salmonid eggs. *Dis. Aquat. Org.* **19,** 15–19.

Yousif, A. N., Albright, L. J., and Evelyn, T. P. T. (1994b). Purification and characterization of a galactose-specific lectin from the eggs of coho salmon, *Oncorhynchus kisutch,* and its interaction with bacterial fish pathogens. *Dis. Aquat. Org.* **20,** 127–136.

Yousif, A. N., Albright, L. J., and Evelyn, T. P. T. (1995). Interaction of coho salmon, *Oncorhynchus kisutch,* egg lectin with the fish pathogen *Aeromonas salmonicida. Dis. Aquat. Org.* **21,** 193–199.

9

INFECTION AND DISEASE

TREVOR P. T. EVELYN

I. INTRODUCTION

It is rather remarkable that of the many hundreds of bacterial taxa that finfish must encounter during their lives, relatively few are capable of causing systemic infections culminating in disease (Austin and Austin, 1993). Of these, even fewer, the so-called obligate fish pathogens, depend entirely on the tissues of living fish for their "livelihood." Considering that the generation time of bacteria is very small relative to that of their potential finfish hosts, one might expect that a much larger proportion of the microbes encountered by fish would have developed mutations permitting them to circumvent the fish's defenses. Why this has not happened is not completely understood but it is probably related to the fact that a successful parasite or pathogen must possess a number of attributes that permit it to attach to, enter, survive in, and multiply within living hosts. The chances of a single bacterial cell mutating to possess all of these necessary attributes is thus exceedingly small. In addition, if replication of this cell is poor or impossible outside of a living host, then its chances of contacting a suscepti-

THE FISH IMMUNE SYSTEM:
ORGANISM, PATHOGEN, AND ENVIRONMENT

ble finfish would be essentially nil. Such a bacterium is, thus, only rarely likely to persist in nature, hence the paucity of "obligate" fish pathogens. On the other hand, if the mutant can multiply outside of the living host, its chances of contacting a susceptible finfish host would be enhanced, and its persistence in the environment as an "opportunist" fish pathogen would be considerably greater than that of the "obligate" fish pathogen. The laws of chance, therefore, likely explain why opportunist pathogens outnumber obligate fish pathogens and why more representatives of both types do not occur in the aquatic world of the fish.

The purpose of this chapter is to examine how microbes that have achieved the attributes necessary to warrant the label "fish pathogen" manage to attach to and penetrate the fish's epithelial barriers, and how they manage to persist and multiply within fish despite the antimicrobial substances and processes available to the fish. The chapter is not intended to be a comprehensive review of virulence factors possessed by all fish pathogens. Its goal is more modest. The intention is to focus on some of the more important and better-studied bacterial fish pathogens. The pathogens selected for discussion all cause systemic infections and are responsible for major losses in a number of commercially valuable and farmed finfishes. They are the following:

1. *Vibrio anguillarum,* causative agent of vibriosis, is the best studied of the six or so species of vibrios pathogenic to fish. It affects a wide range of fishes, primarily marine and anadromous species, including salmonids, producing, in acute cases, a hemolytic anemia associated with severe destruction of the lymphoid, hematopoietic, and other tissues, and an extensive focal liquefactive necrosis of the skeletal musculature.
2. *Aeromonas salmonicida* (typical strain), causative agent of furunculosis, affects primarily salmonids in both fresh- and seawater; it may occur in acute or chronic form. The acute form is usually seen in young fish and is so rapidly fatal that gross clinical signs are usually lacking except perhaps for some hemorrhaging. In the chronic form, the fish become anemic and all visceral organs may be severely affected; the kidney tissues often appear liquefied, and cavernous lesions ("furuncles"), which give the disease its name, may occur in the skeletal musculature.
3. *Renibacterium salmoninarum,* causative agent of bacterial kidney disease, affects salmonids in both fresh- and seawater, normally causing a slowly progressive and often fatal infection that is characterized by the presence of granulomatous lesions in the kidney, spleen, and liver, anemia, and occasionally, cavernous lesions in the skeletal musculature.

4. *Yersinia ruckeri,* causative agent of enteric redmouth disease, is a disease that primarily affects salmonids in freshwater and is best characterized as a hemorrhagic septicemia in which there is massive destruction of various tissues, particularly lymphoid and hematopoietic tissues in the kidney and spleen.
5. *Edwardsiella ictaluri,* causative agent of enteric septicemia of catfish, is a disease that primarily affects catfish (as the name indicates), particularly channel catfish (*Ictalurus punctatus*), and is characterized externally by the number inflamed and hemorrhagic foci on the skin (which often ulcerate), and by an open lesion in the skull between the eyes (in chronic cases), and internally by extensive necrosis of tissues in the kidney, spleen, and liver, accompanied by hemorrhaging and anemia.
6. *Pasteurella piscicida,* causative agent of pasteurellosis, is a disease that affects many species of marine fish but causes the largest problems in farmed yellowtail (*Seriola quinqueradiata*) in Japan. The disease is characterized by the presence of creamy-white granulomatous nodules composed of masses of bacterial cells, epithelial cells, and fibroblasts; the nodules are most prominent in the kidney and spleen, and the infection is accompanied by widespread internal necrosis.

In mammalian medicine, prevention of disease has often resulted from knowledge about a pathogen's virulence factors. For example, the lethal effects of potent toxins produced by some pathogens (such as those causing diphtheria and tetanus) have readily been negated by vaccination. The motivation to identify the virulence factors of fish pathogens is therefore understandably high because it is driven by the possibility that the understanding will translate into methods for preventing costly diseases. Unfortunately, studies on the virulence factors produced by fish pathogens are much more recent than those of mammalian pathogens. Thus, far less is known about how fish pathogens function. Notwithstanding this, considerable progress has been made in recent years and our understanding is rapidly expanding.

II. COLONIZATION AND ENTRY INTO THE HOST

Very little has been done to identify the factors that permit fish pathogens to attach to and colonize their fish hosts, or to establish the importance of attachment in the infection process. Intuitively, however, one would expect that in an aquatic environment, the ability to attach to a host would be an important prerequisite for the successful establishment of an infection, because without it, the pathogen would run the risk of being washed off the host or of being voided from the host's gastrointestinal (GI) tract. Many bacterial pathogens have been shown in *in vitro* tests to be capable of attaching to fish cells and fish mucus. Unfortunately, these studies provide, at best, only very indirect evidence about a bacterium's ability to colonize the surface of its host. Furthermore, tests have shown that possession of putative attachment factors does not necessarily correlate with virulence (Santos *et al.*, 1991).

The most convincing evidence regarding the sites by which bacteria enter fish (presumably these sites represent the initial points of bacterial attachment and colonization) comes from sequential observations on the surface tissues of fish using microscopic, culture, or radioactive tracer techniques following exposure of the fish to various pathogens via water or by feeding. Additional evidence on the sites of entry by bacteria comes from studies on the uptake by fish of killed bacterial cells being used as vaccines, the implication here being that if a killed bacterium can induce its own uptake then its live counterpart should certainly also be capable of doing so. Unfortunately, however, definitive studies have been conducted on only a few fish pathogens and so with most fish pathogens, reasonable speculation is all that is possible concerning this question. As will be illustrated below, the studies provide direct evidence that the gill and intestinal surfaces are important sites of colonization and entry for a number of fish pathogens. However, colonization of the intestine appears to present a problem for some pathogens, apparently because of their susceptibility to the harsh secretions of the stomach which they must encounter before they gain access to the kinder environment of the intestine. Direct evidence for invasion of fish via the intact skin has been convincingly demonstrated with only one bacterial fish pathogen, *V. anguillarum* (see later), and one would hypothesize that in the absence of injury, or without the assistance of another parasite such as a leach or louse, invasion via the skin would be much more difficult to accomplish because, structurally, the skin presents a far more formidable barrier to penetration than does the gill or intestine.

A. Penetration via the Gills

Because the fish's gills are constantly being flushed with water that may contain fish pathogens, are covered with only a thin layer of protective mucus, and are constructed so that only a single layer of fragile cells separates the fish's vascular system from the external environment, it is probably a very important site of entry for pathogens. Indeed, the epithelial cells of the fish gill seem capable of actively taking up particles, for example latex beads (Smith, 1982) and kaolin (Goldes *et al.,* 1986), and they have clearly been shown to be the primary sites of entry for at least three different viral fish pathogens (Ahne, 1978; Chilmonczyk, 1980; Mulcahy *et al.,* 1983). They have also been shown to take up killed and live cells of *V. anguillarum* (Smith, 1982; Nelson *et al.,* 1985; Baudin-Laurencin and Germon, 1987), and probably also killed and live cells of *A. salmonicida* (Tatner *et al.,* 1984; Brune *et al.,* 1986), live cells of *P. piscicida* (Kawahara *et al.,* 1989), and killed cells of *Y. ruckeri* (Zapata *et al.,* 1987). Whether the gill also serves as a portal of entry for the kidney disease bacterium *R. salmoninarum* has not been investigated but entry via the gill would not be surprising because infections are readily established by bath exposure to the bacterium (Murray *et al.,* 1992). Further, as will be discussed later, *R. salmoninarum* appears capable of inducing its own uptake by fish cells not commonly thought of as phagocytic and so it would not be surprising to find that gill uptake also occurs with this bacterium.

The mechanism of particle uptake by the gill appears to involve attachment of the particle to and engulfment by gill epithelial cells followed by transfer of the particle to underlying mononuclear phagocytes, which presumably take part in disseminating internalized particles to other sites in the fish (Goldes *et al.,* 1986; Zapata *et al.,* 1987). The importance of the gill as a site of colonization and entry by fish pathogens may often have been overlooked in studies using routine histology because of the insensitivity of the technique and the rapidity with which the particles such as bacterial cells can disappear from the gill. For example, Alexander *et al.* (1981) found that live *Escherichia coli* entering the gill disappeared from the gill very rapidly, an observation also made by Zapata *et al.* (1987) with respect to killed *Y. ruckeri* cells. Rapid disappearance of bacteria from the gills may explain why Ransom (1978), working with *V. anguillarum,* overlooked the gill as an important entry site. With other bacterial pathogens, for example, *E. ictaluri,* the gill has apparently not been considered as a possible portal of entry, an oversight that should be reexamined in light of the foregoing.

B. Penetration via the Gastrointestinal Tract

In larval fish, infections via the GI tract are quite likely to occur with a wide range of fish pathogens because the lethal secretions in the stomach, present in fish at a later stage of development, may still largely be absent (Olafsen, 1994). In fish beyond the larval stage, however, infection via the GI tract presents problems for some pathogens. For example, although infection via the GI tract has been reported for all of the pathogens under discussion, for two of them, *V. anguillarum* and *A. salmonicida,* the GI tract is obviously a hostile environment (Bogwald *et al.,* 1994), perhaps explaining why the GI tract may not be the usual route of infection for these two pathogens. With *V. anguillarum,* infection via the GI tract, suggested by the results of Ransom (1978), has indeed been shown to occur under experimental conditions (Baudin-Laurencin and Tangtrongpiros, 1980; Watkins *et al.,* 1981; Kanno *et al.,* 1989; Grisez *et al.,* 1996) but large numbers of *V. anguillarum* cells were required to kill fish when administered *per os,* and, in some cases, challenge of susceptible fish by this route failed to cause any mortalities (Chart and Munn, 1980; Horne, 1982; Kodera *et al.,* 1974); mortalities occurred only when the challenge was introduced via the anus to avoid exposing the challenge bacterium to the secretions of the stomach (Chart and Munn 1980). With *A. salmonicida* a similar picture holds true. Rose *et al.* (1989) could establish infections via the GI tract only with large numbers of the bacterium, and various other workers have failed to establish or have reported difficulty in establishing infections by feeding or gastric intubation (Cipriano, 1982; Tatner *et al.,* 1984; Perez *et al.,* 1996). The problem may be partly related to a limited ability of *A. salmonicida* to cross the intestinal epithelium; for example, unlike killed *V. anguillarum* cells (Vigneulle and Baudin-Laurencin, 1991), killed *A. salmonicida* cells have been reported as unable to traverse the intestinal wall (Tatner *et al.,* 1984; Brudeseth and Evensen, 1995). More likely, however, the problem is due to poor survival of the bacterium in the face of stomach secretions because, having managed to enter the intestine intact, the bacterium can apparently survive there to initiate furunculosis in carrier fish when the fish become stressed (Hiney *et al.,* 1994).

With the other pathogens being considered, infection via the GI tract does not appear to present a major problem. Infections with the kidney disease bacterium *R. salmoninarum* have been successfully established by the feeding of infected material (Wood and Wallis, 1955), and in Pacific salmon farmed in sea water, infections by this route probably help to account for the increasing prevalence of bacterial kidney disease during their rearing (Balfry *et al.,* 1996). The mechanism of penetration by the bacterium has not been investigated but, as mentioned before, the bacterium appears capable of inducing its uptake by cells other than phagocytes; thus, it seems likely that its crossing of the intestinal wall may occur in a manner analogous to that described for particles crossing the gill epithelium. With *E. ictaluri,* the intestinal tract is believed to be the primary portal of entry (Shotts *et al.,* 1986; Newton *et al.,* 1989; Baldwin and Newton, 1993), although there is convincing evidence based on histopathology and culture that infections may also occur via the sensory epithelium in the olfactory sac (Miyazaki and Plumb, 1985; Shotts *et al.,* 1986). The mechanism of penetration appeared to involve transport across the intestinal wall via the intestinal epithelium; it was extremely rapid and was not associated with any destruction of the intestinal epithelium (Baldwin and Newton, 1993). With *P. piscicida,* infections via the GI tract apparently occur, because they can be established by the feeding of infected material (Kawahara *et al.,* 1989); in addition, it has been shown, using the fluorescent antibody technique, that killed *P. piscicida* cells are taken up by and cross the intestinal epithelium (Kawahara and Kusuda, 1988). Finally, the important fish pathogen *Y. ruckeri* probably also invades fish via the GI tract, because it is possible to infect fish via the oral route (Ross *et al.,* 1966) and because fish develop demonstrable protection against the live pathogen following vaccination *per os* using killed cells of the pathogen (Anderson and Nelson, 1974; Vigneulle, 1990); in addition, once the bacterium gains access to the intestines, the bacterium can apparently persist there to initiate active infections at a later stage (Busch and Lingg, 1975).

C. Penetration via the Skin

Evidence for invasion of fish via the intact skin has been convincingly demonstrated with only one bacterial fish pathogen, *V. anguillarum.* Funahashi *et al.* (1974) found, using routine histopathological techniques, that in natural and experimental infections of ayu (*Plecoglossus altivelis*), the skin was the first tissue to be colonized, an observation subsequently supported by the results of Muroga and De La Cruz (1987) who used immunohistochemical and viable count methods to investigate this question. These results were consistent with the findings of Kawai *et al.* (1981) who showed that vaccination was effective against vibriosis because it prevented a quantifiable colonization of the skin by the pathogen. The most direct evidence for the skin as a portal of entry, however, has been provided by Kanno *et al.* (1989) who showed that placing a *V. anguillarum*–containing patch of paper on the fish's skin for one minute resulted in a quantifiable colonization of the skin at the contact site followed by mortalities due to vibriosis. Similar type experiments have not been performed with the other fish pathogens under discussion but the propensity for one of them, *A. salmonicida,* to occur in skin mucus (Cipriano *et al.,* 1992, 1994) suggests that this route of infection should be considered for this pathogen.

D. Penetration via Miscellaneous Routes

It has already been mentioned that the sensory epithelium in the olfactory sac of channel catfish likely serves as a point of entry for the bacterial fish pathogen *E. ictaluri.* Other surface tissues that have been noted by various workers as points of attachment for other fish pathogens include those covering the thymus and pseudobranch. However, except for the study by Flano *et al.* (1996) with the thymus, no follow-up studies have been done to assess the importance of these tissues in the uptake of fish pathogens. Flano *et al.* (1996) found that the pharyngeal epithelium overlying the thymus was not penetrated by *R. salmoninarum.* Infections of salmonids with *R. salmoninarum* occurring via the eyes and skin have been reported (Hendricks and Leek, 1975; Hoffman *et al.,* 1984) but it seems likely that these infections were precipitated by superficial injuries (Hendricks and Leek, 1975). The only other important route of infection is via the egg. Among the fish-pathogenic bacteria, the kidney disease bacterium *R. salmoninarum* is unique because it is transmitted from parent to progeny via the egg (Evelyn *et al.,* 1986). Because the bacterium has been shown to occur intraovum (Evelyn *et al.,* 1984), it clearly has the ability to induce uptake by the ovum. The only other microbial fish pathogen for which there is convincing evidence of egg transmission is the infectious pancreatic necrosis virus. However, there is some disagreement about whether the transfer is accomplished by virus particles located within the ovum (Fijan and Giorgetti, 1978; Dorson and Torchy, 1985) or external to it (Ahne and Negele, 1985).

III. POSTENTRY EVENTS (SPREAD)

In the nonimmune host, the six pathogens under consideration are all capable of causing, and indeed frequently cause, systemic infections following their entry into the host. From the site of entry, which is affected by the route of the challenge, the pathogens are spread to the various organs and tissues chiefly via the blood. Spread via the lymphatic system probably also occurs but this system in fish has received very little study. With all of the pathogens, the spread is apparently mediated by free cells but with at least three of them, *E. ictaluri* (Baldwin and Newton, 1993), *R. salmoninarum* (Bruno, 1986), and *P. piscicida* (Noya *et al.,* 1995), spread via infected phagocytes, mainly macrophages, may also be important.

The speed with which the various organs and tissues of nonimmune hosts are colonized by the various pathogens depends on a number of factors, including the infectious dose actually entering the fish and the rate at which the pathogen can multiply in the host tissues. With the slow-growing kidney disease bacterium *R. salmoninarum,* the time to death may be several weeks (Bruno, 1986; Flano *et al.,* 1996) for injected fish to many months (Murray *et al.,* 1992) for fish infected by more natural methods. Ultimately, however, virtually every tissue and organ in the fish is colonized (sometimes including organs normally considered refractory to invasion such as brain and thymus) (Speare *et al.,* 1993; Flano *et al.,* 1996). By the time death occurs, large numbers of *R. salmoninarum* cells, many of them located within host cells, are present in the tissues, the counts ranging from an average of 10^3 g^{-1} in the somatic muscle to 10^8 or 10^9 g^{-1} in spleen and kidney (T. P. T. Evelyn, unpublished data).

With the rapidly growing pathogens such as *V. anguillarum, A. salmonicida, Y. ruckeri, E. ictaluri,* and *P. piscicida,* the spread can be rapid, and lethal effects may occur within a few days of challenge, affected tissues containing large numbers of the pathogens. For example, following exposure to water-borne *V. anguillarum,* counts of viable cells in the blood and tissues reached 10^8 ml^{-1} and 10^7–10^9 cells g^{-1}, respectively, in the terminal stages of the infection, and the tissues involved included kidney, liver, spleen, cardiac muscle, somatic muscle, visceral peritoneum, gills, digestive tract, and skin (Ransom *et al.*, 1984; Muroga and De La Cruz, 1987). With *A. salmonicida,* a bacteremia rapidly developed following challenge via the intramuscular route, and by 72 h postchallenge, large numbers of the bacterium (10^6–10^9 cells g^{-1}) occurred in the organs tested (kidney, liver, spleen, heart) (Munn and Trust, 1984); fish dead of the pathogen contained on average 10^5 *A. salmonicida* cells per gram of somatic muscle (T. P. T. Evelyn, unpublished data). With *Y. ruckeri,* progression of the acute form of the disease is characterized by colonization of highly vascularized tissues such as kidney, liver, spleen, heart, gills, and muscle (Busch, 1978). With *E. ictaluri,* challenge of channel catfish by various routes resulted in colonization of the digestive tract, kidney, liver, spleen, brain, muscle, and skin (Newton *et al.,* 1989; Shotts *et al.,* 1986; Baldwin and Newton, 1993; Ciembor *et al.,* 1995), the numbers of *E. ictaluri* cells in the tissues of the digestive tract, liver, and kidney ranging from approximately 10^5 to 10^7 g^{-1} by four days after gastric intubation (Baldwin and Newton, 1993). With *P. piscicida,* the bacterium occurred in the gills, stomach, pyloric caeca, intestine, liver, spleen, kidney, and heart by 72 h after water-borne challenge of yellowtail (Kawahara *et al.,* 1989). Nelson *et al.* (1989) also indicated its presence in the muscle (but not brain) of naturally infected specimens. The tissue distribution observed by Kawahara *et al.* (1989) in yellowtail was very similar to that observed in injection-challenged seabream by Noya *et al.* (1995). Interestingly, as with the slow-growing *R. salmoninarum,* both *E. ictaluri* and *P. piscicida* cause infections in which a large proportion of their cells occur within host cells (macrophages).

IV. MECHANISMS OF SURVIVAL IN THE FISH HOST

The ability of the six pathogens under discussion to invade nonimmune fish and to spread to various tissues and organs via the blood indicates that these pathogens must be able to survive in the blood and phagocytes of such fish. The question is: How do these pathogens accomplish this?

A. Survival in the Face of Humoral Factors

Survival in the presence of blood or, more accurately, in the serum of nonimmune fish has been documented for at least five of these pathogens: *V. anguillarum* (Trust *et al.*, 1981), *A. salmonicida* (Munn *et al.*, 1982; Sakai and Kimura, 1985), *E. ictaluri* (Ourth and Bachinski, 1987; Ourth *et al.*, 1991), *Y. ruckeri* (Davies, 1991a; Ourth and Bachinski, 1987), and *P. piscicida* (Magarinos *et al.*, 1994). The mechanisms underlying the serum resistance are not known in all cases. In *V. anguillarum* the factors responsible for serum resistance were not identified but they appeared to be chromosome specified. It seems likely, however, that serum resistance in *V. anguillarum* and also in *Y. ruckeri* is related to the presence of cell-surface proteins (discussed later). In *Y. ruckeri,* all virulent strains tested proved serum resistant. However, some avirulent strains also proved serum resistant, suggesting that virulence in *Y. ruckeri* is related to an additional factor. In *P. piscicida,* virulent strains survived in unheated normal serum but avirulent strains survived only when the serum was heated to inactivate complement (Magarinos *et al.*, 1994).

In *A. salmonicida,* serum resistance appears to be related to a protein layer (the A layer) that envelopes the cell, and to the long side chains of the *A. salmonicida* lipopolysaccharide (LPS) that traverse the A layer. These structures presumably serve to prevent access of complement to susceptible target sites on *A. salmonicida* cells because loss of the structures results in sensitivity to killing by normal serum (Munn *et al.*, 1982; Sakai and Kimura, 1985). Another possibility is that the presence of large quantities of sialic acid on *A. salmonicida* cells serves to protect them by preventing the activation of complement via the alternate pathway (Ourth and Bachinski, 1987). However, because only one virulent strain of *A. salmonicida* has been tested to date for its sialic acid content, further studies are warranted to investigate whether sialic acid is indeed a factor in the serum resistance of all virulent strains of *A. salmonicida.*

Siliac acid–induced suppression of activation of the alternate complement pathway has been proposed as the mechansim explaining the survival of *E. ictaluri* in normal serum of channel catfish (Ourth and Bachinski, 1987). However, these authors also noted that two other systemic fish pathogens tested, *V. anguillarum* and *Y. ruckeri,* lacked sialic acid but yet were completely resistant to normal fish serum. Factors other than sialic acid, perhaps cell surface proteins, must therefore explain the survival of *V. anguillarum* and *Y. ruckeri* in normal serum. Certainly, the strains of *V.*

anguillarum, Y. ruckeri, and *E. ictaluri* studied by Ourth and Bachinski (1987) all had cell surface proteins as demonstrated by the Congo red method. Further, in the case of *E. ictaluri,* virulent strains had greater amounts of surface proteins and polysaccharide capsular material than avirulent strains (Stanley *et al.,* 1994). Presumably, these surface components prevent bactericidal substances in normal serum such as complement and lysozyme from coming into contact with vulnerable sites on the bacterial cell surface.

B. Survival in Phagocytes

On entry into a nonimmune host, a pathogen must not only be able to survive in the host's body fluids, as discussed above, but it must also be capable of avoiding being killed by the host's phagocytic cells. The involvement of salmonid phagocytes, presumably primarily macrophages, in eliminating *in vivo* bacterial infections has been clearly illustrated by Olivier *et al.* (1985) who demonstrated that fish show markedly reduced resistance to bacterial challenges if phagocyte function is reduced by phagocyte-blockading or -killing particles.

In fish, the phagocytic cells consist of two main types, the granulocytes (mainly neutrophils) and mononuclear phagocytes (tissue macrophages and circulating monocytes). Both groups of phagocytes appear to be capable of bacterial uptake (Ellis, 1982; Ainsworth, 1992), but macrophages appear to be far more active in this function, particularly in salmonids. Further, whereas macrophages are clearly capable of killing ingested bacteria, considerable doubt still exists about the extent to which neutrophils effect intracellular bacterial killing. For example, data on channel catfish neutrophils suggest that, like macrophages, they are capable of producing a respiratory burst response. However, bactericidal activity of these neutrophils appeared to be feeble (as measured using virulent *E. ictaluri*) and was exerted only on the extracellular bacterium, and only in the presence of unheated serum (see Ainsworth and Dexiang, 1990; Waterstrat *et al.,* 1991). Recent studies with Atlantic salmon (*Salmo salar*) neutrophils have shown that they are also capable of ingesting bacteria (*A. salmonicida,* particularly the avirulent strain) and that upon activation by *A. salmonicida* opsonized with complement or specific antibody, they showed a respiratory burst response, became swollen, and then apparently lysed (Lamas and Ellis, 1994a). The activated neutrophils also appeared to kill ingested virulent and avirulent *A. salmonicida* but the practical significance of this with respect to virulent *A. salmonicida* is uncertain because it was not actively phagocytosed (Lamas and Ellis, 1994b). However, in view of the tendency of activated neutrophils to swell, show degranulation, and lyse, and considering that neutrophils appear to be less actively phagocytic than macrophages, it is possible that their bactericidal effects are mediated primarily on extracellular targets following lysis and release of their cell contents.

Phagocytic cells in the form of macrophages are widely distributed in fish tissues and they occur in high concentrations in tissues underlying epithelial barriers such as intestine and gills, and in the kidney, spleen, and atrium of the heart. In these locations, they are in a position to monitor the blood stream for foreign entities.

The mechanisms by which the pathogens under consideration resist phagocytosis and phagocytic killing are incompletely understood. With *E. ictaluri* and *P. piscicida,* histological evidence shows that they are phagocytosed and that they apparently survive and multiply within the phagocytes of the nonimmune host (see Miyazaki and Plumb, 1985; Stanley, 1991; and Baldwin, 1992 with respect to *E. ictaluri;* see Kubota *et al.,* 1970; Nelson *et al.,* 1989; and Noya *et al.,* 1995 regarding *P. piscicida*). In the case of *E. ictaluri,* the virulent strain attached to phagocytes more readily than the avirulent strain, and, in contrast to the latter, appeared to multiply within the phagocytes (Stanley, 1991). Stanley *et al.* (1994) therefore speculated that the polysaccharide capsule and surface proteins found on virulent strains of *E. ictaluri* facilitate attachment and protect against the acidic environment and hydrolytic enzymes within phagolysosomes. Waterstrat *et al.* (1991) also indicate that *E. ictaluri* LPS suppresses the respiratory burst response in channel catfish neutrophils, an observation that, if also true for catfish macrophages, could help to explain why this pathogen is refractory to intracellular killing in channel catfish phagocytes. Obviously, further studies are needed to determine the mechanisms involved in phagocyte resistance in these two pathogens, particularly with *P. piscicida. Pasteurella piscicida* is readily taken up by macrophages, in which it can grow to large numbers, eventually killing the macrophages. However, the factors permitting these events have not been studied (Kubota *et al.,* 1970). Further studies are also needed with *Y. ruckeri* and *V. anguillarum.* The former appears to resist phagocytic killing (Griffin, 1983) but the mechanism involved is unknown. It may, however, be partly related to the ability of the more virulent strains to elicit a reduced respiratory burst response (Stave *et al.,* 1987). With *V. anguillarum,* the paucity of information on its interaction with phagocytes probably stems from the fact that protection in the immune animal appears to be effectively mediated by the humoral arm of the immune system and because immunity to vibriosis is very easily accomplished (Evelyn, 1988). It is just as well that *V. anguillarum* infections in the immune host are arrested by circulating antibodies because the results of Lamas *et al.* (1994) indicate that macrophages offer virtually no defense against the bacterium in the nonimmune host. In such hosts, the bacterium was indeed observed within macrophages. However, in the absence of circulating anti–*V. anguillarum* antibodies, the bacterium multiplied extremely rapidly in the blood, producing toxic extracellular products that caused extensive tissue destruction. The consequence was that any potential effectiveness of the macrophages appeared to be neutralized by the amount of host tissue debris that they had to deal with—a result that was similar to that obtained *in vivo* by Olivier *et al.* (1985) using known phagocyte-blockading agents in the presence of two other bacterial fish pathogens. It seems likely that the success of *V. anguillarum* as a pathogen depends on the foregoing strategy.

With *A. salmonicida, in vitro* (Sakai and Kimura, 1985; Olivier *et al.*, 1986; Daly *et al.*, 1994) and *in vivo* (Munn and Trust, 1984; Sakai and Kimura, 1985) studies have shown that avirulent strains of the bacterium are susceptible to ingestion and killing by macrophages whereas virulent strains are not. These effects appeared to be mediated by the A layer because virulent strains possess an A layer whereas avirulent strains do not. Even though virulent strains of *A. salmonicida* attach to macrophages by virtue of their A layer (Garduno and Kay, 1992), they were less readily phagocytosed in the presence of normal serum (Sakai and Kimura, 1985) and they elicited a lower chemotactic response (Weeks-Perkins and Ellis, 1995) than avirulent strains. In addition, virulent strains of *A. salmonicida* tended to be more resistant to killing by superoxide anion, a product of the respiratory burst response (Karczewski *et al.*, 1991). Further, in *in vitro* studies, *A. salmonicida* exerted a toxic effect on salmonid phagocytes when present in high enough numbers (Olivier *et al.*, 1992; Garduno and Kay, 1992), the effect being more pronounced with virulent than avirulent strains (Lamas and Ellis, 1994a). This effect is consistent with earlier *in vivo* observations on *A. salmonicida* infections that showed the infections to cause severe phagocytic depletion (Klontz *et al.*, 1966), a depletion almost certainly mediated by one of the two important extracellular toxins produced by the bacterium (Ellis, 1991). This toxin, a 25-kDa phospholipase (glycerophospholipid: cholesterol acyltransferase or GCAT), exhibits high leukocytolytic activity, especially when complexed with *A. salmonicida* LPS.

The interactions of *R. salmoninarum* with fish phagocytes deserve special mention because this bacterium tends to occur intracellularly *in vivo,* and the histological evidence strongly suggests that it survives and multiplies within phagocytes (Young and Chapman, 1978; Bruno, 1986). Certainly, the cell wall, which is a chemically unique and formidable structure (Kusser and Fiedler, 1983; Fiedler and Draxl, 1986), may contribute to intracellular survival (Gutenberger, 1993). In addition, intracellular survival may, in part, be a function of a capsule (Dubreil *et al.,* 1990b). However, the existence of a capsule needs to be confirmed because other electron microscopic studies on the bacterium have not noted such a structure (Gutenberger, 1993).

Contrary to the results of Rose and Levine (1992), *R. salmoninarum* is readily ingested by rainbow trout (*Oncorhynchus mykiss*) macrophages without opsonization by complement (Bandin *et al.*, 1993, 1995; Gutenberger, 1993). Cells ingested without opsonization survived in the macrophages for several days, and even apparently replicated, before showing declines in numbers. Ingested cells induced a respiratory burst response in macrophages but apparently survived the response to be killed by other factors (Bandin *et al.*, 1993). This finding suggests that the bacterium is resistant to reactive oxygen products resulting during the respiratory burst but the protective factors involved have not been identified. The coincident ingestion by macrophages of melanin granules, which are known to protect against such products, may contribute to the protection (Gutenberger, 1993). Also, the p57 surface protein (henceforth referred to as p57) on *R. salmoninarum* cells may help to account for the bacterium's survival in the face of reactive oxygen products because it has been shown to be a potent inhibitor of the respiratory burst response (Kaattari *et al.*, 1988). Based on the findings of Gutenberger (1993), the killing factors may have been the products of phagolysomal fusion because the walls of some *R. salmoninarum* cells remaining in the phagosomes showed signs of destruction. On the other hand, many of the ingested *R. salmoninarum* cells appeared to escape from the phagosomes into the macrophage cytoplasm (Gutenberger, 1993), in which location they would likely have escaped the products of phagolysomal fusion and, based on their resistance to lysozyme (Fryer and Sanders, 1981; Yousif *et al.*, 1994), would likely have withstood the effects of any cytoplasmic lysozyme present. Escape into the cytoplasm suggests the existence of an *R. salmoninarum* factor capable of lysing phagosome membranes.

The foregoing studies of Bandin *et al.* (1993) and Gutenberger (1993), although highly informative, did not fully represent the situation one would expect to occur *in vivo* where invading *R. salmoninarum* cells would likely have been bathed for some time in serum (normal or immune) prior to ingestion by macrophages. When an experiment was conducted using *R. salmoninarum* cells opsonized with complement, specific antibody, or both, survival of the pathogen in macrophages was increased above that of cells not exposed to serum factors. Maximum survival occurred when both complement and antibody were present. In fact, when opsonization occurred in the presence of both complement and antibody, no intracellular killing occurred, and cells of the ingested bacterium actually grew faster than those incubated in the presence of lysed macrophages (the control system). Interestingly, this effect on survival occurred only if the exposure to unheated immune serum was prolonged (16 h instead of 3 h), suggesting that the elaboration of a protective factor (or factors) still to be identified was involved. In mammals, the complement-mediated uptake of bacterial cells by macrophages results in suppression of macrophage respiratory burst activity (Wright and Silverstein, 1983) but Bandin *et al.* (1995) did not determine whether opsonization of *R. salmoninarum* with complement and antibody suppressed this response. However, it is assumed that because the bacterium resists killing via the respiratory burst response (Bandin *et al.,* 1993), the enhanced survival observed following prolonged exposure to complement and antibody was due to some other protective mechanism, perhaps to the elaboration of protective factors that facilitate escape from phagolysosomes, that interferes with phagosome–lysosome fusion, or that somehow neutralizes the effects of lysosomal contents.

V. VIRULENCE FACTORS PERMITTING GROWTH WITHIN THE HOST

Successful pathogens must be capable of deriving all of their growth requirements from their hosts. To do this they must produce a number of enzymes and other factors that ensure that this requirement is satisfied.

One of the best-studied nutrient acquisition systems among the bacterial fish pathogens is that involved in the acquisition of iron. Iron is needed for bacterial growth but it is normally in short supply in animal tissues because of the presence of host glycoproteins, such as transferrin and lactoferrin, that bind iron very tightly. If bacteria are to grow in such situations, they must produce iron-gathering substances capable of competing for iron with molecules like transferrin. Many bacteria lack such iron-scavenging systems. With these bacteria, transferrin acts as an effective bacteriostat. In a study with *V. anguillarum,* Crosa *et al.* (1980) showed that when the bacterium was "cured" of a suspected virulence plasmid it also lost its virulence. This study confirmed the relationship between the plasmid and virulence. The relationship between virulence and the possession of an iron-sequestering system was then established in another study (Crosa, 1980). In this study, the loss of the ability of the plasmid-cured strain of *V. anguillarum* to produce disease was reversed by providing exogenous iron in amounts sufficient to result in free iron in the tissues of the test fish (Crosa, 1980). The genes coding for iron scavenging in *V. anguillarum* were located on a plasmid in some isolates (Crosa, 1980) and on the chromosome in others (Toranzo *et al.,* 1983). The iron-acquiring system studied by Crosa and colleagues in *V. anguillarum* consisted of a diffusable iron-sequestering molecule, a siderophore (Actis *et al.,* 1986), and an iron-regulated outer membrane protein (IROMP) that served as the receptor for the siderophore bearing scavenged iron (Actis *et al.,* 1985). Using mutants deficient in the ability to produce the siderophore or the IROMP, it was shown that both were needed for iron uptake and that an inability to produce either component resulted in a loss of virulence in *V. anguillarum* (Tomalski and Crosa, 1984; Wolf and Crosa, 1986).

Iron aquisition has been studied in four of the other bacterial pathogens under consideration in this chapter. Iron acquisition in *A. salmonicida* was effected in some isolates by the two-component system described above for *V. anguillarum;* in other isolates it was mediated by a siderophore-independent mechanism involving direct contact of the bacterium with the glycoprotein–iron complex and the transfer of the iron from the complex to the bacterium (Chart and Trust, 1983; Hirst *et al.,* 1991). In *Y. ruckeri,* a rather similar situation occurs; the two-component system was present in some isolates (Romalde *et al.,* 1991); in others, siderophores could not be demonstrated, and iron uptake was postulated to occur as a result of direct contact of IROMPs on the bacterial surface with the iron-bearing glycoproteins (Davies, 1991b). With *P. piscicida,* iron uptake was mediated via the two-component system described for *V. anguillarum,* avirulent strains of the bacterium being unable to grow in heat-inactivated fish serum or *in vivo* unless iron supplements in the form of hemin, hemoglobin, or ferric ammonium citrate were provided (Magarinos *et al.,* 1994). With *R. salmoninarum,* no siderophores or IROMPs could be demonstrated, but polysaccharide molecules of uncertain function were produced under conditions of iron deprivation. In addition, under such conditions, enhanced levels of iron reductase were produced, suggesting that iron acquisition in *R. salmoninarum* is at least partly accomplished via this enzyme (Grayson *et al.,* 1995a). The reductase functions by converting glycoprotein-bound ferric iron to the less tightly bound ferrous iron, thus making the iron more available to the bacterial cell. The iron acquisition systems present in *A. salmonicida, Y. ruckeri,* and *R. salmoninarum* doubtless contribute to their virulence. However, unequivocal proof of this awaits studies to determine whether mutants defective in the production of the various iron acquisition components lose their virulence. Recently, Hirst and Ellis (1994) showed that antibodies raised against the IROMPs of *A. salmonicida* were protective against furunculosis in Atlantic salmon. This finding suggested that the antibodies might function by attaching to the IROMPs, thus blocking iron uptake via siderophores. However, additional testing by these workers suggests that the protection by these antibodies was mediated by another route: complement-mediated killing of *A. salmoninarum* cells opsonized with anti-IROMP antibodies.

Four of the pathogens being considered in this chapter produce a wide array of extracellular products (ECPs) that very likely contribute to their success as pathogens and that help to explain the lesions that they cause in their hosts. For example, *V. anguillarum* produces a number of proteases, hemolysins, cytotoxins, and dermatotoxins; *A. salmonicida* produces, in addition, a phospholipase; while *Y. ruckeri* produces all of the foregoing products except for the cytotoxins (Toranzo and Barja, 1993). In addition, *Y. ruckeri* produces a heat labile factor (perhaps a lipoprotein) which, in the isolates tested, was clearly correlated with virulence (Furones *et al.*, 1990). Vaccination of fish with preparations containing this factor resulted in enhanced immunity to *Y. ruckeri,* suggesting that this component is an important virulence factor in this pathogen. The ECPs of *P. piscicida* contained proteolytic, hemolytic, cytotoxic, and phospholipase activity, the amount of proteolytic activity depending on the strain tested (Magarinos *et al.*, 1992) and on whether the strain was grown under conditions of iron deprivation (Magarinos *et al.,* 1994). Injection of the ECPs from these four pathogens into fish resulted in death, often with clinical signs resembling those of the disease caused by the pathogen serving as the ECP source (Toranzo and Barja, 1993; Magarinos *et al.,* 1992; Noya *et al.,* 1995). The ECPs probably serve to impair host defense mechanisms, lyse cells, and hydrolyze host tissues, thus facilitating the spread of the pathogens throughout the host and the release of nutrients required by the pathogens.

One of the other pathogens under discussion, *E. ictaluri,* appears to produce a more limited range of tissue damaging products. For example, it seems to produce no remarkable proteolytic or cytotoxic activity, thus leading Thune *et al.* (1993) to suggest that the tissue damage caused by *E. ictaluri* in channel catfish is an indirect result of the acute inflammatory response to the pathogen. However, *E. ictaluri* produces chondroitinase activity that correlates well with virulence (Stanley *et al.,* 1994) and that probably accounts for the "hole-in-the-head" syndrome often observed in *E. ictaluri* infections.

Unfortunately, ECP components have only been adequately purified and shown to be important virulence factors in a few cases. One such component occurs in the ECPs of *A. salmonicida* and is a 70-kDa protease (caseinase). The protease is resistant to all serum protease inhibitors except α2-macroglobulin, causes extensive tissue liquefaction, activates blood clotting, and is lethal for fish at 2.4 $\mu g\ g^{-1}$ fish. It shows broad-spectrum protease activity and its function *in vivo* appears to be the provision of amino acids for growth of *A. salmonicida* (Ellis, 1991). The value of this protease to the bacterium *in vivo* has been established by Coleman *et al.* (1992), who found that when used (in the form of a genetically engineered fusion protein) to vaccinate Atlantic salmon, it conferred measurable protection against challenge with *A. salmonicida.* Another ECP component of *A. salmonicida* that has been particularly well studied is its phospholipase, referred to earlier as GCAT. This 25-kDa enzyme normally occurs complexed *in vivo* with high molecular weight LPS, which not only stabilizes it but also enhances its activity. The complex is extremely hemolytic for fish erythrocytes and, based on *in vitro* studies, is leukocytolytic. This latter property, expressed *in vivo,* likely helps the bacterium to overwhelm the cellular defenses of the nonimmune host and probably accounts for the lack of leukocytes seen in advanced cases of furunculosis (Klontz *et al.,* 1966). The complex is extremely lethal (45 ng g^{-1} fish), by mechanisms still not understood, and in conjunction with the above-mentioned protease, accounts for the hemorrhagic and liquefactive lesions (furuncles) seen in fish suffering from furunculosis (Ellis, 1991).

The factors accounting for virulence in *R. salmoninarum* are gradually being uncovered. Unlike most of the other pathogens discussed so far, this pathogen produces no acutely lethal toxins. It is not surprising, therefore, that the ECPs of *R. salmoninarum* examined by Bandin *et al.* (1991, 1992) contained no detectable (or appreciable) proteolytic, hemolytic, or cytotoxic activity. In fact, this feature of the bacterium serves it well because it means that fish can survive to spawn while carrying the large numbers of *R. salmoninarum* cells that favor egg infection, thus increasing the bacterium's chances of being transmitted to the next generation. Although the ECPs of *R. salmoninarum* are not lethal when injected into fish (Bandin *et al.,* 1991, 1992), they do in fact contain readily detectable amounts of the previously mentioned protein, p57 (Getchell *et al.,* 1985), a 100-kDa protease capable of hydrolyzing p57 (and presumably host proteins) (Rockey *et al.,* 1991), and a dermatotoxic factor of uncertain function (Bandin *et al.,* 1992). The protein (p57) is a major surface component on *R. salmoninarum* cells, and appears to occur there in the form of fimbriae (Dubreil *et al.,* 1990a). During *in vivo* and *in vitro* growth, the bacterium produces considerable amounts of p57, much of it being released extracellularly in the process (Turaga *et al.,* 1987a; Wiens and Kaattari, 1989). The protein appears to play a role in the granulomatous lesion formation characteristic of bacterial kidney disease. It elicits an antibody response in fish but because it is produced in such large amounts, it tends to "mop up" the antibodies. Some of the resulting antibody–p57 complexes persist in fish tissues, including glomerular tissues, and contribute to a tissue-destructive hypersensitivity response culminating in the granulomas (Kaattari, 1989). The role of the above-mentioned protease and of two *R. salmoninarum* cytolysins described by Evenden *et al.* (1990) and Grayson *et al.* (1995b) in the pathogenesis of bacterial kidney disease requires further study. These enzymes could help to account, however, for the tissue destruction and reduced hematocrits characteristic of the disease.

In *R. salmoninarum,* the loss of the ability to produce p57 is correlated with a loss of virulence (Bruno, 1988), indicating that the protein plays a major role in the virulence of the pathogen. The protein is responsible for the hydrophobicity and autoagglutinating properties of *R. salmoninarum* cells (Bruno, 1988) and apparently accounts for their ready attachment to host cells, for example, leukocytes (Wiens and Kaattari, 1991) and sperm (Daly and Stevenson, 1989). This propensity for attaching to cells may account for *R. salmoninarum's* ability to enter host cells not normally considered to be phagocytic, for example ova (Evelyn *et al.,* 1984), thrombocytes (Lester and Budd, 1979), and endothelial cells (Bruno, 1986). The advantage of becoming intraovum has already been mentioned. However, there may also be advantages to entering cells such as thrombocytes and endothelial cells because, while there, the bacterium may be "hidden" from the immune system of the host, thus reducing exposure to any protective serum factors present and minimizing the chances of eliciting an immune response. Another advantage to cell attachment, particularly attachment to leukocytes, is that it may facilitate their destruction, either directly by the pathogen or by causing them to become targets for phagocytosis and destruction by macrophages (Gutenberger, 1993). The attachment ability may therefore result in immunosuppression by removing leukocytes involved in antibody production and macrophage activation. Support for this statement is provided by Wiens and Kaattari (1991) who found that leukocytes bearing surface immunoglobulin and leukocytes lacking surface immunoglobulin were both agglutinated by *R. salmoninarum* cells and by the cell-free p57.

Finally, there is strong additional evidence that p57 is an important virulence factor because, *in vitro,* it results in significant suppression of both antibody production (Turaga *et al.,* 1987b) and the respiratory burst response (Kaattari *et al.,* 1988), and *in vivo* it appears to be immunotolerogenic under certain conditions (Brown *et al.,* 1996). Suppression of antibody formation was observed with head kidney lymphocytes taken from fish with severe *R. salmoninarum* infections as well as with lymphocytes, taken from healthy fish, and then exposed to p57; the immunosuppression was prevented if p57 was destroyed with the 100-kDa proteinase mentioned above (Rockey *et al.,* 1991). Suppression of the respiratory burst activity by p57 (Kaattari *et al.,* 1988) has already been mentioned but new information (L. L. Brown, pers. comm.) indicates that *R. salmoninarum* continues to produce the protein following its ingestion by macrophages. It seems likely, therefore, that suppression of the response would be maintained by the bacterium during its intracellular phase. Immunotolerance (as measured in terms of a decreased ability to survive challenge with *R. salmoninarum*) was demonstrated in young fish that were derived from eggs injected before fertilization with p57 (Brown *et al.*, 1996). Early exposure of embryos to this antigen via the egg, which is entirely feasible when the eggs are derived from adults with severe *R. salmoninarum* infections, can thus immunlogically compromise all of the resulting progeny, making them good sources of the infection for their uninfected cohorts.

VI. CONCLUDING REMARKS

In this chapter, the focus has been on the factors that permit bacteria to circumvent the defenses of their fish hosts and, thus, qualify for the designation "fish pathogen." For practical reasons, the discussions were limited to six pathogens that have caused enormous losses in a number of farmed fishes. Additional information on these and other bacterial fish pathogens are contained in other recent reviews (Evenden *et al.*, 1993; Furones *et al.*, 1993; Kusuda and Salati, 1993; Thune *et al.*, 1993; Toranzo and Barja, 1993).

It should be obvious from the material presented in this chapter that although there is still much to learn about the mechanisms employed by fish pathogens to avoid the host's defenses, a considerable body of knowledge on this topic is beginning to accumulate. What is particularly remarkable about this is that this information, prompted by an interest in obtaining vaccines for controlling fish diseases, has essentially all been obtained in the last 20 years. During that time, commercially available vaccines have been developed for controlling diseases caused by three of the six pathogens (*V. anguillarum, A. salmonicida,* and *Y. ruckeri*) dealt with in this chapter, and, what is more, a fair understanding exists about why these vaccines function so well. Interestingly, the three pathogens for which vaccines are still not commercially available are the ones that spend much of their time intracellularly in the host. It is no wonder then that a major thrust in fish immunology these days is in the direction of cellular immunity, particularly on ways of "up-regulating" bacterial killing by phagocytes (Secombes and Fletcher, 1992). The considerable interest in this area is evident in a number of recent publications on this topic (for example, Jorgensen *et al.*, 1993; Sharp and Secombes, 1993; Francis and Ellis, 1994; Hardie *et al.*, 1994) that show the potential value of this approach for dealing with pathogens such as *A. salmonicida* and *R. salmoninarum* which have long been regarded as "difficult" pathogens.

REFERENCES

Actis, L. A., Potter, S. A., and Crosa, J. H. (1985). Iron-regulated outer membrane protein OM2 of *Vibrio anguillarum* is encoded by virulence plasmid pJM1. *J. Bacteriol.* **161,** 736–742.

Actis, L. A., Fish, W., Crosa, J. H., Kellerman, K., Ellenberger, S. R., Hauser, S. M., Sanders-Loehr, F. M. J. (1986). Characterization of anguibactin, a novel siderophore from *Vibrio anguillarum* 775 (pJM1). *J. Bacteriol.* **167,** 57–65.

Ahne, W. (1978). Uptake and multiplication of spring viremia of carp virus in carp, *Cyprinus carpio* L. *J. Fish Dis.* **1,** 265–268.

Ahne, W., and Negele, R. D. (1985). Studies on the transmission of infectious pancreatic necrosis virus via eyed eggs and sexual products of salmonid fish. *In* "Fish and Shellfish Pathology" (A. E. Ellis, ed.), p. 261–269. Academic Press, London.

Ainsworth, A. J. (1992). Fish granulocytes: Morphology, distribution, and function. *Annu. Rev. Fish Dis.* **2,** 123–148.

Ainsworth, A. J., and Dexiang, C. (1990). Differences in the phagocytosis of four bacteria by channel catfish neutrophils. *Dev. Comp. Immunol.* **14,** 201–209.

Alexander, J. B., Bowers, A., and Shamshoon, S. M. (1981). Hyperposmotic infiltration of bacteria into trout: Route of entry and fate of the infiltrated bacteria. *Dev. Biol. Stand.* **49,** 441–445.

Anderson, D. P., and Nelson, J. S. (1974). Comparison of protection in rainbow trout (*Salmo gairdneri*) inoculated with and fed Hagerman redmouth bacterins. *J. Fish. Res. Bd. Can.* **31,** 214–216.

Austin, B., and Austin, D. A. (1993). "Bacterial Fish Pathogens: Disease in Farmed and Wild Fish," 2nd ed. Ellis Horwood, New York.

Baldwin, T. J. (1992). *Edwardsiella ictaluri:* Pathogenic mechanisms and immunogenic antigens. Ph.D. Thesis, Louisiana State University, Baton Rouge, Lousiana.

Baldwin, T. J., and Newton, J. C. (1993). Pathogenesis of enteric septicemia of channel catfish, caused by *Edwardsiella ictaluri:* Bacteriologic and light and electron microscopic findings. *J. Aquat. Anim. Health* **5,** 189–198.

Balfry, S. K., Albright, L. J., and Evelyn, T. P. T. (1996). Horizontal transfer of *Renibacterium salmoninarum* among farmed salmonids via the fecal–oral route. *Dis. Aquat. Org.* **25,** 63–69.

Bandin, I., Santos, Y., Bruno, D. W., Raynard, R. S., Toranzo, A. E., and Barja, J. L. (1991). Lack of biological activities in the extracellular products of *Renibacterium salmoninarum. Can. J. Fish. Aquat. Sci.* **48,** 421–425.

Bandin, I., Santos, Y., Toranzo, A. E., and Barja, J. L (1992) Detection of a vascular permeability factor in the extracellular products of *Renibacterium salmoninarum. Microb. Path.* **13,** 237–241.

Bandin, I. Ellis, A. E., Barja, J. L., and Secombes, C. J. (1993). Interaction between rainbow trout macrophages and *Renibacterium salmoninarum in vitro. Fish Shellfish Immunol.* **3,** 25–33.

Bandin, I., Rivas, C., Santos, Y., Secombes, C. J., Barja, J. L., and Ellis, A. E. (1995). Effect of serum factors on the survival of *Renibacterium salmoninarum* within rainbow trout macrophages. *Dis. Aquat. Org.* **23,** 221–227.

Baudin-Laurencin, F., and Germon, E. (1987). Experimental infection of rainbow trout, *Salmo gairdneri* R., by dipping in suspensions of *Vibrio anguillarum:* Ways of bacterial penetration; influence of temperature and salinity. *Aquaculture* **67,** 203–205.

Baudin-Laurencin, F., and Tangtrongpiros, J. (1980). Some results of vaccination against vibriosis in Brittany. *In* "Fish Diseases, Third COPRAQ Session" pp. 60–68. Springer-Verlag, Berlin, Heidelberg, New York.

Bogwald, J., Stensvag, K., Stuge, T. B., and Jorgensen, T. O. (1994). Tissue localization and immune responses in Atlantic salmon, *Salmo salar* L., after oral administration of *Aeromonas salmonicida, Vibrio anguillarum,* and *Vibrio salmonicida* antigens. *Fish Shellfish Immunol.* **4,** 353–368.

Brown, L. L., Iwama, G. K., and Evelyn, T. P. T. (1996). The effect of early exposure of coho salmon (*Oncorhynchus kisutch*) eggs to the p57 protein of *Renibacterium salmoninarum* on the development of immunity to the pathogen. *Fish Shellfish Immunol.* **6,** 149–165.

Brudeseth, B., and Evensen, O. (1995). Uptake of soluble and particulate antigens of *Aeromonas salmonicida* by the gut of Atlantic salmon (*Salmo salar*). p. 32 *In* Abstracts Book, European Association of Fish Pathologists Seventh International Conference "Diseases of Fish and Shellfish", 10-15 September, 1995, Palma de Mallorca, Spain.

Bruno, D. W. (1986). Histopathology of bacterial kidney disease in laboratory infected rainbow trout, *Salmo gairdneri* Richardson, and Atlantic salmon, *Salmo salar* L., with reference to naturally infected fish. *J. Fish Dis.* **9,** 523–537.

Bruno, D. W. (1988). Relationship between autoagglutination, cell surface hydrophobicity, and virulence of the fish pathogen *Renibacterium salmoninarum. FEMS Microbiol. Lett.* **51,** 135–140.

Bruno, D. W., Munro, A. L. S., and Needham, E. A. (1986). Gill lesions caused by *Aeromonas salmonicida* in sea-reared Atlantic salmon, *Salmo salar* L. *ICES CM* **1986/F,** 6.

Busch, R. A. (1978). Enteric redmouth disease (Hagerman strain). *Mar. Fish. Rev.* **40(3),** 42–51.

Busch, R. A., and Lingg, A. J. (1975). Establishment of an asymptomatic carrier state of enteric redmouth disease in rainbow trout (*Salmo gairdneri*). *J. Fish Res. Bd. Can.* **32,** 2429–2432.

Chart, H., and Munn, C. B. (1980). Experimental vibriosis in the eel (*Anguilla anguilla*). *In* "Fish Diseases, Third COPRAQ Session" (W. Ahne, ed.), pp. 39–44. Springer-Verlag, Berlin, Heidelberg, New York.

Chart, H., and Trust, T. J. (1983). Acquisition of iron by *Aeromonas salmonicida. J. Bacteriol.* **156,** 758–764.

Chilmonczyk, K. (1980). Some aspects of trout gill structure in relation to Egtved virus infection and defence mechanisms. *In* Fish Diseases, Third COPRAQ session (W. Ahne, ed.), pp. 18–22. Springer-Verlag, Berlin, Heidelberg, New York.

Ciembor, P. G., Blazer, V. S., Dawe, D., and Shotts, E. B. (1995). Susceptibility of channel catfish to infection with *Edwardsiella ictaluri:* Effect of exposure method. *J. Aquat. Anim. Health* **7,** 132–140.

Cipriano, R. C. (1982). Furunculosis in brook trout: Infection by contact exposure. *Prog. Fish Cult.* **44,** 12–14.

Cipriano, R. C., Ford, L. A., Teska, J. D., and Hale, L. E. (1992). Detection of *Aeromonas salmonicida* in the mucus of salmonid fishes. *J. Aquat. Anim. Health* **4,** 114–118.

Cipriano, R. C., Ford, L. A., Schacte, J. H., and Petrie, C. (1994). Evaluation of mucus as a valid site to isolate *Aeromonas salmonicida* among asymptomatic populations of lake trout (*Salvelinus namaycush*). *Biomed. Lett.* **49,** 229–233.

Coleman, G., Bennett, A. J., Whitby, P. W., and Bricknell, I. R. (1992). A 70 kDa *Aeromonas salmonicida* serine protease-β-galactosidase hybrid protein as an antigen and its protective effect on Atlantic salmon (*Salmo salar* L.) against virulent *A. salmonicida* challenge. *Biochem. Soc. Trans.* **21,** 49S.

Crosa, J. H. (1980). A plasmid associated with virulence in the marine fish pathogen *Vibrio anguillarum* specifies an iron-requiring sequestering system. *Nature* **284,** 566–568.

Crosa, J. H., Hodges, L. L., and Schiewe, M. H. (1980). Curing of a plasmid is correlated with an attenuation of the marine fish pathogen *Vibrio anguillarum. Infect. Immun.* **18,** 509–513.

Daly, J. G., and Stevenson, R. M. W. (1989). Agglutination of salmonid spermatozoa by *Renibacterium salmoninarum. J. Aquat. Anim. Health* **1,** 163–164.

Daly, J. G., Moore, A. R., and Olivier, G. (1994). Bacteriocidal activity of brook trout (*Salvelinus fontinalis*) peritoneal macrophages against avirulent strains of *Aeromonas salmonicida. Fish Shellfish Immunol.* **4,** 273–283.

Davies, R. L. (1991a). Virulence and serum-resistance in different clonal groups and serotypes of *Yersinia ruckeri. Vet. Microbiol.* **29,** 289–297.

Davies, R. L. (1991b). *Yersinia ruckeri* produces four iron-regulated other membrane proteins but does not produce detectable siderophores. *J. Fish Dis.* **14,** 563–570.

Dorson, M., and Torchy, C. (1985). Experimental transmission of infectious pancreatic necrosis virus via the sexual products. *In* "Fish and Shellfish Pathology" (A. E. Ellis, ed.), pp. 251–260. Academic Press, London

Dubreil, J. D., Jacques, M., Graham, L., and Lallier, R. (1990a). Purification and biochemical and structural characterization of a fimbrial haemagglutinin of *Renibacterium salmoninarum. J. Gen. Microbiol.* **136,** 2443–2448.

Dubreil, J. D., Lallier, R., and Jacques, M. (1990b). Immunoelectron microscopic demonstration that *Renibacterium salmoninarum* is encapsulated. *FEMS Microbiol. Lett.* **66,** 313–316.

Ellis, A. E. (1982). Differences between the immune mechanisms of fish and higher vertebrates. *In* "Microbial Diseases of Fish" (R. J. Roberts, ed.), pp. 1–29. Academic Press, New York.

Ellis, A. E. (1991). An appraisal of the extracellular toxins of *Aeromonas salmonicida* ssp. *salmonicida. J. Fish Dis.* **14,** 265–277.

Evelyn, T. P. T. (1988). Vibrio vaccines for salmonids. pp. 459–469. *In:* Congress Proceedings, Aquaculture International Congress and Exposition, Vancouver Trade Centre, Vancouver, British Columbia, Canada, September 6–9, 1988.

Evelyn, T. P. T., Ketcheson, J. E., and Prosperi-Porta, L. (1984). Further evidence for the presence of the kidney disease bacterium (*Renibacterium salmoninarum*) in the salmonid egg and for the failure of iodine, in the form of povidone-iodine, to reduce the intra-ovum infection rate in water-hardened eggs. *J. Fish Dis.* **7,** 173–182.

Evelyn, T. P. T., Prosperi-Porta, L., and Ketcheson, J. E. (1986). Experimental intra-ovum infection of salmonid eggs with *Renibacterium salmoninarum* and vertical transmission of the pathogen with such eggs despite their treatment with erythromycin. *Dis. Aquat. Org.* **1,** 197–202.

Evenden, A. J., Gilpin, M. L., and Munn, C. B. (1990). The cloning and expression of a gene encoding haemolytic activity from the fish pathogen *Renibacterium salmoninarum. FEMS Microbiol. Lett.* **71,** 31–34.

Evenden, A. J., Grayson, T. H., Gilpin, M. L., and Munn, C. B. (1993). *Renibacterium salmoninarum* and bacterial kidney disease—the unfinished jigsaw. *Annu. Rev. Fish Dis.* **3,** 87–104.

Fiedler, F., and Draxl, R. (1986). Biochemical and immunological properties of *Renibacterium salmoninarum. J. Bacteriol.* **168,** 799–804.

Fijan, N. N., and Giorgetti, G. (1978). Infectious pancreatic necrosis: Isolation of virus from eyed eggs of rainbow trout *Salmo gairdneri* Richardson. *J. Fish Dis.* **1,** 269–270.

Flano, E., Kaattari, S. L., Razquin, B., and Villena, A. J. (1996). Histopathology of the thymus of coho salmon, *Oncorhynchus kisutch,* experimentally infected with *Renibacterium salmoninarum. Dis. Aquat. Org.* **26,** 11–18.

Francis, C. H., and Ellis, A. E. (1994). Production of a lymphokine (macrophage activating factor) by salmon (*Salmo salar*) leukocytes stimulated with other membrane protein antigens of *Aeromonas salmonicida. Fish Shellfish Immunol.* **4,** 489–497.

Fryer, J. L., and Sanders, J. E. (1981). Bacterial kidney disease of salmonid fish. *Annu. Rev. Microbiol.* **35,** 273–298.

Funahashi, N., Miyazaki, T., Kodera, K., and Kubota, S. (1974). Histopathological studies on vibriosis in ayu. *Fish Pathol.* **8,** 136–143.

Furones, M. D., Gilpin, M. J., Alderman, D. J., and Munn, C. B. (1990). Virulence of *Yersinia ruckeri* serotype I strains is associated with a heat sensitive factor (HSF) in cell extracts. *FEMS Microbiol. Lett.* **66,** 339–344.

Furones, M. D., Rodgers, C. J., and Munn, C. B. (1993). *Yersinia ruckeri,* the causal agent of enteric redmouth disease (ERM) in fish. *Annu. Rev. Fish Dis.* **3,** 105–125.

Garduno, R. A., and Kay, W. W. (1992). Interaction of the fish pathogen *Aeromonas salmonicida* with rainbow trout macrophages. *Infect. Immun.* **60,** 4612–4620.

Gettchell, R. G., Rohovec, J. S., and Fryer, J. L. (1985). Comparison of *Renibacterium salmoninarum* isolates. *Fish Pathol.* **20,** 149–159.

Goldes, S. A., Ferguson, H. W., Daoust, P. Y., and Moccia, R. D. (1986). Phagocytosis of the inert suspended clay kaolin by the gills of rainbow trout, *Salmo gairdneri* Richardson. *J. Fish Dis.* **9,** 147–151.

Grayson, T. H., Bruno, D. W., Evenden, A. J., Gilpin, M. L., and Munn, C. B. (1995a). Iron acquisition by *Renibacterium salmoninarum;* contribution by iron reductase. *Dis. Aquat. Org.* **22,** 157–162.

Grayson, T. H., Evenden, A. J., Gilpin, M. L., and Munn, C. B. (1995b). Production of a *Renibacterium salmoninarum* hemolysin fusion protein in *Escherichia coli* K12. *Dis. Aquat. Org.* **22,** 153–156.

Griffin, B. R. (1983). Opsonic effect of rainbow trout (*Salmo gairdneri*) antibody on phagocytosis of *Yersinia ruckeri* by trout leukocytes. *Dev. Comp. Immunol.* **7,** 253–259.

Grisez, L., Chair, M., Sorgeloose, P., and Ollevier, F. (1996). Mode of infection and spread of *Vibrio anguillarum* in turbot (*Scophthalmus maximus*) larvae after oral challenge through live feed. *Dis. Aquat. Org.* **26,** 181–187.

Gutenberger, S. K. (1993). Phylogeny and intracellular survival of *Renibacterium salmoninarum.* Ph.D. Thesis, Oregon State University, Corvallis, Oregon.

Hardie, L. J., Ellis, A. E., and Secombes, C. J. (1994). *In vitro* activation of rainbow trout macrophages stimulates killing of *Renibacterium salmoninarum.* p. W-3.1. *In:* Program and Abstracts, International Symposium on Aquatic Animal Health, Seattle, Washington, September 4–8, 1994.

Hendricks, J. D., and Leek, S. L. (1975). Kidney disease postorbital lesions in spring chinook salmon (*Oncorhynchus tshawytscha*). *Trans. Am. Fish. Soc.* **104,** 805–807.

Hiney, M. P., Kilmartin, J. J., and Smith, P. R. (1994). Detection of *Aeromonas salmonicida* in Atlantic salmon with asymptomatic furunculosis infections. *Dis. Aquat. Org.* **19,** 161–167.

Hirst, I. D., and Ellis, A. E. (1994). Iron-regulated outer membrane proteins of *Aeromonas salmonicida* are important protective antigens in Atlantic salmon against furunculosis. *Fish Shellfish Immunol.* **4,** 29–45.

Hirst, I. D., Hastings, T. S., and Ellis, A. E. (1991). Siderophore production by *Aeromonas salmonicida. J. Gen. Microbiol.* **137,** 1185–1192.

Hoffmann, R., Popp, W., and Van De Graaff, S. (1984). Atypical BKD predominantly causing ocular and skin lesions. *Bull. Eur. Assoc. Fish Pathol.* **4,** 7–9.

Horne, M. T. (1982). The pathogenicity of *Vibrio anguillarum. In* "Microbial Diseases of Fish." Special Publication of the Society for General Microbiology (R. J. Roberts, ed.), pp. 171–187. Academic Press, London, New York.

Jorgensen, J. B., Sharp, G. J., Secombes, C. J., and Robertsen, B. (1993). Effect of a yeast-cell-wall glucan on the bactericidal activity of rainbow trout macrophages. *Fish Shellfish Immunol.* **3,** 267–277.

Kaattari, S. (1989). Development of a vaccine for bacterial kidney disease in salmon. Final Report. U.S. Department of Energy, Bonneville Power Administration, Division of Fisheries and Wildlife, Portland, Oregon.

Kaattari, S., Chen, D., Turaga, P., and Wiens, G. (1988). Development of a vaccine for bacterial kidney disease in salmon. Annual Report FY 1987. U.S. Department of Energy, Bonneville Power Administration, Division of Fish and Wildlife, Portland, Oregon.

Kanno, T., Nakai, T., and Muroga, K. (1989). Mode of transmission of vibriosis among ayu *Plecoglossus altivelis. J. Aquat. Anim. Health* **1,** 2–6.

Karczewski, J. M., Sharp, G. J. E., and Secombes, C. J. (1991). Susceptibility of strains of *Aeromonas salmonicida* to killing by cell-free generated superoxide anion. *J. Fish Dis.* **14,** 367–373.

Kawahara, E., and Kusuda, R. (1988). Location of *Pasteurella piscicida* antigens in tissues of yellowtail *Seriola quinqueradiata* vaccinated by immersion. *Nip. Suisan Gakkaishi* **54,** 1101–1105.

Kawahara, E., Kawai, K., and Kusuda, R. (1989). Invasion of *Pasteurella piscicida* in tissues of experimentally infected yellowtail *Seriola quinqueradiata. Nip. Suisan Gakkaishi* **55,** 499–501.

Kawai, K., Kusuda, R., and Itami, T. (1981). Mechanisms of protection in ayu orally vaccinated for vibriosis. *Fish Pathol.* **15,** 257–262.

Klontz, G. W., Yasutake, W. T., and Ross, A. J. (1966). Bacterial diseases of the Salmonidae in the western United States: Pathogenesis of furunculosis in rainbow trout. *Am. J. Vet. Res.* **27,** 1455–1460.

Kodera, K., Funahashi, N., Miyazaki, T., and Kubota, S. (1974). Studies on vibriosis of ayu (*Plecoglossus altivelis*). II. Experimental infection with *Vibrio anguillarum* isolated from diseased fish. *Fish Pathol.* **8,** 185–189.

Kubota, S., Kimura, T., and Egusa, S. (1970). Studies of a bacterial tuberculosis of yellowtail. I. Symptomatology and histopathology. *Fish Pathol.* **4,** 111–118.

Kusser, W., and Fiedler, F. (1983). Murein type and polysaccharide composition of cell walls from *Renibacterium salmoninarum. FEMS Microbiol. Lett.* **20,** 391–394.

Kusuda, R., and Salati, F. (1993). Major bacterial diseases affecting mariculture in Japan. *Annu. Rev. Fish Dis.* **3,** 69–85.

Lamas, J., and Ellis, A. E. (1994a). Atlantic salmon (*Salmo salar*) neutrophil responses to *Aeromonas salmonicida. Fish Shellfish Immunol.* **4,** 201–219.

Lamas, J., and Ellis, A. E. (1994b). Electron microscopic observations of phagocytosis and subsequent fate of *Aeromonas salmonicida* by Atlantic salmon neutrophils *in vitro. Fish Shellfish Immunol.* **4,** 539–546.

Lamas, J., Santos, Y., Bruno, D., Toranzo, A. E., and Anadon, R. (1994). A comparison of the pathological changes caused by *Vibrio anguillarum* and its extracellular products in rainbow trout (*Oncorhynchus mykiss*). *Fish Pathol.* **29,** 79–89.

Lester, R. J. G., and Budd, J. (1979). Some changes in the blood of diseased coho salmon. *Can. J. Zool.* **57,** 1458–1464.

Magarinos, B., Santos, Y., Romalde, J. L., Rivas, C., Barja, J. L., and Toranzo, A. E. (1992). Pathogenic activities of live cells and extracellular products of the fish pathogen *Pasteurella piscicida. J. Gen. Microbiol.* **138,** 2491–2498.

Magarinos, B., Romalde, J. L., Lemos, M. L., Barja, J. L., and Toranzo, A. E. (1994). Iron uptake by *Pasteurella piscicida* and its role in pathogenicity for fish. *Appl. Environ. Microbiol.* **60,** 2990–2998.

Miyazaki, T., and Plumb, J. A. (1985). Histopathology of *Edwardsiella ictaluri* in channel catfish, *Ictalurus punctatus. J. Fish Dis.* **8,** 389–392.

Mulcahy, D., Pascho, R. J., and Jenes, C. K. (1983). Detection of infectious haematopoietic necrosis virus in river water and demonstration of waterborne transmission. *J. Fish Dis.* **6,** 321–330.

Munn, C. B., and Trust, T. J. (1984). Role of additional protein in virulence of *Aeromonas salmonicida. In* "Fish Diseases, Fourth COPRAQ Session" (Acuigrup, eds.), pp. 69–75. Editora ATP, Madrid.

Munn, C. B., Ishiguro, E. E., Kay, W. W., and Trust, T. J. (1982). Role of surface components in serum resistance of virulent *Aeromonas salmonicida. Infect. Immun.* **36,** 1069–1075.

Muroga, K., and De La Cruz, M. C. (1987). Fate and location of *Vibrio anguillarum* in tissues of artificially infected ayu (*Plecoglossus altivelis*). *Fish Pathol.* **22,** 99–103.

Murray, C. B., Evelyn, T. P. T., Beacham, T. D., Barner, L. W., Ketcheson, J. E., and Prosperi-Porta, L. (1992). Experimental induction of bacterial kidney disease in chinook salmon by immersion and cohabitation challenges. *Dis. Aquat. Org.* **12,** 91–96.

Nelson, J. S., Rohovec, J. S., and Fryer, J. L. (1985). Location of *Vibrio anguillarum* in tissues of infected rainbow trout (*Salmo gairdneri*) using the fluorescent antibody technique. *Fish Pathol.* **20,** 229–235.

Nelson, J. S., Kawahara, E., Kawai, K., and Kusuda, R. (1989). Macrophage infiltration in pseudotuberculosis of yellowtail, *Seriola quinqueradiata. Bull. Mar. Sci. Fish. Kochi Univ.* **11,** 17–22.

Newton, J. C., Wolfe, L. G., Grizzle, J. M., and Plumb, J. A. (1989). Pathology of experimental enteric septicaemia in channel catfish, *Ictalurus punctatus* (Rafinesque), following immersion-exposure to *Edwardsiella ictaluri. J. Fish Dis.* **12,** 335–347.

Noya, M., Magarinos, B., Toranzo, A. E., and Lamas, J. (1995). Sequential pathology of experimental pasteurellosis in gilthead seabream *Sparus aurata.* A light- and electron-microscopic study. *Dis. Aquat. Org.* **21,** 177–186.

Olafsen, J. (1994). Marine animals as hosts for fish-pathogenic vibrios, (Abstract p-92). In: Third International Marine Biotechnology Conference: Program, Abstracts and List of Participants. Tromsoe University, Tromsoe, Norway.

Olivier, G., Evelyn, T. P. T., and Lallier, R. (1985). Immunity to *Aeromonas salmonicida* in coho salmon (*Oncorhynchus kisutch*) induced by modified Freund's complete adjuvant: Its non-specific nature and the possible role of macrophages in the phenomenon. *Dev. Comp. Immunol.* **9,** 419–432.

Olivier, G., Eaton, A., and Campbell, N. (1986). Interaction between *Aeromonas salmonicida* and peritoneal macrophages of brook trout, *Salvelinus fontinalis. Vet. Immunol. Immunopathol.* **12,** 253–265.

Olivier, G., Moore, A. R., and Fildes, J. (1992). Toxicity of *Aeromonas salmonicida* cells to Atlantic salmon, *Salmo salar,* peritoneal macrophages. *Dev. Comp. Immunol.* **16,** 49–61.

Ourth, D. D., and Bachinski, L. M. (1987). Bacterial sialic acid modulates the alternative complement pathway of channel catfish (*Ictalurus punctatus*). *Dev. Comp. Immunol.* **11,** 511–564.

Ourth, D. D., Ratts, V. D., and Parker, N. C. (1991). Bactericidal complement activity and concentrations of immunoglobulin M, transferrin, and protein at different ages of channel catfish. *J. Aquat. Anim. Health* **3,** 274–280.

Perez, M. J., Fernandez, A. I. G., Rodriguez, L. A., and Nieto, T. P. (1996). Differential susceptibility to furunculosis of turbot and rainbow trout and release of the furunculosis agent from furunculosis-affected fish. *Dis. Aquat. Org.* **26,** 133–137.

Ransom, D. P. (1978). Bacteriologic, immunologic, and pathologic studies of *Vibrio* sp. pathogenic to salmonids. Ph.D. Thesis, Oregon State University, Corvallis, Oregon.

Ransom, D. P., Lannon, C. N., Rohovec, J. S., and Fryer, J. L. (1984). Comparison of histopathology caused by *Vibrio anguillarum* and *V. ordalii* in three species of Pacific salmon. *J. Fish Dis.* **7,** 107–115.

Rockey, D. D., Turaga, P. S. D., Wiens, G. D., Cook, B. A., and Kaattari, S. L. (1991). Serine proteinase of *Renibacterium salmoninarum* digests a major autologous extracellular and cell-surface protein. *Can. J. Microbiol.* **37,** 758–763.

Romalde, J. L., Conchas, R. F., and Toranzo, A. E. (1991). Evidence that *Yersinia ruckeri* possesses a high affinity iron uptake system. *FEMS Microbiol. Lett.* **80,** 121–126.

Rose, A. S., and Levine, R. P. (1992). Complement-mediated opsonization and phagocytosis of *Renibacterium salmoninarum. Fish Shellfish Immunol.* **2,** 223–240.

Rose, A. S., Ellis, A. E., and Munro, A. L. S. (1989). The infectivity by different routes of exposure and shedding rates of *Aeromonas salmonicida* subsp. *salmonicida* in Atlantic salmon, *Salmo salar* L., held in sea water. *J. Fish Dis.* **12,** 573–578.

Ross, A. J., Rucker, R. R., and Ewing, W. H. (1966). Description of a bacterium associated with redmouth disease of rainbow trout (*Salmo gairdneri*). *Can. J. Microbiol.* **12,** 763–770.

Sakai, D. K., and Kimura, T. (1985). Relationship between agglutinative properties of *Aeromonas salmonicida* strains isolated from fish in Japan and their resistance to mechanisms of host defense. *Fish Pathol.* **20,** 9–21.

Santos, Y., Bandin, I., Nieto, T. P., Barja, J. L., Toranzo, A. E., and Ellis, A. E. (1991). Cell-surface-associated properties of fish pathogenic bacteria. *J. Aquat. Anim. Health* **3,** 279–301.

Secombes, C. J., and Fletcher, T. C. (1992). The role of phagocytes in the protective mechanisms of fish. *Annu. Rev. Fish Dis.* **2,** 53–71.

Sharp, G. J. E, and Secombes, C. J. (1993). The role of reactive oxygen species in the killing of the bacterial fish pathogen *Aeromonas salmonicida* by rainbow trout macrophages. *Fish Shellfish Immunol.* **3,** 119–129.

Shotts, E. B., Blazer, V. S., and Waltman, W. D. (1986). Pathogenesis of experimental *Edwardsiella ictaluri* infections in channel catfish (*Ictalurus punctatus*). *Can. J. Fish. Aquat. Sci.* **43,** 36–42.

Smith, P. D. (1982). Analysis of the hyperosmotic and bath methods for fish vaccination—comparison of uptake of particulate and nonparticulate antigens. *Dev. Comp. Immunol. Suppl.* **2,** 181–186.

Speare, D. J., Ostland, V. E., and Ferguson, H. W. (1993). Pathology associated with meningoencephalitis during bacterial kidney disease of salmonids. *Res. Vet. Sci.* **54,** 25–31.

Stanley, L. A. (1991). Characteristics of pathogenic mechanisms of *Edwardsiella ictaluri* in channel catfish, *Ictalurus punctatus.* Ph.D. Thesis, Clemson University, Clemson, South Carolina.

Stanley, L. A., Hudson, J. S., Schwedler, T. E., and Hayasaka, S. S. (1994). Extracellular products from virulent and avirulent strains of *Edwardsiella ictaluri* from channel catfish. *J. Aquat. Anim. Health* **6,** 36–43.

Stave, J. W., Cook, T. M., and Roberson, B. S. (1987). Chemiluminescent responses of striped bass, *Morone saxatilis* (Walbaum), phagocytes to strains of *Yersinia ruckeri. J. Fish Dis.* **10,** 1–10.

Tatner, M. F., Johnson, C. M., and Horne, M. T. (1984). The tissue localization of *Aeromonas salmonicida* in rainbow trout, *Salmo gairdneri* Richardson, following three methods of administration. *J. Fish Biol.* **25,** 95–108.

Thune, R. L., Stanley, L. A., and Cooper, R. K. (1993). Pathogenesis of Gram-negative bacterial infections in warmwater fish. *Annu. Rev. Fish Dis.* **3,** 37–68.

Tomalsky, M. E., and Crosa, J. H. (1984). Molecular cloning and expression of genetic determinants for the iron uptake system mediated by *Vibrio anguillarum* plasmid pJM1. *J. Bacteriol.* **160,** 860–866.

Toranzo, A. E., and Barja, J. L. (1993). Virulence factors of bacteria pathogenic for cold water fish. *Annu. Rev. Fish Dis.* **3,** 5–36.

Toranzo, A. E., Barja, J. L., Potter, S. A., Colwell, R. A., Hetrick F. M., and Crosa, J. H. (1983). Molecular factors associated with virulence of marine vibrios isolated from striped bass in Chesapeake Bay. *Infect. Immun.* **39,** 1220–1227.

Turaga, P. S. D., Wiens, G. D., and Kaattari, S. L. (1987a). Analysis of *Renibacterium salmoninarum* antigen production *in situ. Fish. Pathol.* **22,** 209–214.

Turaga, P. S. D., Wiens, G. D., and Kaattari, S. L. (1987b). Bacterial kidney disease: The potential role of soluble protein antigen(s). *J. Fish Biol. Suppl. A* **31,** 191–194.

Trust, T. J., Courtice, I. D., Khouri, A., Crosa, J. H., and Schiewe, M. H. (1981). Serum resistance and haemagglutinating ability of marine vibrios pathogenic for fish. *Infect. Immun.* **34,** 702–707.

Vigneulle, M. (1990). Yersiniose des salmonides: Etudes comparee des differents modes de vaccination. *Ichtyophysiol. Acta* **13,** 43–58.

Vigneulle, M., and Baudin-laurencin, F. (1991). Uptake of *Vibrio anguillarum* bacterin in the posterior intestine of rainbow trout *Oncorhynchus mykiss,* sea bass *Dicentrarchus labrax,* and turbot *Scophthalmus maximus* after oral administration and anal intubation. *Dis. Aquat. Org.* **11,** 85–92.

Waterstrat, P. R., Ainsworth, A. J., and Capley, G. (1991). In vitro responses of channel catfish, *Ictalurus punctatus,* neutrophils to *Edwardsiella ictaluri. Dev. Comp. Immunol.* **15,** 53–63.

Watkins, W. D., Wolke, R. E., and Cabelli, V. J. (1981). Pathogenicity of *Vibrio anguillarum* for juvenile winter flounder *Pseudopleuronectes americanus. Can. J. Fish. Aquat. Sci.* **38,** 1045–1051.

Weeks-Perkins, B. A., and Ellis, A. E. (1995). Chemotactic responses of Atlantic salmon (*Salmo salar*) macrophages to virulent and attenuated strains of *Aeromonas salmonicida. Fish Shellfish Immunol.* **5,** 313–323.

Wiens, G. D., and Kaattari, S. L. (1989). Monoclonal antibody analysis of common surface proteins of *Renibacterium salmoninarum. Fish Pathol.* **24,** 1–7.

Wiens, G. D., and Kaattari, S. L. (1991). Monoclonal antibody characterization of a leukoagglutinin produced *Renibacterium salmoninarum. Infect. Immun.* **59,** 631–637.

Wolf, M. K., and Crosa, J. H. (1986). Evidence for the role of a siderophore in promoting *Vibrio anguillarum* infections. *J. Gen. Microbiol.* **132,** 2949–2952.

Wood, J. W., and Wallis, J. (1955). Kidney disease in adult chinook salmon and its transmission by feeding to young chinook salmon. Research Briefs 6, pp. 32–40. Fisheries Commission of Portland, Oregon.

Wright, S. D., and Silverstein, S. C. (1983). Receptors for C3b and C3bi promote phagocytosis but not release of O_2^- from human phagocytes. *J. Exp. Med.* **158,** 2016–2023.

Young, C. L., and Chapman, G. B. (1978). Ultrastructural aspects of the causative agent and renal histopathology of bacterial kidney disease in brook trout (*Salvelinus fontinalis*). *J. Fish Res. B. Can.* **35,** 1234–1248.

Yousif, A. N., Albright, L. J., and T. P. T. Evelyn (1994). *In vitro* evidence for the antibacterial role of lysozyme in salmonid eggs. *Dis. Aquat. Org.* **19,** 15–19.

Zapata, A. G., Torroba, M., Alvarez, F., Anderson, D. P., Dixon, D. W., and Wisniewski, M. (1987). Electron microscopic examination of antigen uptake by salmonid gill cells after bath immunization with a bacterin. *J. Fish Biol. Suppl. A* **31,** 209–217.

INDEX

D

E

F

G

J

K

L

M

N

O

P

Q

R

T